The Evolution of Protein Structure and Function

A SYMPOSIUM IN HONOR OF PROFESSOR EMIL L. SMITH

UCLA FORUM IN MEDICAL SCIENCES

Professor Emil L. Smith

UCLA FORUM IN MEDICAL SCIENCES

NUMBER 21

The Evolution of Protein Structure and Function

A SYMPOSIUM IN HONOR OF PROFESSOR EMIL L. SMITH

Edited by

DAVID S. SIGMAN

Department of Biological Chemistry
School of Medicine
University of California Los Angeles
Los Angeles, California

MARY A. B. BRAZIER

Department of Anatomy
School of Medicine
University of California Los Angeles
Los Angeles, California

1980

ACADEMIC PRESS

A Subsidiary of Harcourt Brace Jovanovich, Publishers

New York London Toronto Sydney San Francisco

ACADEMIC PRESS, INC.
111 Fifth Avenue, New York, New York 10003

United Kingdom Edition published by
ACADEMIC PRESS, INC. (LONDON) LTD.
24/28 Oval Road, London NW1 7DX

Library of Congress Cataloging in Publication Data

Main entry under title:
Evolution of protein structure and function.

(UCLA forum on medical sciences; no. 21)
Held Dickson Art Auditorium, University of California, Los Angeles, June 28-29, 1979.
"Representative publications of Emil L. Smith (1935–1979)": p.
Includes indexes.
1. Proteins – Congresses. 2. Chemical evolution––Congresses. 3. Biochemorphology–– Congresses. 4. Smith, Emil L. I. Smith, Emil L., Date II. Sigman, D. S. III. Brazier, Mary Agnes Burniston, Date IV. Series.
QP551.E94 574.19'245 80-18140
ISBN 0–12–643150–7

PRINTED IN THE UNITED STATES OF AMERICA

80 81 82 83 9 8 7 6 5 4 3 2 1

Contents

List of Contributors and Participants . . . xiii
Preface . . . xv
Other Titles in the Series . . . xvii

Introductory Review to a Symposium Concerned with Forty Years of Research on Proteins

STANFORD MOORE

Amino Acid Analysis . . . 2
Ribonuclease . . . 5
Ribonuclease Revisited . . . 8
The Ribonuclease Inhibitor from the Human Placenta . . . 11
Conclusion . . . 14
References . . . 14

ENZYMOLOGY

On the Role of the recA Protein of *Escherichia coli* in General Recombination

I. ROBERT LEHMAN

recA Protein: A DNA-Dependent Nucleoside Triphosphatase . . . 24
Annealing of Complementary Single Strands of DNA Catalyzed by recA Protein . . . 25
Binding of recA Protein to Duplex DNA in the Presence of a Nucleoside Triphosphate . . . 28
Assimilation of Homologous Single Strands into Duplex DNA by recA Protein . . . 30
Discussion . . . 34
References . . . 36

Experimental Enzyme Evolution

BRIAN S. HARTLEY

Evolution of Enzyme Families 40
Experimental Evolution 40
Evolution of Xylitol Dehydrogenases 41
Superproduction of Ribitol Dehydrogenase 42
Ribitol Dehydrogenase mRNA 42
Transfer of Pentitol Operons to *Escherichia coli* K12 43
Specialized Transducing Phages Containing the Pentitol Operons 44
Evolution of the Pentitol Operons 44
DNA Sequences of the Pentitol Operons 46
Conclusion 47
References 47

β-Galactosidase and the Lactose Operon

IRVING ZABIN

Text 49
References 60

Glycosyltransferases in Analysis of Oligosaccharide Structure and Function

ROBERT L. HILL

Introduction 63
Purification of Glycosyltransferases 66
Glycosyltransferases in Oligosaccharide Biosynthesis 69
Glycosyltransferases in the Analysis of Oligosaccharide Structure and Function 74
References 83

Metalloenzymes

BO G. MALMSTRÖM

Introduction 87
Carbonic Anhydrase 88
"Blue" Copper Proteins 89
Cytochrome *c* Oxidase 92
The Mechanism of Reduction of Dioxygen 92
Conclusion 94
References 95

PROTEIN STRUCTURE AND FUNCTION

Glycoprotein Hormones: Similar Molecules with Different Functions

JOHN G. PIERCE AND THOMAS F. PARSONS

Text 99
References 112

Thymus Hormones: Diverse Molecules Producing Similar Biological Actions

ABRAHAM WHITE

Historical Background 119
Criteria for Designation of a Preparation as a Thymic Hormone 121
Purified Molecules with Thymic Hormone-like Activity 122
Protein Fraction from Human Plasma with Thymic Hormone-like Activity 123
Biological Activities of Fraction 3 126
Conclusion 129
References 130

Chromosome Biology and Chemistry

JAMES BONNER

Introduction 133
Primary Structure of the Histones 135
Histone Sequencing—A Growth Industry 136
Histone Biology 137
The Expressed Portion of the Genome 141
What Makes Chromatin Template Active? 144
High-Mobility Group and Related Proteins 146
A General Picture 147
References 147

The Evolution of Histones in Relationship to Recent Advances in Elucidating Chromatin Structure

ROBERT J. DeLANGE

Text 151
References 156

Thermal Stability and Protein Structure

PATRICK ARGOS, MICHAEL G. ROSSMAN, ULRICH M. GRAU, HERBERT ZUBER, GERHARD FRANK, AND JON DÜRI TRATSCHIN

Introduction 159
Method 160
Stabilizing Effects of the Preferred Exchanges 162
Useful Exchanges for Evolving Stability 166
Conclusions 167
References 167

EVOLUTION OF PROTEINS

The Cytochromes *c*: An Exercise in Scientific Serendipity

RICHARD E. DICKERSON

Electron Transport in Bacteria 174
Cytochrome Sequence and Structure 176
Confusion from Lateral Gene Transfer? 181
Evolution in the Photosynthetic Bacteria 192
Taxonomy within the Rhodospirillaceae 196
References 199

Neutral Changes Revisited

THOMAS H. JUKES

Relation of the Genetic Code to Amino Acid Composition of Proteins 204
One Gene, Two Proteins 207
The Evolutionary Clock 208
Silent Changes in mRNA Comparisons 210
Discussion and Summary 216
References 218

Structure and Evolution of Photosynthetic Accessory Pigment Systems with Special Reference to Phycobiliproteins

ALEXANDER N. GLAZER

Introduction 221
The "Accessory Light-Harvesting Domain" in Cyanobacteria and Red Algae 225
Phycobiliproteins—The Light-Harvesting Proteins of Cyanobacteria (Blue-Green Algae) and Red Algae 227
Cryptomonad Phycobiliproteins—A Case of Interspecific Gene Transfer? 231
Energy Transfer 235
Phycobilisome Morphology and Its Relation to Function 237
References 239

Structure and Evolution of Chloroplast- and Bacterial-Type Ferredoxins

H. MATSUBARA, T. HASE, S. WAKABAYASHI, AND K. WADA

Chloroplast-Type Ferredoxins: Amino Acid Sequences and Three-Dimensional Structure 246
Molecular Evolution and Gene Duplication 253
Bacterial-Type Ferredoxins: Amino Acid Sequences, Characteristics, and Molecular Evolution 257
References 262

Amino Acid Sequences of Globin Chains and Their Use in Phylogenetic Divergence Point Estimations

E. O. P. THOMPSON

Introduction 267
Myoglobins and Hemoglobins 268
The Myoglobin Fold 269
Tetrameric Hemoglobins 271
Contact Residues and Sites of Cooperative Interactions 271
Studies on Marsupial Globins 279
Studies on Monotreme Globins 281
Rates of Evolution 282
Studies on Shark Globins 284
Studies on Mollusk Globins 288
References 293

Evolutionary Adaptation of Mitochondrial Cytochrome *c* to Its Functional Milieu

E. MARGOLIASH

Neutral Mutations versus Selection in the Evolution of Cytochrome *c* 299
The Reaction of Cytochrome *c* with Mitochondrial Electron Donors and Acceptors 305
Do *in Vitro* Tests of Function Reflect Physiologic Activities in the Organism? 316
References 317

Appendix I: Tributes to Professor Emil L. Smith by Symposium Participants

Tribute by Robert L. Hill 323
Tribute by Bo G. Malmström 324
Tribute by Thomas H. Jukes 326

Appendix II: Representative Publications of Emil L. Smith (1935–1979) 327

Author Index **332**
Subject Index **347**

List of Contributors and Participants

Names of contributors to this volume are marked with an asterisk.

***Patrick Argos**, Department of Biological Sciences, Purdue University, West Lafayette, Indiana 47907

Elkan R. Blout, Department of Biological Chemistry, School of Medicine, Harvard University, Boston, Massachusetts 02115

***James Bonner**, Division of Biology, California Institute of Technology, Pasadena, California 91125

Paul D. Boyer, Molecular Biology Institute, University of California Los Angeles, Los Angeles, California 90024

***Robert J. DeLange**, Department of Biological Chemistry, School of Medicine, and Molecular Biology Institute, University of California Los Angeles, Los Angeles, California 90024

***Richard E. Dickerson**, Division of Chemistry and Chemical Engineering, California Institute of Technology, Pasadena, California 91125

***Gerhard Frank**, Institut für Molekularbiologie und Biophysik, ETH-Hönggerberg, CH-8093 Zürich, Switzerland

***Alexander N. Glazer**, Department of Microbiology and Immunology, University of California Berkeley, Berkeley, California 94720

***Ulrich M. Grau**, Department of Biological Sciences, Purdue University, West Lafayette, Indiana 47907

Philip Handler, National Academy of Sciences, Washington, D.C. 20418

***Brian S. Hartley**, Department of Biochemistry, Imperial College of Science and Technology, London SW7 2AZ England

***T. Hase**, Department of Biology, Faculty of Science, Osaka University, Toyonaka, Osaka 560, Japan

***Robert L. Hill**, Department of Biology, School of Medicine, Duke University, Durham, North Carolina 27710

*Thomas H. Jukes, Department of Biophysics and Medical Physics, University of California Berkeley, Berkeley, California 94720

*I. Robert Lehman, Department of Biology, School of Medicine, Stanford University, Stanford, California 94305

*Bo G. Malmström, Department of Biology and Biophysics, Chalmers University of Technology, S-412 96 Göteborg, Sweden

*E. Margoliash, Departments of Biochemistry and Molecular Biology, Northwestern University, Evanston, Illinois 60201

*H. Matsubara, Department of Biology, Faculty of Science, Osaka University, Toyonaka, Osaka 560, Japan

Sherman Mellinkoff, School of Medicine, University of California Los Angeles, Los Angeles, California 90024

*Stanford Moore, The Rockefeller University, New York, New York 10021

*Thomas F. Parsons, Department of Biological Chemistry, School of Medicine, University of California Los Angeles, Los Angeles, California 90024

*John G. Pierce, Department of Biological Chemistry, School of Medicine, University of California Los Angeles, Los Angeles, California 90024

Gerhard W. E. Plaut, Department of Biology, School of Medicine, Temple University, Philadelphia, Pennsylvania 19140

George J. Popjak, Departments of Biological Chemistry and Psychiatry, School of Medicine, University of California Los Angeles, Los Angeles, California 90024

*Michael G. Rossmann, Department of Biological Sciences, Purdue University, West Lafayette, Indiana 47907

David S. Sigman, Department of Biological Chemistry, School of Medicine, University of California Los Angeles, Los Angeles, California 90024

*Emil L. Smith, Department of Biological Chemistry, School of Medicine, University of California Los Angeles, Los Angeles, California 90024

*E. O. P. Thompson, School of Biochemistry, University of New South Wales, Kensington, New South Wales 2033, Australia

*Jon Düri Tratschin, Institut für Molekularbiologie und Biophysik, ETH-Hönggerberg, CH-8093 Zürich, Switzerland

*K. Wada, Department of Biology, Faculty of Science, Osaka University, Toyonaka, Osaka 560, Japan

*S. Wakabayashi, Department of Biology, Faculty of Science, Osaka University, Toyonaka, Osaka, Japan

*Abraham White,[1] Institute of Biological Sciences, Syntex Research, Palo Alto, California, and Department of Biology, School of Medicine, Stanford University, Stanford, California 94304

*Irving Zabin, Department of Biological Chemistry, School of Medicine, and Molecular Biology Institute, University of California Los Angeles, Los Angeles, California 90024

*Herbert Zuber, Institut für Molekularbiologie und Biophysik, ETH-Hönggerberg, CH-8093 Zürich, Switzerland

[1]Deceased

Preface

The symposium "Evolution of Protein Structure and Function" was held June 28–29, 1979, in the Dickson Art Auditorium, University of California Los Angeles (UCLA). Its objective was to honor Professor Emil L. Smith on the occasion of his retirement as Professor and Chairman, Department of Biological Chemistry, School of Medicine, UCLA. The papers presented by Emil's colleagues, friends, and students from all phases of his long and varied scientific career provided a valuable review of enzymology, protein chemistry, and biochemical evolution, and prompted the publication of these proceedings as a volume in the UCLA Forum in Medical Sciences series.

The warm personal reminiscences interwoven into the contributions and some others reproduced in Appendix I obviate the need for a formal biographical sketch. These remarks, as well as the enthusiasm and interest of all those participating or attending the symposium, clearly convey the high esteem and regard felt toward Emil. Representative publications, reflective of the broad range of his research interests, are listed in Appendix II.

The symposium was made possible by the generous financial support of the UCLA School of Medicine and the Calbiochem–Behring Corporation. Dean Sherman Mellinkoff and Associate Dean A. Fred Rasmussen, Jr., provided immediate support and warm encouragement for the entire project.

Special thanks are due Douglas Brown and Katherine Kanamori for assistance in conducting the symposium. Assistant Dean Byron Backlar was instrumental in arranging for the publication of the proceedings in the present form. Melody Horner, Assistant Editor of the UCLA Forum series, played a central role in the editing and organization of the manuscripts.

David S. Sigman

Mary A. B. Brazier

UCLA FORUM IN MEDICAL SCIENCES

Other Titles in the Series

1. Brain Function: Cortical Excitability and Steady Potentials; Relations of Basic Research to Space Biology. *Edited by Mary A. B. Brazier (1963)*
2. Brain Function: RNA and Brain Function. Memory and Learning. *Edited by Mary A .B. Brazier (1964)*
3. Brain and Behavior: The Brain and Gonadal Function. *Edited by Roger A. Gorski and Richard E. Whalen (1966)*
4. Brain Function: Speech, Language, and Communication. *Edited by Edward C. Carterette (1966)*
5. Gastrin. *Edited by Morton I. Grossman (1966)*
6. Brain Function: Brain Function and Learning. *Edited by Donald B. Lindsley and Arthur A. Lumsdaine (1967)*
7. Brain Function: Aggression and Defense. Neural Mechanisms and Social Patterns. *Edited by Carmine D. Clemente and Donald B. Lindsley (1967)*
8. The Retina: Morphology, Function, and Clinical Characteristics. *Edited by Bradley R. Straatsma, Raymond A. Allen, Frederick Crescitelli, and Michael O. Hall (1969)*
9. Image Processing in Biological Science. *Edited by Diane M. Ramsey (1969)*
10. Pathophysiology of Congenital Heart Disease. *Edited by Forrest H. Adams, H. J. C. Swan, and V. E. Hall (1970)*
11. The Interneuron. *Edited by Mary A. B. Brazier (1969)*
12. The History of Medical Education. *Edited by C. D. O'Malley (1970)*
13. Cardiovascular Beta Adrenergic Responses. *Edited by Albert A. Kattus, Jr., Gordon Ross, and Rex N. MacAlpin (1970)*
14. Cellular Aspects of Neural Growth and Differentiation. *Edited by Daniel C. Pease (1971)*
15. Steroid Hormones and Brain Function. *Edited by Charles Sawyer and Roger A. Gorski (1971)*
16. Multiple Sclerosis: Immunology, Virology, and Ultrastructure. *Edited by Frederick Wolfgram, George Ellison, Jack Stevens, and John Andrews (1972)*
17. Epilepsy: Its Phenomena in Man. *Edited by Mary A. B. Brazier (1973)*
18. Brain Mechanisms in Mental Retardation. *Edited by Nathaniel A. Buchwald and Mary A. B. Brazier (1975)*
19. Amyotrophic Lateral Sclerosis: Recent Research Trends. *Edited by John M. Andrews, Richard T. Johnson, and Mary A. B. Brazier (1976)*
20. Prevention of Neural Tube Defects: The Role of Alpha-Fetoprotein. *Edited by Barbara F. Crandall and Mary A. B. Brazier (1978)*
21. The Evolution of Protein Structure and Function: A Symposium in Honor of Professor Emil L. Smith. *Edited by David S. Sigman and Mary A. B. Brazier (1980)*

In preparation

22. The Regulation of Muscle Contraction: Excitation–Contraction Coupling. *Edited by Alan D. Grinnell and Mary A. B. Brazier (1981)*
23. Cellular Basis of Chemical Messengers in the Digestive System. *Edited by Morton I. Grossman, Mary A. B. Brazier, and Juan Lechago (1981)*

Introductory Review to a Symposium Concerned with Forty Years of Research on Proteins

STANFORD MOORE

The Rockefeller University
New York, New York

The contributors to this symposium spoke of advances in protein chemistry during four decades in the career of one of the most productive contemporary investigators of the subject. This occasion, in honor of Emil L. Smith, took my thoughts back to when I first met him when he joined Max Bergmann's laboratory at The Rockefeller Institute for Medical Research in 1940. I respected him as an experienced scientist who had already written an impressive series of papers on photosynthesis and chlorophyll–protein complexes (e.g., refs. 1–3) from the laboratory of Selig Hecht at Columbia University and in cooperation with David Keilin at the Molteno Institute in Cambridge. He had publications on seed globulins (e.g., ref. 4) with Hubert Vickery of the Connecticut Agricultural Experiment Station and proceeded to publish with Bergmann a pioneering paper (5) on the peptidases of intestinal mucosa (Fig. 1), a communication that marked the beginning of his experimental concern with enzymes; the bibliography of the 1944 article provides a glimpse into the European heritage of the chemistry of peptides and peptidases in that epoch; it began with the name of K. Linderstrøm-Lang and closed with that of E. Abderhalden. The staff of The Rockefeller Institute in 1940 reflected the same tradition; we recall lunching on occasion with Karl Landsteiner, Leonor Michaelis, or P. A. Levene. Today the bibliographies of current papers in our field reflect the truly worldwide growth of biochemistry in the intervening years, an enterprise in which Emil Smith has been a major contributor as an experimentor and educator and through his concern for science at the international level. The breadth of participation in this symposium warmly illustrated the scope of his scientific missions.

ISBN 0-12-643150-7

THE JOURNAL OF BIOLOGICAL CHEMISTRY
Vol. 153, No. 2, May, 1944

THE PEPTIDASES OF INTESTINAL MUCOSA

BY EMIL L. SMITH AND MAX BERGMANN

(*From the Laboratories of The Rockefeller Institute for Medical Research, New York*)

(Received for publication, February 28, 1944)

BIBLIOGRAPHY

1. Linderstrøm-Lang, K., *Z. physiol. Chem.*, **182**, 151 (1929); **188**, 48 (1930).

Etc.

28. Abderhalden, E., and Komm, E., *Z. physiol. Chem.*, **145**, 308 (1925).

Fig. 1. From Emil Smith's first paper on enzymes (5) in 1944. (By permission.)

During these four decades Emil Smith and his associates have explored a remarkable number of aspects of protein chemistry; the list of proteins studied in depth includes aminopeptidases, carboxypeptidases, γ-globulins, cytochromes, papains, subtilisins, histones, and glutamic dehydrogenases. Through direct participation, and through the influence of example, Emil is appreciated by protein chemists working on a very wide variety of researches.

John Pierce has asked me, in the spirit of the following addresses, to focus this essay in considerable part on current experiments from our laboratory, with a measure of history as to how our researches have evolved since Emil and I were among the apprentices in the Bergmann laboratory; that is the limited theme of the following paragraphs. The breadth of this symposium rests on the sum of its contributions.

AMINO ACID ANALYSIS

In a report with Vickery on the amino acid composition of seed globulins in 1941 (4), Emil Smith and his co-authors undertook to determine three amino acids, arginine, tyrosine, and tryptophan; the methods used were gravimetric or colorimetric. The state of the analytical art at that time was not encouraging. For example, in those years, William H. Stein and I had been assigned by Bergmann the task of seeing whether we could develop new methods for determining several amino acids, including glycine and leucine, that did not have unique functional groups. We were struggling with the design of a new approach to gravimetric analysis (6,7), with precipitating agents pioneered by Bergmann. The aim, in the tradition of organic chemistry, was to be able to express the complete composition of a protein in terms of the constituent amino acids, but the goal seemed far away. Erwin Brand, at Columbia, was making a major effort to do this for β-lactoglobulin by the combined use of chemical, microbiological, and gravimetric methods (8). When Stein and I returned to amino acid analysis after the war, the renaissance in chromatography stimulated by Martin and Synge (9) in England was under way. Their concept of partition chromatography, with starch or cellulose as the stationary phase, was rendering the chromatographic method applicable to many classes of water-soluble compounds. Also, Lyman Craig, who was our neighbor on the fifth floor of Flexner Hall, had introduced countercurrent liquid–liquid distribution (10) as a separation method that further evidenced the potential resolving power of multiplate systems. Stein and I decided to see whether we could develop quantitative chromatographic methods for amino acids. We were very happy in those days, when, after considerable developmental work, which included the design of a drop-counting fraction collector (11) and a quantitative photometric ninhydrin method for analysis of the effluent (12), we were able to analyze an acid hydrolysate of serum albumin or β-lactoglobulin in 2 weeks by running three starch columns, with alcohol–water eluents, to resolve all overlaps (13,14). The first user of our starch column procedure, outside of our laboratory, was John Pierce, then working across the street with Vincent du Vigneaud at Cornell University Medical College (15); they thus determined the composition of a hydrolysate of purified beef oxytocin (Fig. 2). At about this time synthetic ion exchange resins came on the market and the time for a complete amino acid analysis was reduced to one week by developing aqueous buffer systems for the elution of amino acids from sulfonated polystyrene columns (16). Emil Smith's laboratory at the University of Utah was one of the first users of the ion exchange, fraction-collector procedure; it was Joe Kimmel who assumed primary responsibility for placing the method in operation for the determination of the amino acid composition of crystalline papain (Fig. 3) in 1954 (17). I recall one special feature of the early analyses in Salt Lake City. It was

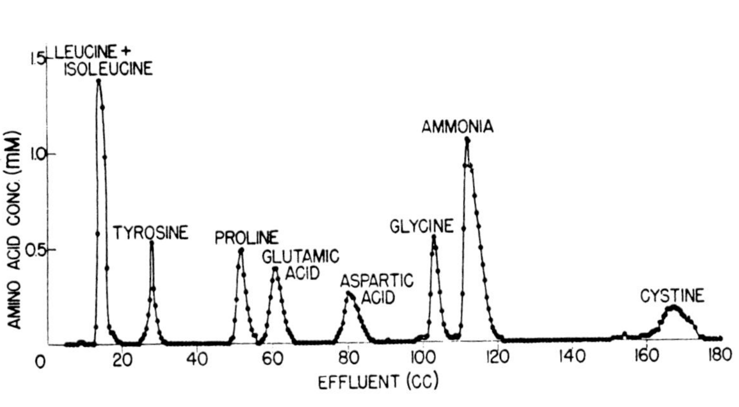

Fig. 2. Separation of the amino acids from a hydrolysate of bovine oxytocin on a starch column [Pierce and du Vigneaud (15), by permission]. A second column was run to separate leucine and isoleucine.

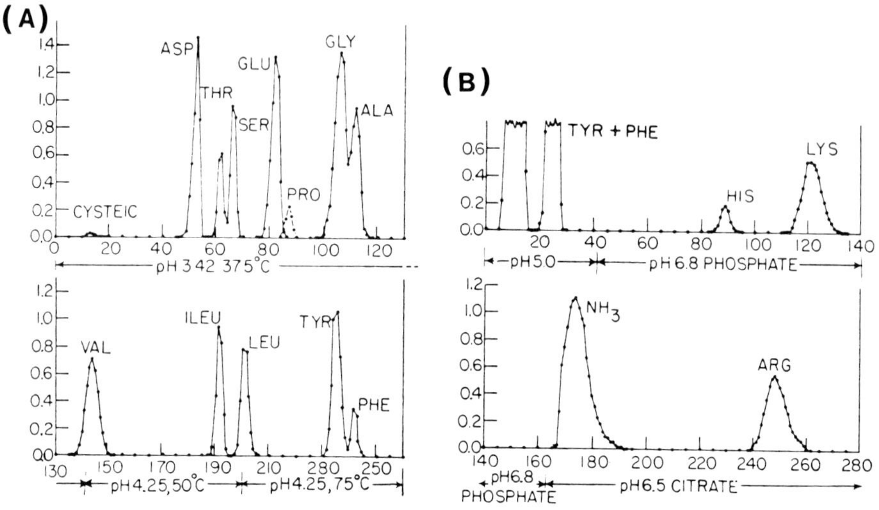

Fig. 3. Elution curves obtained with a 70-hr acid hydrolysate of papain on (A) a 0.9 × 100 cm column and (B) a 0.9 × 15 cm column of Dowex 50. [Smith *et al.* (17), by permission.]

customary, in those days, to hydrolyze proteins by refluxing them with "constant boiling" HCl, which boils at 110° at "atmospheric pressure." Twenty hours of hydrolysis in 6 *N* HCl at 110° generally yields a hydrolysate that contains no measurable amounts of ninhydrin-positive peptides; in Utah they occasionally observed some extra peaks on the chromatographs attributable to peptides. The determining factor was the altitude at Salt Lake City, at which "constant boiling" HCl refluxes at about 105°; hydrolysis was not close enough to completion in 20 hours. Hydrolysis in thermostated sealed tubes at 110° permits reproducible results worldwide.

With the utility of the chromatographic method established and the resolution extended to include analysis of physiologic fluids (18–22), we then decided to improve the system (23) and to bypass the fraction collection step by automating the procedure to measure the ninhydrin color on the flowing stream (24). The key associate on this project was Darrel Spackman, who came to our laboratory in 1954 as a postdoctoral fellow after finishing his thesis on leucine aminopeptidase with Emil Smith. In cooperation with Spackman, an automatic recording instrument was built and the analysis time was reduced to an overnight run.

Emil Smith's laboratory had one of the first commercial models of the automatic amino acid analyzer. In those years, the late 1950s, one of the pioneers in the temporary buildings on a Utah hillside was Robert Hill, who shared with Joe Kimmel the responsibility of seeing that the analyzer delivered results and not too many blue thumbs. I always enjoyed the opportunities to talk shop with Emil and his colleagues in those busy days, with the scene of those conversations soon shifting to the University of California at Los Angeles (UCLA).

Improvements developed in the subsequent 20 years in both academic and industrial laboratories have reduced the time for analysis of an acid hydrolysate to about 1 hr (25) and the sample size to a few micrograms of protein. Special procedures have been designed to meet the needs for the determinations of the dozens of less common amino acids (26) and amino acids that require alkaline hydrolysis [tryptophan (27) and γ-carboxyglutamic acid (28)]. The development of fluorescent detection (29,30) offers the important possibility of further increase in the sensitivity. In cooperation with James Manning, the chromatographic method has been extended to the determination of D- and L-amino acids as D-L and L-L dipeptides (31).

Ribonuclease

When Emil and I started with Max Bergmann in the late 1930s, the determination of the chemical structure of a protein seemed far in the future. Frederick Sanger and his associates (32) changed the picture. The sequence of beef insulin was realized largely through the use of Sanger's dinitrophenyl end-group method, qualitative paper chromatography, and partial acid hydrolysis. The quantitative chromatographic methods, which we were developing concurrently, were designed to increase the feasibility of being able to go up the scale to the study of larger proteins. In looking for a protein on which to test our approach, we wished to select an enzyme, the chemical structure of which would open the way to studies on what makes a protein a catalyst.

We surveyed the literature in 1950 for an enzyme that would be about twice the size of insulin and that gave promise of being obtainable in homogeneous form and in reasonable quantity. Bovine pancreatic ribonuclease (MW 14,000), which had been crystallized by Moses Kunitz at The Rockefeller Institute in 1940 (33), was the protein that we selected for study. Insulin, conveniently, had lacked methionine and tryptophan; ribonuclease had methionine but no tryptophan, and thus represented a one-step increase in complexity in terms of labile amino acids. In cooperation with Werner Hirs, our first postdoctoral associate, we decided to see whether there was any chance of using ion exchange chromatography to purify a protein. It turned out that ribonuclease could be chromatographed on a polymethacrylic acid resin, Amberlite IRC-50, with a finite distribution coefficient (Fig. 4) (34,35), just as if it were an amino acid (34). Sober and Peterson (36), in 1954, vastly widened the scope of ion exchange chromatography of proteins by the key introduction of ion exchangers with a hydrophilic matrix, such as cellulose.

With chromatographically homogeneous ribonuclease on hand, the next problem was how to split the protein into appropriate peptide segments. Relatively random hydrolysis by acid of a protein of this size, as Sanger had pointed out (37), would probably give unduly complicated mixtures of peptides. There was serious concern at that time as to whether enzymic cleavage

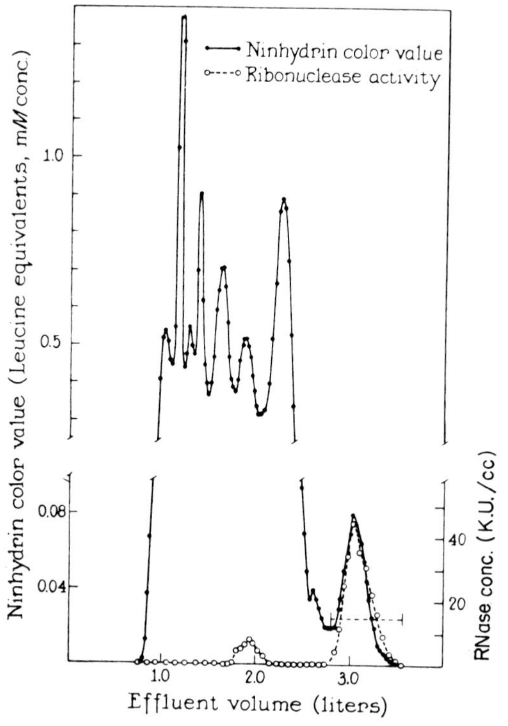

Fig. 4. Isolation of RNase A by chromatography of a 0.25 *N* sulfuric acid extract of comminuted beef pancreas. Preparative scale column, 7.5 × 60 cm [Hirs *et al.* (34), by permission]. The smaller active peak was later characterized as a glycosylated derivative of RNase A by Plummer and Hirs (35).

could be generally used because of the hazard of production of artifacts by transpeptidation. Fortunately, it has turned out that this phenomenon has not interfered with the use of trypsin, chymotrypsin, and pepsin in protein structure work; all three enzymes were used on performic acid-oxidized ribonuclease with consistent results (38–40). The peptides thus obtained, primarily by Werner Hirs, were separated quantitatively by ion exchange chromatography on Dowex 50. The use of pepsin on the intact protein at pH 2, to minimize disulfide interchange, was the key step in Darrel Spackman's location of the disulfide bridges in ribonuclease (41). The chemical reaction that made possible most of the sequencing of the peptides was, of course, the invaluable stepwise degradation with phenylisothiocyanate introduced by Edman in 1950 (42), with important assistance from two enzymes extensively studied in Emil Smith's laboratory, leucine aminopeptidase (43) and carboxypeptidases (44).

As is frequently true in science, we were not alone in our endeavor on RNase. The timeliness of research on pancreatic ribonuclease also occurred to Christian Anfinsen at the National Institutes of Health who began similar structural studies in the early 1950s (45–47). Cordial and free interchange of results of experiments in progress expedited the solution of the problem. The final sequence (Fig. 5), obtained in cooperation with Derek Smyth in 1963 (48,49), was also dependent on the introduction by Gross and Witkop (50) in 1961 of the specific cleavage of proteins at methionine residues by cyanogen bromide, a method that provided definitive information on the sequence around Met-13.

With the sequence known, various experiments were undertaken to define the residues most directly concerned with the catalytic function. Iodoacetate had been shown (51) to inactivate RNase and at that time iodoacetate was thought of as a reagent for SH-groups. When we found that RNase does not have any SH-groups, reexamination of the iodoacetate reaction was in order. Research by Barnard and W. D. Stein (52,53) and by our laboratory (54–56) led to the identification of 1-carboxymethyl-His-119 and 3-carboxymethyl-His-12 (55) as the products of the reaction. Our conclusion that these two residues were near the active center of the enzyme was confirmed when the three-dimensional structure of RNase S was realized through the X-ray analyses of Richards and Wyckoff (57). In a study of the effects of denaturing reagents on the special reactivity toward iodoacetate of these two histidine residues, George Stark tried the iodoacetate reaction in 8 *M* urea at 40° (58). Amino acid analyses showed the presence of homocitrulline formed by carbamylation of the ϵ-NH_2 groups of lysine residues in the protein by cyanate present as a contaminant in the urea. The result served as one of the several reminders that the equilibrium between ammonium cyanate and urea, first studied by Wöhler (59) in 1828, continues to be of practical concern to biochemists (Fig. 6). Stark's observations on cyanate in urea provided part of the basis for Anthony Cerami's idea, explored in cooperation with James

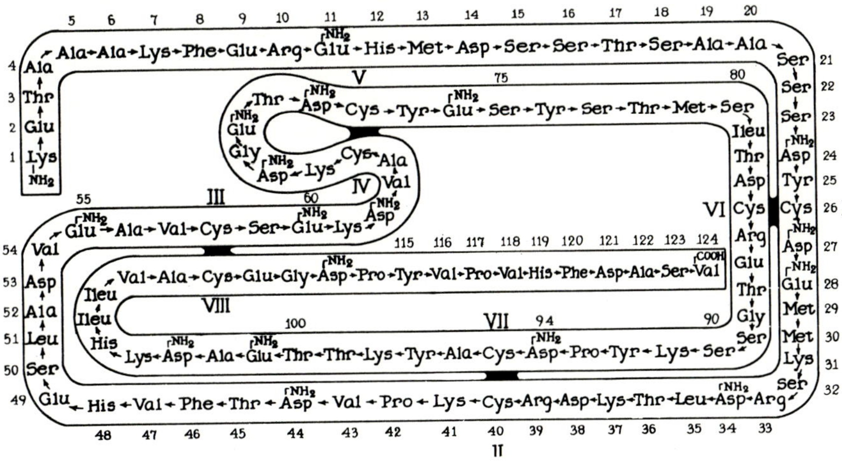

Fig. 5. The sequence of amino acid residues in bovine pancreatic ribonuclease A [from Smyth *et al.* (49), by permission, based on Hirs *et al.* (38–41), Anfinsen *et al.* (45–48), and Gross and Witkop (50)].

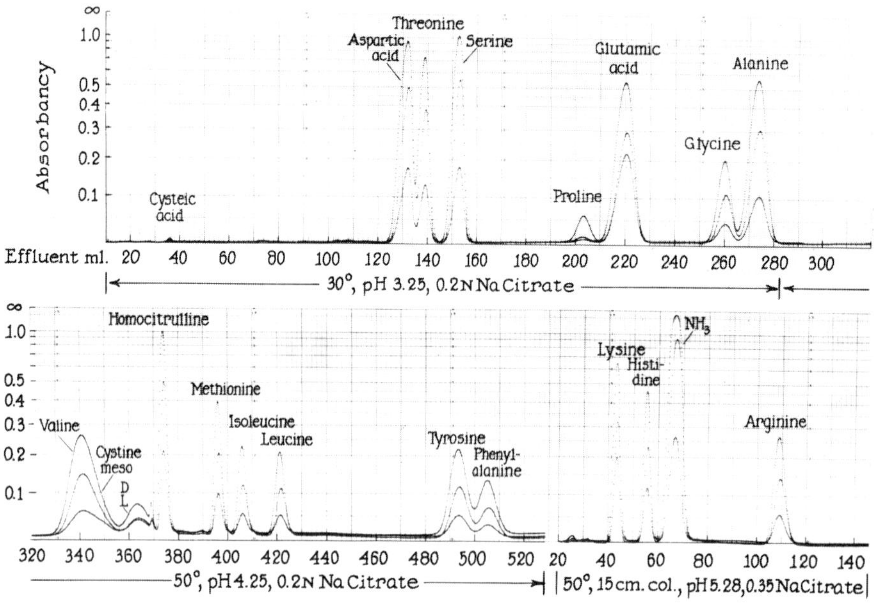

Fig. 6. Amino acid analysis of ribonuclease A that had been exposed to cyanate. [Stark *et al.* (58), by permission.]

Manning (60), of trying cyanate as an inhibitor of the sickling of erythrocytes carrying hemoglobin S (61). The renewed interest in the simple chemistry of the cyanate reaction proved useful in Emil Smith's laboratory in study of the pK_a of the ϵ-NH_2 group of lysine-97 in glutamate dehydrogenase (62).

From ribonuclease we moved on to pancreatic deoxyribonuclease (63,64), a protein that is twice the size of ribonuclease. The sequence study was completed (65,66) in cooperation with Ta-hsiu Liao, who came to our laboratory as a postdoctoral fellow after receiving his biochemical training at UCLA.

Today the dozen years that it took to arrive at the chemical structure of ribonuclease seem very long in the light of the sequence explosion that has brought protein chemistry into a new era, and X-ray crystallography has added a new dimension. A host of enzymic, chemical, and physical methods have been brought to bear on protein sequence studies, a number pioneered by Emil Smith and by speakers at this symposium. The texts of this volume will reflect the extent of progress with numerous proteins of diverse functions. In the interest of continuity in this introduction, I am going to focus on ribonuclease, as of 1979, to illustrate how the subject has grown in the 40 years since the crystallization of the enzyme by Kunitz.

Ribonuclease Revisited

When we and others began our chemical studies on pancreatic ribonuclease, we thought that we were working on an enzyme of admittedly limited physiologic function. The protein was considered as one of the catalysts of the digestive tract. Species variations in the occurrence and in the structures of pancreatic ribonucleases have received detailed study in the laboratories of Barnard (67) and Beintema (68). The first broadening of the picture to include chemical studies on neutral ribonucleases from other mammalian sources came from the researches of Leone and D'Alessio and their associates in Naples. They found (69) that about 3% of the total protein of bovine seminal plasma is a ribonuclease, which makes the fluid the richest known source of a ribonuclease. They proceeded to purify the seminal enzyme and to determine its chemical structure (70). The molecule, which has an enzymic specificity similar to that of the pancreatic enzyme, turned out to be a dimer (MW 28,000) cross-linked by two —S—S— bonds. The sequence of the 124-residue subunit proved to be homologous with but not identical to the sequence of the pancreatic enzyme of the same species. Key features of RNase A (71) were maintained; the four intramolecular disulfide bonds, His-12, His-119, and Lys-41, are in exactly the same positions. However, there are about 23 substitutions in noncrucial parts of the chain. The protein is made by a different gene from that which makes the pancreatic enzyme in the same species. Thus nature makes more than one RNase to the same catalytic model but with different accessories by different genes in different tissues of the body. (We will return to the breadth of the tissue distribution of RNase-A-like enzymes in the following discussion of ribonuclease inhibitors.) A special difference in this instance is the substitution of two cysteine residues at positions 31 and 32 to permit the formation of the two disulfide bonds that yield the dimer. Libonati and Floridi (72) then made the observation that the dimeric seminal enzyme, at pH 7.5 in 0.14 *M* salt, is more active than the pancreatic monomer toward double-stranded substrates such as poly(A)·poly(U).

In the early studies on RNase A by Stein and myself, and the number of young scholars with whom we had the pleasure of collaboration in that work, one of the approaches was to see whether we could *lower* the activity of the enzyme by specific derivatizations and thereby plot the positions of residues at or near the active center of the catalyst. It is also informative to see whether you can take an enzyme and make it *more* active than the native molecule. With this aim in view, the findings of Libonati and Floridi prompted us to see whether, if we were to take two molecules of the monomeric pancreatic enzyme and chemically cross-link them, we would obtain a covalently linked, nondissociable dimer that would simulate the seminal enzyme in enzymic properties. In cooperation with Dalton Wang and Glynn Wilson (73), conditions were developed for cross-linking at ϵ-NH_2-groups of lysine residues with dimethyl suberimidate that would favor maximum yield of dimeric products with optimum activity toward double-stranded substrates. A 20% yield was obtained of a preparation that possessed eight times the activity of the pancreatic monomer toward the U strand of the poly(A)·poly(U).

An eightfold activation is only modest. How might we do better? From studies of Libonati and Beintema on the activities of ribonucleases of different species, particularly whale pancreatic ribonuclease (74), it was concluded that the determining characteristic for activity toward double-stranded substrates may be the number of basic charges on the enzyme. Therefore we decided to see what would happen if we cross-linked ribonuclease and spermine, a polyamine with special affinity for nucleic acids. In cooperation with Wang (75), we obtained a product with about eight spermine residues, mainly in one chain, attached via dimethyl suberimidate; the polyspermine ribonuclease had 115 times the activity of the monomer toward poly (A) · poly (U) at low substrate concentrations. Wang's kinetic data (76) showed that the binding to poly (A) · poly (U) has been made 100 times as strong and that V_{max} has been increased sixfold. Under similar conditions the activity toward the rU strand of the hybrid substrate, poly(rU)·poly(dA), was increased 380-fold. A natural double-stranded RNA from reovirus 3 could be completely hydrolyzed to oligonucleotides in 10 min by the polyspermine-RNase A with 1 part of enzyme to 120 parts of substrate at physiologic pH and salt concentration. We conclude that synthetic hybrids offer a number of possibilities for the design of enzymes with new biochemical properties.

Experiments along this line have included the chemical coupling of carbohydrates to RNase A to alter its biological properties. Several lines of research are indicating that dimeric ribonucleases and their derivatives, for reasons that are not yet defined, are more cytostatic than the monomer toward tumor cells (77,78), and Bartholeyns and Baudhuin (79) have observed the inhibition of hepatoma cells in culture. With the thought of possibly aiming the cross-linked RNase dimer for the liver, Glynn Wilson (80) has undertaken the coupling of galactose residues to RNase dimer. The biological approach, in this instance, stems from the findings of Ashwell, Morell, and co-workers (81) that the exposure of galactose as the terminal sugar on serum glycoproteins provides a specific recognition signal for receptor-mediated uptake into hepatocytes, and the demonstration by Rogers and Kornfeld (82) that the asialoglycopeptide of fetuin (which contains terminal galactose residues), when covalently linked to either albumin or lysozyme, caused rapid selective uptake of these proteins from the blood by the liver.

Reductive lactosamination with cyanoborohydride, according to the procedure of Gray (83), has worked well with RNase A dimer. The small amount of Schiff base formed at any one time between the aldehydic moiety of the disaccharide and the ϵ-NH_2 groups of a lysine residue is reduced to give a stable alkyl derivative (Fig. 7). The disaccharide lactose is used in our experiments in order to yield a product that retains the galactopyranose ring, which is essential for liver uptake. The product with eight galactose residues attached per molecule of dimer retains 80% of the native activity toward yeast RNA. For the tracer experiments in rats (Fig. 8), the protein derivatives were treated with [^{14}C]iodoacetate, which carboxymethylates His-12 or His-119 and largely inactivates the enzyme (53). The top diagram shows the uptake of nonglycosylated RNase A dimer after intravenous injection in the rat. At 30

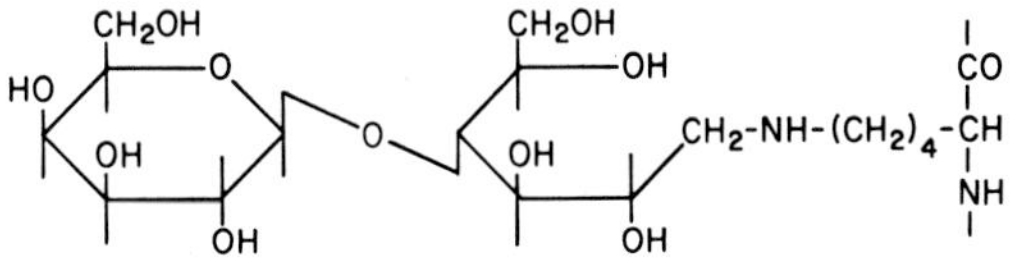

Fig. 7. The product [an N^{ϵ}-1-(deoxylactitolyl-lysyl residue)] of the coupling of lactose to the ϵ-NH_2 group of a lysine residue by reductive lactosamination with cyanoborohydride according to Gray (83).

min, more than 90% of the injected labeled dose was removed from the blood and taken up by the kidneys. Only 4% was taken out by the liver. By contrast, with lactose attached, the clearance from the blood was more rapid, and at 10 min 69% of the protein was in the liver. This result shows that reductive lactosamination provides a simple method for directing a protein of possible therapeutic interest to the hepatocytes, even when the protein is of sufficiently low molecular weight to be subject to rapid uptake in the kidney.

In a cooperative study, Ann L. Hubbard at Yale University (84) has shown by electron microscope autoradiography that the lactosaminated derivative was exclusively localized in the hepatocytes; the similarly prepared mannobiosaminated dimer has been prepared by Wilson (85) and was taken up by the endothelial and Kuppfer cells of the liver (84).

The Ribonuclease Inhibitor from the Human Placenta

In parallel with studies on the enhancement of the activity or the cellular uptake of the ribonuclease molecule, we have become concerned with the ways in which nature inhibits RNase. Earlier studies, particularly by J. S. Roth (86), have shown that there is a protein in rat liver that will inhibit pancreatic ribonuclease. Through the finding that *p*-hydroxymercuribenzoate will inactivate the inhibitor, Roth demonstrated that there are small but significant amounts of normally inhibited neutral RNases of the pancreatic type in most mammalian tissues. In order to study in more detail the possible roles of such nucleases and their inhibitors in the maintenance of messenger RNA during protein biosynthesis, we undertook the isolation of the pure inhibitor for chemical characterization.

Since the placenta is a tissue of considerable biosynthetic activity, we decided to explore the RNase–inhibitor equilibrium in that tissue. Affinity chromatography is clearly the technique of choice for the isolation of such an inhibitor, since pancreatic RNase provides a readily available protein to which it binds strongly. There have been many attempts to isolate ribonuclease inhibitors from mammalian tissues, but none of the previous methods have yielded a pure, stable protein. There are several features of a method developed by Peter Blackburn and Glynn Wilson (87) that contribute to the procedure's success (Table 1). Human placentas obtained within hours of normal term delivery were homogenized at pH 7.5 in 0.25 *M* sucrose, 5 m*M*

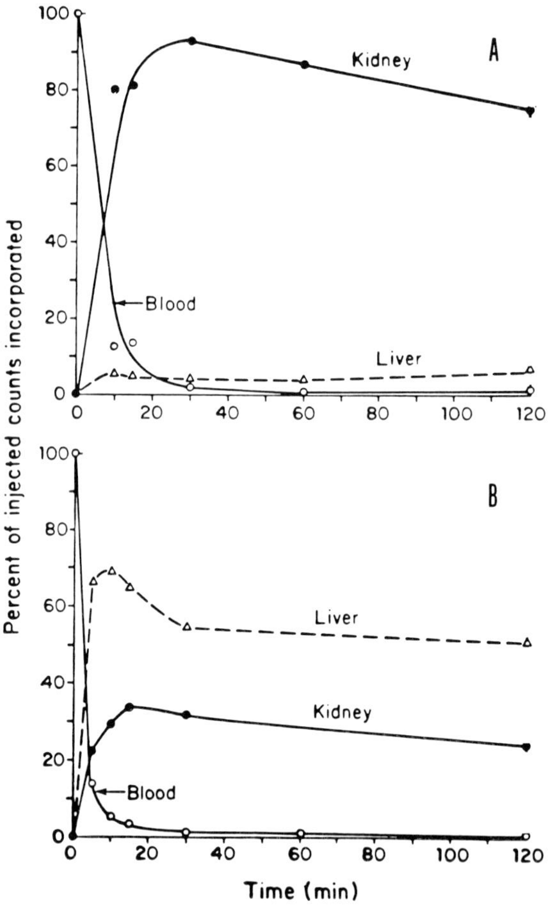

Fig. 8. Tissue uptake of (A) ^{14}C-labeled carboxymethylated cross-linked RNase dimer and (B) its lactosylated derivative after injection into the tail vein of rats. [Wilson (80), by permission].

TABLE 1
PURIFICATION OF RNASE INHIBITOR FROM HUMAN TERM PLACENTA[a]

Step	Total protein (mg)	Total inhibitor (units)	Specific activity (units/mg)	Recovery of activity (%)
1. Initial extract	16,000	400,000	25	100
2. 35–50% ammonium sulfate precipitate	2327	320,000	138	80
3. 48,000 g, 1 hr	1646	320,000	194	80
4. DEAE-cellulose chromatography	308	240,000	780	60
5. Sepharose-RNase A affinity chromatography	1.8	180,000	100,000	45

[a] From Blackburn *et al.* (87).

in dithiothreitol, and 1 mM in EDTA. The inhibitor is a protein with free SH-groups that are essential for activity and it takes an excess of protecting thiol to prevent oxidative inactivation, particularly in the initial extract in which oxidative loss of dithiothreitol is accelerated by some component of the homogenate. The key step is the affinity chromatography on RNase A coupled to Sepharose (Fig. 9). The active inhibitor was successfully eluted by dropping the pH from 6.5 to 5.0 and making the eluent 3 M in NaCl and 15% in glycerol. The yield was about 2 mg per placenta of inhibitor that is fully stable for months at −10°. Blackburn (88) has subsequently simplified the purification and assay. The product is homogeneous by SDS gel electrophoresis (MW ca. 50,000) and by isoelectric focusing (pI = 4.6). When the results of amino acid analysis were expressed in terms of MW 51,000, the major amino acid was leucine (85 residues) and of the 30 half-cystine plus cysteine residues, 8 were in the SH-form, as evidenced by reaction of the native protein with iodoacetic acid to give carboxymethylcysteine residues.

A double-reciprocal plot of $1/v$ versus $1/s$ at different inhibitor concentrations (87) showed that the inhibition is noncompetitive; a Dixon plot of $1/v$ versus inhibitor concentration gave a binding constant of 3×10^{-10} M. In a study of a number of aspects of the nature of the interaction of enzyme and inhibitor, Blackburn and Jailkhani (89) have shown that modification of the arginine residues of the enzyme by 1,2-cyclohexanedione, as studied for

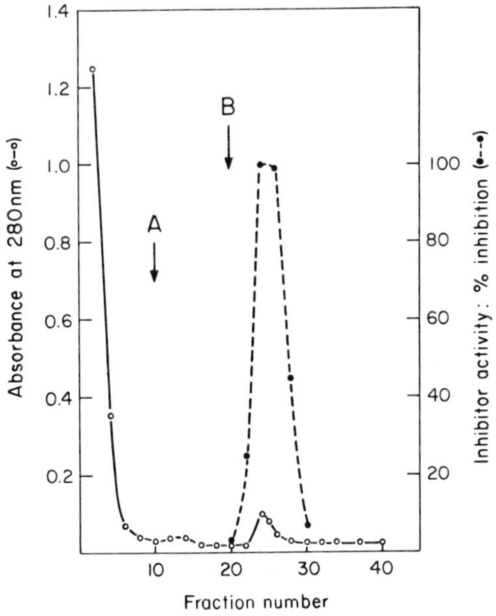

Fig. 9. Affinity chromatography of human placental RNase inhibitor on RNase-Sepharose. At A the 45 mM phosphate buffer at pH 6.4 was made 0.5 M in NaCl. At B the buffer was changed to 20 mM acetate, pH 5.0, 3.0 M in NaCl and 15% in glycerol. All eluents were 5 mM in dithiothreitol and 1 mM in EDTA. [Blackburn *et al.* (87), by permission].

ribonuclease in 1975 by Patthy and Smith (90), does not markedly alter the binding; lysine residues in the enzyme, on the other hand, are crucial.

With sufficient pure inhibitor on hand, experiments are currently being conducted in cooperation with investigators at The Rockefeller University to study the effect of adding the inhibitor to *in vitro* translation systems. George Scheele and Blackburn (91) have found that the placental inhibitor at 10 μg/ml in a mixture of wheat germ extract plus translation components increased the amounts of proteins synthesized via mRNA from dog pancreas or from influenza virus-infected cells. Analysis of the translation products by sodium dodecyl sulfate (SDS)-gel electrophoresis indicated that the presence of the inhibitor preferentially increased the yield of the larger molecular weight proteins, the messenger RNAs for which present larger targets for an endonuclease; yields of the preprocarboxypeptidases were increased 4.5-fold and yields of preamylase were increased 15-fold. Robbi and Lazarow (92), at The Rockefeller University, have shown that the inhibitor permits a dramatic improvement in the translation of rat liver catalase mRNA by wheat germ extract. Scheele and Blackburn have also shown that the inhibitor can be very helpful in the preparation of rough microsomes and detached polysomes from rat liver for the purposes of translation studies.

Conclusion

After 40 years of research on the chemistry of pancreatic ribonuclease, it might be thought that the subject would be reaching a stage of diminishing returns. Yet, as is true of so many aspects of biochemistry, there is exponential growth as the interrelationships in the intricate synergisms of living systems come into fuller view. When Emil Smith and I entered science in the 1930s, progress was slow. Today the tools of the trade and the data bank of biochemical knowledge on which the experimenter can draw provide a rich heritage. Students who are entering biochemistry in the 1980s can look forward to the satisfaction of seeing a far deeper impact of our science on medicine and human welfare than we have seen in our lifetimes. I hope that the young scholars who read the essays to follow will sense the enthusiasm that Emil Smith and contemporary biochemists share.

REFERENCES

1. Smith, E. L., The induction period in photosynthesis. *J. Gen. Physiol.*, 1937, **21**:151–163.
2. Smith, E. L., Limiting factors in photosynthesis: Light and carbon dioxide. *J. Gen. Physiol.*, 1938, **22**:21–35.
3. Smith, E. L., The chlorophyll-protein compound in the green leaf. *J. Gen. Physiol.*, 1941, **24**:565–582.
4. Vickery, H. B., Smith, E. L., Hubbell, R. B., and Nolan, L. S., Cucurbit seed

globulins. I. Amino acid composition and preliminary tests of nutritive value. *J. Biol. Chem.*, 1941, **140**:613–624.
5. Smith, E. L., and Bergmann, M., The peptidases of intestinal mucosa. *J. Biol. Chem.*, 1944, **153**:627–651.
6. Bergmann, M., and Stein, W. H., A new principle for the determination of amino acids and its application to collagen and gelatin. *J. Biol. Chem.*, 1939, **128**:217–232.
7. Moore, S., and Stein, W. H., Determination of amino acids by the solubility product method. *J. Biol. Chem.*, 1943, **150**:113–130.
8. Brand, E., Saidel, L. J., Goldwater, W. H., Kassell, B., and Ryan, F. J., The empirical formula of β-lactoglobulin. *J. Am. Chem. Soc.*, 1945, **67**:1524–1532.
9. Martin, A. J. P., and Synge, R. L. M., Analytical chemistry of the proteins. *Adv. Protein Chem.* 1945, **2**:1–83.
10. Craig, L. C., Identification of small amounts of organic compounds by distribution studies. II. Separation by counter-current distribution. *J. Biol. Chem.*, 1944, **155**:519–534.
11. Stein, W. H., and Moore, S., Chromatography of amino acids on starch columns. Separation of phenylalanine, leucine, isoleucine, methionine, tyrosine, and valine. *J. Biol. Chem.*, 1948, **176**:337–365.
12. Moore, S., and Stein, W. H., Photometric ninhydrin method for use in the chromatography of amino acids. *J. Biol. Chem.*, 1948, **176**:367–388.
13. Moore, S., and Stein, W. H., Chromatography of amino acids on starch columns. Solvent mixtures for the fractionation of protein hydrolysates. *J. Biol. Chem.*, 1949, **178**:53–77.
14. Stein, W. H., and Moore, S., Amino acid composition of β-lactoglobulin and bovine serum albumin. *J. Biol. Chem.*, 1949, **178**:79–91.
15. Pierce, J. G., and du Vigneaud, V., Preliminary studies on the amino acid content of a high potency preparation of the oxytocic hormone of the posterior lobe of the pituitary gland. *J. Biol. Chem.*, 1949, **182**:359–366.
16. Moore, S., and Stein, W. H., Chromatography of amino acids on sulfonated polystyrene resins. *J. Biol. Chem.*, 1951, **192**:663–681.
17. Smith, E. L., Stockell, A., and Kimmel, J. R., Crystalline papain. III. Amino acid composition. *J. Biol. Chem.*, 1954, **207**:551–561.
18. Stein, W. H., A chromatographic investigation of the amino acid constituents of normal urine. *J. Biol. Chem.*, 1953, **201**:45–48.
19. Moore, S., and Stein, W. H., Procedures for the chromatographic determination of amino acids on four per cent cross-linked sulfonated polystyrene resins. *J. Biol. Chem.*, 1954, **211**:893–906.
20. Moore, S., and Stein, W. H., A modified ninhydrin reagent for the photometric determination of amino acids and related compounds. *J. Biol. Chem.*, 1954, **211**:907–913.
21. Stein, W. H., and Moore, S., The free amino acids of human blood plasma. *J. Biol. Chem.*, 1954, **211**:915–925.
22. Tallan, H. H., Moore, S., and Stein, W. H., Studies on the free amino acids and related compounds in the tissues of the cat. *J. Biol. Chem.*, 1954, **211**:927–939.
23. Moore, S., Spackman, D. H., and Stein, W. H., Chromatography of amino acids

on sulfonated polystyrene resins. An improved system. *Anal. Chem.*, 1958, **30**:1185–1190.
24. Spackman, D. H., Stein, W. H., and Moore, S., Automatic recording apparatus for use in the chromatography of amino acids. *Anal. Chem.*, 1958, **30**:1190–1206.
25. Moore, S., The precision and sensitivity of amino acid analysis. In: *Chemistry and Biology of Peptides* (J. Meienhofer, ed.). Ann Arbor Sci. Publ., Ann Arbor, Michigan, 1972:629–653.
26. Hamilton, P. B., Elution behavior of naturally occurring ninhydrin-positive compounds during ion exchange chromatography. In: *Handbook of Biochemistry* (H. A. Sober, ed.). Chemical Rubber Publishing Co., Cleveland, Ohio, 1968:B49–B54.
27. Hugli, T. E., and Moore, S., Determination of the tryptophan content of proteins by ion exchange chromatography of alkaline hydrolysates. *J. Biol. Chem.*, 1972, **247**:2828–2834.
28. Hauschka, P. V., Quantitative determination of γ-carboxyglutamic acid in proteins. *Anal. Biochem.*, 1977, **80**:212–223.
29. Roth, M., Fluorescence reaction for amino acids. *Anal. Chem.*, 1971, **43**:880–882.
30. Udenfriend, S., Stein, S., Böhlen, P., Dairman, W., Leimgruber, W., and Weigele, M., Fluorescamine: A reagent for assay of amino acids, peptides, proteins, and primary amines in the picomole range. *Science*, 1972, **178**:871–872.
31. Manning, J. M., and Moore, S., Determination of D- and L-amino acids by ion exchange chromatography as L-D and L-L dipeptides. *J. Biol. Chem.*, 1968, **243**:5591–5597.
32. Sanger, F., Chemistry of insulin. *Science*, 1959, **129**:1340–1344.
33. Kunitz, M., Crystalline ribonuclease. *J. Gen. Physiol.*, 1940, **24**:15–32.
34. Hirs, C. H. W., Moore, S., and Stein, W. H., A chromatographic investigation of pancreatic ribonuclease. *J. Biol. Chem.*, 1953, **200**:493–506.
35. Plummer, T. H., Jr., and Hirs, C. H. W., The isolation of ribonuclease B, a glycoprotein, from bovine pancreatic juice. *J. Biol. Chem.*, 1963, **238**:1396–1401.
36. Sober, H. A., and Peterson, E. A., Chromatography of proteins on cellulose ion-exchangers. *J. Am. Chem. Soc.*, 1954, **76**:1711–1712.
37. Sanger, F., The arrangement of amino acids in proteins. *Adv. Protein Chem.*, 1952, **7**:1–67.
38. Hirs, C. H. W., Moore, S., and Stein, W. H., Peptides obtained by tryptic hydrolysis of performic acid-oxidized ribonuclease. *J. Biol. Chem.*, 1956, **219**:623–642.
39. Hirs, C. H. W., Stein, W. H., and Moore, S., Peptides obtained by chymotryptic hydrolysis of performic acid-oxidized ribonuclease. A partial structural formula for the oxidized protein. *J. Biol. Chem.*, 1956, **221**:151–169.
40. Bailey, J. L., Moore, S., and Stein, W. H., Peptides obtained by peptic hydrolysis of performic acid-oxidized ribonuclease. *J. Biol. Chem.*, 1956, **221**:143–150.
41. Spackman, D. H., Stein, W. H., and Moore, S., The disulfide bonds of ribonuclease. *J. Biol. Chem.*, 1960, **235**:648–659.
42. Edman, P., Method for the determination of the amino acid sequence in peptides. *Acta Chem. Scand.*, 1950, **4**:283–293.

43. Spackman, D. H., Smith, E. L., and Brown, D. M., Leucine aminopeptidase. IV. Isolation and properties of the enzyme from swine kidney. *J. Biol. Chem.*, 1955, **212**:255–269.
44. Smith, E. L., The specificity of certain peptidases. *Adv. Enzymol.*, 1951, **12**:191–257.
45. Anfinsen, C. B., Redfield, R. R., Choate, W. L., Page, J., and Carroll, W. R., Studies on the gross structure, cross-linkages, and terminal sequences in ribonuclease. *J. Biol. Chem.*, 1954, **207**:201–210.
46. Redfield, R. R., and Anfinsen, C. B., The structure of ribonuclease. II. The preparation, separation, and relative alignment of large enzymically produced fragments. *J. Biol. Chem.*, 1956, **221**:385–404.
47. Potts, J. T., Berger, A., Cooke, J., and Anfinsen, C. B., A reinvestigation of the sequence of residues 11 to 18 in bovine pancreatic ribonuclease. *J. Biol. Chem.*, 1962, **237**:1851–1855.
48. Smyth, D. G., Stein, W. H., and Moore, S., On the sequence of residues 11 to 18 in bovine pancreatic ribonuclease. *J. Biol. Chem.*, 1962, **237**:1845–1850.
49. Smyth, D. G., Stein, W. H., and Moore, S., The sequence of amino acid residues in bovine pancreatic ribonuclease: revisions and confirmations. *J. Biol. Chem.*, 1963, **238**:227–234.
50. Gross, E., and Witkop, B., Nonenzymic cleavage of peptide bonds: The methionine residues in bovine pancreatic ribonuclease. *J. Biol. Chem.*, 1962, **237**:1856–1860.
51. Zittle, C. A., Ribonucleinase. III. The behavior of copper and calcium in the purification of nucleic acid and the effect of these and other reagents on the activity of ribonucleinase. *J. Biol. Chem.*, 1946, **163**:111–117.
52. Barnard, E. A., and Stein, W. D., The histidine residue in the active centre of ribonuclease. I. A specific reaction with bromoacetic acid. *J. Mol. Biol.*, 1959, **1**:339–349.
53. Stein, W. D., and Barnard, E. A., The histidine residue at the active centre of ribonuclease. II. The position of this residue in the primary protein chain. *J. Mol. Biol.*, 1959, **1**:350–358.
54. Gundlach, H. G., Stein, W. H., and Moore, S., The nature of the amino acid residues involved in the inactivation of ribonuclease by iodoacetate. *J. Biol. Chem.*, 1959, **234**:1754–1760.
55. Crestfield, A. M., Stein, W. H., and Moore, S., Alkylation and identification of the histidine residues at the active site of ribonuclease. *J. Biol. Chem.*, 1963, **238**:2413–2420.
56. Crestfield, A. M., Stein, W. H., and Moore, S., Properties and conformation of the histidine residues at the active site of ribonuclease. *J. Biol. Chem.*, 1963, **238**:2421–2428.
57. Richards, F. M., and Wyckoff, H. W., Bovine pancreatic ribonuclease. In: *The Enzymes* (P. B. Boyer, ed.), 3rd ed., Vol. 4. Academic Press, New York, 1971:647–806.
58. Stark, G. W., Stein, W. H., and Moore, S., Reactions of cyanate present in aqueous urea with amino acids and proteins. *J. Biol. Chem.*, 1960, **235**:3177–3181.
59. Wöhler, F., On the artificial formation of urea. *Pogg. Ann. Phys. Chem. (Leipzig)* [2], 1828, **12**:253–256.

60. Cerami, A., and Manning, J. M., Potassium cyanate as an inhibitor of the sickling of erythrocytes *in vitro*. *Proc. Natl. Acad. Sci. U.S.A.*, 1971, **68**:1180–1183.
61. Manning, J. M., and Nigen, A. M., Principles derived from studies on the carbamylation of hemoglobin S. In: *Proceedings of the Symposium on Molecular and Cellular Aspects of Sickle Cell Disease* (J. I. Hercules, G. L. Cottam, M. R. Waterman, and A. N. Schechter, eds.), DHEW Publ. (NIH) 76-1007. National Institutes of Health, Bethesda, Maryland, 1976:361–377.
62. Veronese, F. M., Piszkiewicz, D., and Smith, E. L., Inactivation of bovine glutamate dehydrogenase by carbamyl phosphate and cyanate. *J. Biol. Chem.*, 1972, **247**:754–759.
63. Price, P. A., Liu, T.-Y., Stein, W. H., and Moore, S., Properties of chromatographically purified bovine pancreatic deoxyribonuclease. *J. Biol. Chem.*, 1969, **244**:917–923.
64. Salniknow, J., Moore, S., and Stein, W. H., Comparison of the multiple forms of bovine pancreatic deoxyribonuclease. *J. Biol. Chem.*, 1970, **245**:5685–5990.
65. Salnikow, J., Liao, T.-H., Moore, S., and Stein, W. H., Bovine pancreatic deoxyribonuclease A. Isolation, composition, and amino acid sequences of the tryptic and chymotryptic peptides. *J. Biol. Chem.*, 1973, **248**:1480–1488.
66. Liao, T.-H., Salnikow, J., Moore, S., and Stein, W. H., Bovine pancreatic deoxyribonuclease A. Isolation of cyanogen bromide peptides; complete covalent structure of the polypeptide chain. *J. Biol. Chem.*, 1973, **248**:1489–1495.
67. Barnard, E. A., Ribonucleases. *Annu. Rev. Biochem.* 1969, **38**:677–732.
68. van den Berg, A., and Beintema, J. J., Non-constant evolution rates in amino acid sequences of guinea pig, chinchilla, and coypu pancreatic ribonucleases. *Nature (London)*, 1975, **253**:207–210.
69. D'Alessio, G., Floridi, A., De Prisco, R., Pignero, A., and Leone, E., Bull semen ribonucleases. 1. Purification and physicochemical properties of the major component. *Eur. J. Biochem.*, 1972, **26**:153–161.
70. D'Alessio, G., Malorni, M. C., and Parente, A., Dissociation of bovine seminal ribonuclease into catalytically active monomers by selective reduction and alkylation of the intersubunit disulfide bridges. *Biochemistry*, 1975, **14**:1116–1122.
71. Moore, S., and Stein, W. H., Chemical structures of pancreatic ribonuclease and deoxyribonuclease. *Science*, 1973, **180**:458–464.
72. Libonati, M., and Floridi, A., Breakdown of double-stranded RNA by bull semen ribonuclease. *Eur. J. Biochem.*, 1969, **8**:81–87.
73. Wang, D., Wilson, G., and Moore, S., Preparation of cross-linked dimers of pancreatic ribonuclease. *Biochemistry*, 1976, **15**:660–665.
74. Libonati, M., Furia, A., and Beintema, J. J., Basic charges on mammalian ribonuclease molecules and the ability to attack double-stranded RNA. *Eur. J. Biochem.*, 1976, **69**:444–451.
75. Wang, D., and Moore, S., Polyspermine-ribonuclease prepared by cross-linkage with dimethyl suberimidate. *Biochemistry*, 1977, **16**:2937–2941.
76. Wang, D., Observations on kinetics of action of polyspermine-ribonuclease on poly A·poly U. *Biochim. Biophys. Acta*, 1979, **568**:488–490.
77. Matousek, J., The effect of bovine seminal ribonuclease on cells of Crocker tumor virus. *Experientia*, 1973, **29**:858–859.
78. Tarnowski, G. S., Kassel, R. L., Mountain, I. M., Blackburn, P., Wilson, G.,

and Wang, D., Comparison of antitumor activities of pancreatic ribonuclease and its cross-linked dimer. *Cancer Res.*, 1976, **36**:4074–4078.
79. Bartholeyns, J., and Baudhuin, P., Inhibition of tumor cell proliferation by dimerized ribonuclease. *Proc. Natl. Acad. Sci. U.S.A.*, 1976, **73**:573–576.
80. Wilson, G., Effect of reductive lactosamination on the hepatic uptake of bovine pancreatic ribonuclease A dimer. *J. Biol. Chem.*, 1978, **253**:2070–2072.
81. Ashwell, G., and Morell, A. G., The role of surface carbohydrates in the hepatic recognition and transport of circulating glycoproteins. *Adv. Enzymol.*, 1974, **41**:99–128.
82. Rogers, J. C., and Kornfeld, S., Hepatic uptake of proteins coupled to fetuin glycopeptide. *Biochem. Biophys. Res. Commun.*, 1971, **45**:622–629.
83. Gray, G. R., The direct coupling of oligosaccharides to proteins and derivatized gels. *Arch. Biochem. Biophys.* 1974, **163**:426–428.
84. Hubbard, A. L., Wilson, G., Ashwell, G., and Stukenbrok, R., An electron microscope autoradiographic study of the carbohydrate recognition systems in rat liver. *J. Cell Biol.*, 1979, **83**:47–64.
85. Wilson, G., Selective uptake of synthetic glycoproteins: Mannosaminated ribonuclease A dimer and serum albumin. *J. Gen. Physiol.*, 1979, **74**:495–509.
86. Roth, J. S., Some observations on the assay and properties of ribonucleases in normal and tumor tissues. *Methods Cancer Res.*, 1967, **3**:151–243.
87. Blackburn, P., Wilson, G., and Moore, S., Ribonuclease inhibitor from human placenta. Purification and properties. *J. Biol. Chem.*, 1977, **252**:5904–5910.
88. Blackburn, P., *J. Biol. Chem.*, 1979, **254**:12484–12487.
89. Blackburn, P., and Jailkhani, B. L., *J. Biol. Chem.*, 1979, **254**:12488–12493.
90. Patthy, L., and Smith, E. L., Identification of functional arginine residues in ribonuclease A and lysozyme. *J. Biol. Chem.*, 1975, **250**:565–569.
91. Scheele, G., and Blackburn, P., *Proc. Natl. Acad. Sci. U.S.A.*, 1979, **76**, 4898–4902.
92. Robbi, M., and Lazarow, P. B., Synthesis of catalase in two cell-free protein-synthesizing systems and in rat liver. *Proc. Natl. Acad. Sci. U.S.A.*, 1978, **75**:4344–4348.

ENZYMOLOGY

On the Role of the recA Protein of *Escherichia coli* in General Recombination

I. ROBERT LEHMAN

Department of Biochemistry
Stanford University School of Medicine
Stanford, California

The process of genetic recombination permits an organism to sample many arrangements of its genes in search of a genotype that confers on it a significant selective advantage. It is therefore a major driving force in evolution.

In 1961 Meselson and Weigle (11) and Kellenberger, Zichichi, and Weigle (6) provided the first major insight into the molecular events that underlie this important process when they demonstrated that recombination in bacteriophage lambda occurs by the breaking and rejoining of DNA molecules. This discovery was followed several years later by the isolation by Clark and Margulies (2) of mutant strains of *Escherichia coli* that were specifically defective in their ability to carry out general recombination. Three genetic loci were identified—*recA*, *recB*, and *recC*—whose function was essential for normal recombination; mutations in any of these genes resulted in a profound loss in recombination proficiency.

Genetic and biochemical analysis of the *recB* and *recC* genes have demonstrated that they specify the two subunits of a complex deoxyribonuclease (1,4). In contrast, attempts to isolate and analyze the *recA* gene product have, until recently, been unsuccessful. Furthermore, a variety of genetic and physiologic studies have made it clear that the product of the *recA* gene is involved not only in recombination but in many other aspects of DNA metabolism as well. This point is emphasized by the extreme pleiotropy of *recA* mutations; *recA* mutants, in addition to being unable to perform general (as opposed to site specific) recombination are deficient in prophage induction and in the induction of an error-prone mode of repair of damage to DNA.

THE EVOLUTION OF PROTEIN STRUCTURE AND FUNCTION

ISBN 0-12-643150-7

Such mutants are also defective in recombination repair of ultraviolet-induced lesions, and, in fact, rapidly degrade their DNA in response to such treatments. There are, in addition, a number of other cellular processes and responses that are altered in cells lacking normal *recA* function [see Witkin (14), for review.]

Several years ago the product of the *recA* gene was identified as a protein with a subunit molecular weight of approximately 40,000 (9). Although normally present in relatively small amounts, treatments that damage DNA lead to the accumulation of several thousand molecules of recA protein per cell (5,7,8).

Our attempts to elucidate the structure and function of the recA protein began with the development of a purification procedure for the protein. The strain that served as a source of protein bore a multicopy plasmid containing the *recA* gene as well as a mutation at the *lexA* locus that allows high-level constitutive expression of *recA* (14). Under appropriate conditions extracts prepared from this strain contained recA protein that amounted to 5% of the total protein. With such enriched extracts purification of the recA protein to homogeneity could be achieved by a simple three-step procedure (13).

Analysis of the homogeneous recA protein has shown it to possess a variety of activities, as might perhaps have been expected of a protein that plays a crucial role in so many physiologically significant processes. What follows is a review of these activities (10,13) and an attempt to incorporate them into a plausible model for the function of recA protein in general recombination.

recA Protein: A DNA-Dependent Nucleoside Triphosphatase

recA Protein catalyzes the hydrolysis of nucleoside triphosphates to their corresponding nucleoside diphosphates. The enzyme is most active on ATP, dATP, UTP, and dUTP, although some hydrolysis of CTP can be detected (Table 1). In contrast, GTP, dGTP, and dTTP are inactive as substrates. The hydrolysis reaction has an absolute requirement for single-stranded DNA, which is satisfied by denatured linear duplex DNAs, poly(dT), and circular single-stranded viral DNAs. On the other hand, duplex DNAs are not active in stimulating hydrolysis under these conditions. The reaction requires a divalent cation and is inhibited by *N*-ethylmaleimide (5 m*M*) and the ATP analog, adenosine 5′-(γ-thio)triphosphate ATP[γ-S]. Under optimal conditions the recA protein catalyzes the hydrolysis of approximately 30 mol of ATP per mol of protein per min at 37°C.

The DNA-dependent nucleoside triphosphatase activity of the recA protein isolated from a cold-labile *recA* mutant (*recA*629) is abnormally cold labile (Table 2), demonstrating that this activity is an intrinsic property of the recA protein.

TABLE 1
HYDROLYSIS OF NUCLEOSIDE TRIPHOSPHATES CATALYZED BY THE RECA PROTEIN[a]

NTP	% NDP formed
ATP	60
dATP	66
UTP	41
dUTP	36
CTP	27
dCTP	8
GTP	4
dGTP	6
dTTP	3

[a] The recA protein (0.4 μg) was incubated for 30 min at 37°C with 80 μ*M* NTP and 10 m*M* $MgCl_2$, and formation of NDP was measured. In the absence of DNA, <2% hydrolysis was observed. All values have been corrected for NDP initially present in the NTP (usually less than 5%).

ANNEALING OF COMPLEMENTARY SINGLE STRANDS OF DNA CATALYZED BY RECA PROTEIN

Denatured nicked ColE1 or P22 DNA, incubated with recA protein, Mg^{2+}, and ATP are converted to a form that remains at the top of an agarose gel during electrophoresis (Fig. 1). The product is unaffected by Sarkosyl (1%) or sodium dodecyl sulfate (1%) but is converted to a form that enters the gel by brief incubation in 0.1 *M* NaOH or by heating at 100°C for 1 min. Treatment with the single-stranded DNA-specific S1 nuclease also converts it to a form that migrates into the agarose gel. When examined in the electron microscope, DNA that had been incubated with the recA protein, Mg^{2+}, and ATP is found predominantly in the form of complex aggregates (Fig. 2a and b). The presence of double-stranded DNA in these products was confirmed by

TABLE 2
COLD LABILITY OF ATPASE ACTIVITY OF WILD-TYPE AND COLD-SENSITIVE RECA PROTEINS[a]

recA Protein	Protein added (μg)	ATP hydrolyzed (%) 28°C	ATP hydrolyzed (%) 37°C	Ratio 37°C/28°C
recA$^+$	0.4	8	31	3.9
*recA*629 (cold sensitive)	1.0	3	32	11.0

[a] Measurements of ATP hydrolysis were performed as described in Table 1.

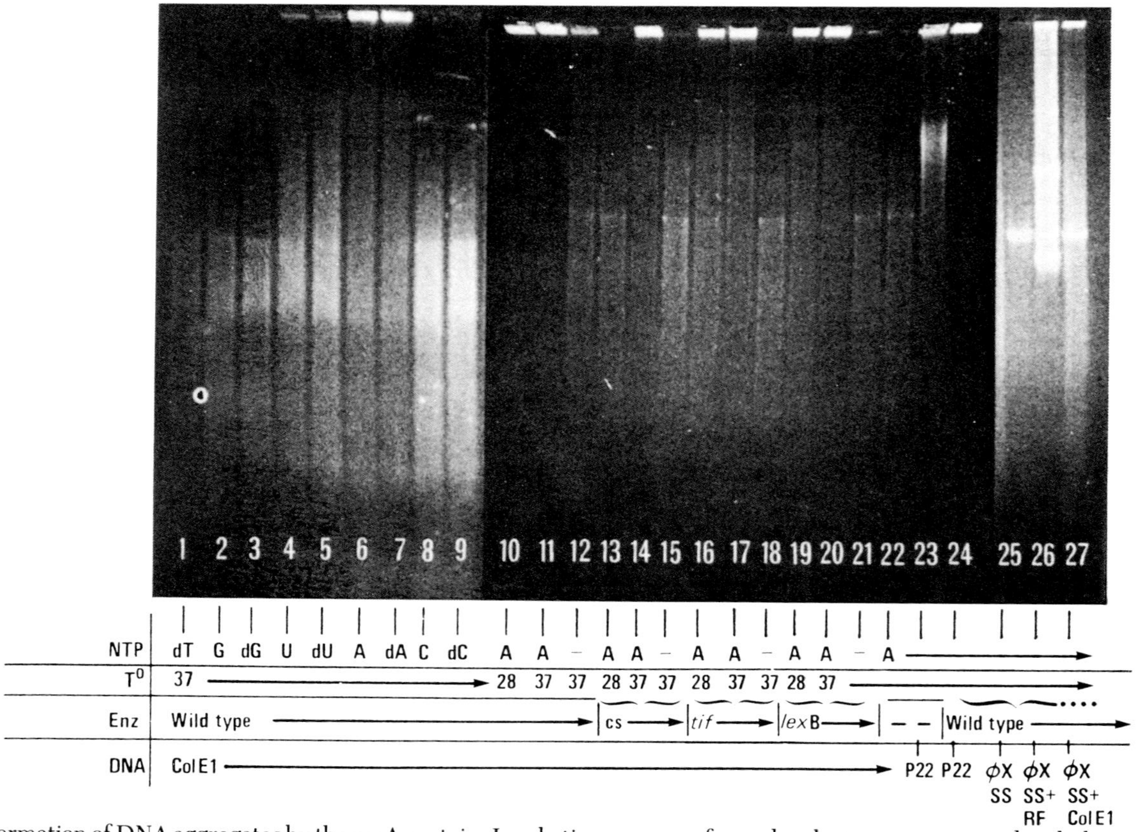

Fig. 1. Formation of DNA aggregates by the recA protein. Incubations were performed at the temperatures and with the components indicated; samples were electrophoresed in 0.8% agarose gels following deproteinization with 1% Sarkosyl. ColE1, heat-denatured, nicked ColE1 DNA; XSS, single-stranded circular X174 DNA; X RFII, the circular duplex replicative form of X174 DNA with a nick in one strand; cs, recA protein from the cold-sensitive *recA*629 mutant; *tif*, recA protein from the *tif-1* mutant of *recA*; *lexB*, recA protein from the *lexB* mutant of *recA*.

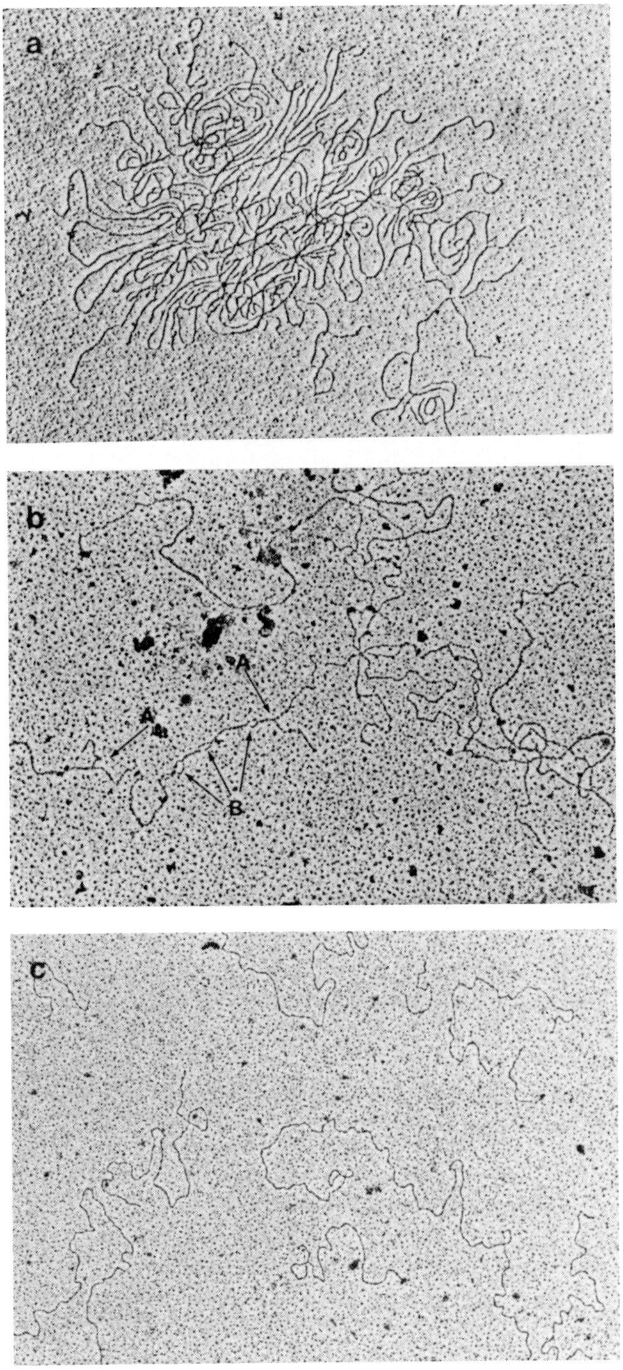

Fig. 2. Electron micrographs of DNA aggregates produced by the recA protein. Heat-denatured P22 or ColE1 DNA were incubated with recA protein, ATP, and Mg^{2+} for 20 min at 37°C. The reaction products were spread by the formamide method of Davis *et al.* (3). (a) ColE1 DNA incubated with recA protein; (b) P22 incubated with recA protein; (c) P22 DNA incubated without recA protein. In (b), branches (A) and single-stranded bubbles (B) are indicated by the arrows.

digestion with S1 nuclease. Incubation of denatured ColE1 or P22 DNA in the absence of recA protein does not result in formation of DNA aggregates; instead, extended single-stranded DNA molecules are seen by electron microscopy (Fig. 2c).

In contrast to the results obtained with denatured duplex DNA, neither intact duplex P22 DNA nor ColE1 DNA (supercoiled or nicked) are altered in their electrophoretic mobility. Furthermore, single-stranded DNA from ϕX174 phage which lacks a complementary strand does not undergo the recA protein-promoted aggregation (Fig. 1). Formation of DNA aggregates requires ATP(dATP) or UTP(dUTP); other nucleoside triphosphates are inert. The aggregation reaction is inhibited by the analog ATP[γ-S], suggesting that ATP hydrolysis is required for aggregate formation. Finally, the recA protein purified from the cold-sensitive strain is inactive in the aggregation reaction at 37°C but is active at 28°C (Fig. 1).

The extent of duplex DNA formed during the aggregation reaction may be determined by measuring conversion of single-stranded P22 DNA to a form insusceptible to S1 nuclease digestion. Incubation of denatured DNA with recA protein produces S1 nuclease-resistant material (up to 50% of the total), and the reaction requires ATP. In contrast, incubation of single-stranded DNA from phage ϕX174 with recA protein and ATP produces no S1 nuclease-resistant material.

The formation of duplex DNA by recA protein as measured by resistance to S1 nuclease has the same requirements as aggregate formation determined by agarose gel electrophoresis. Thus, it requires ATP and is inhibited by ATP[γ-S]. Furthermore, the reaction catalyzed by the cold-sensitive *recA*629 mutant protein is abnormally cold labile. Thus the formation of duplex DNA and the aggregation of complementary single strands of DNA, like the DNA-dependent nucleoside triphosphatase activity, are intrinsic features of the recA protein.

Binding of recA Protein to Duplex DNA in the Presence of a Nucleoside Triphosphate

In the presence of a nucleoside triphosphate, recA protein interacts directly with duplex DNA, as measured by binding of the DNA to nitrocellulose filters. Retention of ^{3}H-labeled duplex P22 DNA on filters requires recA protein but not single-stranded DNA and is completely eliminated by treatment with detergents (1% Sarkosyl) or proteinases (proteinase K). Inasmuch as the recA protein–duplex DNA complexes are retained on alkali-treated nitrocellulose filters that do not bind single-stranded DNA, retention of the labeled duplex DNA is a consequence of the interaction between bound recA protein and the nitrocellulose. The requirements for binding of recA protein to duplex DNA are given in Table 3. Binding requires a nucleoside triphosphate that need not be hydrolyzed. Thus GTP and ATP[γ-S]

TABLE 3
REQUIREMENTS FOR BINDING OF RECA PROTEIN TO LINEAR DUPLEX DNA[a]

Reaction mixture	Duplex DNA retained %
Complete	(100)
−recA protein	2
−ATP	3
−ATP + GTP	140
−ATP + ATP[γ-S]	130
−ATP + ADP[NH]P	4
−ATP + ADP[CH_2]P	4
+*N*-Ethylmaleimide (10 m*M*)	10
+NaCl (100 m*M*)	3

[a] Complete reaction mixtures contained 175 pmole of recA protein, 1.45 m*M* (nucleotide) of P22 duplex ^{3}H-DNA, and 500 μ*M* ATP or other ribonucleoside triphosphate. Incubation was at 37°C for 15 min. ADP[NH]P, adenyl-5′-yl imidodiphosphate; ADP[CH_2]P, adenyl-5′-yl methylene diphosphate.

that are not hydrolyzed by recA protein are as effective as ATP in the binding reaction.

Binding kinetics (Fig. 3) indicate that there are qualitative differences between ATP- and GTP-promoted binding of recA protein to duplex DNA. In the presence of 500 μ*M* ATP, recA protein binds rapidly to duplex DNA. However, on further incubation there is a rapid disappearance of the recA protein–duplex DNA complexes. Addition of ATP after 60 min of incubation restores binding, indicating that the recA protein has not been inactivated during the incubation. Furthermore, increasing the initial ATP concentration to 1.5 m*M* reduces the rate of disappearance of the recA protein–DNA complexes (Fig. 3). These results are plausibly accounted for by hydrolysis of ATP catalyzed by recA protein in the presence of duplex DNA.

In contrast to the results obtained with ATP, binding of recA protein to duplex DNA in the presence of the nonhydrolyzable nucleoside triphosphates GTP and ATP[γ-S] results in the formation of a recA protein–duplex DNA complex (Fig. 3). These complexes are extremely stable at 37°C. Thus, although initial formation of complexes is inhibited by 0.1 *M* NaCl, once formed in the presence of GTP, the recA protein–DNA complex cannot be dissociated by incubation in 0.5 *M* NaCl at 37°C for 60 min.

Under conditions that promote binding of recA protein to duplex DNA, recA protein catalyzes the hydrolysis of ATP to ADP. As in the case of single-stranded DNA, recA protein does not hydrolyze GTP in the presence of duplex DNA.

The binding of recA protein to duplex DNA in the presence of ATP or GTP

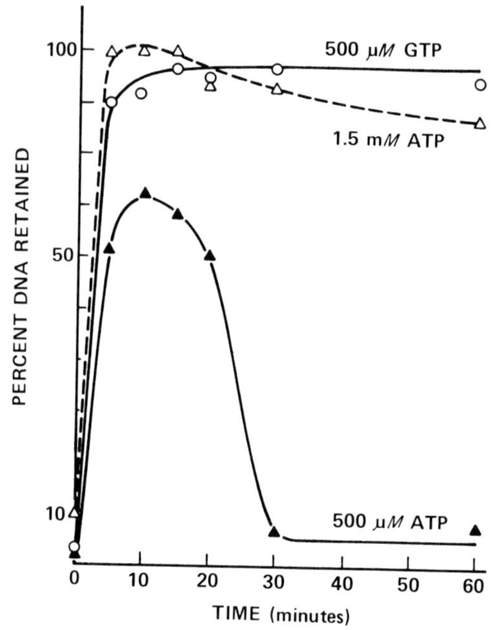

Fig. 3. Kinetics of recA protein binding to duplex DNA. recA Protein was incubated with ^{3}H-labeled duplex P22 DNA in the presence of 500 μM GTP, 500 μM ATP, or 1.5 mM ATP. Aliquots were taken at the times indicated and retention of the DNA on nitrocellulose filters determined.

is cooperative (Fig. 4). This is in contrast to the linear dependence of ATP hydrolysis on recA protein concentration in the presence of single-stranded DNA.

Assimilation of Homologous Single Strands into Duplex DNA by recA Protein

It is clear that the recA protein can catalyze at least two reactions: the hybridization of complementary single strands of DNA coupled to the hydrolysis of ATP and the binding to duplex DNA in the presence of ATP possibly producing a region of localized denaturation. The latter does not require hydrolysis of the ATP. These two reactions can be combined to generate a recombinant molecule in which a complementary single strand is assimilated into a restricted region of duplex DNA.

In the presence of ATP, Mg^{2+}, and complementary (unlabeled) single-stranded P22 DNA, the recA protein converts ^{3}H-labeled duplex P22 DNA into a form that is retained on nitrocellulose filters. Retention of these complexes is eliminated by treatment with the single-stranded DNA-specific S1 nuclease but is resistant to detergents or protease treatment. Retention of the labeled duplex DNA is also greatly reduced by brief heating (55°C for 5 min). These results suggest that the recA protein promotes the assimilation of unlabeled single strands into the labeled duplex to form D-loops that are retained on nitrocellulose filters by virtue of their single-stranded regions.

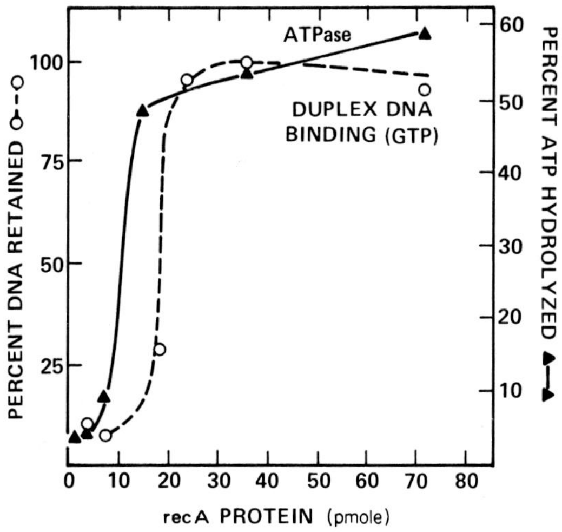

Fig. 4. Effect of recA protein concentration on ATP hydrolysis and binding to duplex DNA. The indicated amounts of recA protein were incubated with 500 μM GTP for 15 min at 37°C. Aliquots from each reaction were taken for the filter retention assay. Hydrolysis of ATP was determined after 30 min of incubation at 37°C with the indicated amounts of recA protein.

The products of the assimilation reaction can in fact be visualized directly by electron microscopic studies. These reveal D-loop-containing molecules such as are shown in Fig. 5A and B. D-loops are also observed when recA protein is incubated with supercoiled M13 RFI DNA and homologous single-stranded DNA fragments (Fig. 5C).

Single-strand assimilation requires recA protein, ATP, and homologous single-stranded DNA (Table 4). Heterologous calf thymus or ColE1 single strands cannot substitute for P22 DNA, although single-stranded phage λ DNA does produce a small but significant amount of binding. The interaction of λ DNA with the P22 DNA duplex may reflect the limited homology between these DNAs. The ATP requirement is not satisfied by any other ribonucleoside triphosphate, including UTP and the ATP[γ-S] derivative, which inhibits both recA protein-catalyzed DNA renaturation and ATP hydrolysis. The assimilation reaction, like the renaturation reaction, is sensitive to *N*-ethylmaleimide. The requirements for uptake of single strands into superhelical DNA are identical to those for linear duplex DNA.

The kinetics of single-strand assimilation into linear P22 duplex DNA are shown in Fig. 6. After a lag of about 2 min the reaction is complete in 15 min, with nearly 50% of the labeled DNA being retained. The extent is unaffected by the addition of more recA protein or ATP after 20 min of incubation. Similar kinetics are observed for the assimilation of single-stranded M13 fragments into supercoiled M13 RFI DNA.

As might be anticipated, the assimilation reaction shows the same cooperative dependence on recA protein concentration as the binding reaction (see Fig. 7).

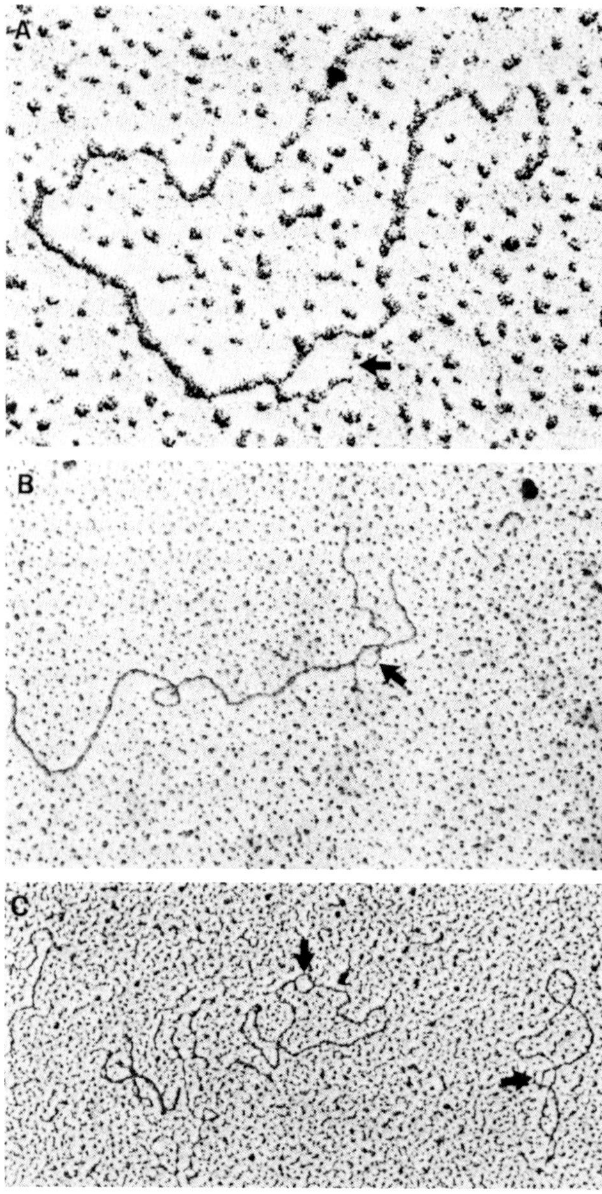

Fig. 5. recA Protein-catalyzed formation of D-loops in linear P22 DNA (A and B) and the covalently closed circular duplex form (RFI) of M13 DNA (C). D-loops are indicated by the arrows.

TABLE 4
REQUIREMENTS FOR SINGLE-STRAND ASSIMILATION INTO LINEAR AND SUPERCOILED DUPLEX DNA[a]

Duplex DNA	Reaction mixture	Complexes formed %
P22	Complete	(100)
	−recA protein	<1
	−ATP	<1
	−ATP + GTP	<1
	−ATP + ATP[γ-S]	<1
	−P22 single strands	2
	−P22 single strands + duplex P22 DNA	1
	−P22 single strands + linear ColE1 single strands	2
	−P22 single strands + λ single strands	6
	+*N*-Ethylmaleimide (10 m*M*)	1
	+NaCl (100 m*M*)	1
M13 RF1	Complete	(100)
	−recA protein	1
	−ATP	2
	−ATP + GTP	2
	−M13 single strands + P22 single strands	2

[a] The complete reaction mixture for P22 DNA contained 350 pmole of recA protein, 1.45 nmole of P22 duplex ^{3}H-DNA, 650 pmole of heat-denatured P22 DNA, and 500 μ*M* ATP or other ribonucleoside triphosphate. The complete reaction mixture for M13 DNA contained 350 pmol of M13 single-stranded DNA fragments and 2.0 nmole ^{3}H-M13 RFI DNA. After incubation at 37°C for 30 min, reactions were stopped by addition of Sarkosyl; retention on nitrocellulose filters was then measured. Forty-five percent of the input P22 duplex ^{3}H-DNA and 40% of the M13 RFI ^{3}H-DNA were retained on filters in their respective complete reaction mixtures.

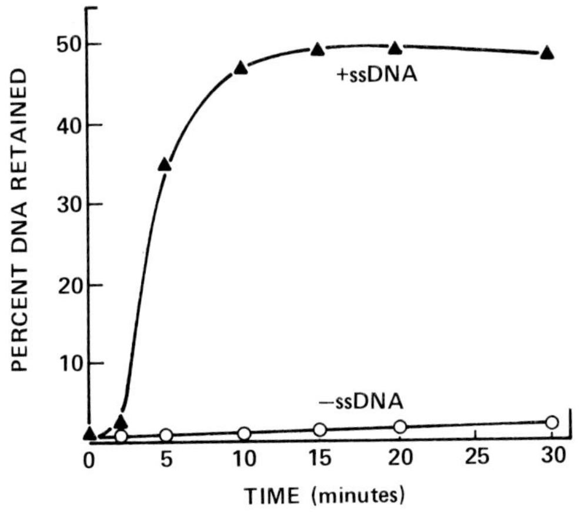

Fig. 6. Kinetics of single-stranded DNA (ssDNA) assimilation into duplex P22 DNA. Binding to nitrocellulose filters after removal of protein was measured at the times indicated.

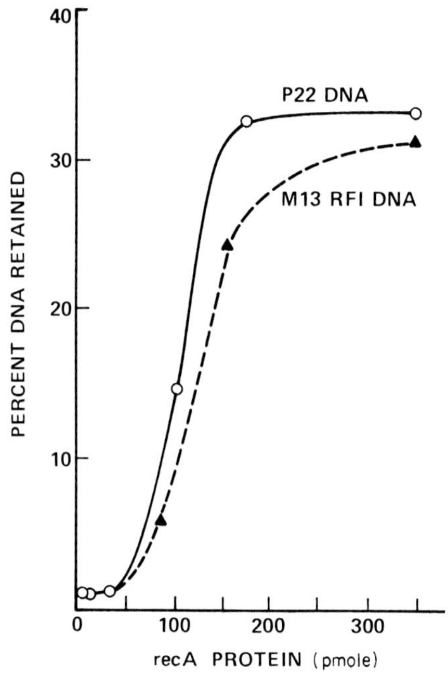

Fig. 7. Effect of recA protein concentration on single-stranded DNA assimilation. recA Protein was incubated with either ^{3}H-labeled duplex P22 DNA and unlabeled heat denatured P22, or M13 RFI DNA and M13 single-stranded DNA fragments for 30 min at 37°C. Binding to nitrocellulose filters was measured after removal of protein.

Discussion

The homogeneous recA protein promotes assimilation of single-stranded DNA into homologous duplex DNA regions *in vitro*. The products of this reaction are recombinant DNA molecules in which one strand of the duplex is hydrogen bonded to an exogenous single strand (D-loop). We have detected efficient D-loop formation both in linear and supercoiled duplex DNA, suggesting that superhelicity is not required for strand uptake promoted by the recA protein. The stability of D-loops in linear DNA, in contrast to superhelical DNA, is extremely sensitive to ionic strength and temperature. At high temperature and low salt content, the single strands are rapidly lost because of branch migration. In order to detect single-strand assimilation into linear duplex P22 DNA we have used filtration conditions (high salt and low temperature) that retard branch migration. *In vivo*, D-loops might be stabilized in relaxed DNA by other binding proteins or nucleases that could cleave the displaced single strand and block branch migration.

A model for the single-strand assimilation reaction and its relationship to single-stranded DNA renaturation catalyzed by recA protein is depicted in Fig. 8. Binding of recA protein to duplex DNA is cooperative and requires a nucleoside triphosphate (ATP, GTP, ATP[γ-S]) but does not require its hydrolysis (A). We presume that the recA protein–DNA intermediate that is formed contains partially denatured regions. recA Protein might unwind the duplex directly; alternatively, it might bind to single-stranded regions formed during "breathing" of the duplex, destabilize adjacent duplex regions, and promote further binding of recA protein, thus accounting for the observed cooperativity. ATP, when present as a cofactor in the binding reaction, can be hydrolyzed by the recA protein–DNA complex (presumably because single-stranded DNA is generated after recA protein binding), leading to loss of the complex and release of the recA protein (B).

Binding of recA protein to single-stranded DNA is not cooperative and does not require a nucleoside triphosphate (C) (unpublished results). ATP is hydrolyzed by recA protein in the absence (D) or presence of a complementary single strand(E). In the latter reaction, recA protein promotes the renaturation of the DNA. When homologous duplex and single-stranded DNAs are incubated together with recA protein and ATP, strand assimilation occurs (F). The hybridization reactions in single-strand assimilation (F) and between free single strands (E) are probably due to the same activity of the recA protein because the recA protein purified from a cold-sensitive *recA* mutant strain is cold labile for the hybridization of free single strands as well as for assimilation of single strands into duplex DNA *in vitro*.

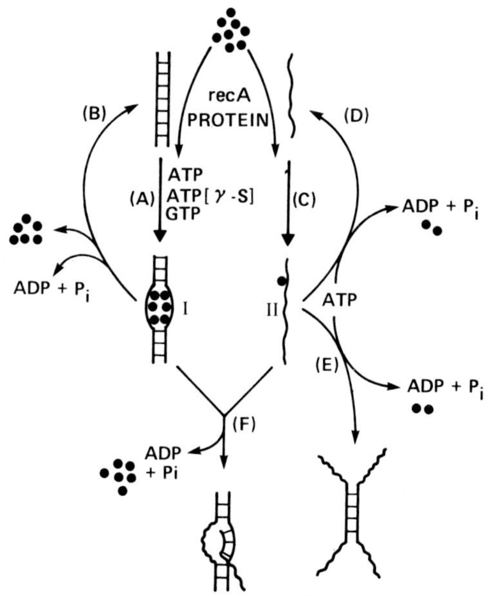

Fig. 8. Model for hybridization reactions catalyzed by recA protein. recA Protein binding to duplex DNA or single-stranded DNA yields protein-DNA complexes I or II, respectively. Binding to duplex DNA results in a partially denatured intermediate (I) that can either renature (B) or hybridize with homologous single-stranded DNA [or recA protein–single-stranded DNA complexes (II)] as in F to produce D-loops. recA Protein bound to a single-stranded DNA (II) can promote renaturation of free complementary strands (E) or dissociate from a single strand (D). Reactions E and D require ATP hydrolysis.

We propose that nucleoside triphosphates induce a conformational change in the recA protein and thus increase its affinity for DNA. In fact, recA protein–DNA complexes formed in the presence of GTP and ATP[γ-S] are extremely stable and are not dissociated at a NaCl concentration (0.5 *M*) that prevents the initial binding step.

When ATP is present in the binding reaction, the formation of recA protein–DNA complexes is followed by dissociation of recA protein and concomitant hydrolysis of ATP. These results suggest that ATP hydrolysis restores recA protein to its original conformation, leading to its release from the DNA because of its lower binding affinity.

The formation of D-loops *in vitro* catalyzed by recA protein is consistent with the role of this protein in recombination *in vivo*. *recA* Function is required at an early step in general recombination *in vivo*, a fact that is most easily explained if the role of the *recA* gene product is to promote the initial strand crossover event required by all recombination pathways (12). Differences in the processing of this early intermediate may distinguish the different recombination pathways.

Although single-strand assimilation *in vitro* requires recA protein levels that probably exceed the intracellular recA protein concentration in uninduced cells, other proteins might assist in duplex DNA unwinding *in vivo*. Alternatively, induction of the recA protein may be required for single-strand assimilation *in vivo*.

ACKNOWLEDGMENTS

The studies reported here were supported by grants from the National Institutes of Health (GM06196) and the National Science Foundation (PCM74-00865) and represent the joint efforts of Drs. Kevin McEntee and George Weinstock, both of whom are Fellows of the Jane Coffin Childs Memorial Foundation for Medical Research.

REFERENCES

1. Barbour, S. D., and Clark, A. J., Biochemical and genetic studies of recombination proficiency in *Escherichia coli* I. Enzymatic activity associated with *recB*$^+$ and *recC*$^+$ genes. *Proc. Natl. Acad. Sci. U.S.A.*, 1970, **65**:955–961.
2. Clark, A. J., and Margulies, A. D., Isolation and characterization of recombination defective mutants of *E. coli* K-12. *Proc. Natl. Acad. Sci. U.S.A.*, 1965 **53**:451–459.
3. Davis, R. W., Simon, M., and Davidson, N., Electron microscope heteroduplex methods for mapping regions of base sequence homology in nucleic acids. In: *Methods in Enzymology* (L. Grossman and K. Moldave, eds.), Vol. 21, Part D. Academic Press, New York, 1971:413–428.
4. Goldmark, P. J., and Linn, S., An endonuclease activity from *Escherichia coli* absent from certain *rec*$^-$ strains. *Proc. Natl. Acad. Sci. U.S.A.*, 1970, **67**:434–441.
5. Gudas, L. J., and Mount, D. W., Identification of the *recA* (*tif*) gene product of *Escherichia coli*. *Proc. Natl. Acad. Sci. U.S.A.*, 1977, **74**:5280–5284.

6. Kellenberger, G., Zichichi, M. L., and Weigle, J. J., Exchange of DNA in the recombination of bacteriophage. *Proc. Natl. Acad. Sci. U.S.A.*, 1961, **47**:869–878.
7. Little, J. W., and Kleid, D. G., *Escherichia coli* protein X is the *recA* gene product. *J. Biol. Chem.*, 1977, **252**:6251–6252.
8. McEntee, K., Protein X is the product of the *recA* gene of *Escherichia coli. Proc. Natl. Acad. Sci. U.S.A.*, 1979, **74**:5275–5279.
9. McEntee, K., Hesse, J. E., and Epstein, W., Identification and radiochemical purification of the recA protein of *Escherichia coli* K-12. *Proc. Natl. Acad. Sci. U.S.A.*, 1976, **73**:3979–3983.
10. McEntee, K., Weinstock, G. M., and Lehman, I. R., Initiation of general recombination catalyzed *in vitro* by the recA protein of *Escherichia coli. Proc. Natl. Acad. Sci. U.S.A.*, 1979, **76**:2615–2619.
11. Meselson, M., and Weigle, J. J., Chromosome breakage accompanying genetic recombination in bacteriophage. *Proc. Natl. Acad. Sci. U.S.A.*, 1961, **47**:857–868.
12. Radding, C. M., Genetic recombination: Strand transfer and mismatch repair. *Annu. Rev. Biochem.*, 1978, **47**:847–880.
13. Weinstock, G. M., McEntee, K., and Lehman, I. R., ATP-dependent renaturation of DNA catalyzed by the recA protein of *Escherichia coli. Proc. Natl. Acad. Sci. U.S.A.*, 1979, **76**:126–130.
14. Witkin, E., Ultraviolet mutagenesis and inducible DNA repair in *Escherichia coli. Bacteriol. Rev.*, 1976, **40**:869–907.

Experimental Enzyme Evolution

BRIAN S. HARTLEY

Department of Biochemistry
Imperial College of Science and Technology
London, England

For most of us who have been engaged in protein chemistry over the last three decades, a request to speak at a symposium in honor of Emil Smith is not so much an invitation as a summons. History will certainly mark him as one of the great figures in the development of our subject, and I am pleased that this volume will help historians to get things right.

However, in the climate of our time, to be acknowledged as a great protein chemist is a little less flattering than it might seem. Younger biochemists tend to view protein chemists as rather dull and plodding people: excellent and necessary technologists, of course, but somewhat lacking in biological imagination. What other sort of person would put up with the tedious labor of purifying a protein to homogeneity and grinding out its complete amino acid sequence?

To anyone who has known Emil Smith, this sort of picture is laughably absurd. His knowledge and his imagination are vast; indeed his friends know that the latter needs to be sat on at regular intervals. But this is not surprising, since in the 1950s only a person of boundless optimism and imagination would have tackled the job of purifying proteins and determining their complete amino acid sequences. One knew, of course, that even the complete sequence would not unlock all the secrets of protein function. One speculated that perhaps certain rules of protein folding would emerge that would allow rational model building. One hoped, but did not always believe, that X-ray crystallographers would someday come up with a method for translating their diffraction patterns into three-dimensional structures. But meanwhile protein chemists pushed on with the sequencing, convinced that only more solid chemical data would bring to life the protein structures of which they dreamed.

ISBN 0-12-643150-7

EVOLUTION OF ENZYME FAMILIES

My own dreams were crystallized into reality with the determination of the tertiary structures of three pancreatic proteases: chymotrypsin (2), elastase (19), and trypsin (20). It has been clear for some time that these enzymes were a family with a common catalytic mechanism but different specificities (4). Moreover, sequence comparisons pointed to an evolutionary origin by divergence from a common ancestor (9,21). However, the tertiary structures revealed not only the catalytic mechanism but also an amazing conservation of conformation of the polypeptide chains, which are largely superimposable in three dimensions. The surfaces of these proteins differ considerably, because of amino acid substitutions, insertions or deletions at external loops of the predominant β-structure, but the internal architecture and most of the interactions of internal side chains are faithfully conserved. The considerable differences in specificity are due to only one or two amino acid replacements in the substrate binding pocket. Hence it appears that new enzyme specificities could be readily grafted onto an existing enzyme by relatively trivial substitutions in the environment of the catalytic site, without significant changes in conformation of the polypeptide chain. But why, therefore, are there so many other changes in the surface residues in these proteins? Here we join the battle between the "neutralists" and the "selectionists" and a prudent protein chemist might retire. However, I have previously advanced the argument that the selective pressures involved in the evolution of an enzyme family might differ from those applying to a single protein in different species (5). Gene duplication is obviously the first step in evolving an enzyme family. Mutations that destroy the function of one copy might then be conserved within the population by the product of the remaining viable copy. Mutations would thereafter accumulate in the "silent gene," and the population would become heterogeneous in its progeny. A change that restored the capacity of the protein to fold into a stable tertiary structure would bring with it the mutations acquired during the silent phase of its ancestry, and some of these could contribute to new enzyme specificities.

EXPERIMENTAL EVOLUTION

Like so many evolutionary arguments, this is little more than a molecular "Just So Story" (10), but experimentalists tend to be unsatisfied with such arguments; they want to see evolution in action. New enzyme activities are clearly evolving before our very eyes in the microbial world, and it might be possible to set up model systems that would mimic the evolution of new enzymes.

Our own efforts have concentrated on applying selective pressure to a particular enzyme to change its substrate specificity. The pioneering studies of Mortlock (13) and of Lin (11) and their colleagues on pentitol metabolism in *Klebsiella aerogenes* allowed us to devise a suitable system. This organism metabolizes ribitol and D-arabitol as follows:

$$\text{Ribitol} \xrightarrow{\text{RDH}} \text{D-ribulose} \xrightarrow{\text{RK}} \text{D-ribulose-5-phosphate}$$

$$\text{D-Arabitol} \xrightarrow{\text{ArDH}} \text{D-xululose} \xrightarrow{\text{XK}} \text{D-xululose-5-phosphate}$$

where RDH = ribitol dehydrogenase, ArDH = D-arabitol dehydrogenase, RK = ribulokinase, and XK = xylulokinase. The ribitol pathway is induced by D-ribulose and the D-arabitol pathway by D-arabitol. *Klebsiella aerogenes* will not grow on xylitol or L-arabitol, but mutants constitutive for RDH grow slowly on these pentitols by virtue of a weak side specificity of RDH that converts them to D-xylulose, which then produces sufficient D-arabitol to induce D-xylulokinase (22). Hence growth on xylitol in a chemostat should be controlled by the intracellular activity of RDH for this substrate, and mutants with increased xylitol dehydrogenase activity should be selected by virtue of their faster growth rates—a reasonable analog of a natural selection process.

Evolution of Xylitol Dehydrogenases

We found, as we had hoped, that after 200–1000 generations of growth on xylitol in a chemostat, strains arose that gave significantly higher steady-state biomass (17). All of these showed higher xylitol dehydrogenase activity, confirming that this activity was indeed growth limiting. Most of these mutants proved to have increased levels of RDH (RDH superproducers), amounting to up to 30% of the total soluble protein in some cases. However, after mutagenesis, or in some cases spontaneously, mutants were selected that had an altered RDH with an improved specific activity for xylitol (XDH mutants).

We have investigated the kinetic properties, stability, and electrophoretic mobility of several of these improved xylitol dehydrogenases from *K. aerogenes* (7) and also from strains of *E. coli* that contain similar pentitol operons (3a). To our surprise, all the 13 enzymes that we have investigated are different in one or more properties. This implies that there are many different amino acid changes that can improve the specificity of RDH for xylitol. We have determined the complete amino acid sequence of the wild-type ribitol dehydrogenase (RDHA) and of one of these "XDH mutants" (RDHD), where the only change is Pro-194 for Ala-194 in the wild type. Unfortunately we have so far failed to crystallize the wild-type enzyme and have not given priority to identifying the amino acid replacements in other enzymes of this class, since there is now some prospect of sequencing the genes themselves (see the following discussion).

One can, however, make some preliminary observations on this family of "xylitol dehydrogenases" that we have evolved from RDHs. In most cases neither the catalytic activity toward ribitol nor that toward L-arabitol (another poor substrate for the enzyme) is significantly affected by the increased activity for xylitol; nor do most of the mutations affect the stability of the protein. This suggests that there are alternative quaternary structures for this tetrameric enzyme: an R-conformation that attacks ribitol and X- or A-conformations that act on xylitol and L-arabitol, respectively. The latter would be minor components in an equilibrium mixture, but several mutations might cause interactions that favor the X-conformation at little expense to the R- or A-conformations. Where catalytic sites lie at the interface between discrete protein "domains" one can imagine that there would be fruitful opportunities for dramatic shifts in substrate specificity with minor changes in tertiary structure (6).

Superproduction of Ribitol Dehydrogenase

Most of the *K. aerogenes* mutants selected by continuous culture on xylitol contained vastly elevated levels of unchanged RDH. The wild type (strain A) was constitutive for this enzyme, making RDH about 1% of its soluble protein. Chemostat selection typically gave strains yielding RDH as 5% of soluble protein, and subsequent steps yielded 10–15%, followed by 20–30%. Our early experiments (17) indicated that gene multiplication might often be a reason for this phenomenon. Most of our superproducing strains show a high frequency of reversion (10^{-2} to 10^{-3}) to normal constitutive levels of production of RDH; this is typical of tandem gene duplications, because of loss by recombination. Moreover, an experiment designed to measure gene dosage by relative frequencies of mutation of superproducers to strains lacking RDH activity supported this conclusion (17).

However, these experiments are by no means conclusive, and there are many other conceivable explanations for enzyme superproduction (e.g., reduced catabolite repression, "up-promoter" mutations, improved translation efficiency or stability of the mRNA, etc.). The availability of a large family of mutants that superproduce RDH in response to a "natural" selective pressure gives us an opportunity to evaluate the relative importance of such hypothetical mechanisms.

Ribitol Dehydrogenase mRNA

Isolating a pure prokaryotic mRNA has proved to be difficult because of the instability of such molecules. However, we felt that a strain that produced massive quantities of a particular enzyme might lend itself to such an exercise. Figure 1 shows that in all strains RDH synthesis goes hand in hand with cell growth until early log phase, probably because of competition with other bacterial mRNAs, resulting in a steady-state level of RDH. Thereafter there is a sudden switch-off of most mRNA synthesis, but RDH synthesis rises to the very high levels of specific activity indicated. Hence late log phase mRNA is likely to be rich in RDH–mRNA. We were able to label specifically polysomes engaged in RDH synthesis with rabbit antibody monospecific to RDH in the presence of heparin. The profile of such polysomes suggests that RDH–mRNA is a major species after the "switch-point" mentioned earlier. Hence by labeling polysomes from late log phase cells with rabbit anti-RDH,

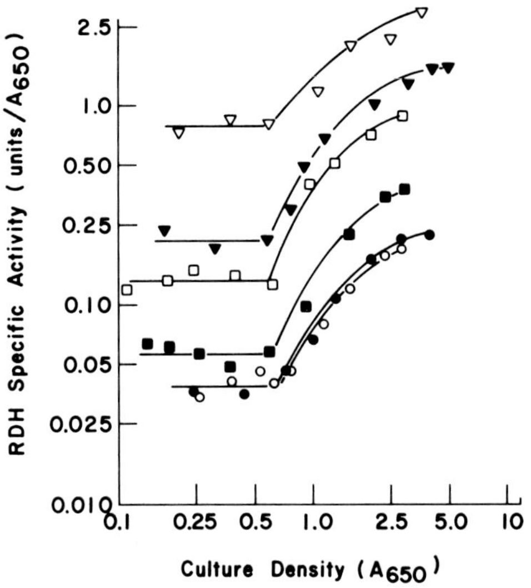

Fig. 1. Intracellular levels of ribitol dehydrogenase (RDH) in various strains of *K. aerogenes* during batch culture on 2% (w/v) casamino acids medium (1a). The wild-type strain FG5 (O), which is *gua*, was supplemented with 0.25 m*M* guanine and 0.2% (w/v) ribitol to induce the RDH. Strain A (●) is a mutant of this constitutive for RDH synthesis. Strains A1 (■), A11 (□), A111 (▼), and A112 (▽) are RDH superproducers evolved from strain A by continuous culture on xylitol (17). Samples of the cultures were taken at the densities indicated and assayed for total RDH activity after ultrasonication. Control experiments showed that RDH specific activity can be expressed equivalently as either units/ml/A_{650} or units/mg soluble protein.

and precipitating with sheep antirabbit IgG, we were able to isolate essentially pure RDH-mRNA (1a). This translated an order of magnitude more efficiently than Qβ mRNA in a crude *E. coli* cell-free system for protein synthesis and yielded intact RDH monomer as the only major product. Hence in superproducing strains RDH-mRNA is indeed made in large amounts and translated very efficiently. We suspected at first that mutations to stability and/or translation efficiency of RDH-mRNA may have been involved in some of the superproducing strains (8). However, direct comparisons of the stability of wild-type and superproducing RDH-mRNA after rifampicin treatment of the cells now make this unlikely. Gene multiplication remains the favorite explanation for the RDH superproducing strains selected by continuous culture on xylitol.

Transfer of Pentitol Operons to *Escherichia coli* K12

The questions we were now asking demanded an answer at the level of the DNA sequence of the genes responsible for pentitol metabolism. Charnetzky and Mortlock (3) showed that these genes are tightly clustered in *K. aerogenes* in the order:

-*rbtK* - *rbt-D* - *rbtC* - *dalB* - *rbtB* - *dalC* - *dalD* - *dalK*-

where *rbtK* encodes D-ribulokinase; *dalK*, D-xylulokinase; *rbtD*, ribitol dehydrogenase; and *dalD*, D-arabitol dehydrogenase. *rbtB*, *rbtC*, *dalB*, and *dalC* are loci identified by control mutations in the appropriate operons.

E. coli K12 does not grow on pentitols, but we constructed a hybrid by Pl transduction from our *K. aerogenes* strain constitutive for RDH synthesis (18). This showed a similar pattern of pentitol metabolism to the parental *K. aerogenes* strain and the *rbt* and *dal* operons mapped at 45′ on the *E. coli* chromosome. Independently, Reiner (16) discovered that natural *E. coli C* strains contain similar *rbt* and *dal* operons that map in an identical position. Altosaar and Hartley (1) showed that the sequence of this *E. coli C* ribitol dehydrogenase is 95% homologous with that from *K. aerogenes*. It seems that natural and artificial genetic engineering have achieved the same results!

Chemostat selection on xylitol was carried out with both the *E. coli* K12–*K. Aerogenes* hybrid and the *E. coli C* strain mentioned earlier. The results were similar to those observed with *K. aerogenes*, i.e., either superproducers of unchanged ribitol dehydrogenase or strains with an improved xylitol dehydrogenase were selected (7). Hence the evolutionary response of the gene does not greatly depend on the organism that contains it.

Specialized Transducing Phages Containing the Pentitol Operons

The next step was to transfer the pentitol genes of *K. aerogenes* from the *E. coli* K12 hybrid to λ bacteriophage. We achieved this by transferring the genes to a host deleted for the λ attachment site and selecting for lysogens at secondary sites (14). Phage from such lysogens were then screened for their ability to transduce the capacity to grow on ribitol into suitable *E. coli* K12 hosts. One such phage, λ p*rbt*, proved to be a plaque-forming phage that transduced constitutive *rbtD* and *rbtK* but not *dalD* or *dalK*. By lysogenizing this into an *E. coli* host deleted in the λ attachment site and containing the *rbt* and *dal* genes, it was easy to obtain λ p*rbt dal*, which contains both pentitol operons.

Figure 2 shows a restriction endonuclease map of λ p*rbt* and λ p*rbt dal* (14). By separating the two strands of DNA and hybridizing with *rbt*-mRNA or *dal*-mRNA we showed that the two operons are transcribed in bipolar fashion. The transcribed regions were mapped by hybridizing restriction endonuclease fragments with radioactive *rbt*- or *dal*-mRNA.

Evolution of the Pentitol Operons

The structure of these two pentitol operons clearly invites speculation about their evolutionary origin. ***rbtB*** and ***dalB*** presumably encode repressors (both operons are under negative control) and *rbtC* and *dalC* are probably operator regions. The genetic analysis of Charnetzky and Mortlock

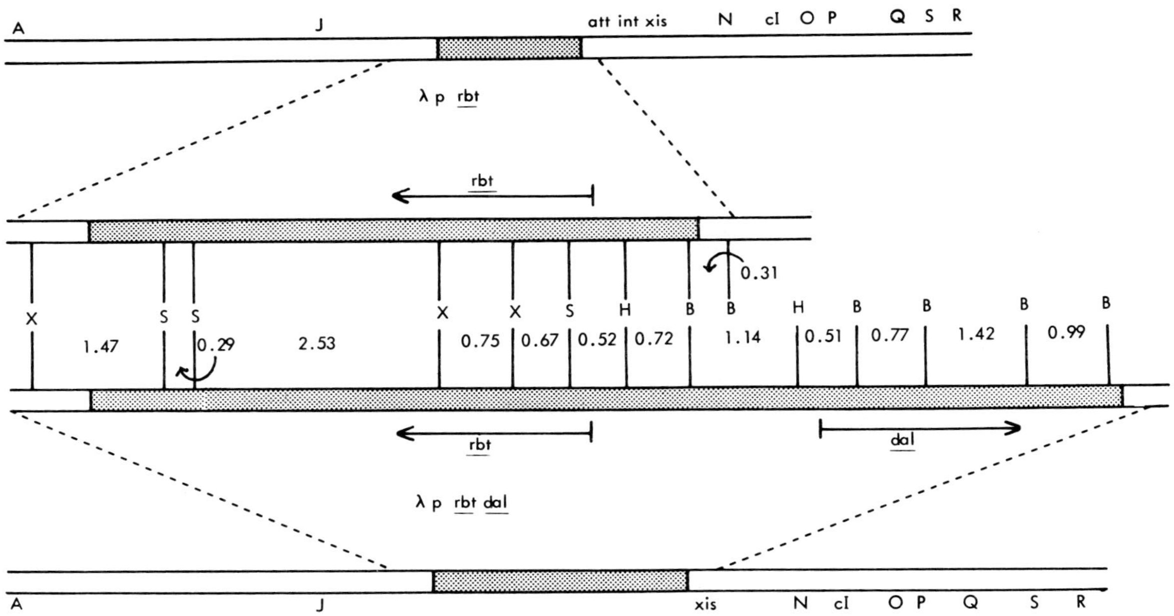

Fig. 2. Structure of λ p*rbt* and λ p*rbt dal* (14). The upper and lower diagrams indicate the position of the inserted *K. aerogenes* DNA relative to λ genes. The middle diagrams show some restriction endonuclease cleavages in these inserted regions: B = *Bst*I, X = *Xma*I, H = *Hin*dIII, and S = *Sal*I; the distances are in kilobase pairs. The arrows show the approximate locations of the *rbt* and *dal* transcriptions.

(2) suggests that these regions are interdigitated. For the structural genes, *rbtD*, *rbtK*, *dalD*, and *dalK*, an evolutionary origin involving an invert gene duplication can be guessed. Invert gene duplications have the advantage that they would be immune to elimination by recombination, such as is observed for tandem duplications. This would give more time for divergence of structure and function.

Clearly it would be interesting to compare the sequences of *rbtD* and dalD, and *rbtK* and *dalK*. This can be done at the level of amino acid sequence, and to this end we have constructed strains that superproduce D-arabitol dehydrogenase. It has then been relatively easy to purify the protein in good yield (15). It is a monomer of MW 46,000. In contrast, RDH is a tetramer of 27,000 dalton subunits, so there can be at best partial homology.

We have also attempted to screen for weak homology between appropriate parts of the *rbt* and *dal* operons by hybridizing λ p*rbt* to restriction endonuclease fragments from the *dal* region under various salt conditions. This revealed no weak homology; instead a considerable stretch of repeated sequence was disclosed that appears to lie at the promoter distal regions of both operons, probably just outside the transcribed regions.

This repeated sequence also invites evolutionary speculations. One observes that the whole of pentitol metabolism and its control is neatly packaged into a small region of DNA. This region is absent in *E. coli* B and *E. coli* K12 but present in *E. coli C* and *K. aerogenes* in similar locations on their chromosomes (23). In the latter species the two RDHs show high sequence homology (1). This structure bears the hallmark of a "metabolic plasmid" (such as have been recognized in *Pseudomonas* species) that has become stably integrated into the genome of some species of enterobacter. We have discovered two plasmids in our strain of *K. aerogenes* but neither are implicated in pentitol metabolism (14a). Nevertheless interspecies transfer of these pentitol operons via phages or plasmids remains an attractive hypothesis for their evolutionary history.

DNA Sequences of the Pentitol Operons

λ p*rbt* and λ p*rbt dal* have proved to be useful tools in screening strains of *K. aerogenes* that have putative gene duplications. Restriction endonuclease digests of the chromosomal DNA from such strains were run on gels and hybridized with probes of radioactive λ p*rbt* or λ p*rbt dal*. In this way a partial duplication was revealed in *K. aerogenes* strain A3, which is a "first stage superproducer" of RDH, selected spontaneously after growth on xylitol in a chemostat. The restriction endonuclease map shows that the duplication embraces *rbtD* and *dalD* and part of *rbtK* and *dalK*. We hope to determine the DNA sequence at the "joint" of this duplication in order to throw light on the mechanism of gene duplication.

DNA sequencing may also throw light on the evolution of the pentitol operons themselves. As mentioned earlier, the different subunit sizes of

RDH and D-arabitol dehydrogenase and lack of immunological cross reaction (15) make it improbable that they share a common ancestor. However, we have recently purified D-ribulokinase and D-xylulokinase and determined their subunit structure and N-terminal sequences (14b). They are both dimers of 60,000 and 54,000 dalton subunits, respectively. If DNA sequencing reveals homology between the kinases and not the dehydrogenases, it would imply that the kinases arose by an invert gene duplication and subsequently recruited the dehydrogenases. Another possibility is that there may be homology between the *rbt* repressor and *dal* repressor proteins. We have therefore begun to clone interesting parts of the DNA of the two operons into *E. coli* plasmids and to sequence relevant parts of these.

We have also isolated small amounts of the four enzymes of pentitol metabolism and sequenced their N-termini and several peptides by rapid mass-spectrometric techniques (12). In conjunction with DNA sequencing this should help us quickly to achieve further data of evolutionary interest.

Conclusion

This symposium speaks to the fact that great contributions to evolutionary biology have been made by protein chemists, in general, and Emil Smith, in particular. The new techniques for isolating, mapping, and sequencing DNA combined with the arts of the protein chemist promise a further explosion of hard data to discipline our speculations about the mechanisms of evolution.

REFERENCES

1. Altosaar, I., and Hartley, B. S., Comparison of ribitol dehydrogenase from *E. coli C* and *K. aerogenes*. *Proc. Int. Congr. Biochem., 10th, 1976* Abstr. 04-6-319, p. 200.

1a. Bahramian, M. B., and Hartley, B. S., In preparation.

2. Birktoft, J. J., and Blow, D. M., The atomic structure of tosyl α-chymotrypsin at 2Å resolution. *J. Mol. Biol.*, 1972, **68**:187–240.

3. Charnetzky, W. T., and Mortlock, R. P., Close genetic linkage of the determinants of the ribitol and D-arabitol catabolic pathways in *Kiebsiella aerogenes*. *J. Bacteriol.*, 1974, **119**:176–182.

3a. Dothie, J. M., Silcox, M. A., and Hartley, B. S., In preparation.

4. Hartley, B. S., Proteolytic enzymes. *Annu. Rev. Biochem.*, 1960, **29**:45–72.

5. Hartley, B. S., Enzymes are proteins. *Adv. Sci.*, 1966, 47–54.

6. Hartley, B. S., Enzyme families. In: *Evolution in the Microbial World* (M. J. Carlile and J. J. Skehel, eds.). Cambridge University Press, London and New York, 1974:151–182.

7. Hartley, B. S., Altosaar, I., Dothie, J. M., and Neuberger, M. S., Experimental evolution of a xylitol dehydrogenase. In: *Structure-Function Relationships in Proteins* (R. Markham and R. W. Horne, eds.). North-Holland Publishers, Amsterdam, 1976:191–200.

8. Hartley, B. S., Bahramian, M. B., and Neuberger, M. S., Mechanisms of enzyme superproduction. In: *Protein: Structure, Function and Industrial Applications* E. Hoffman *et al.*, eds.). Pergamon, Oxford, 1979:7–14.
9. Hartley, B. S., Brown, J. L., Kauffman, D. L., and Smillie, L. B., Evolutionary similarities between pancreatic proteolytic enzymes. *Nature (London)*, 1965, **207**:1157–1159.
10. Kipling, R., *Just So Stories for Children*. Macmillan, New York, 1902.
11. Lerner, S. A., Wu, T. T., and Lin, E. C. C., Evolution of a catabolic pathway in bacteria. *Science*, 1964, **146**:1313–1315.
12. Morris, H. R., Williams, D. M., Midwinter G. G. and Hartley, B. S., A mass-spectrometric sequence study of the enzyme ribitol dehydrogenase from *K. aerogenes*. *Biochem. J*, 1974, **141**:701–713.
13. Mortlock, R. P., Fossitt, D. D., and Wood, W. A., A basis for utilization of unnatural pentoses and pentitols by *Aerobacter aerogenes*. *Proc. Natl. Acad. Sci. U.S.A.*, 1965, **54**:572–579.
14. Neuberger, M. S., and Hartley, B. S., Investigations into the *K. aerogenes* pentitol operons using specialised transducing phages λ p*rbt* and λ p*rbt dal*. *J. Mol. Biol*, 1979, **132**:435–470.
14a. Neuberger, M. S., and Hartley, B. S., Minicircular Col El-related DNA in strains of *K. aerogenes* selected for fast growth on xylitol. *J. Gen. Microbiol* ,1980, **118**: 171–177.
14b. Neuberger, M. S., Hartley B. S., and Walker, J. D., D-Ribulokinase and D-xylulokinase of *K. aerogenes*: properties of the purified enzymes. *Biochem J.*, in press.
15. Neuberger, M. S., Patterson, R. A., and Hartley, B. S., Purification and properties of *K. aerogenes* D-arabitol dehydrogenase. *Biochem. J.*, 1979, **183**:31–42.
16. Reiner, A. M., Genes for ribitol and D-arabitol metabolism in *E. coli:* Their loci in C strains and absence in K-12 and B strains. *J. Bacteriol.*, 1975, **123**:530–536.
17. Rigby, P. W. J., Burleigh, B. D., and Hartley, B. S., Gene duplication in experimental enzyme evolution. *Nature* (*London*), 1974, **251**:200–204.
18. Rigby, P. W. J., Gething, M. J., and Hartley, B. S., Construction of intergeneric hybrids using bacteriophage PICM: Transfer of the *K. aerogenes* ribitol dehydrogenase gene to *E. coli*. *J. Bacteriol.*, 1976, **125**:728–738.
19. Shotton, D. M., and Watson, H. C., Three dimensional structure of tosyl-elastase. *Nature* (*London*), 1970, **225**:811–816.
20. Stroud, R. M., Kay, L. M., and Dickerson, R. E., The crystal and molecular structure of DIP-inhibited bovine trypsin at 2.7Å resolution. *Cold Spring Harbor Symp. Quant. Biol.*, 1971, **36**:125–140.
21. Walsh, K. A., and Neurath, H., Trypsinogen and chymotrypsinogen as homologous proteins. *Proc. Natl. Acad. Sci. U.S.A.*, 1964, **52**:884–889.
22. Wu, T. T., Lin, E. E. C., and Tanaka, S., Mutants of *Aerobacter aerogenes* capable of utilising xylitol as a novel carbon substrate. *J. Bacteriol*, 1968, **96**:447–456.

β-Galactosidase and the Lactose Operon

IRVING ZABIN

Department of Biological Chemistry
UCLA School of Medicine
and Molecular Biology Institute
University of California Los Angeles
Los Angeles, California

In this chapter I shall discuss β-galactosidase of *Escherichia coli* and present some of the highlights of this enzyme's existence as a scientific subject, with a greater emphasis on recent work carried out in my laboratory at the University of California, Los Angeles (UCLA).

I begin by presenting three views of β-galactosidase. One is the crystalline enzyme first prepared in 1957 by Wallenfels and co-workers (37) from ammonium sulfate-sodium chloride suspensions of the protein (Fig. 1). A second view (Fig. 2) is one of a number of electron micrographs of β-galactosidase negatively stained with uranyl acetate. These photographs were taken at UCLA in 1964 (21). A molecular weight of roughly 500,000 daltons can be calculated from the dimensions of the molecule and the tetrameric structure of the protein can be seen. The third view is the amino acid sequence of the subunit of β-galactosidase (Fig. 3). The determination of the primary structure was carried out by Audrée Fowler and myself and was completed in 1977 (10,11). The polypeptide chain contains 1021 amino acids for a molecular weight of 116,349, which corresponds to 465,396 for the whole tetramer.

The earliest report that I could find of any enzyme in *Escherichia coli* that catalyzes the hydrolysis of lactose was made by Karstrom in 1930 (22). Karstrom and his colleagues spoke of this enzyme as a lactase. The name β-galactosidase was not used commonly until about 1950, when Lederberg introduced the chromogenic substrate *o*-nitrophenyl-β-D-galactoside to simplify the assay (26). He measured some of the enzymic parameters of the *E. coli* β-galactosidase in crude extracts including kinetics, effect of metal ions, and substrate specificity. The first purified preparation was obtained by Cohn and Monod (7).

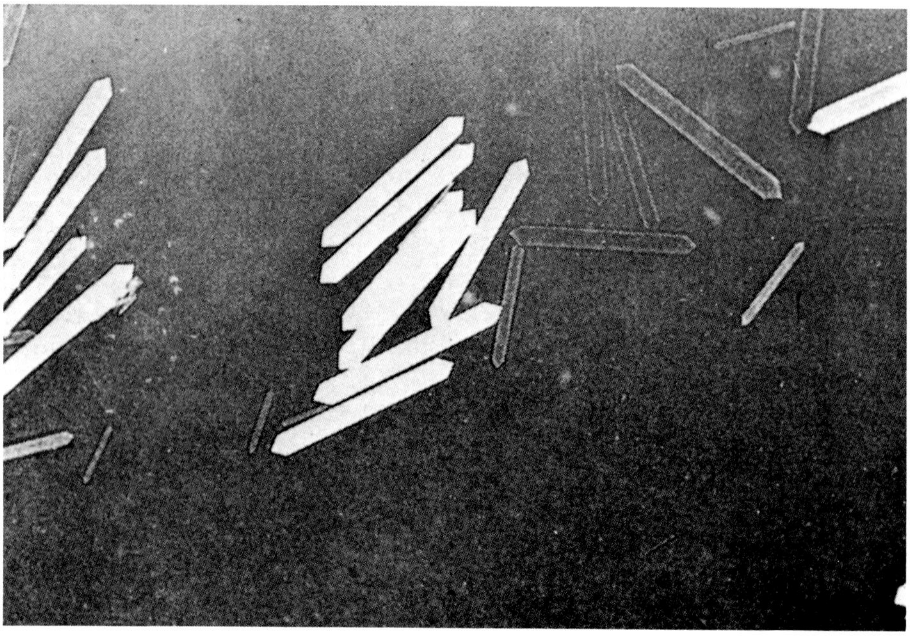

Fig. 1. Crystals of β-galactosidase.

These early experiments set the stage for the great achievement by the Pasteur school over the next decade, the development of the general concept of operon structure, expression and regulation. β-Galactosidase was a central subject. I will mention a few of the important experiments. The appearance of β-galactosidase activity in a culture of growing wild-type *E. coli* following addition of a galactoside to the medium was found to be *de novo* synthesis of β-galactosidase from amino acids (16). It was not necessary for the galactoside inducer to be a substrate (29). Lactose permease was discovered (33) and later, thiogalactoside transacetylase was demonstrated (40). Genetic experiments showed that the synthesis of these induced enzymes was under negative control and implied the existence of a *lac* repressor specified by a regulatory gene (32). The structural genes defining β-galactosidase, lactose permease, and transacetylase were controlled by an operator locus, and all three proteins are expressed coordinately. This set of genes is the lactose operon, which is described in the classic paper of Jacob and Monod, "Genetic Regulatory Methods in the Synthesis of Proteins," published in 1961 (19).

I would like to explain how my laboratory began work on β-galactosidase. During my sabbatical leave at the Pasteur Institute in 1959, Kepes, Monod, and I discovered the transacetylase in extracts of bacteria induced for β-galactosidase (40). The reaction catalyzed by this enzyme is the transfer of an acetyl group from acetyl-CoA to a thiogalactoside acceptor to form a 6-0 acetyl ester. It was of interest to investigate the properties of the protein and to purify it to homogeneity. When this was accomplished (38) it became possible to compare relative amounts of *lac* operon proteins produced during expression of the lactose operon. But to do this required information on the subunit structure of both β-galactosidase and transacetylase. This was not clear for β-galactosidase and, in fact, was controversial (39). Therefore, since it was important for the understanding of operon function to define the substructure of β-galactosidase, studies on end groups and peptide mapping of β-galactosidase were carried out (4,23). The results obtained indicated that β-galactosidase was in fact a tetramer composed of four identical, but quite long, polypeptide chains (39).

One of the big advantages for work on a bacterial enzyme is the availability of mutant strains that produce enzymically inactive protein. It was well known that proteins from missense mutants that contain amino acid substitutions can be detected by their immunologic cross-reactivity. At about the time we were engaged in end-group studies, nonsense mutants mapping in *LacZ* were isolated (31). I wondered whether end groups could be demonstrated in the corresponding β-galactosidase prematurely terminated chains

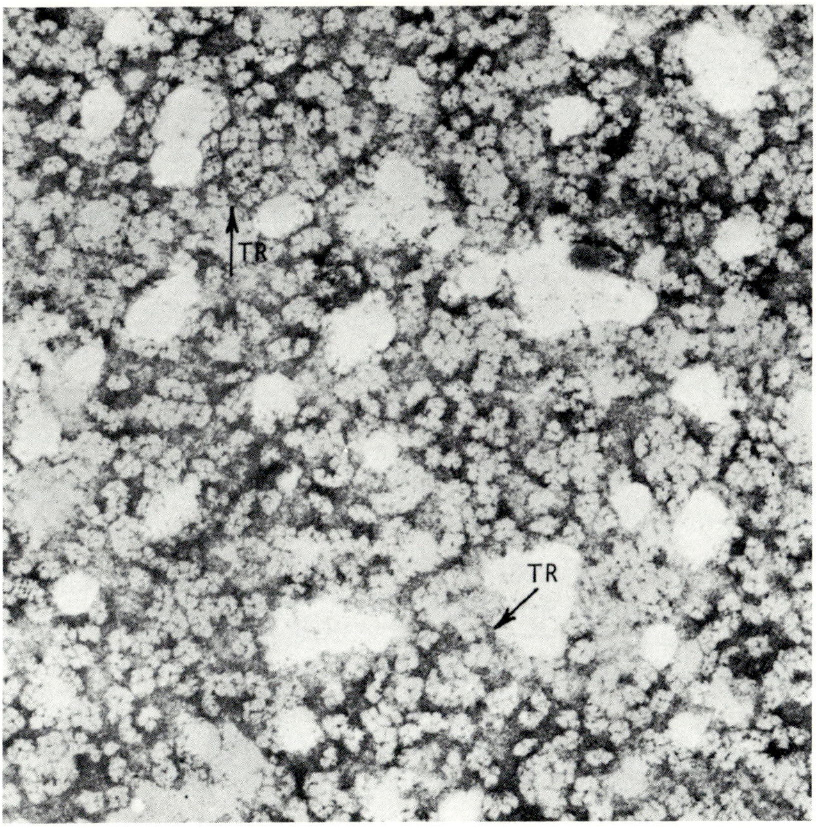

Fig. 2. Formaldehyde-treated preparation of β-galactosidase after staining with uranyl acetate. TR = tetrameric structure.

10 20 30
Thr-Met-Ile-Thr-Asp-Ser-Leu-Ala-Val-Val-Leu-Gln-Arg-Arg-Asp-Trp-Glu-Asn-Pro-Gly-Val-Thr-Gln-Leu-Asn-Arg-Leu-Ala-Ala-His-
CNBr1 T1

40 50 60
Pro-Pro-Phe-Ala-Ser-Trp-Arg-Asn-Ser-Glu-Glu-Ala-Arg-Thr-Asp-Arg-Pro-Ser-Gln-Gln-Leu-Arg-Ser-Leu-Asn-Gly-Glu-Trp-Arg-Phe-
CNBr2

70 80 90
Ala-Trp-Phe-Pro-Ala-Pro-Glu-Ala-Val-Pro-Glu-Ser-Trp-Leu-Glu-Cys-Asp-Leu-Pro-Glu-Ala-Asp-Thr-Val-Val-Val-Pro-Ser-Asn-Trp-
T8

100 110 120
Gln-Met-His-Gly-Tyr-Asp-Ala-Pro-Ile-Tyr-Thr-Asn-Val-Thr-Tyr-Pro-Ile-Thr-Val-Asn-Pro-Pro-Phe-Val-Pro-Thr-Glu-Asn-Pro-Thr-

130 140 150
Gly-Cys-Tyr-Ser-Leu-Thr-Phe-Asn-Val-Asp-Glu-Ser-Trp-Leu-Gln-Glu-Gly-Gln-Thr-Arg-Ile-Ile-Phe-Asp-Gly-Val-Asn-Ser-Ala-Phe-
CNBr3

160 170 180
His-Leu-Trp-Cys-Asn-Gly-Arg-Trp-Val-Gly-Tyr-Gly-Gln-Asp-Ser-Arg-Leu-Pro-Ser-Glu-Phe-Asp-Leu-Ser-Ala-Phe-Leu-Arg-Ala-Gly-

190 200 210
Glu-Asn-Arg-Leu-Ala-Val-Met-Val-Leu-Arg-Trp-Ser-Asp-Gly-Ser-Tyr-Leu-Glu-Asp-Gln-Asp-Met-Trp-Arg-Met-Ser-Gly-Ile-Phe-Arg-
CNBr4 CNBr5
T13 T14 T15
C43

220 230 240
Asp-Val-Ser-Leu-Leu-His-Lys-Pro-Thr-Thr-Gln-Ile-Ser-Asp-Phe-His-Val-Ala-Thr-Arg-Phe-Asn-Asp-Asp-Phe-Ser-Arg-Ala-Val-Leu-
CNBr6

250 260 270
Glu-Ala-Glu-Val-Gln-Met-Cys-Gly-Glu-Leu-Arg-Asp-Tyr-Leu-Arg-Val-Thr-Val-Ser-Leu-Trp-Gln-Gly-Glu-Thr-Gln-Val-Ala-Ser-Gly-
T18-19

280 290 300
Thr-Ala-Pro-Phe-Gly-Gly-Glu-Ile-Ile-Asp-Glu-Arg-Gly-Gly-Tyr-Ala-Asp-Arg-Val-Thr-Leu-Arg-Leu-Asn-Val-Glu-Asn-Pro-Lys-Leu-

310 320 330
Trp-Ser-Ala-Glu-Ile-Pro-Asn-Leu-Tyr-Arg-Ala-Val-Val-Glu-Leu-His-Thr-Ala-Asp-Gly-Gln-Leu-Ile-Glu-Ala-Gly-Thr-Cys-Asp-Phe-
CNBr7

340 350 360
Arg-Glu-Val-Arg-Ile-Glu-Asn-Gly-Leu-Leu-Leu-Leu-Asn-Gly-Lys-Pro-Leu-Leu-Ile-Arg-Gly-Val-Asn-Arg-His-Gln-His-His-Pro-Leu-

370 380 390
His-Gly-Gln-Val-Met-Asp-Glu-Gln-Thr-Met-Val-Gln-Asp-Ile-Leu-Leu-Met-Lys-Gln-Asn-Asn-Phe-Asn-Ala-Val-Arg-Cys-Ser-His-Tyr-
CNBr8 CNBr9
T28-30
C38

400 410 420
Pro-Asn-His-Pro-Leu-Trp-Tyr-Thr-Leu-Cys-Asp-Arg-Tyr-Gly-Leu-Tyr-Val-Val-Asp-Glu-Ala-Asp-Ile-Glu-Thr-His-Gly-Met-Val-Pro-
CNBr10 CNBr11
T31-32
C2

430 440 450
Met-Asn-Arg-Leu-Thr-Asp-Asp-Pro-Arg-Trp-Leu-Pro-Ala-Met-Ser-Glu-Arg-Val-Thr-Arg-Met-Val-Gln-Arg-Asp-Arg-Asn-His-Pro-Ser-
CNBr12 CNBr13
T34 T36

460 470 480
Val-Ile-Ile-Trp-Ser-Leu-Gly-Asn-Glu-Ser-Gly-His-Gly-Ala-Asn-His-Asp-Ala-Leu-Tyr-Arg-Trp-Ile-Lys-Ser-Val-Asp-Pro-Ser-Arg-
CNBr14

490 500 510
Pro-Val-Gln-Tyr-Glu-Gly-Gly-Gly-Ala-Asp-Thr-Thr-Ala-Thr-Asp-Ile-Ile-Cys-Pro-Met-Tyr-Ala-Arg-Val-Asp-Glu-Asp-Gln-Pro-Phe-
T39-46

Fig. 3. Amino acid sequence of β-galactosidase. The letters CNBr indicate cyanogen bromide peptides; T refers to tryptic peptides; and C to chymotryptic peptides.

520 530 540
Pro-Ala-Val-Pro-Lys-Trp-Ser-Ile-Lys-Lys-Trp-Leu-Ser-Leu-Pro-Gly-Glu-Thr-Arg-Pro-Leu-Ile-Leu-Cys-Glu-Tyr-Ala-His-Ala-Met-
CNBr15
T39-46

550 560 570
Gly-Asn-Ser-Leu-Gly-Gly-Phe-Ala-Lys-Tyr-Trp-Gln-Ala-Phe-Arg-Gln-Tyr-Pro-Arg-Leu-Gln-Pro-Gly-Phe-Val-Trp-Asp-Trp-Val-Asp-
CNBr16

580 590 600
Gln-Ser-Leu-Ile-Lys-Tyr-Asp-Glu-Asn-Gly-Asn-Pro-Trp-Ser-Ala-Tyr-Gly-Gly-Asp-Phe-Gly-Asp-Thr-Pro-Asp-Asp-Arg-Gln-Phe-Cys-
T47-52
C8

610 620 630
Met-Asn-Gly-Leu-Val-Phe-Ala-Asp-Arg-Thr-Pro-His-Pro-Ala-Leu-Thr-Glu-Ala-Lys-His-Gln-Gln-Gln-Phe-Phe-Gln-Phe-Arg-Leu-Ser-
CNBr17

640 650 660
Gly-Gln-Thr-Ile-Glu-Val-Thr-Ser-Glu-Tyr-Leu-Phe-Arg-His-Ser-Asp-Asn-Glu-Leu-Leu-His-Trp-Met-Val-Ala-Leu-Asp-Gly-Lys-Pro-

670 680 690
Leu-Ala-Ser-Gly-Glu-Val-Pro-Leu-Asp-Val-Ala-Pro-Gln-Gly-Lys-Gln-Leu-Ile-Glu-Leu-Gln-Glu-Leu-Pro-Gln-Pro-Glu-Ser-Ala-Gly-
CNBr18
T53-55

700 710 720
Pro-Leu-Trp-Leu-Thr-Val-Arg-Val-Val-Gln-Pro-Asn-Ala-Thr-Ala-Trp-Ser-Glu-Ala-Gly-His-Ile-Ser-Ala-Trp-Gln-Gln-Trp-Arg-Leu-

730 740 750
Ala-Glu-Asn-Leu-Ser-Val-Thr-Leu-Pro-Ala-Ala-Ser-His-Ala-Ile-Pro-His-Leu-Thr-Thr-Ser-Glu-Met-Asp-Phe-Cys-Ile-Glu-Leu-Gly-
T56-57

760 770 780
Asn-Lys-Arg-Trp-Gln-Phe-Asn-Arg-Gln-Ser-Gly-Phe-Leu-Ser-Gln-Met-Trp-Ile-Gly-Asp-Lys-Lys-Gln-Leu-Leu-Thr-Pro-Leu-Arg-Asp-
CNBr19
T59-60
C70

790 800 810
Gln-Phe-Thr-Arg-Ala-Pro-Leu-Asp-Asn-Asp-Ile-Gly-Val-Ser-Glu-Ala-Thr-Arg-Ile-Asp-Pro-Asn-Ala-Trp-Val-Glu-Arg-Trp-Lys-Ala-
CNBr20

820 830 840
Ala-Gly-His-Tyr-Gln-Ala-Glu-Ala-Ala-Leu-Leu-Gln-Cys-Thr-Ala-Asp-Thr-Leu-Ala-Asp-Ala-Val-Leu-Ile-Thr-Thr-Ala-His-Ala-Trp-

850 860 870
Gln-His-Gln-Gly-Lys-Thr-Leu-Phe-Ile-Ser-Arg-Lys-Thr-Tyr-Arg-Ile-Asp-Gly-Ser-Gly-Gln-Met-Ala-Ile-Thr-Val-Asp-Val-Glu-Val-
T68-70

880 890 900
Ala-Ser-Asp-Thr-Pro-His-Pro-Ala-Arg-Ile-Gly-Leu-Asn-Cys-Gln-Leu-Ala-Gln-Val-Ala-Glu-Arg-Val-Asn-Trp-Leu-Gly-Leu-Gly-Pro-
CNBr21

910 920 930
Gln-Glu-Asn-Tyr-Pro-Asp-Arg-Leu-Thr-Ala-Ala-Cys-Phe-Asp-Arg-Trp-Asp-Leu-Pro-Leu-Ser-Asn-Met-Tyr-Thr-Pro-Tyr-Val-Phe-Pro-
T72-74

940 950 960
Ser-Glu-Asn-Gly-Leu-Arg-Cys-Gly-Thr-Arg-Glu-Leu-Asn-Tyr-Gly-Pro-His-Gln-Trp-Arg-Gly-Asp-Phe-Gln-Phe-Asn-Ile-Ser-Arg-Tyr-
CNBr22

970 980 990
Ser-Gln-Gln-Gln-Leu-Met-Glu-Thr-Ser-His-Arg-His-Leu-Leu-His-Ala-Glu-Glu-Gly-Thr-Trp-Leu-Asn-Ile-Asp-Gly-Phe-His-Met-Gly-
CNBr23
T78 T79

1000 1010 1021
Ile-Gly-Gly-Asp-Asp-Ser-Trp-Ser-Pro-Ser-Val-Ser-Ala-Glu-Phe-Gln-Leu-Ser-Ala-Gly-Arg-Tyr-His-Tyr-Gln-Leu-Val-Trp-Cys-Gln-Lys
CNBr24

Fig. 3. *(Continued)*

and whether some of these might cross-react with anti-β-galactosidase. The answer was yes, and it was possible to demonstrate colinearity of β-galactosidase with its gene as well as to add further support for the conclusion that *LacZ* specified a single polypeptide chain. The molecular weight of fragments of β-galactosidase, detected by cross-reaction, was found to be in good agreement with the size predicted by genetic mapping (Table 1, ref. 9).

At first, interest in my laboratory in the primary structure of β-galactosidase was limited to the sequence at the amino terminus. There was some question as to whether the operator locus, adjacent to *LacZ* in the lactose operon, specified part of the protein. Examination of an operator deletion mutant showed that it did not (1a). For a variety of reasons, a conscious decision was then made to determine the sequence of β-galactosidase. One important reason was to extend gene-protein correlations further, and a second was to provide a basis for structure-function studies. It became evident also that information on the primary structure might help to explain the origin of *LacZ*. Another reason, probably, was finally to lay to rest any lingering doubts about the subunit structure of β-galactosidase. Still another was the challenge; no polypeptide chain containing more than about 200 amino acid residues had been sequenced up to that time. Certainly a strong psychological as well as physical support was the presence down the hall of Dr. Emil Smith and his experienced, active, and expert protein research group.

The hard work involved in determining the sequence of β-galactosidase was begun about 1968. Conventional techniques using trypsin and chymotrypsin digestions were used at first, but the problem was considerably more difficult because of the complexity of the mixture. Of key importance was the study of cyanogen bromide peptides. The polypeptide contains 23 methionine residues and yields on cleavage 8 small peptides of less than 15 residues each and 16 large peptides. Each peptide has been obtained in pure form. The separation of the large peptides was accomplished by chromatog-

TABLE 1
SPECIFIC ACTIVITIES OF CROSS-REACTING MATERIAL (CRM), SEDIMENTATION CONSTANTS, AND APPROXIMATE MOLECULAR WEIGHTS OF INCOMPLETE POLYPEPTIDE CHAINS PRODUCED BY β-GALACTOSIDASE NONSENSE MUTANTS

Mutant type	*NG 125* *amber*	*NG 200* *amber*	*U 366* *amber*	*X-90* *ochre*
Units CRM per mg protein	38	31	38	52
$S_{20,w}$ by sucrose density gradient centrifugation	4.2	5.5	6.0	6.6
MW calculated from $S_{20,w}$	60,000	91,000	103,000	122,000
MW calculated from genetic mapping	58,000	87,000	96,000	110,000

raphy on carboxymethyl-cellulose columns in 8 *M* urea followed by additional steps of purification when necessary (8).

The sequence determination was completed by using the whole protein, but it was aided by taking advantage of a number of strains with mutations in *LacZ*. Prematurely terminated chains from a number of nonsense mutants were isolated and characterized, and one of these, containing the amino terminal half of the polypeptide, was treated with cyanogen bromide (11,36). Therefore each isolated cyanogen bromide peptide from this mutant protein could be placed in the first half of the chain. Ideally, the sequence determination might be carried out by examining prematurely terminated fragments of different lengths from nonsense mutants mapping further and further along the gene. However, protein fragments produced by mutant strains are degraded by proteolysis during growth; consequently, yields of material available for sequencing are low. In the experiment shown in Fig. 4, wild-type and various nonsense mutant strains were induced, washed free of inducer, and sampled during a second period of growth. β-Galactosidase, represented by the top horizontal line, is stable, but fragments disappear at varying rates (27).

Mutant strains producing complementing fragments were also useful. It had been observed by Ullmann and co-workers (35) that an extract made from a strain called M15, which had no β-galactosidase activity, could have its enzyme activity restored by addition of an extract from various strains carrying nonsense mutations. M15 contains a deletion in the beginning part of *LacZ*. Subsequently it was found that a cyanogen bromide digest of whole β-galactosidase could substitute for the second extract, the nonsense mutant

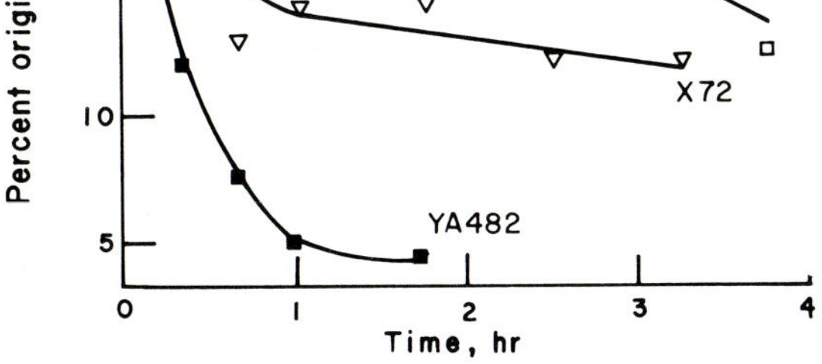

Fig. 4. Degradation of β-galactosidase fragments produced by termination mutants. The polypeptides in exponentially growing cells after removal of inducer were assayed by α-donor activity. (W.T.) = wild-type.

extract (28). The complementation test was then used to monitor purification of the specific complementing peptide in the cyanogen bromide digest. This was isolated and shown to be derived from residues 3–92 of β-galactosidase (24). The M15 protein produced by the *LacZ* deletion mutant was obtained in pure form and was found to lack residues 11–41 of the polypeptide (25).

Immunologic methods were also utilized for the sequence determination. Peptides obtained from cyanogen bromide and tryptic digests were used to make antibodies (6), and some of these were used to search for specific peptides in complex mixtures. One of the last overlapping peptides required for ordering two cyanogen bromide peptides was isolated from a tryptic digest by monitoring its purification with a radioimmunoassay. The tryptic peptide overlapping the two CNBr peptides and hence containing an identical antigenic determinant was measured by inhibition of the reaction between one of the CNBr peptides labeled with ^{125}I and its antibody (2).

The challenge after unfolding a protein and determining its structure is to fold it and ask what its shape is. One way we have used to probe the topography of β-galactosidase is with antibodies. These were prepared from 18 tryptic and cyanogen bromide peptides, which ranged in size from 15 to 96 amino acid residues, and represented more than 80% of the polypeptide chain. An active antibody was obtained in each case (Table 2, ref. 6). It can be seen that these antibodies had respectable titers against the homologous antigen. Each antibody was then tested for binding to the native protein. Nine antibodies gave positive results. Figure 5 shows the Ab reaction with the whole protein plotted against the position in the chain of the peptide fragment used to elicit that antibody (6). Therefore the corresponding segments of the polypeptide chain or part of each, all from central and carboxyl-terminal regions, are exposed on the surface of the molecule.

The segments corresponding to antibodies that did not bind to native β-galactosidase may very well be buried within the molecule. This suggestion is supported by experiments with the dimeric protein from deletion mutant strain M15. Several antibodies made from peptides near the amino terminus that do not react with β-galactosidase were found to react with M15 protein (5). One of these was anti-CNBr2, which also interferes with complementation (Fig. 6). In addition, probes with iodoacetic acid of the native and mutant proteins are consistent. Certain cysteine residues, at positions 498 and 1019, react with the reagent in both forms, but additional residues are available in the dimer (20). Since the results with antibodies are confirmed by the chemical probe, they appear valid not only to give a preliminary picture of the topography of the protein but to point to intersubunit binding regions as well.

β-Galactosidase has been recently in both the public and the scientific eye in an interesting new way. Bacterial strains have been prepared that contain a large part of the *LacZ* gene spliced to a chemically prepared DNA fragment specifying either the hormone somatostatin or insulin (13,18). The protein product produced consists of a long β-galactosidase polypeptide lacking the

TABLE 2
ANTIPEPTIDE ANTIBODIES

Peptide			Antibody	
Designation	Size (amino acid residues)	Position in β-galactosidase (residue number)	Binding capacity (nmoles antigen per ml serum)	Avidity (liters per mole)
CNBr2	90	3–92	1.3	5×10^7
T8	81	60–140	1.0	3×10^7
CNBr3	95	93–187	2.3	1×10^6
CNBr4	15	188–202	0.4	2.4×10^6
T16	20	211–230	4.0	5×10^7
T28-30	36	351–386	8.0	1×10^7
CNBr10	41	378–418	1.0	2×10^8
CNBr14	59	442–500	0.45	8×10^7
CNBr15	40	501–540	2.2	7×10^7
CNBr16	61	541–601	1.7	4×10^6
CNBr18	90	654–743	5.9	3.6×10^6
CNBr19	23	744–766	1.4	4×10^5
CNBr20	96	767–862	11.9[a]	4.6×10^6
CNBr20B	62	801–862	13.3	4×10^6
CNBr21	61	863–923	3.5	4.4×10^7
CNBr22	43	924–966	6.7	3.3×10^5
CNBr23	23	967–989	0.6	2.5×10^7
CNBr24	32	990–1021	5.0	2×10^6

[a] Binding capacity and avidity were tested against CNBr20B. [Adapted from Celada *et al.* (6). By permission.]

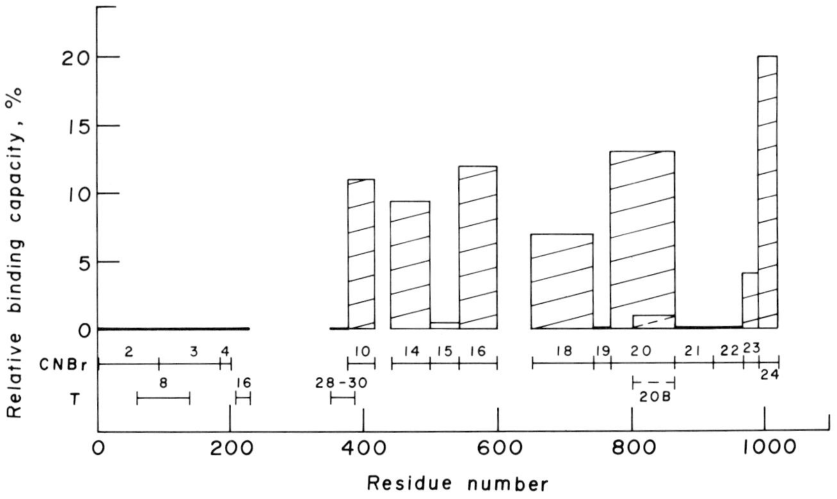

Fig. 5. Relative binding of β-galactosidase by antipeptide antibodies. The figure shows a linear map of the polypeptide chain of β-galactosidase and the cyanogen bromide (CNBr) and tryptic (T) peptides that were used to prepare antibodies. Solid lines and hatched segments represent the binding of antibodies to β-galactosidase as compared with the homologous peptide. [From Celada *et al.* (6). By permission.]

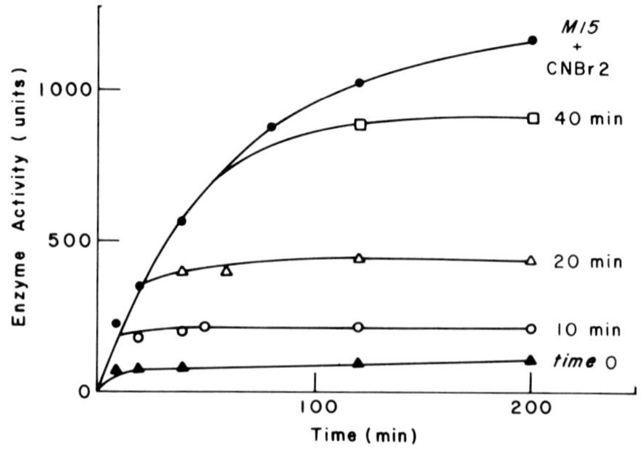

Fig. 6. Effect of anti-CNBr2 on α-complementation. Equal volumes of 10^{-7} *M* CNBr2 and 10^{-7} *M* M15 protein were mixed and anti-CNBr2 diluted 1:2 with normal serum was added to separate portions of the mixture at times 0, 10, 20, and 40 min after the initial mixing. The control tube received normal serum at the time of mixing. [From Celada *et al.* (6). By permission.]

carboxyl terminal 17 amino acids and containing instead the sequence corresponding to the hormone. The hybrid protein contains a trace at most of β-galactosidase enzyme activity, but the hormone derived from the hybrid is biologically active.

By genetic methods, a number of other hybrid or chimeric proteins have been prepared in which a foreign sequence has been substituted for the native β-galactosidase sequence at the other end, the amino terminus. In these cases the β-galactosidase retains complete or nearly complete enzyme activity. One such protein from a strain in which *LacI* and *LacZ* are fused (30) is a monster containing 1353 amino acids in its polypeptide chain. The joining CNBr peptide was detected by radioimmunoassay and was analyzed. The hybrid protein contains the *lac* repressor's polypeptide minus the carboxyl 4 amino acids and β-galactoside lacking its amino terminal 23 amino acids (3). Leu at the joining position is common to both proteins. This hybrid retains both activities. The β-galactosidase portion is tetrameric. In collaboration with J. Beckwith and M. Schwartz and their co-workers, we are now studying hybrids containing a portion of other proteins linked in the same manner to β-galactosidase. These include the periplasmic proteins alkaline phosphatase and the maltose binding protein and several maltose operon membrane proteins (1, 28a, 33a). In certain cases, the fused protein retains all or part of its "leader" sequence, allowing determination of its structure. Such studies might help to tell us why a protein is directed to a specific part of a cell.

What about the origin of the β-galactosidase gene and of the lactose operon? How did such a large protein evolve, and what evolutionary relationship might there be among the different proteins of the lactose operon? By comparison of amino acid sequences, evidence has been obtained that certain genes arose by gene duplication followed by fusion (15). When the sequence of β-galactosidase was compared to itself, a number of striking similarities were seen. Specifically, eight pairs of segments, each separated by about 390 residues, have a large number of amino acid identities, ranging from about 25 to 35% (17). The overall number of identities was found to be about 25%. This is considerably more than expected by chance, which is about 10 or 11%. The results suggest that the first three-fourths of *LacZ* was derived by duplication and fusion of a smaller gene. For control comparisons, a number of other proteins were also examined. Curiously, another protein from *E. coli*, the small dihydrofolate reductase containing 156 residues, was found to be homologous to the carboxyl-terminal fourth of β-galactosidase (17). The results clearly imply that fusion to an earlier dihydrofolate reductase gene occurred at some time in *LacZ*'s evolutionary history. A model for the origin of *LacZ* is shown in Fig. 7. These results also suggest that further detailed comparisons of amino acid sequences within a species may reveal interesting ancestral relationships.

Comparison of β-galactosidase and *lac* repressor showed no similarity in sequence greater than expected on a random basis. The primary structure determination of thiogalactoside transacetylase is now in progress, and the sequence of about two-thirds of its 260 amino acids is known. It is not homologous to β-galactosidase. Therefore no evidence has been obtained so far to indicate that any of the proteins of the lactose operon are evolutionarily related to each other.

For any enzyme protein, important questions are, "How does it act?" and "How does it fold?" For the first, some progress has been made in defining the active site of β-galactosidase. The active site reagent β-D-galactopyranosylmethyl *p*-nitrophenyltriazene binds to methionine residue 500 of the protein (12). Studies of the reaction mechanism suggest that tyrosine is part of the catalytic site (34). It is unlikely that methionine is involved in catalysis, but there is a tyrosine residue at position 501. In regard to the pathway of folding, complementation studies and amino acid comparisons suggest the presence in each subunit of three domains. It seems most likely that these are formed during rather than following synthesis of the polypeptide chain. It was shown that anti-β-galactosidase antibody, which reacts with native but not denatured enzyme, reacts with nascent β-galactosidase on ribosomes (14). The steps in polymerization and the sites of intersubunit binding bear further

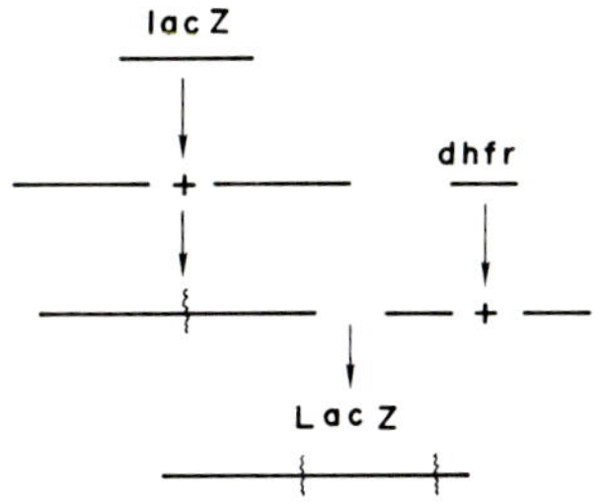

Fig. 7. Model for origin of *LacZ* from earlier *lacZ* and dihydrofolate reductase (dhfr) genes.

study. Polypeptide chains produced by nonsense mutants and lacking a segment at the carboxyl-terminus are monomeric. Presumably, therefore, the monomer–monomer contact points are in this part of the chain. Recently the structure of a protein from deletion mutant strain M112 has been determined (37a). This lacks only nine amino acids, residues 23–31; is a dimer; and can also be complemented with CNBr2. Therefore this segment is a dimer–dimer binding area.

An X-ray diffraction analysis of β-galactosidase awaits the preparation of suitable crystals. In the meantime, antibody probes offer a means to extend and refine the preliminary view of the shape of the molecule already obtained. Most of the antipeptide antibodies used so far are very likely directed against multiple antigenic determinants. By purification of these antibodies, it should be possible to obtain a more highly resolved view of the topography of the enzyme.

Finally, what are the forces that cause a protein to assume a particular shape? Complementation studies allow one approach to this problem. α-Complementation with the 90-residue peptide CNBr2 allows the restoration of enzyme activity to deletion mutant proteins. CNBr2 containing different amino acids at various positions can be obtained from suppressed nonsense mutant strains mapping in the CNBr2 area. What is the effect of these changes on the complementation reaction? And what parts of the 90-residue peptide actually interact with the acceptor protein? β-Galactosidase has been a rich experimental subject and there is every reason to believe it will continue to be.

ACKNOWLEDGMENTS

It is a pleasure to acknowledge the contributions of the following collaborators: George Collins, Paulette Osborne, Val H. Schaeffer, Joseph Welply, Drs. Anthony J. Brake, Franco Celada, Joyce Hamlin, Hans Jörnvall, Jon Katze, Keith E. Langley, Shin Lin, Merna R. Villarejo, and particularly, Audrée V. Fowler.

Work performed in this laboratory was supported in part by U.S. Public Health Service Grants AI-04181 and GM07104 and National Science Foundation Grant PCM-7819974.

REFERENCES

1. Bedouelle, H., Bassford, P. J. Jr., Fowler, A.V., Zabin, I., and Beckwith, J. Mutations which alter the function of the signal sequence of the maltose binding protein of *Escherichia coli*, (1980) *Nature* **285**:78–81.

1a. Bhorjee, J. S., Fowler, A. V., and Zabin, I., Biochemical evidence that the

2. Brake, A. J., Celada, F., Fowler, A. V., and Zabin, I., An immunochemical aid to sequence determination of proteins. *Anal. Biochem.*, 1977, **80**:108–115.

3. Brake, A. J., Fowler, A. V., Zabin, I., Kania, J., and Müller-Hill, B., β-Galactosidase chimeras: Primary structure of a *lac*-repressor-β-galactosidase protein. *Proc. Natl. Acad. Sci. U.S.A.*, 1978, **75**:4824–4827.

4. Brown, J. L., Koorajian, S., Katze, J., and Zabin, I., β-Galactosidase, amino- and carboxyl-terminal studies. *J. Biol. Chem.*, 1966, **241**:2826–2831.

5. Celada, F., and Zabin, I., A dimer-dimer binding region in β-galactosidase. *Biochemistry*, 1979, **18**:404–407.

6. Celada, F., Fowler, A. V., and Zabin, I., Probes of β-galactosidase structure with antibodies. Reaction of anti-peptide antibodies against native enzyme. *Biochemistry*, 1978, **17**:5156–5160.
7. Cohn, M., and Monod, J., Purification et propriétés de la β-galactosidase (lactase) d'*Escherichia coli*. *Biochim. Biophys. Acta*, 1951, **7**:153–174.
8. Fowler, A. V., Amino acid sequence of β-galactosidase. V. Isolation of the 24 cyanogen bromide peptides. *J. Biol. Chem.*, 1978, **253**:5499–5504.
9. Fowler, A. V., and Zabin, I., Colinearity of β-galactosidase with its gene by immunological detection of incomplete polypeptide chains. *Science*, 1966, **154**:1027–1029.
10. Fowler, A. V., and Zabin, I., The amino acid sequence of β-galactosidase. *Proc. Natl. Acad. Sci. U.S.A.*, 1977, **74**:1507–1510.
11. Fowler, A. V., and Zabin, I., Amino acid sequence of β-galactosidase. XI. Peptide ordering procedures and the complete sequence. *J. Biol. Chem.*, 1978, **253**:5521–5525.
12. Fowler, A. V., Zabin, I., Sinnott, M. L., and Smith, P. J., Methionine 500, the site of covalent attachment of an active site-directed reagent of β-galactosidase. *J. Biol. Chem.*, 1978, **75**:4824–4827.
13. Goeddel, D. V., Kleid, D. G., Bolivar, F., Heyneker, H. L., Yansura, D. G., Crea, R., Hirose, T., Kraszewski, A., Itakura, K., and Riggs, A. D., Expression in *Escherichia coli* of chemically synthesized genes for human insulin. *Proc. Natl. Acad. Sci. U.S.A.*, 1979, **76**:106–110.
14. Hamlin, J., and Zabin, I., Immunological activity of ribosome-bound, growing polypeptide chains. *Proc. Natl. Acad. Sci. U.S.A.*, 1972, **69**:412–416.
15. Hartley, B. S., Enzyme families. *Symp. Soc. Gen. Microbiol.*, 1974, **24**:151–182.
16. Hogness, D. S., Cohn, M., and Monod, J., Studies on the induced synthesis of β-galactosidase in *Escherichia coli:* The kinetics and mechanism of sulfur incorporation. *Biochim. Biophys. Acta*, 1955, **16**:99–116.
17. Hood, J. M., Fowler, A. V., and Zabin, I., On the evolution of β-galactosidase. *Proc. Natl. Acad. Sci. U.S.A.*, 1978, **75**:113–116.
18. Itakura, K., Hirose, T., Crea, R., Riggs, A. D., Heyneker, H. L., Bolivar, F., and Boyer, H. W., Expression in *Escherichia coli* of a chemically synthesized gene for the hormone somatostatin. *Science*, 1977, **198**:1056–1063.
19. Jacob, F., and Monod, J., Genetic regulatory mechanisms in the synthesis of proteins. *J. Mol. Biol.*, 1961, **3**:318–356.
20. Jörnvall, H., Fowler, A. V., and Zabin, I., Probe of β-galactosidase structure with iodoacetate. Differential reactivity of thiol groups in wild-type and mutant forms of β-galactosidase. *Biochemistry*, 1978, **17**:5160–5164.
21. Karlsson, U., Koorajian, S., Zabin, I., Sjöstrand, F. S., and Miller, A., High resolution electron microscopy on highly purified β-galactosidase from *Escherichia coli. J. Ultrastruct. Res.*, 1964, **10**:457–469.
22. Karstrom, H., Über die Enzymbildung in Bakterien, *Ann. Acad. Sci. Fenn. Ser. A*, 1930, **33**:1–147.
23. Koorajian, S., and Zabin, I., Carboxypeptidase studies on β-galactosidase: Detection of one C-terminal lysine per monomer. *Biochem. Biophys. Res. Commun.*, 1965, **18**:384–388.
24. Langley, K. E., Fowler, A. V., and Zabin, I., The amino acid sequence of β-galactosidase. IV. Sequence of an α-complementing cyanogen bromide peptide, residues 3–92. *J. Biol. Chem.*, 1975, **250**:2587–2592.

25. Langley, K. E., Villarejo, M. R., Fowler, A. V., Zamenhof, P. J., and Zabin, I., Molecular basis of β-galactosidase α-complementation. *Proc. Natl. Acad. Sci. U.S.A.*, 1975, **72**:1254–1257.
26. Lederberg, K., The beta-D-galactosidase of *Escherichia coli*, strain K-12. *J. Bacteriol.*, 1950, **60**:381–392.
27. Lin, S., and Zabin, I., β-Galactosidase: Rates of synthesis and degradation of incomplete chains. *J. Biol. Chem.*, 1972, **247**:2205–2211.
28. Lin, S., Villarejo, M., and Zabin, I., β-galactosidase, α-complementation of a deletion mutant with cyanogen bromide peptides. *Biochem. Biophys. Res. Commun.*, 1970, **40**:249–254.
28a. Moreno, F., Fowler, A. V., Hall, M., Silhavy, T. J., Zabin, I., and Schwartz, M., A signal sequence is not sufficient to lead β-galactosidase out of the cytoplasm, (1980) *Nature* **286**:356–359.
29. Monod, J., Cohen-Bazire, G., and Cohn, M., Sur la biosynthèse de la β-galactosidase (lactase) chez *Escherichia coli*. La spécificité de l'induction. *Biochim. Biophys. Acta*, 1951, **7**:585–599.
30. Müller-Hill, B., and Kania, J., Lac repressor can be fused to β-galactosidase. *Nature (London)*, 1974, **249**:561–563.
31. Newton, W. A., Beckwith, J. R., Zipser, D., and Brenner, S., Nonsense mutants and polarity in the *lac* operon of *Escherichia coli*. *J. Mol. Biol.*, 1965, **14**:290–296.
32. Pardee, A. B., Jacob, F., and Monod, J., The genetic control and cytoplasmic expression of inducibility in the synthesis of β-galactosidase of *E. coli*. *J. Mol. Biol.*, 1959, **1**:165–178.
33. Rickenberg, H. W., Cohen, G. N., Buttin, G., and Monod, J., La galactoside-permease d'*Escherichia coli*. *Ann. Inst. Pasteur, Paris*, 1956, **91**:829–857.
33a. Sarthy, A., Fowler, A., Zabin, I., and J., Beckwith, Use of gene fusions to determine a partial signal sequence of alkaline phosphatase, (1979) *J. Bact.* **139**:932–939.
34. Sinnott, M. L., and Smith, P. J., Affinity labelling with a deaminatively generated carbonium ion. *Biochem. J.*, 1978, **175**:525–538.
35. Ullmann, A., Jacob, F., and Monod, J., Characterization by in vitro complementation of a peptide corresponding to an operator-proximal segment of the β-galactosidase structural gene of *Escherichia coli*. *J. Mol. Biol.*, 1967, **24**:339–343.
36. Villarejo, M. R., and Zabin, I., β-galactosidase from termination and deletion mutant strains. *J. Bacteriol.*, 1974, **120**:466–474.
37. Wallenfels, K., and Zarnitz, M. L., Kristallization der β-galaktosidase aus *Escherichia coli*. *Angew. Chem.*, 1957, **69**:482.
37a. Welply, J.K.,Fowler, A.V., Beckwith, J. R., and Zabin, I., Positions of early nonsense and deletion mutations in *lacZ*, (1980) *J. Bact.* **142**:732–734.
38. Zabin, I., Crystalline thiogalactoside transacetylase. *J. Biol. Chem.*, 1963, **238**:3300–3306.
39. Zabin, I., and Fowler, A. V., β-galactosidase and thiogalactoside transacetylase, In: *The Lactose Operon* (J. R. Beckwith and D., Zipser, eds.). Cold Spring Harbor Laboratory, Cold Spring Harbor, New York, 1970:27–47.
40. Zabin, I., Kepes, A., and Monod, J., On the enzymic acetylation of isopropyl β-D-thiogalactoside and its association with galactoside permease. *Biochem. Biophys. Res. Commun.*, 1959, **1**:289–292.

Glycosyltransferases in Analysis of Oligosaccharide Structure and Function

ROBERT L. HILL

The Department of Biochemistry
Duke University Medical Center
Durham, North Carolina

Introduction

Ten different monosaccharides (sialic acid, fucose, galactose, mannose, glucose, xylose, *N*-acetylglucosamine, *N*-acetylgalactosamine, glucuronic acid, and iduronic acid) are found in the oligosaccharide prosthetic groups of mammalian glycoproteins and form a wide variety of different glycosidic linkages with one another, as well as with certain side chains of amino acids. The number and types of linkages vary considerably from one monosaccharide to another, as illustrated by examples of typical structures shown in Fig. 1. Thus xylose occurs in only one structure, linking the glycosaminoglycan groups of certain proteoglycans to seryl residues in a polypeptide chain, as in chondroitin 4-sulfate. In contrast, *N*-acetylglucosamine, which occurs in a larger number of different linkages than any of the other sugars, forms at least 16 different linkages with other monosaccharides and one linkage with the side chain of asparaginyl residues in glycoproteins. Together the 10 sugars are known to form at least 80 different types of linkages in the oligosaccharides of mammalian glycoproteins. Since new types of linkages continue to be reported periodically in the current literature, the exact number of naturally occurring linkages in oligosaccharides is presently unknown.

It is generally acknowledged that each glycosidic linkage between a monosaccharide and another structure is formed by the catalytic action of a single, highly specific glycosyltransferase (13). This one-enzyme–one-linkage concept suggests that there are as many glycosyltransferases as there are different types of glycosidic linkages. Notwithstanding this great diversity of the glycosyltransferases, each catalyzes a reaction of the following general form.

ISBN 0-12-643150-7

Glycosylnucleotide + acceptor → Glycosyl—acceptor + nucleotide (1)
(donor substrate) (acceptor substrate)

The donor substrate is cytidine monophosphosialic acid (CMPSia) for the sialyltransferases, guanosine diphosphomannose (GDPMan) and guanosine diphosphofucose (GDPFuc) for the mannosyl and the fucosyltransferases, respectively, and a uridine diphosphomonosaccharide (UDPGlyc) for all of the other transferases. Table 1 lists some typical glycosyltransferase catalyzed reactions and illustrates the diversity of acceptor substrates they employ. Iduronic acid is an exception, in that it is not directly incorporated into glycosidic linkage by a transferase but is formed from glucuronic acid residues in an oligosaccharide by the action of a glucuronosyl-5-epimerase (22).

The first glycosyltransferase to be purified to homogeneity was the *N*-acetylglucosaminide β-1,4-galactosyltransferase (43) that catalyzes the synthesis of the Galβ1,4GlcNAc sequence frequently found in glycoproteins (reaction 7a, Table 1). It also catalyzes synthesis of lactose (reaction 7b, Table

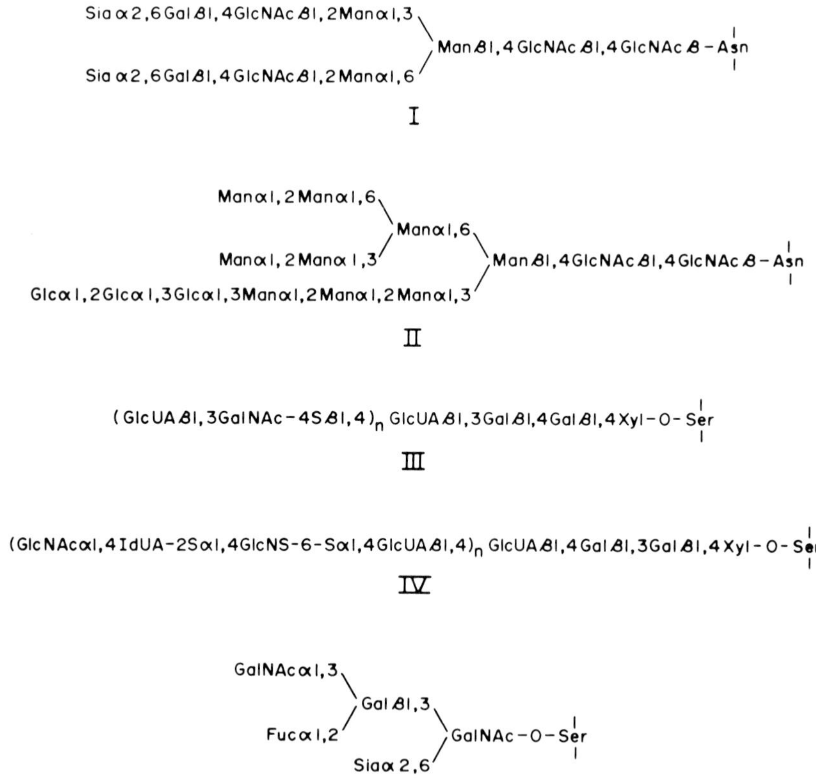

Fig. 1. Some typical structures of N- and O-linked oligosaccharides from mammalian glycoproteins that illustrate the diversity of linkages among the 10 monosaccharides found in these complex sugars. Structures were taken from I (38); II (10); III, IV, and V (16).

TABLE 1

GLYCOSYLTRANSFERASES THAT HAVE BEEN HIGHLY PURIFIED, SOME TO HOMOGENEITY[a]

Sialyltransferases

1. CMPSia + Galβ1,4GlcNAc- → Siaα2,6Galβ1,4GlcNAc- + CMP
2. CMPSia + Galβ1,3GalNAc- → Siaα2,3Galβ1,4GlcNAc- + CMP
3. CMPSia + Galβ1,3GalNAc- → Galβ1,3\\ GalNAc- + CMP / Siaα2,6

Fucosyltransferases

4. GDPFuc + Galβ- → Fucα1,2Galβ- + GDP
6. GDPFuc + Galβ1,4GlcNAc- → Galβ1,4\\ GlcNAc- + GDP / Fucα1,3
6. GDPFuc + Galβ1,3Glc → Galβ1,3\\ Glc + GDP / Fucα1,4

Galactosyltransferases

7a. UDPGal + GlcNAc → Galβ1,4GlcNAc + UDP

7b. UDPGal + Glc —(α-lactalbumin)→ Galβ1,4Glc + UDP

8. UDPGal + Fucα1,3Galβ- → Galα1,3\\ Galβ- + UDP / Fucα1,2

9. UDPGal + Hyl- → Galβ-*O*-Hyl- + UDP

N-Acetylgalactosaminyltransferases

10. UDPGalNAc + Fucα1,2Galβ- → GalNAcα1,3\\ Galβ- + UDP / Fucα1,2

N-Acetylglucosaminyltransferases

11. UDPGlcNAc + Manα1,3\\ Manα1,6 / Manα1,6\\ Man-R / Manα1,3 → Manα1,3\\ Manα1,6 / Manα1,6\\ Man-R / GlcNAcβ1,2Manα1,3 + UDP

12. UDPGlcNAc + Manα1,6\\ Man-R / GlcNAcβ1,2Manα1,3 → GlcNAcβ1,2Manα1,6\\ Man-R / GlcNAcβ1,2Manα1,3 + UDP

Glucosyltransferases

13. UDPGlc + Galβ-0-Hyl- → Glcα1,2Galβ-0-Hyl + UDP

Xylosyltransferases

14. UDPXyl + Ser- → Xyl-β-0-Ser + UDP

Glucuronyltransferases

15. UDPGlcUA + Xylβ-0-Ser → GluUAβ1,4Xylβ-0-Ser

---

[a] Names of the transferases are as follows: (1) β-galactoside α-2,6-sialyltransferase, (2) β-galactoside α-2,3-sialyltransferase, (3) *N*-acetylgalacrosaminide α-2,6-sialyltransferase, (4) β-galactoside α-1,2-fucosyltransferase, (5) *N*-acetylglucosaminide α-1,3-fucosyltransferase, (6) glucoside α-1,3-fucosyltransferase, (7a) *N*-acetylglucosaminide β-1,4-galactosyltransferase, (7b) lactose synthase, (8) fucosyl α-1,2-galactoside α-1,3-galactosyltransferase, (9) collagen hydroxylysyline β-galactosyltransferase, (10) fucosyl α-1,2-galactoside α-1,3-*N*-acetylgalactosaminyltransferase, (11) α-1,3-mannoside β-1,2-*N*-acetylglucosaminyltransferase, (12) α-1,6-mannoside β-1,2-*N*-acetylglucosaminyltransferase, (13) collagen galactosylhydroxylysine α-1,2-glucosyltransferase, (14) proteoglycan serine β-xylosyltransferase and (15) β-xyloside β-1,4-glucoronyltransferase. A discussion of these transferases is given in the recent review by Beyer *et al.* (6), in which the references to the publications describing each transferase listed may be found.

1) in the presence of α-lactalbumin, a regulatory protein made uniquely in the mammary gland (7). It appears to be the only transferase that catalyzes formation of two different glycosidic bonds and is the sole known exception to the one-enzyme–one-linkage concept. In the eight years since this transferase was obtained in homogeneous form, the 14 other transferases that are listed in Table 1 have been highly purified, some to homogeneity. The experience gained in purification of these enzymes along with knowledge of some of their enzymic and physicochemical properties has given considerable insight into this important class of enzymes, including their roles in oligosaccharide biosynthesis. Moreover, the availability of these well-characterized enzymes has permitted some to be used as reagents for probing the structures and the structure–function relationships of oligosaccharides, especially those on cell surfaces. It is the purpose of this chapter to discuss aspects of these properties of some of the glycosyltransferases, especially those published from the author's laboratory.

Purification of Glycosyltransferases

Although the glycosyltransferases are widely distributed among mammalian tissues and can be assayed easily if a specific acceptor substrate is available, the development of methods for their purification has presented problems until recently (6, 33). Most transferases are bound to membranes of either the endoplasmic reticulum or the Golgi apparatus and must be solubilized prior to fractionation, usually with aqueous solutions of neutral detergents. They are present in rather small amounts, and large quantities of a tissue may be required to obtain even small quantities of a pure enzyme. After partial purification, many transferases are unstable, especially in dilute aqueous solutions. Inactivation in many instances occurs by adsorption on glass surfaces, and plastic ware or siliconized glassware must be used during purification to prevent loss of enzyme. Metal cofactors in some instances stabilize the enzymes as do nucleotide co-factors, but the latter usually inhibits the enzymes as well. The transferases may also be stabilized by the aqueous solutions of detergents in which they are solubilized, but often the presence of the detergent may preclude use of many conventional purification procedures. The enzymically homogeneous transferases have often been found to be structurally heterogeneous. All appear to be glycoproteins, and the different electrophoretic forms that are frequently observed under non-denaturing conditions may be due to differences in sugar content, particularly in sialic acid. Some enzymically homogeneous preparations also contain species with different molecular weights, indicating differences in the size of their polypeptide chains. Size heterogeneity may be caused by proteolysis during isolation, although this has been demonstrated directly only with a galactosyltransferase (19,31).

In view of the foregoing general properties of glycosyltransferases it is now possible to make some generalizations about methods for their purification. First, essential for the purification of any enzyme is a simple, specific assay.

This is especially important for the glycosyltransferases, since there are often several transferases in a single tissue that employ the same donor substrate, as illustrated in Table 1. Because these enzymes differ in the nature of their acceptor substrates, it is essential to have a chemically pure acceptor substrate. It is beyond the scope of this chapter to describe how appropriate acceptors, such as those listed in Table 1, have been obtained and the reader is referred elsewhere (6) for more detailed discussions of this problem, but the importance of a specific assay in the purification of a transferase cannot be underestimated. Second, extensive purification of each transferase has required one or more steps by affinity chromatography (6). A wide variety of specific adsorbents has been used since the purification of the first transferase, galactosyltransferase, by affinity chromatography on α-lactalbumin bound covalently to cyanogen bromide-activated agarose (2,43). This affinity adsorbent can only be used with the galactosyltransferase, but, in principle, structural analogs of either of the two reactants or the two products could serve if they can be coupled to an insoluble support (12,30). Derivatives of acceptor substrates have been used with some transferases, including the *N*-acetylglucosaminide β-1,4-galactosyltransferase (Table 1, reaction 7), the xylosyltransferase (Table 1, reaction 14) that forms the linkage to protein in proteoglycans, the galactosyltransferase (Table 1, reaction 9), and the glucosyltransferase (Table 1, reaction 13) involved in synthesis of the disaccharide groups of collagen, and the B blood group specific galactosyltransferase (Table 1, reaction 8) of human serum (6). Some of the best adsorbents, however, are formed with hexanolamine nucleotide phosphates [P^1-(6-amino-1-hexyl)-P^2-(5^1-nucleoside) pyrophosphates] with the following general structure:

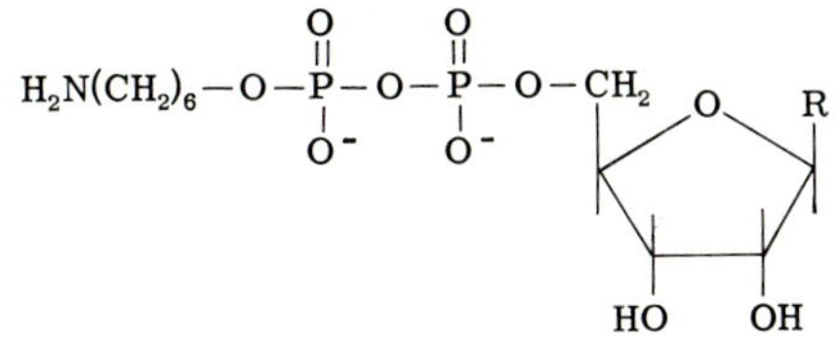

where *R* may be either uridine, cytidine, guanine, or adenine (3,6). The exact nucleotide to be used depends on the type of sugar nucleotide employed as the donor substrate by the transferase. These derivatives are easily synthesized in reasonable yields and can be coupled to agarose to give adsorbents containing from 1 to 15 μmoles of ligand per ml of adsorbent. The adsorbents may be used for batch adsorption or in columns and can be used repeatedly over many months. They have a high degree of specificity; for example, a sialyl, but not a galactosyltransferase, adsorbs to CDP-hexanolamine agarose, whereas the galactosyl but not the sialyltransferase adsorbs to UDP-hexanolamine-agarose (23). In addition, if more than one transferase of the same type binds to an adsorbent, it may be possible to separate the different enzymes by gradient elution with either sodium chloride or the appropriate nucleotide (36). The utility of these adsorbents is indicated by several other observations. Only enzymically active transferases bind to the adsorbent, and their binding and elution are largely independent of the charge or the size of the enzymes, both of which may vary because of the charge and size heterogeneity of a given transferase. If a metal cofactor is required for transferase activity, it is usually needed for adsorption to the nucleotide-agarose, and its removal may aid desorption. The adsorbents can be used in the dilute solutions of aqueous detergent that are required not only for solubilization of transferases but also often for stabilization during purification. Moreover, transferases are usually quite stable when bound to the affinity adsorbents. Finally, the adsorbents are ideally suited for adsorbing the enzymes from the large volumes of solutions that may be unavoidably encountered during purification. Once adsorbed, the transferase can usually be eluted in a small volume.

The reader is referred elsewhere for more detailed discussions of the affinity chromatography of glycosyltransferases (6), but Figs. 2 and 3 give some details for the purification of a fucosyltransferase on GDPhexanolamine-agarose and illustrate the utility of nucleotide-hexanolamine-agarose adsorbents.

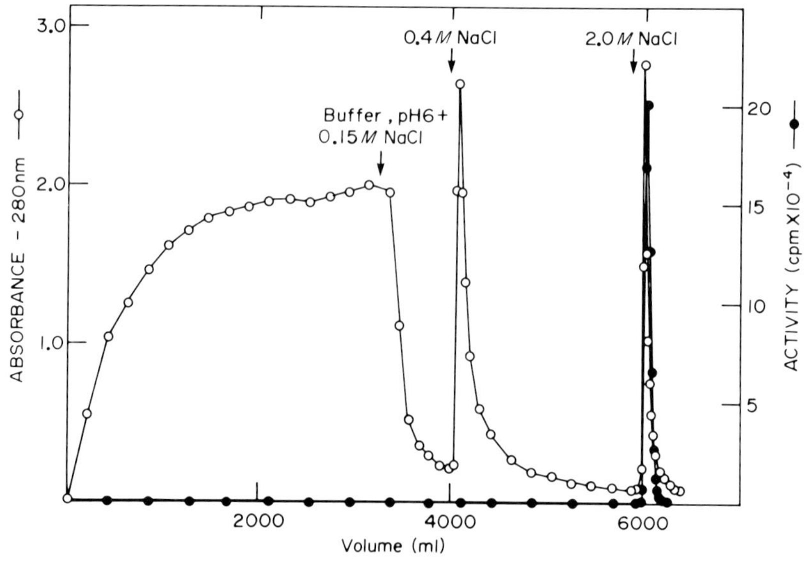

Fig. 2. Affinity chromatography of β-galactoside α-1,2-fucosyltransferase on GDP-hexanolamine-agarose. The transferase was extracted from a particulate fraction of porcine submaxillary glands with an aqueous solution of Triton X-100, the extract absorbed onto sulfopropyl-Sephadex in batches and then eluted (14). The eluate (40) was then applied to a column (3.7 × 15 cm) of GDP-hexanolamine-agarose. The column was eluted with buffered (0.025 *M* sodium cacodylate, pH 6) 0.4 *M* sodium chloride as indicated and then a gradient of sodium chloride (0.4–2.0 *M*). The enzyme was purified 62-fold on this column, and overall was purified 1125-fold after this step. Taken from Beyer *et al.* (5).

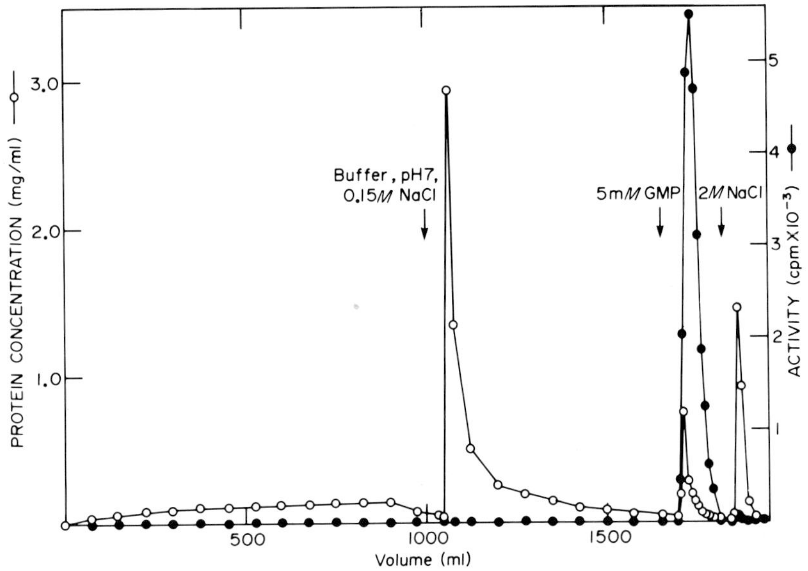

Fig. 3. Affinity chromatography of -galactoside -1,2-fucosyltransferase on GDP-hexanolamine-agarose. The enzyme from the column shown in Fig. 2 was desalted by gel filtration and then applied (925 ml) to a column (1.5 × 6.8 cm) of GDP-hexanolamine-agarose. The column was washed at A with buffered 0.15 *M* sodium chloride and then at B, the enzyme was eluted with buffered 0.005 *M* GMP. At C, the column was washed with 2 *M* sodium chloride to remove inert protein. The enzyme was purified 19-fold on the column to give an overall purification of 21,000-fold. The inert protein remaining in this preparation was removed by gel filtration to bring the transferase to homogeneity (5).

Glycosyltransferases in Oligosaccharide Biosynthesis

Current knowledge of the biosynthesis of the N- and O-linked oligosaccharides of mammalian glycoproteins has been obtained largely by examination of the glycosyltransferase activities and the products they form in whole cells, tissue extracts, or tissue homogenates. Although this approach was necessary and gave valuable information, the low and variable activities of the complex mixture of transferases present in such crude systems often did not permit elucidation of the substrate specificities of individual transferases, and thus a complete understanding of their potential roles in oligosaccharide synthesis. Now that several glycosyltransferases have been highly purified and free of other transferase activities, it is possible to study their substrate specificities in detail. These studies have largely confirmed the high degree of specificity predicted by the one-enzyme–one-glycosidic bond concept. But perhaps more important, they have also shown how certain transferases can act sequentially or in concert to catalyze the formation of several known or theoretically possible oligosaccharide structures (4).

The experimental details used to examine the sequential action of certain glycosyltransferases have been reported earlier (4) and need not be repeated here. In essence, the actions of six different pure transferases have been studied and the reactions examined for each are shown in Figs. 4 and 5.

As shown in Fig. 4, three of the transferases use the structure Galβ1,3GalNAc-O-Thr/Ser as an acceptor substrate and form products that are often found at the nonreducing termini of O-linked oligosaccharides. Moreover, in principle, the product of one enzyme is a substrate for the other two enzymes, and the products of one of the fucosyltransferases is a potential substrate for a fourth enzyme, an *N*-acetylgalactosaminyltransferase that forms the A blood group antigenic determinant.

Figure 5 shows the action of three transferases that employ another structure, Galβ1,4GlcNAc-, as an acceptor substrate to form products that are also often found at the nonreducing termini of N-linked oligosaccharides. As in Fig. 4, the product of one transferase is potentially a substrate for the other two enzymes and the products of one of the enzymes are potential substrates for a fourth transferase, the *N*-acetylgalactosaminyltransferase.

The sequential actions of the glycosyltransferases shown in Figs. 4 and 5 reveal several interesting points. First, the actions of some transferases may be mutually exclusive, since the combined action of two enzymes does not lead to the formation of theoretically possible structures. This is the case for the sialyl- and fucosyltransferases that could potentially form structures *f* (Fig. 4) and *l* (Fig. 5).

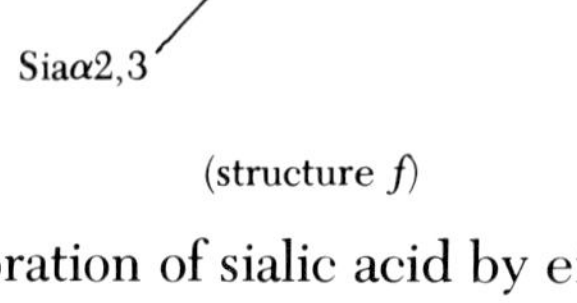

(structure *f*)

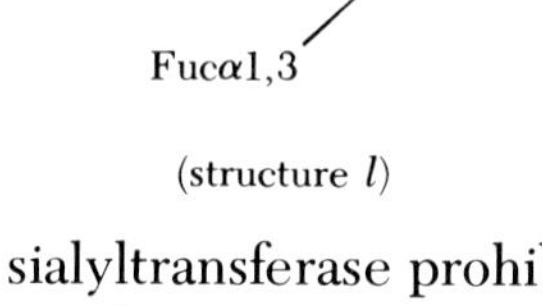

(structure *l*)

Thus incorporation of sialic acid by either sialyltransferase prohibits incorporation of fucose by the fucosyltransferases, and vice versa. Second, the synthesis of some products requires that the transferases act sequentially in a specific order. This was the case for formation of structures *o* and *p* (Fig. 5).

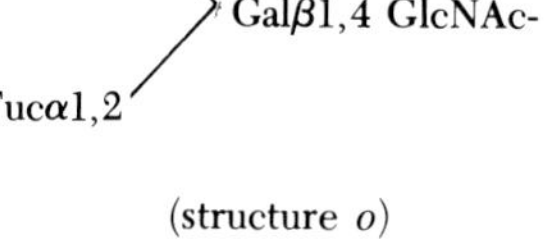

(structure *o*)

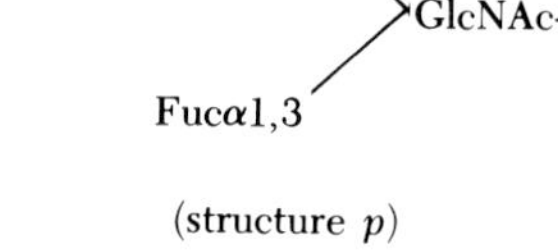

(structure *p*)

In each case, the Fucα1,2Gal structure must first be formed, since structure *k* [Siaα2,6Galβ1,4GlcNAc-] and structure *l* [Galβ1,4(Fucα1,3)GlcNAc-] are not acceptor substrates for the β-galactoside α1,2-fucosyltransferase. Structures *h* and *q* could also be formed sequentially only when the fucosyltransferase acted first, and the *N*-acetylgalactosaminyltransferase acted next.

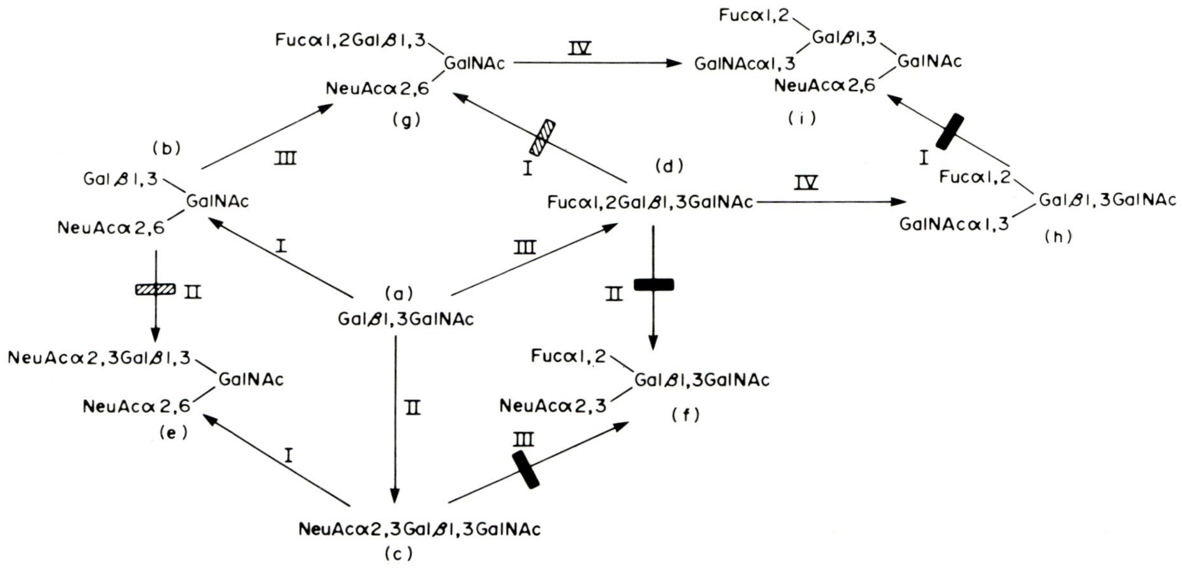

Fig. 4. The sequential reactions of two sialyltransferases, a fucosyltransferase, and a *N*-acetylgalactosaminyltransferase. Antifreeze glycoprotein, whose disaccharide groups are indicated by structure *a*, was the substrate from which all other structures were derived. Structures *b*, *c*, *d*, *e*, *g*, and *h* were synthesized by reacting the appropriate acceptor substrate and nucleotide sugar with excess enzyme and then isolated from reactants chromatographically. The rates of reaction of each of the structures with the enzyme indicated were measured at equivalent enzyme and substrate concentrations. Reactions that could not proceed are indicated by a solid bar across the arrow, and those that proceeded very slowly are indicated by the hatched bars. (I) *N*-acetylgalactosaminide α-2,6-sialyltransferase; (II) β-galactoside α-2,3-sialytransferase; (III) β-galactoside α-1,2-fucosyltransferase; (IV) α-1,2-fucosylgalactoside α-1,3-*N*-acetylgalactosaminyltransferase. [Taken from Beyer *et al.* (4).]

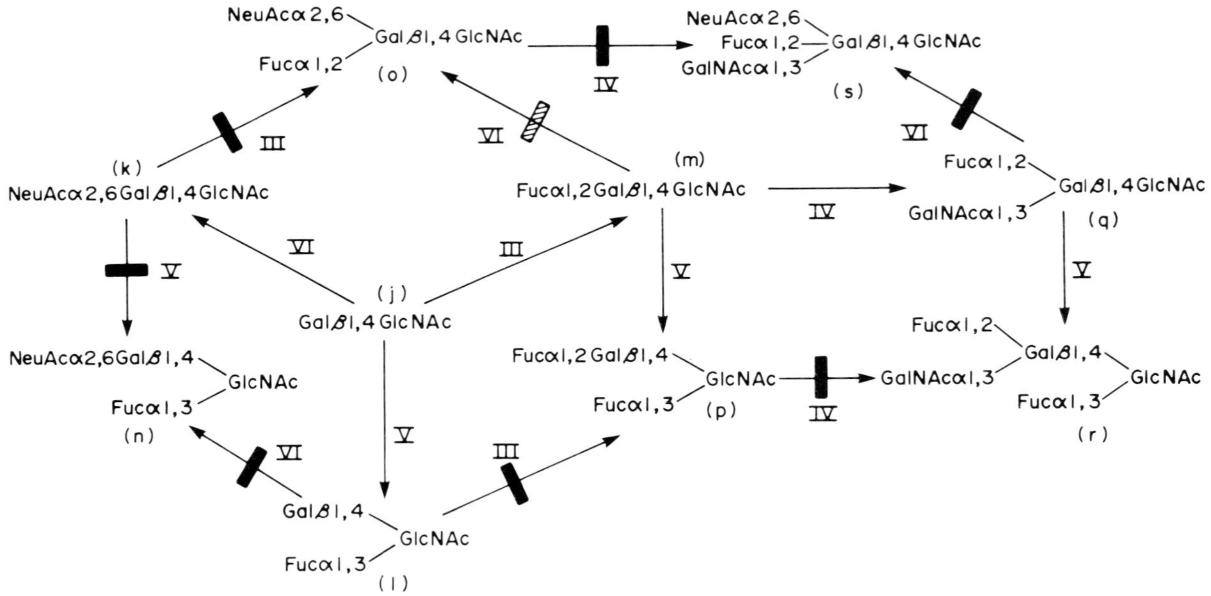

Fig. 5. The sequential reactions of two fucosyltransferases, a sialyltransferase, and a *N*-acetylgalactosaminyltransferase. Asialotransferrin, prepared by treatment of human transferrin with neuraurinidase, with nonreducing termini as shown for structure *j*, was the substrate from which all other structures were derived. Structures *k*, *l*, *m*, *o*, *q*, and *r* were synthesized, and the rates of each in the reaction shown were measured at equivalent enzyme and substrate concentrations as described in Fig. 4. The symbols are the same as in Fig. 4. (III) β-galactoside α-1,2-fucosyltransferase; (IV) α-1,2-fucosylgalactoside α-1,3-*N*-acetylgalatosaminyltransferase; (V) *N*-acetylglucosaminide α-1,3-fucosyltransferase; (VI) β-galactoside α-2,6-sialytransferase. [Taken from Bever *et el.* (4).]

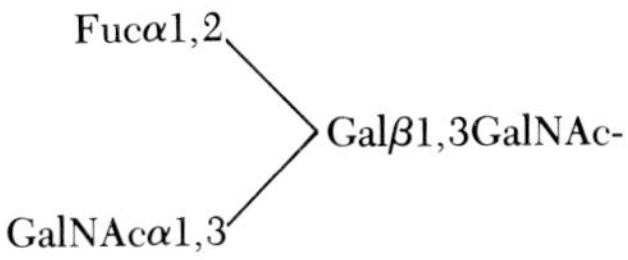

(structure *h*)

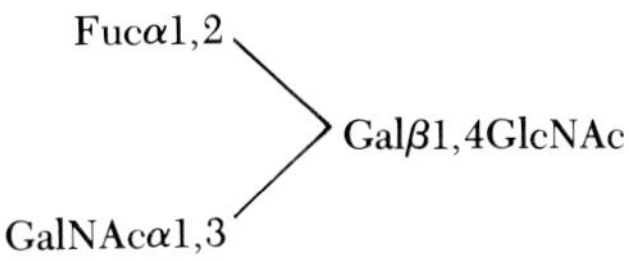

(structure *q*)

The requirement for sequential action of transferases was also noted for structures *i* (Fig. 4) and *r* (Fig. 5).

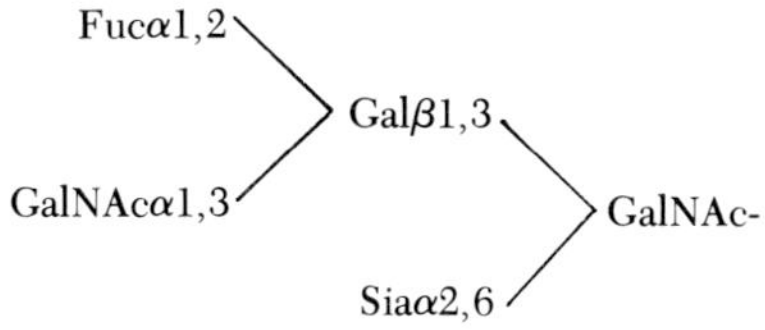

(structure *i*)

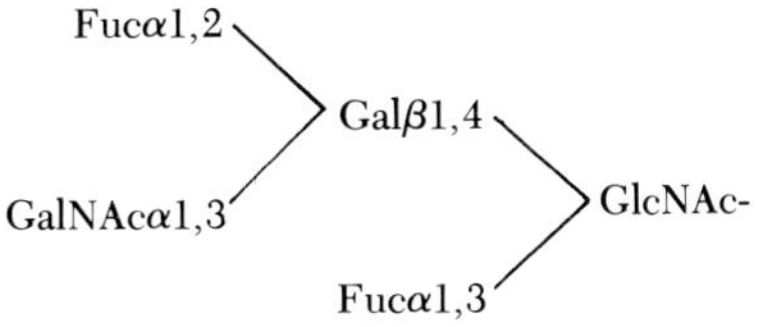

(structure *r*)

In structure *i* the Fucα1,2Gal linkage must be formed initially before the *N*-acetylgalactosaminyltransferase can form the GalNAcα1,3Gal linkage. However, formation of structure *r* requires just the opposite order of action, that is, the transferase forming the GalNAcα1,3Gal linkage must act prior to that forming the Fucα1,3GlcNAc linkage. Third, when the same structure can be formed by two alternate pathways, one of the two pathways may be highly preferred. This is illustrated for the synthesis of structures *e* and *g* (Fig. 4).

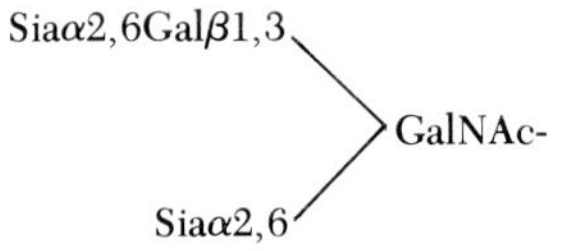

(structure *e*)

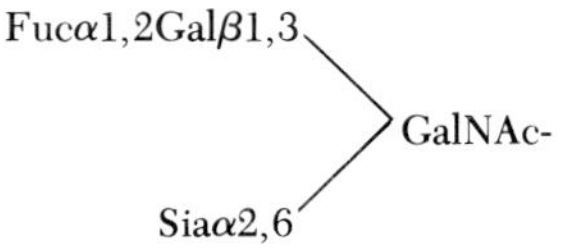

(structure *g*)

For structure *e* the rate of formation of the Siaα2,6GalNAc linkage is about 30 times faster if the Sia α2,3Gal linkage is formed initially, and for structure *g* the rate of formation of the Fucα1,2Gal linkage is much more rapid if the Siaα2,6GalNAc linkage is formed first.

Thus the pathways in Figs. 4 and 5 indicated that the prior action of one transferase may completely or partially inhibit the action of another transferase. This suggests that the combined actions of the transferases may account for the structural heterogeneity often observed at the nonreducing termini of oligosaccharides in glycoproteins. For example, bovine lactoferrin contains an N-linked oligosaccharide with the following sequences at its nonreducing termini (40).

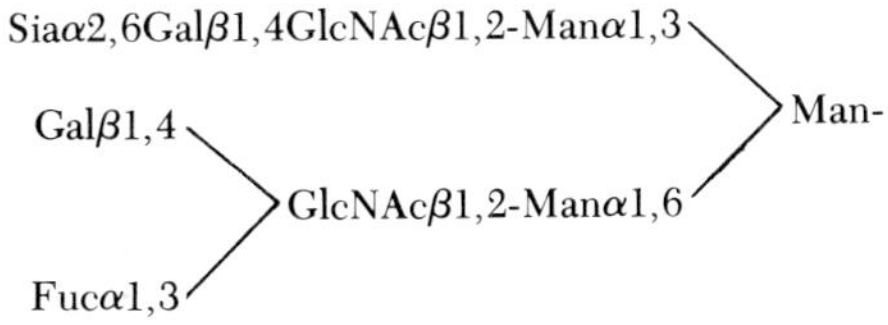

Clearly, the presence of the Siaα2,6Gal linkage in one branch would prevent formation of the Fucα1,3GlcNAc linkage by the fucosyltransferase in the same branch. Similarly, if the Fucα1,3GlcNAc linkage were formed initially, then the Siaα2,6Gal linkage could not be formed in that branch. The structure of the lactoferrin oligosaccharide as well as the fact that the structure Siaα2,6Galβ1,4(Fucα1,3)GlcNAc- has never been reported to exist in oligosaccharides lend support to the view that the glycosyltransferases act *in vivo* as they do *in vitro* and, as a result, help to explain the structural heterogeneity of oligosaccharides. It is interesting, however, that the structure Siaα2,3Galβ1,4(Fucα1,3)GlcNAc- has been reported in rat brain (17), suggesting that the sialyltransferase forming the Siaα2,3Gal linkage and the fucosyltransferase forming the Fucα1,3GlcNAc linkage can act together, although in which order is unknown.

The structural heterogeneity of oligosaccharides could also be partly accounted for if one of two alternate paths that lead to the synthesis of the same oligosaccharide is highly preferred. As indicated in Fig. 4, the preferred path for synthesis of structure *e* is *a*→*c*→*e* rather than *a*→*b*→*e*, since *c* is a far better substrate for the *N*-acetylgalactosaminide α-2,6-sialyltransferase than structure *b* for the β-galactoside α-2,3-sialyltransferase. Thus if structure *b* is formed it may not be further modified and could be considered a "dead end" product. Of course, several factors other than the strict substrate specificity could aid regulation of oligosaccharide biosynthesis, but at present no allosteric controls appear to occur, nor have multicellular complexes of transferases been demonstrated, even though they have been postulated to exist (15).

Finally, the same studies used to construct the pathways shown in Fig. 4 revealed the most likely pathways for the biosynthesis of the A blood groups antigenic pentasaccharide of porcine mucin. The five transferases that are involved could act in several alternative orders to give the desired structure, as shown in Fig. 6, but only one appears to be preferred, since at least three structures, if formed, are dead-end products. The preferred path deduced from these studies differs in two steps from that suggested from studies with very impure enzyme preparations (37). It is also of interest that the oligosaccharides of porcine mucins are structurally heterogeneous (1,21) and contain significant amounts of some of the structures shown in Fig. 6 that are considered to be dead end products. This supports the view that the order in which the glycosyltransferases act accounts at least in part for the structural heterogeneity of glycoprotein oligosaccharides.

Glycosyltransferases in the Analysis of Oligosaccharide Structure and Function

Glycosidases have been used for many years to determine the structures and the biological functions of oligosaccharides. However, it is only recently that well-characterized, highly purified glycosyltransferases have become

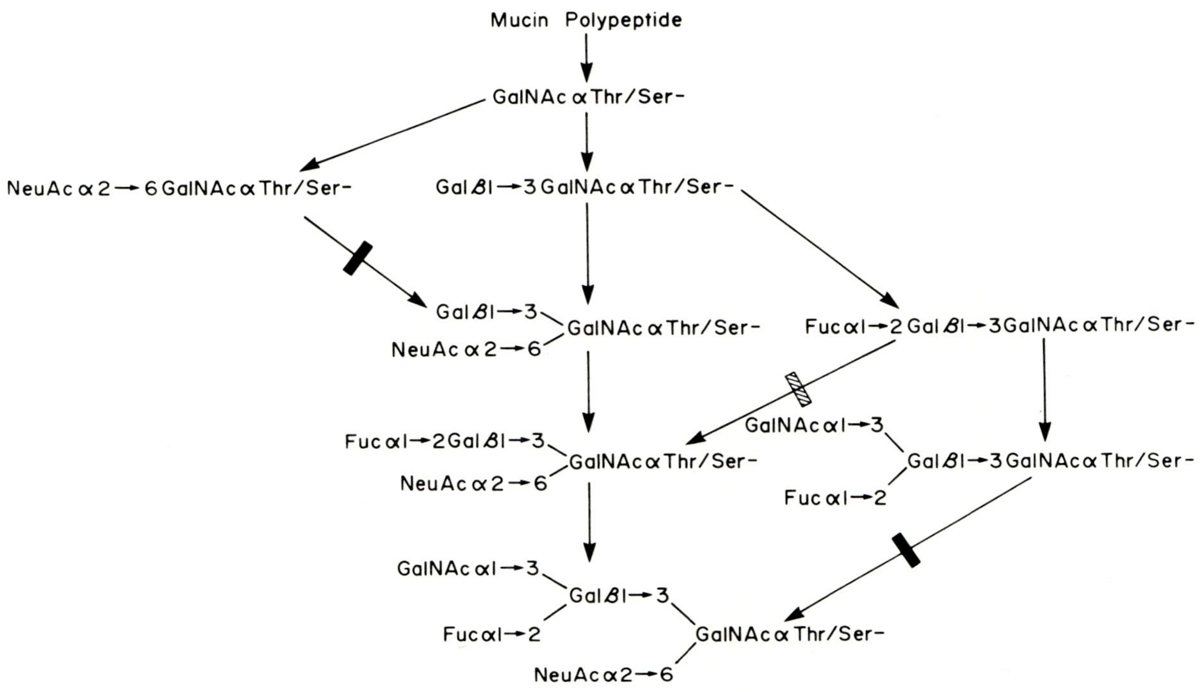

Fig. 6. The proposed biosynthetic pathway for the A blood group antigenic oligosaccharide of porcine mucin as deduced from the studies summarized in Fig. 4. [Taken from Beyer *et al.* (4).]

available for these purposes. Before employing glycosyltransferases in structure-function studies, extensive knowledge of their acceptor substrate specificities is required, including an understanding of their permitted sequential actions of the kinds summarized in Figs. 4 and 5. With this knowledge the transferases can be used in seveal ways to obtain considerable information from rather simple experiments. Either native or glycosidase-treated glycoproteins in nanomolar amounts can be reacted with a given transferase and radiolabeled donor substrate, and the extent of incorporation of labeled sugar can be measured with time. If label is incorporated, then it is highly likely that the nonreducing terminal groups of the oligosaccharide have the same structure as those of the single, best acceptor substrate of the transferase. For example, consider use of the three pure sialyltransferases. One, the β-galactoside α-2,6-sialytltransferase (reaction 1, Table 1) will incorporate sialic acid only if the sequence Galβl,4GlcNAc- is present. Another, the β-galactoside α-2,3-sialyltransferase (reaction 2, Table l) incorporates sialic acid if the sequence Galβ1,3GalNAc- is present. Thus the two enzymes permit identification of not only the terminal residue but also the penultimate residue and the nature of the glycosidic bond linking the two. The third transferase, the α-*N*-acetylgalactosaminide α-2,6-sialyltransferase (reaction 3, Table 1), uses about equally well as acceptors either GalNAcα- or Galβ1,3GalNAcα- linked to the hydroxyl groups of serine or threonine. However, the two structures can be distinguished from one another, since the α-2,3-sialyltransferase will not use GalNAcα- linked to hydroxyl groups in glycoproteins as an acceptor. Thus, since the number of acceptor groups per molecule can be estimated from the specific activity of the products, considerable *quantitative* structural information becomes available with the use of only small amounts of substrates and enzyme. Other transferases may not be as specific as the sialyltransferases, and aid in identification of only the terminal sugar. For example, the β-galactoside α-1,2-fucosyltransferase (reaction 4, Table 1) uses any of a number of β-galactosides and the *N*-acetylglucosaminide β-1,4-galactosyltransferase (reaction 7a, Table 1) uses *N*-acetylglucosamine linked in a variety of ways to other sugars.

Examples of the use of glycosyltransferases in the structural analysis of glycoproteins have been reported recently and need only be mentioned briefly here.

1. The extent of hydrolysis of α_1-acid glycoprotein by the sequential action of a neuraminidase, a β-1,4-galactosidase and a β-*N*-acetylglucosaminidase was readily measured with aid of transferases (24). Removal of sialic acid generates acceptor sites for the β-galactoside α-2,6-sialyltransferase (reaction 1, Table 1), whereas removal of sialic acid and galactose destroys the acceptor sites for the sialyltransferase but generates acceptor sites for the *N*-acetylglucosaminide β-1,4-galactosyltransferase (reaction 7a, Table 1). Removal of *N*-acetylglucosamine then destroys the acceptor sites for the galactosyltransferase.

2. The β-galactoside α-1,2-fucosyltransferase (reaction 4, Table 1) was used to detect the presence of Galβ1,3GalNAc- groups in ovine mucin (36). Only 0.5% of all the oligosaccharides in ovine mucin have this structure, and their removal by an endo-α-*N*-acetylgalactosamidase was monitored by loss of acceptor sites for the fucosyltransferase.

3. With the aid of the β-galactoside α-2,6-sialyltransferase, it was demonstrated that the galactose residues in the galactoside-binding protein of rabbit liver were in the sequence Galβ1,4GlcNAc, rather than in the sequence Galβ1,3GlcNAc (25).

4. The galactosyltransferase was used to show that 65% of the galactose residues in human Factor VIII/von Willebrand factor were in the sequence Galβ1,4GlcNAc (39).

5. The fucosyl α-1,2-galactoside α-1,3-*N*-acetylgalactosaminyltransferase (reaction 10, Table 1) was used to demonstrate the H blood group antigenic determinant (Fucα1,2Gal) in human Factor VIII/von Willebrand factor, the first instance of this determinant on a human plasma protein (39).

The glycosyltransferases have also been used to synthesize small oligosaccharides; thus the disaccharide *N*-acetyllactosamine (Galβ1,4GlcNAc) was obtained in 54% yield (1.1 g) with the *N*-acetylglucosaminide β-1,4-galacotosyltransferase (28). Lacto-*N*-tetraose(Galβ1,3GlcNacβ1,3Galβ1,4 Glc) and lacto-*N*-neotetraose (Galβ1,4G-lcNAcβ1,3Galβ1,4Glc) were also prepared with aid of the galactosyltransferase (27). These two milk oligosaccharides are very difficult to separate from one another. Thus a mixture of the two (80 mg) was hydrolyzed with a galactosidase that specifically cleaves Galβ1,4-linked galactosides, and the resulting trisaccharide (13 mg) was readily separated from the resistant lacto-*N*-tetraose (50 mg) by paper chromatography. Lacto-*N*-neotetraose was then synthesized in 50% yield with aid of the galactosyltransferase. Since UDPgalactose is rather expensive, it was generated *in situ* during these syntheses from the less expensive UDPglucose and UDPgalactose-4-epimerase. Use of UDPgalactose would likely have given higher yields of products in these syntheses.

The glycosyltransferases were also used effectively in the synthesis of fucose-containing ligands that are required to detect hepatic receptors that bind oligosaccharides in glycoproteins containing fucose (14,32). These studies also illustrate the ability of the transferases in assessing a biological function of fucose in glycoproteins. Fucose receptors in liver were first suspected from measurements of the rate of clearance of intravenously injected human lactoferrin in mice. Clearance was thought to be mediated via the fucosyl α-1,3-*N*-acetylglucosaminyl groups in the one branch of the biantennary oligosaccharide of human lactoferrin (see structure 1, page 70). In order to evaluate this possibility, the clearance of another protein that will not clear from blood rapidly, does not contain fucose, but in which fucose could be incorporated by fucosyltransferases was examined. Human transferrin, which contains two oligosaccharide groups per molecule that are identical to structure I in Fig. 1, was chosen and treated as shown in Fig. 7A. Sialic acid was quantitatively removed with neuraminidase to give asialotransferrin, which was then used as a substrate for the *N*-acetylglucosamide α-1,3-fucosyltransferase (reaction 5, Table 1) and the galactoside α-1,2-fucosyltransferase (reaction 4, Table 1). Fucose was incorporated quantitatively into α-1,3 linkage and 64% of the acceptor sites for the α-1,2-fucosyltransferase were fucosylated. The α-1,3-fucosylated derivative was also treated with a fucosidase from almond emulsin that specifically hydrolyzes fucose linked α-1,3 to *N*-acetylglucosamine in glycoproteins (46). The clearance of the ^{125}I-linked derivatives of each type of transferrin from the blood of mice was then measured. As shown in Fig. 8, only the derivative containing fucose-linked α-1,3 to *N*-acetyllactosamine was cleared, and its clearance was returned to that of asialotransferrin on removal of the fucose.

Other studies (14) revealed that α-1,3 fucosylated asialotransferrin rapidly cleared from the circulation by the liver but was not taken up rapidly, presumably by endocytosis after binding to receptors. The basis for this conclusion was the observation that after clearance, the ^{125}I-label of the fucosylated asialotransferrin reappeared in the circulation up to an hour after clearance, if a protein such as unlabeled fucosylated asialotransferrin, as well as others, was injected. Presumably, the unlabeled protein displaced the labeled protein from the receptors. Since this lack of uptake was perhaps related to the fact that there were too few fucoside groups per molecule, or the groups were not in favorable locations on the molecule, it was decided to examine the clearance and uptake of neoglycoproteins (18), which are extensively glycosylated derivatives of bovine serum albumin (BSA) that could be modified by glycosyltransferases (Fig. 7B) to give a series of derivatives whose clearance could be measured. Figure 9 shows that these derivatives were rapidly cleared as expected. Moreover, they appear to be endocytosed by the liver, since 10 min after clearance they could not be displaced into the circulation by injection of unlabeled neoglycoproteins, in contrast to fucosylated asialotransferrin.

The various neoglycoproteins have also been used to examine fucose-dependent binding to isolated membrane fractions from mouse liver, as shown in Fig. 10. Since the ligands used to measure clearance (Fig. 8) also contain terminal, nonreducing galactose residues, it is essential to distinguish binding through the galactose via the hepatic galactoside receptors first described by Ashwell and binding via fucose. For this purpose the neoglycoprotein fucosyl-BSA also proved to be useful, since it does not bind to other known carbohydrate-specific binding proteins of liver, including the galactoside binding protein. The studies in Fig. 10 show that binding is specific for fucose, but most important, the binding of fucosides has a different dependence on pH than that involved in the binding of galactosides.

The glycosyltransferases can also be used to modify oligosaccharide structures on intact cells and thereby provide greater understanding of the role of

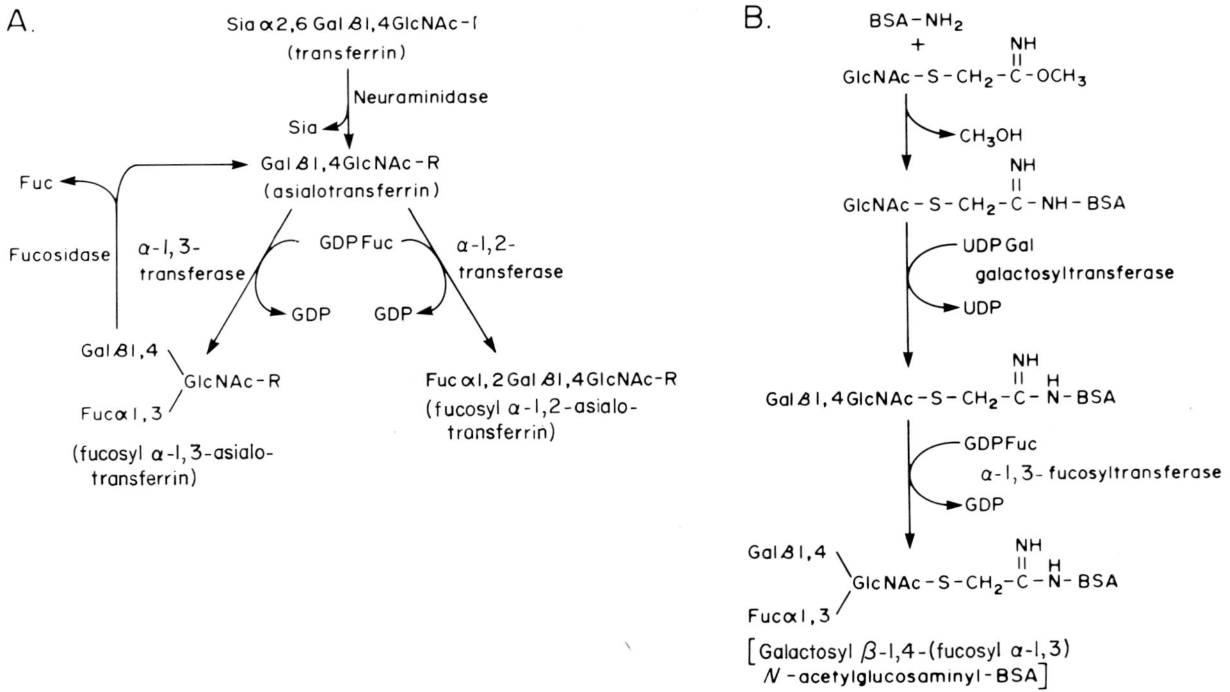

Fig. 7. Modification of glycoproteins by glycosyltransferase catalyzed reactions. (A) Human asialotransferrin. (B) Bovine serum *N*-acetylglucosaminylalbumin.

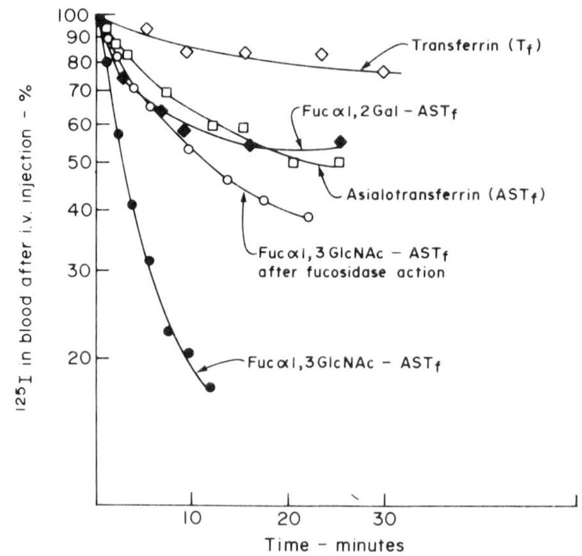

Fig. 8. The rate of clearance from blood of ^{125}I-labeled derivatives (Fig. 7A) of asialotransferrin intravenously injected into mice. The experimental details were described earlier (32).

specific kinds of carbohydrate linkages in certain biological phenomena. Sialyltransferases to incorporate sialic acid into desialized human erythrocytes has been particularly useful in clarifying the role of sialic acid in determining the molecular basis of the M/N blood group activity (34). It has been

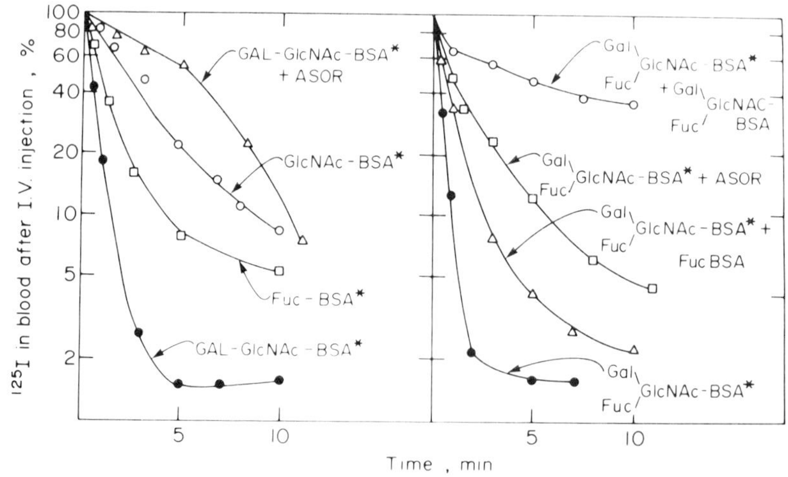

Fig. 9. The rate of clearance from blood of ^{125}I-labeled neoglycoproteins intravenously injected into mice. The derivatives were synthesized as shown in Fig. 7B. Where indicated, a 1000-fold excess of unlabeled ligand was used to test the inhibition of clearance of the ligand indicated. ASOR = asialoorosomucoid. The experimental details were described earlier (32). *, ^{125}I-labeled.

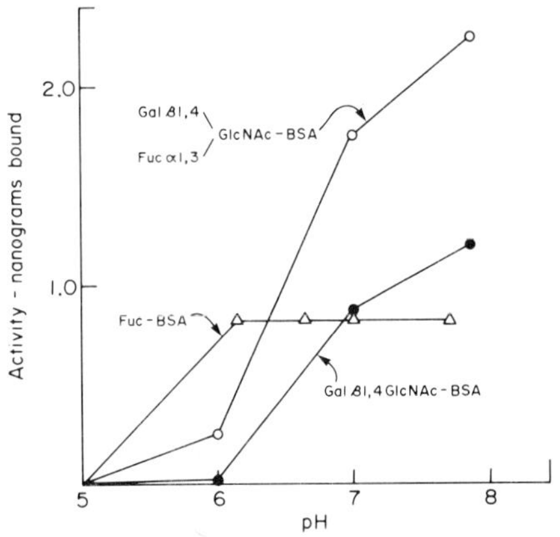

Fig. 10. The binding of ^{125}I-labeled neoglycoproteins to membrane fractions from mouse liver. Plasma membranes (75 μg protein) from mouse liver in 0.5 ml containing BSA (1 mg/ml); 10 m*M* $CaCl_2$; O.15 *M* NaCl; 50 m*M* solium acetade buffer, pH 5, 6, or 7, or 50 m*M* disodium maleate, pH 5, 6, or 7, or 50 m*M* Tris-C1, pH 7.8, and 10 ng of ^{125} I-labeled neoglycoprotein were incubated for 60 min at 37°, then centrifuged, and the pellets were counted. Blanks were the same except that 10 μg of unlabeled ligand were also present in the assay. (Unpublished observations, M. A. Lehrman, M. J. Imber, S. V. Pizzo, and R. L. Hill.)

known for some time that erythrocytes treated with neuraminidase to remove their surface sialic acid groups lose their ability to be hemagglutinated by either anti-M or anti-N antisera (20,41). Moreover, removal of sialic acid from glycophorin, the erythrocyte cell surface glycoprotein that carries the majority of the cell surface sialic acid, results in a loss of its interaction with either anti-M or anti-N antisera. This has led to the suggestion that sialic acid in two different linkage types in the oligosaccharide groups of glycophorin, that have the structure

Siaα2,3Galβ1,3
GalNAcα-0-Ser
Siaα2,6

are essential antigenic determinants for the M/N blood groups. Indeed, it was proposed that sialyltransferases are specified by the M/N genes and that the product of one gene acts on the sequence Galβ1,3GalNAcα-O to give rise to a monosialyled N antigen, which is then acted on by another genetically determined sialyltransferase to give the M antigen (45). The experimental support for this proposal is unconvincing, but more important, structural analysis (11) of glycophorins from erythrocytes of individuals homozygous for either M or N blood groups showed that there were two glycophorin types, each with identical sequences, except for the first and fifth residues from the amino terminus, as follows:

Glycophorin A (M cells)	H_2N-*Ser*-$\overset{*}{\text{Ser}}$-$\overset{*}{\text{Thr}}$-$\overset{*}{\text{Thr}}$-*Gly*-
Glycophorin B (N cells)	H_2N-*Leu*-$\overset{*}{\text{Ser}}$-$\overset{*}{\text{Thr}}$-$\overset{*}{\text{Thr}}$-*Glu*-

The asterisks indicate that 3 of the 15 disialyltetrasaccharides of the glycophorins are attached to residues 2, 3, and 4. Considerable evidence suggests that the sequence differences are the basis for the M and N blood group activities (9,41), yet the loss of these activities on removal of sialic acid remained unexplained.

The three pure sialyltransferases (reactions 1, 2, and 3, Table 1) were used to clarify the role of sialic acid in M/N blood group activities (34). Each of the transferases incorporated sialic acid into neuraminidase-treated erythrocytes, although only those that introduced sialic acid into α-2,3 linkage to galactose or α-2,6 linkage to *N*-acetylgalactosamine in the glycophorin oligosaccharides restored blood group activity. Only M activity was regenerated on sialylation of M cells, and only N activity from N cells, thus the antigenicity of the reasialyzed cells was always the same as that of the native cells from which they were prepared. Moreover, there was no indication that one transferase gave rise to cells of one type and the other, a second type. Thus these results indicate that specific sialic acid groups are important for recognition by the antisera that are used for determining the M and N blood groups, but that the specific M and N determinants reside in the amino acid sequence of the glycophorin, not in sialic acids. These studies also revealed considerable variation in the specificity of M and N antisera, in accord with earlier observations. Moreover, antisera used to detect these blood groups are usually treated with asialo-erythrocytes to remove M/N antibodies against cells that do *not* require sialic acid. Since sialic acid independent antibodies are occasionally reported, it would appear that the role of sialic acid in M/N blood group activity is just another instance of the requirement for multiple groups on an antigen for combination with an antibody.

The roles of sialic acid in viral adherence to erythrocytes has also been clarified with aid of the three sialyltransferases (8,29). Several ortho- and paramyxoviruses (29) and the polyoma oncogenic virus (8) adhere to host cells as well as to erythrocytes by direct binding to cell surface sialic acid, since removal of sialic acid from the cells by neuraminidase prevents adherence. Selective replacement of sialic acid by the β-galactoside α-2,3-sialyltransferases that form the Siaα2,3Galβ1,3GalNAc sequence restored adherence of Sendai virus, Newcastle disease virus, an equine influenze virus, a human influenze virus, and polyomavirus. Restoration of the Siaα2,6Galβ1,4GlcNAc, and Siaα2,6GalNAc groups, however, restored adherence of only the human influenza virus, and none of the others. Thus the transferases are useful for defining the molecular basis for viral adherence.

The foregoing studies with erythrocytes were designed in part to ascertain whether intact cells could be modified by glycosidases with loss of a biological function, and whether that function could be restored by reglycosylation of the cells with glycosyltransferases. This objective was achieved and demonstrates the potential of the glycosyltransferases for revealing the structural basis of cell surface oligosaccharide function. The erythrocyte, however, is an unusually hardy cell with a rather rigid membrane and a considerably altered metabolism. Thus it was of interest to see whether glycosyltransferases could be used with more typical cells that are capable of cell division and have a more complete complement of metabolic activities than the erythrocyte, including the ability to restore sugars that have been removed from its cell surface. Recent studies (26) demonstrated that, indeed, this type of cell could be used with sialyltransferases to study cell surface adherence of Sendai virus. Confluent monolayers of Madin-Darby bovine kidney cells lost their ability to adsorb Sendai virus when treated with neuraminidase, as judged by exposure of treated and untreated cells to virus for 10–60 min, then incubation of the cells for 48 hr and measurement of virus particles that appeared in the culture medium as the result of adherence, infection of the cell by virus, and then viral replication and cell lysis. Desialyzed cells spontaneously restored demonstrable cell surface sialic acid on incubation for 90 min at 37°C, and regained their complete sialic acid complement in 6–8 hr, as judged by adherence and infection with Sendai virus. However, adherence and infection with virus was restored completely within 1 hr on treatment of desialyzed cells with CMPSia and the β-galactoside α-2,3-sialyltransferase. Thus it is likely that conditions can be found in which the glycosyltransferases can be used with any number of cells, thereby aiding understanding of the molecular basis for the many cellular phenomena with which oligosaccharides have been implicated (42).

REFERENCES

1. Aminoff, D., Gathmann, W. D., and Baig, M. M., Glycoproteins and blood group activity: Isolation and characterization of oligosaccharides of H^+ hog submaxillary glycoprotein and their comparison to those found in A^+ and A^-H^- glycoproteins. *J. Biol. Chem.*, 1979, **254**:8909–8913.
2. Andrews, P., Purification of lactose synthetase A protein from human milk and demonstration of its interaction with α-lactalbumin. *FEBS Lett.*, 1970, **9**:297–300.
3. Barker, R., Trayer, I. P., and Hill, R. L., Nucleoside phosphates attached to agarose. In: *Methods in Enzymology* (W. B. Jakoby and M. Wilchek, eds.), Vol. 34, Part B. Academic Press, New York, 1974:479–491.
4. Beyer, T. A., Rearick, J. I., Paulson, J. C., Prieels, J.-P., Sadler, J. E., and Hill, R. L., Biosynthesis of mammalian glycoproteins: glycoproteins pathways in the synthesis of the nonreducing terminal sequences. *J. Biol. Chem.*, 1979, **254**: 12531–12541.
5. Beyer, T. A., Sadler, J. E., and Hill, R. L., Purification to homogeneity of the

H-blood group β-galactoside $\alpha 1 \rightarrow 2$ fucosyltransferases from porcine submaxillary gland. *J. Biol. Chem.*, 1980 **255**:5364–5372.
6. Beyer, T. A., Sadler, J. E., Rearick, J. I., Paulson, J. C., and Hill, R. L., Glycosyltransferases and their use in assessing oligosaccharide structure and structure-function relationships. *Adv. Enzymol.*, 1980 (in press).
7. Brodbeck, U., and Ebner, K. E., Resolution of a soluble lactose synthetase into two protein components and solubilization of microsomal lactose synthetase. *J. Biol. Chem.*, 1966, **241**:762–764.
8. Cahan, L. D., and Paulson, J. C., Polyoma virus adsorbs to specific sialyloligosaccharide receptors on erythrocytes. *Virology*, 1980, **103**:503–509.
9. Dahr, W., Uhlenbruck, G., Janssen, E., and Schmalisch, R., Different N-terminal amino acids in the MN-glycoprotein from MM and NN erythrocytes. *Human. Genet.*, 1977, **35**:335–343.
10. Finne, J., Krusius, T., and Rauvalla, H., Occurrence of disialosyl groups in glycoproteins. *Biochem. Biophys. Res. Commun.*, 1977, **74**:405–410.
11. Furthmayr, H., Structural comparison of glycophorins and immunochemical analysis of genetic variants. *Nature* (*London*), 1978, **271**:519–524.
12. Graves, D. J., and Wu, Y.-T., On predicting the results of affinity procedures. In: *Methods in Enzymology* (W. B. Jakoby and M. Wilchek, eds.), Vol. 34, Part B. Academic Press, New York, 1974:140–163.
13. Hagopian, A., and Eylar, E. H., Glycoprotein biosynthesis: Studies on the receptor specificity of the polypeptidyl: N-acetylgalactosaminyl transferase from bovine submaxillary glands. *Arch. Biochem. Biophys.*, 1968, **128**:422–433.
14. Hill, R. L., Pizzo, S. V., Imber, M., Lehrman, M., Prieels, J.-P., Glasgow, L. R., Guthrow, C. E., and Paulson, J. C., Receptors on hepatocytes that bind ligands containing fucosyl $\alpha 1,3$ N-acetylglucosamine linkages. In: *Symposium on Enzyme Replacement Therapy in Genetic Disease* (R. Desnick, ed.). Birth Defects: Original Article Series, Vol. XVI, No. 1, pp. 85–91, 1980.
15. Ivatt, R. J., Complex formation by sequential glycosyltransferases: Characterization of a novel sugar sequence from rat-brain glycoproteins containing fucose and sialic acid. In: *Glycoconjugates Research* (J. D. Gregory and R. W. Jeanloz, eds.), Vol. 2. Academic Press, New York, 1979:705–711.
16. Kornfeld, R., and Kornfeld, S., Comparative aspects of glycoprotein structure. *Annul. Rev. Biochem.*, 1976, **45**:217–237.
17. Krusuis, T., and Finne, J., Characterization of a novel sugar sequence from rat-brain glycoproteins containing fucose and sialic acid. *Eur. J. Biochem.*, 1978, **84**:395–403.
18. Lee, Y. C., Stowell, C. P., and Krantz, M. J., 2-Imino-2-methoxyethyl 1-thioglycosides: New reagents for attaching sugars to proteins. *Biochemistry*, 1976, **15**:3956–3963.
19. Magee, S. C., Geren, C. R., and Ebner, K. E., Plasmin and the conversion of the molecular forms of bovine milk galactosyltransferase. *Biochim. Biophys. Acta*, 1976, **420**:187–194.
20. Mäkelä, O., and Cantell, K., Destruction of M and N blood group receptors of human red cells by some influenza viruses. *Ann. Med. Exp. Biol. Fenn.*, 1958, **35**:366–374.
21. McGuire, E. J., Biosynthesis of submaxillary mucins. In: *Blood and Tissue Antigens* (D. Aminoff, ed.). Academic Press, New York, 1970:461–474.
22. Molmström, A., Rodén, L., Feingold, D. S., Jacobsson, I., Bäckström, G.,

Höök, M., and Lindahl, U., Biosynthesis of heparin. Partial purification of the uronosyl C-5 epimerase. *J. Biol. Chem.*, 1980, **255**:3878–3883.

23. Paulson, J. C., Beranek, W. E., and Hill, R. L., Purification of a sialyltransferase. *J. Biol. Chem.*, 1977, **252**:2356–2362.
24. Paulson, J. C., Glasgow, L. R., Beyer, T. A., Lowman, C., Holroyde, M. J., and Hill, R. L., Use of glycosyltransferases and glycosidases in structural analysis of oligosaccharides. In: *Glycoconjugate Research* (J. D. Gregory and R. W. Jeanloz, eds.), Vol. 1. Academic Press, New York, 1979:247–249.
25. Paulson, J. C., Hill, R. L., Tanabe, T., and Ashwell, G., Reactivation of asialo-rabbit liver binding protein by resialylation with β-D-galactoside α2→6 sialyltransferase. *J. Biol. Chem.*, 1977, **252**:8624–8628.
26. Paulson, J. C., Markwell, M. A. K., Cahan, L., Higa, H. H., Marxhall, L., and Weinstein, J., The interaction of myxoviruses with sialyloligosaccharide receptors. In: *Glycoconjugates, Proceedings of the Fifth International Symposium* (R. Schaurer, P. Boer, E. Buddecke, M. F. Kramer, J. F. G. Vliegenthart, and H. Wiegandt, eds.). Thieme, Stuttgart, 1979:580–581.
27. Paulson, J. C., Prieels, J.-P., Glasgow, L. R., and Hill, R. L., Sialyl- and fucosyltransferases in the biosynthesis of asparaginyl-linked oligosaccharides in glycoproteins: Mutually exclusive glycosylation by β-galactoside α2→6 sialyltransferase and *N*-acetylglycosaminide α1→3 fucosyltransferase. *J. Biol. Chem.* 1978, **253**:5617–5624.
28. Paulson, J. C., Rearick, J. I., and Hill, R. L., Enzymatic properties of β-D-galactoside α2→6 sialyltransferase from bovine colostrum. *J. Biol. Chem.*, 1977, **252**:2363–2371.
29. Paulson, J. C., Sadler, J. E., and Hill, R. L., Restoration of specific myxovirus receptors to asialoerythrocytes by incorporation of sialic acid with pure sialyltransferases. *J. Biol. Chem.*, 1979, **254**:2120–2124.
30. Porath, J., and Kristiansen, T., Biospecific affinity chromatography and related methods. In: *The Proteins* (H. Neurath and R. L. Hill, eds.), Vol. 1. Academic Press, New York, 1975:95–178.
31. Powell, J. T., and Brew, K., The preparation and characterization of two forms of bovine galactosyl transferase. *Eur. J. Biochem.*, 1974, **48**:217–228.
32. Prieels, J.-P., Pizzo, S. V., Glasgow, L. R., Paulson, J. C., and Hill, R. L., Hepatic receptor that specifically binds oligosaccharides containing fucosylα1→3 *N*-acetylglucosamine linkages. *Proc. Natl. Acad. Sci. U.S.A.*, 1978, **75**:2215–2219.
33. Rodén, L., and Horowitz, M. I., Structure and biosynthesis of connective tissue proteoglycans. In: *The Glycoconjugates* (M. I. Horowitz and W. Pigman, eds.), Vol. 2. Academic Press, New York, 1978:3–71.
34. Sadler, J. E., Paulson, J. C., and Hill, R. L., The role of sialic acid in the expression of human MN blood group antigens. *J. Biol. Chem.*, 1979, **254**:2112–2119.
35. Sadler, J. E., Rearick, J. I., and Hill, R. L., Purification to homogeneity and enzymatic characterization of an α-*N*-acetylgalactosaminide α2→6 sialyltransferase from porcine submaxillary glands. *J. Biol. Chem.*, 1979, **254**:5934–5941.
36. Sadler, J. E., Rearick, J. I., Paulson, J. C., and Hill, R. L., Purification to homogeneity of a β-galactoside α2→3 sialyltransferase and partial purification of

an α-N-acetylgalactosaminide α2→6 sialyltransferase from porcine submaxillary glands. *J. Biol. Chem.* 1979, **254**:4434–4443.
37. Schachter, H., McGuire, E. J., and Roseman, S., Sialic acids. XIII. A uridine diphosphate D-galactose: Mucin galactosyltransferase from porcine submaxillary gland. *J. Biol. Chem.*, 1971, **246**:5321–5328.
38. Schut, B. L., Dorland, L., Haverkamp, J., Vliegenthart, J. F. G., and Fournet, B., Structure of complex asparagine bound glycopeptide from horse pancreatic ribonuclease containing (2→3) and (2→6) linked sialic acid residues. *Biochem. Biophys. Res. Commun.*, 1978, **82**:1223–1228.
39. Sodetz, J. M., Paulson, J. C., Pizzo, S. V. and McKee, P. A., Carbohydrate composition and identification of blood group A, B and H oligosaccharide structures on human factor VIII/ von Willebrand factor. *J. Biol. Chem.*, 1979, **253**:7202–7206.
40. Spik, G., Fournet, B., Cheron, A., Strecker, G., Montreuil, J., Dorland, L., and Vliegenthart, J. F. G., Comparative study of the structure of the glycans isolated from different transferrins. In: *Glycoconjugates: Proceedings of the Fifth International Symposium* (R. Schauer, P. Boer, E. Buddecke, M. F. Kramer, J. F. G. Vliegenthart, and H. Wiegandt, eds.). Georg Thieme Publishers, Stuttgart, 1979:21–22.
41. Springer, G. F., and Ansell, N. J., Inactivation of human erythrocyte agglutinogens M and N by influenza viruses and receptor-destroying enzyme. *Proc. Natl. Acad. Sci. U.S.A.*, 1958, **44**:182–189.
42. Talmadge, K. W., and Burger, M. M., Carbohydrates and cell-surface phenomena. MTP International Review of Science, *Biochemistry (London,) Ser. One*, 1975, **5**:43–93.
43. Trayer, I. P., and Hill, R. L., The purification and properties of the A protein of lactose synthetase. *J. Biol. Chem.*, 1971, **246**:6666–6675.
44. Wasniowska, K., Drzeniek, Z., and Lisowska, E., The amino acids of M and N blood group glycopeptides are different. *Biochem. Biophys. Res. Commun.*, 1977, **76**:385–390.
45. Watkins, W. M., Blood group substances: Their nature and genetics. In: *The Red Blood Cell* (C. Bishop and D. M. Surgenor, eds.). Academic Press, New York, 1964:359–396.
46. Yoshima, H., Takasaki, S., Ito-Mega, S., and Kobata, A., Purification of almond emulsin α-L-fucosidase 1 by affinity chromatography. *Arch. Biochem. Biophys.*, 1979, **194**:394–398.

Metalloenzymes

BO G. MALMSTRÖM

Department of Biochemistry and Biophysics
Chalmers University of Technology
and
University of Göteborg
Göteborg, Sweden

INTRODUCTION

Metalloenzymes and other metalloproteins occupy a prominent position in modern discussions of protein evolution. This is indeed apparent from the other contributors to this volume, as about half of the speakers dealt either specifically with metalloproteins or with those groups of proteins that contain metal ions. The reason for this quantitative prominence is not, of course, that the evolution of metalloproteins entails any specific principle different from that governing the emergence of other proteins. It depends rather on the fact that metalloproteins have been attractive objects for the investigation of protein structure and function. The simplest explanation is that about one-third of all enzymes have metal ions as prosthetic groups or activators and that a large fraction of noncatalytic proteins also depends on metal ions for carrying out their physiologic function. It is not only the prevalence of metalloproteins that has made them common objects of structural investigations; some biochemists realized several decades ago that many specific features of metalloenzymes make them particularly suitable for the elucidation of the relationship between chemical structure and catalytic activity. First, when the metal is directly involved in the catalytic mechanism, it provides a natural label of the active site. Second, the metal ion often has physical properties, such as color or paramagnetism, allowing a direct study of chemical events at this site. In fact, the interpretation of spectroscopic properties has often made it possible for one to surmise structural aspects in the catalytic mechanisms of metalloenzymes long before determination of their complete structures. Investigations during recent years have made it obvious that a combination of detailed spectroscopic and kinetic data for metalloenzymes with extensive structural information can give particularly deep insights into the biological

ISBN 0-12-643150-7

functions of proteins, as I hope to illustrate in subsequent sections of this essay.

I have chosen a rather general title in order to allow for touching on a variety of enzymes with which I have been concerned during my 30 years as a biochemist. These enzymes include enolase, phosphoglycerate kinase, carbonic anhydrase, the so-called blue oxidases, as well as many related copper proteins, and cytochrome *c* oxidase. A major generalization (14) is that metalloenzymes are metal complexes unique in comparison with the simpler systems generally studied by coordination chemists. The unusual coordination imparts properties to the metal ion of paramount importance for its role in the catalytic mechanism. The unique structures have consequently evolved because they possess a functional advantage. In the following pages I will attempt to illustrate this with results of my own or, in some cases, of my students.

I will first briefly review some of the structural studies by P. O. Nyman and L. Rydén on carbonic anhydrase and on several copper proteins. My own work with the same proteins has involved kinetic and spectroscopic measurements aimed at correlating the structure, particularly the metal coordination, with the catalytic function, and I will also present some of my most recent results on oxidase mechanisms. These may not seem directly relevant to the theme of this volume, but in conclusion I will suggest that they relate to certain evolutionary questions.

Carbonic Anhydrase

In 1949 Emil Smith (24) suggested that the "simplest example of our theory concerns its application to carbonic anhydrase," a statement that in 1957 I decided to put to an experimental test. At that time there was, however, no entirely satisfactory purification method available for carbonic anhydrase from any source. When Nyman started work as a graduate student his first task was to purify the enzyme to homogeneity from human erythrocytes (18). At that time several reports on the existence of isoenzymes started to appear (17). I had, for example, reported (13) the separation of several electrophoretically distinct forms of yeast enolase. In general, however, it was not known whether the isoenzymes represented genetically distinct proteins or were the products of epigenetic modifications. Our research group was therefore quite excited when Nyman (18) showed that the human erythrocyte contained two isoenzymes, generally designated carbonic anhydrase B and C, that displayed different kinetic properties, indicating that they had to be structurally distinct proteins. Nyman (19) established this by showing that carbonic anhydrase B and C have significantly different amino acid compositions.

Nyman and his associates (4) reported the complete primary structure of human carbonic anhydrase B in 1972, and the sequence of the C enzyme was reported the following year (11). The two sequences show obvious similarities, the differences being interpretable in terms of a number of single amino acid replacements. Thus the two isoenzymes reflect homology in an evolutionary sense, suggesting that they are derived by a process of gene duplication, followed by divergent changes of the two genes.

The primary structure of carbonic anhydrase from many different biological species has now been determined, and the results as well as their evolutionary implications have been summarized by Tashian (27). A polymorphism similar to that of the human enzyme occurs in a variety of species, which suggests that the gene duplication is not a recent event in evolution. It has furthermore been established that the C-type enzymes from different species display a greater similarity in sequence than the two human enzymes, a conclusion that Nyman and his colleagues (3,20) had arrived at already in 1968 by a comparison of the sequences in the carboxyl-terminal region of the human and bovine carbonic anhydrase. This suggests that the B form has undergone, and perhaps is still undergoing, a faster evolutionary change. The smaller changes in the C enzyme could be explained as the result of a stronger selective pressure against mutational variants, and this form probably represents an older evolutionary type of carbonic anhydrase. This concept is supported by the demonstration of Nyman and collaborators (6) that a bacterial enzyme is of the C type. They stress, however, that their studies are limited to the genus *Neisseria*, whose species occur as parasites in mammalian systems and show an ability to undergo interspecific genetic transformation.

"Blue" Copper Proteins

A number of copper proteins characterized by an intense blue color have been isolated from a variety of organisms spanning the whole evolutionary scale from bacteria to man. They comprise the blue oxidases (16), which are rather large molecules with four to eight copper atoms, as well as a number of small blue proteins (10) containing a single blue copper center. The blue oxidases are laccase found in a variety of fungi and plants, ascorbate oxidase from various plants, and ceruloplasmin from mammalian blood serum. The blue oxidases can, like cytochrome *c* oxidase, utilize the full oxidizing capacity of dioxygen, which is reduced to two molecules of water. The small blue proteins are the azurins and rusticyanin (8) from bacteria, plastocyanins from photosynthetic cells, and the plant proteins stellacyanin and umecyanin. Plastocyanin functions in photosynthetic electron transport, and other small blue proteins probably also participate in electron transfer reactions. Thus these proteins are related to the blue copper in the oxidases in the same way as cytochrome *c* is related to cytochrome *a*.

Quite a number of small blue proteins have been sequenced and the results discussed in evolutionary terms (22,23). Limited primary structure data are also available for two of the blue oxidases, laccase (5a) and ceruloplasmin (12). As these contain several copper atoms per molecule, it is interesting to consider if their separate sites have developed from an ancestral simple protein through gene duplication followed by gene fusion and mutational modification. To be able to consider this question it is first necessary to say a few words about the three types of copper found in the oxidases.

THE FORMS OF COPPER (16)

In general, only half of the copper found in the blue oxidases is paramagnetic and detectable by EPR. In addition, the EPR spectra of the native proteins always consist of contributions from two Cu^{2+} ions in different environments. The two EPR-detectable ions are called type 1 and type 2, respectively, whereas the EPR-nondetectable coppers are called type 3.

It is type 1 Cu^{2+} that is entirely responsible for the intense blue color of these proteins; consequently, the single copper ion in the small proteins has a type 1 character. The extinction coefficient for type 1 Cu^{2+} ions is close to 5000 M^{-1} cm^{-1}, which is about 100 times larger than the coefficients of "normal" coordination compounds of Cu^{2+}. The type 1 ions also have other unusual properties. Thus their hyperfine coupling constants as determined by EPR are about half of the values generally observed. Furthermore, their standard reduction potentials, although variable, are unusually high, with that for fungal laccase being 785 mV.

Type 2 Cu^{2+} has normal spectral parameters but displays an unusual affinity for F^- and other anions, which function as inhibitors of the enzymes.

The type 3 ions were long believed to be in the Cu^+ state, as this is the most common nonparamagnetic form of copper. Later studies, however, have shown that they consist of an antiferromagnetically coupled pair of Cu^{2+} ions.

The structural basis of the unusual coordination properties is well established only in the case of type 1 ions. The first structural models were based on spectroscopic measurements (25), which predicated a tetrahedral distortion, at least two imidazole nitrogen ligands, and one thiol ligand, which functions as the electron donor responsible for the charge-transfer transition that gives rise to the intense blue color. Now this model has been validated by the crystallographic determination of the structure of plastocyanin (7) and azurin (2). The plastocyanin structure is given in Fig. 1. This shows a nearly tetrahedral symmetry with two imidazole nitrogens and one cysteine as well as one methionine sulfur as ligands. The great similarities in spectral properties of all type 1 ions would suggest that they have almost identical coordination. As there are small differences in extinction coefficients and hyperfine structure constants, and a considerable variation in reduction potentials, some variation in structure would be expected. This expectation is borne out by the fact that stellacyanin contains no methionine (5), which, consequently, cannot be a ligand in this protein.

SEQUENCE HOMOLOGY AMONG BLUE PROTEINS

The amino acid sequences for at least eight plastocyanins and nine azurins and for stellacyanin are known, and partial sequences are available for cerulo-

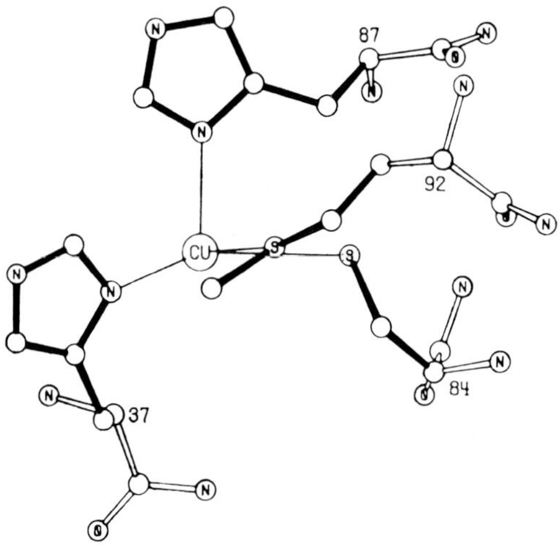

Fig. 1. The copper-binding (CU^{2+}) site of plastocyanin. Ligends: His, His, Cys, Met. [Adapted from Colman *et al.* (7).]

plasmin (12) and laccase. Rydén and Lundgren (22,23) have carried out a statistical comparison of these sequences and have concluded that all the small blue proteins must be derived from a common ancestral protein. In particular, it is striking that, in the plastocyanins and azurins, the histidines and sulfur-containing amino acids that provide the copper ligands are invariable.

Rydén and Lundgren also provide statistical evidence that two regions of ceruloplasmin display sufficient homology with the small proteins for one to conclude that they also have the same origin through divergent evolution. In Fig. 2 the sequence around the ligand cysteine and one of the ligand histidines is compared with the corresponding sequences in plastocyanin and azurin. The figure also gives the corresponding sequence for laccase determined by Briving *et al.* (5a). It is striking that the two oxidases, but not the small proteins, have the ligand cysteine surrounded by two histidine residues. This has led P. O. Nyman (personal communication) to suggest that these two histidines may function as ligands to type 2 copper. As this particular ion can be selectively removed (21), the idea can be tested by chemical modification. Such experiments are in progress in Nyman's laboratory.

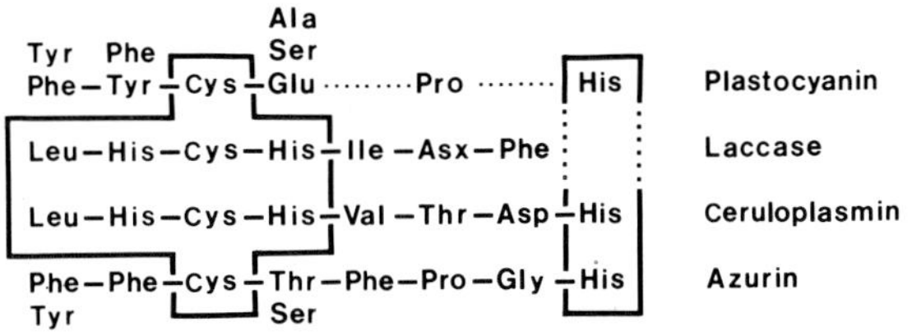

Fig. 2. Comparison of partial sequences of laccase and ceruloplasmin with the sequences around two ligand residues in plastocyanin and azurin.

Even if all blue proteins appear to be monophyletic, there is really no reason to believe that the other copper sites have arisen through further genetic modification in a duplicated gene. In particular, there is one protein, tyrosinase, that contains only type 3 copper (9), and the blue oxidases may equally well have their origin in the fusion of genes for a blue protein and for a tyrosinase ancestor.

Cytochrome *c* Oxidase

Cytochrome *c* oxidase is a complex protein spanning the inner mitochondrial membrane. Although it contains heme iron as well as copper, it shows many similarities to the blue proteins. Thus the minimal functional unit contains four electron acceptors, like laccase, but these are 2 heme Fe^{3+} and 2 Cu^{2+}, rather than 4 Cu^{2+}. Of these components only two, cytochrome *a* and Cu_A, are detectable by EPR. As in the blue oxidases, the two EPR-nondetectable centers, cytochrome a_3 and Cu_B, constitute an antiferromagnetically coupled metal ion pair. Whereas the blue proteins all appear to consist of a single peptide chain, cytochrome oxidase consists of at least seven distinct peptides (28). It is interesting to note that three of the cytochrome oxidase peptides are coded for by mitochondrial DNA, the remaining four being controlled by nuclear DNA.

Several groups are working on the primary structure of the cytochrome oxidase subunits. Of particular interest in the context of this chapter is the report by Steffens and Buse (26) that peptide II shows significant homologies with small blue proteins, as illustrated in Fig. 3. This indicates that one of the cytochrome oxidase peptides shares a common origin with the blue proteins. It is also interesting that peptide II is one of those coded for by mitochondrial DNA, in view of the suggested bacterial origin of mitochondria.

The Mechanism of Reduction of Dioxygen

The oxidases described here have the ability to reduce dioxygen to two molecules of water, and I will now briefly consider the role of some of the structures described in the catalysis of this reaction. One might expect a four-equivalent reduction to involve four consecutive one-electron donations to the dioxygen molecule, but this would require that the enzyme stabilize several intermediates, including highly reactive radicals (OH·). In addition, the relatively high standard reduction potential of these oxidases would tend to make the transfer of one electron from the reduced enzymes to dioxygen with the formation of superoxide an energetically unfavorable process (15). I suggest that the presence of an antiferromagnetically coupled metal pair in these enzymes is the key feature in circumventing these difficulties. In both laccase and cytochrome oxidase these metal pairs constitute the oxygen-reducing unit. In both cases dioxygen does not react with the unit until it has undergone a two-electron reduction. The function of the two EPR-detectable

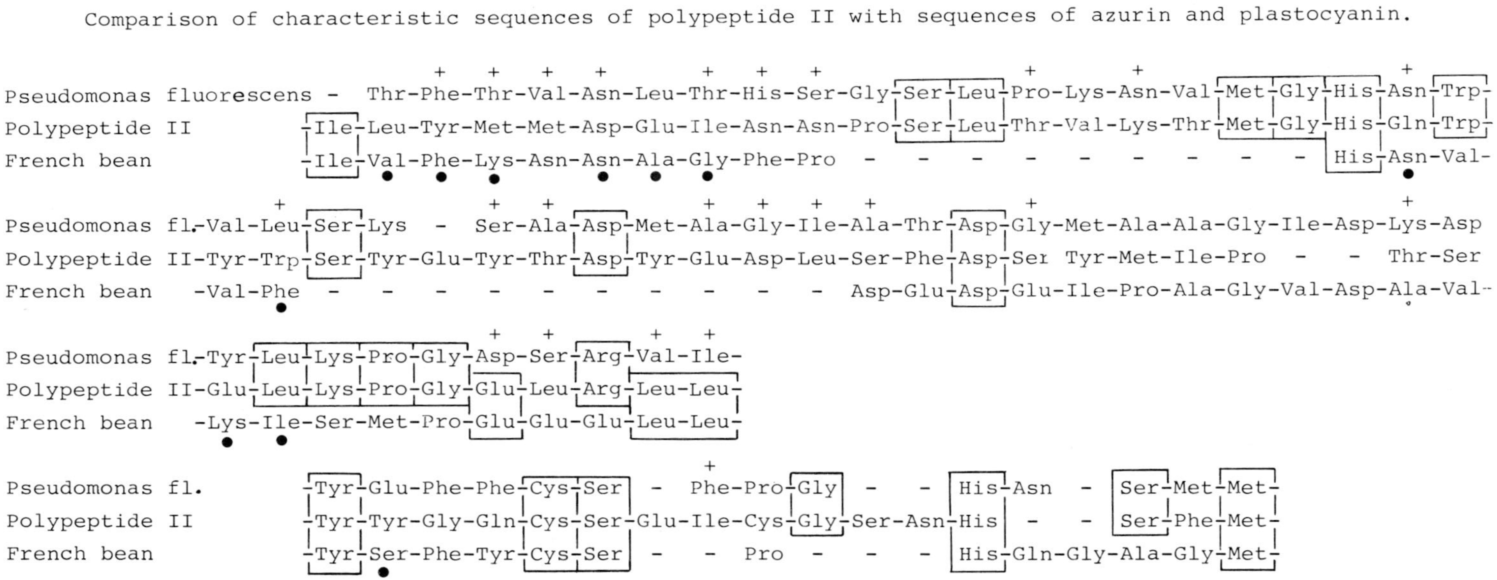

Fig. 3. Comparison of the characteristic sequence of cytochrome oxidase subunit II with sequences of azurin and plastocyanin. +, One-point mutations:azurin↔polypeptide II; ●, one-point mutations:plastocyanin↔polypeptide II.

ions in the enzymes is to provide the two electrons required for this reduction.

Recently, a considerable amount of information on intermediates in the reduction of dioxygen has been derived for laccase as well as for cytochrome oxidase. The first intermediates are formed through the combination of O_2 with the reduced forms of the binuclear units, giving $[Cu_2 \cdot O_2]^{2+}$ in the case of laccase and $[Cu_B a_3 \cdot O_2]^{3+}$ in cytochrome oxidase. It has been shown that in both enzymes the units so formed are further reduced by two successive one-electron donations, so that in the second intermediate the unit must be paramagnetic. This is illustrated for laccase in Fig. 4, which shows the appearance of a new EPR signal during catalytic turnover. With the aid of oxygen isotopes this intermediate has been shown to be an oxygen radical with unusual relaxation properties (1). The unpaired electron is undoubtedly delocalized to the metal nuclei as well, which could stabilize the radical formally at the $OH\cdot$ reduction state. Such interaction with the metals would also be consistent with the short relaxation time. In cytochrome oxidase the coupling has the effect that the radical cannot be detected even around 10 K, but from monitoring the valence state of cytochrome *a* and Cu_A in the intermediates, it is still possible to show that one of them must be paramagnetic (6a).

Conclusion

Only a few enzymes can reduce dioxygen to two molecules of water. I have suggested that the key feature in these enzymes is the presence of binuclear metal complexes with antiferromagnetically coupled ions. These complexes have entirely different structures in the blue oxidases and in cytochrome

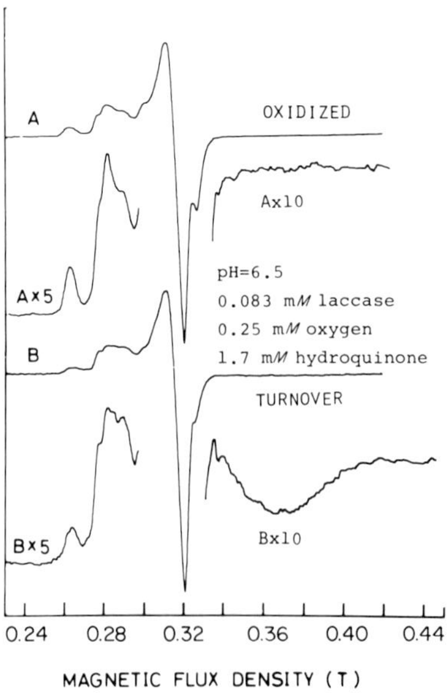

Fig. 4. The EPR spectrum of laccase during turnover (B), showing the presence of the paramagnetic oxygen intermediate (A).

oxidase, however. In the first case they consist of two Cu^{2+} ions, whereas cytochrome oxidase has a heme Fe^{3+}-Cu^{2+} pair. The presence of similar functional units in enzymes with largely different structures has sometimes been taken as evidence for convergent evolution. A membrane protein like cytochrome oxidase, which consists of several polypeptide subunits, can, however, be formed by the assembly of individual polypeptides, each one produced by divergent evolution of genes for ancestral proteins. Thus it is likely that the copper-binding peptides in cytochrome oxidase are derived from ancestral copper proteins, and for one of the copper ions this appears established. Similarly, the two heme peptides are most likely also formed through divergent evolution from primitive cytochromes.

ACKNOWLEDGMENTS

I wish to thank Drs. P. O. Nyman and L. Rydén for helpful discussions and for allowing me to include some of their unpublished material. My original investigations have been supported by grants from the Swedish Natural Science Research Council.

REFERENCES

1. Aasa, R., Deinum, J., Malmström, B. G., Reinhammar, B., and Vänngård, T., A ^{17}O-effect on the EPR spectrum of the intermediate in the dioxygen-laccase reaction. *Biochem. Biophys. Res. Commun.*, 1976, **70**:1204–1209.
2. Adman, E. T., Stenkamp, R. E., Sieker, L. C., and Jensen, L. H., A crystallographic model for azurin at 3 Å resolution. *J. Mol. Biol.*, 1978, **123**:35–47.
3. Andersson, B., Göthe, P. O., Nilsson, T., Nyman, P. O., and Strid, L., Carboxyl-terminal region of human and bovine erythrocyte carbonic anhydrases. *Eur. J. Biochem.*, 1968, **6**:190–195.
4. Andersson, B., Nyman, P. O., and Strid, L., Amino acid sequence of human erythrocyte carbonic anhydrase B. *Biochem. Biophys. Res. Commun.*, 1972, **48**:670–677.
5. Bergman, C., Gandvik, E.-K., Nyman, P. O., and Strid, L., The amino acid sequence of stellacyanin from the lacquer tree. *Biochem. Biophys. Res. Commun.*, 1977, **77**:1052–1059.

5a. Briving, C., Gandvik, E.-K., and Nyman, P. O., Structural studies around cysteine and cystine residues in the "blue" oxidase fungal laccase B. Similarity in amino acid sequence with ceruloplasmin. *Biochem. Biophys. Res. Commun.*, 1980, **93**:454–461.

6. Brundell, J., Falkbring, S. O., and Nyman, P. O., Carbonic anhydrase from *Neisseria sicca*, strain 6021. II. Properties of the purified enzyme. *Biochim. Biophys. Acta*, 1972, **284**:311–323.

6a. Clore, G. M., Andréasson, L.-E., Karlsson, B., Aasa, R., and Malmström, B. G., Characterization of the low-temperature intermediates of the reaction of fully reduced soluble cytochrome oxidase with oxygen by electron-paramagnetic-resonance and optical spectroscopy. *Biochem. J.*, 1980, **185**:139–154.

7. Colman, P. M., Freeman, H. C., Guss, J. M., Murata, M., Norris, V. A., Ramshaw, J. A. M., and Venkatappa, M. P., X-ray crystal structure analysis of plastocyanin at 2.7 Å resolution. *Nature (London)*, 1978, **272**:319–324.
8. Cox, J. C., Aasa, R., and Malmström, B. G., EPR studies on the blue copper protein, rusticyanin. A protein involved in Fe^{2+} oxidation at pH 2.0 in *Thiobacillus ferro-oxidans*. *FEBS Lett.*, 1978, **93**:157–160.

9. Deinum, J., Lerch, K., and Reinhammar, B., An EPR study of *Neurospora* tyrosinase. *FEBS Lett.*, 1976, **69**:161–164.
10. Fee, J. A., Copper proteins—Systems containing the "blue" copper center. *Struct. Bonding (Berlin)*, 1975, **23**:1–60.
11. Henderson, L. E., Henriksson, D., and Nyman, P. O., Amino acid sequence of human erythrocyte carbonic anhydrase C. *Biochem. Biophys. Res. Commun.*, 1973, **52**:1388–1394.
12. Kingston, B., Kingston, B. L., and Putnam, F. W., Complete amino acid sequence of a histidine-rich proteolytic fragment of human ceruloplasmin. *Proc. Natl. Acad. Sci. U.S.A.*, 1979, **76**:1668–1672.
13. Malmström, B. G., The purification of yeast enolase by zone electrophoresis and ion-exchange chromatography, and the existence of several active forms of the enzyme. *Arch. Biochem. Biophys.*, 1957, **70**:58–69.
14. Malmström, B. G., The 'unique' metal-binding properties of metalloenzymes. *Pure Appl. Chem.*, 1970, **24**:393–406.
15. Malmström, B. G., Cytochrome *c* oxidase: Some current biochemical and biophysical problems. *Q. Rev. Biophys.*, 1973, **6**:389–431.
16. Malmström, B. G., Andréasson, L.-E., and Reinhammar, B., Copper-containing oxidases and superoxide dismutase. In: *The Enzymes* (P. D. Boyer, ed.). 3rd ed., Vol. 12. Academic Press, New York, 1975:507–579.
17. Makert, C. L., and Møller, F., Multiple forms of enzymes: Tissue, ontogenetic, and species specific patterns. *Proc. Natl. Acad. Sci. U.S.A.*, 1959, **45**:753–763.
18. Nyman, P. O., Purification and properties of carbonic anhydrase from human erythrocytes. *Biochim. Biophys. Acta*, 1961, **52**:1–12.
19. Nyman, P. O., and Lindskog, S., Amino acid composition of various forms of bovine and human erythrocyte carbonic anhydrase. *Biochim. Biophys. Acta*, 1964, **85**:141–151.
20. Nyman, P. O., Strid, L., and Westermark, G., Carboxylterminal region of human and bovine erythrocyte carbonic anhydrases. 1. Amino acid sequences of terminal cyanogen bromide fragments. *Eur. J. Biochem.*, 1968, **6**:172–189.
21. Reinhammar, B., and Oda, Y., *J. Inorg. Biochem.*, 1979, **11**:115–127.
22. Rydén, L., and Lundgren, J.-O., Homology relationships among the small blue proteins. *Nature (London)*, 1976, **261**:344–346.
23. Rydén, L., and Lundgren, J.-O., On the evolution of blue proteins. *Biochimie*, 1979, **61**:781–790.
24. Smith, E. L., The mode of action of the metal-peptidases. *Proc. Natl. Acad. Sci. U.S.A.*, 1949, **35**:80–90.
25. Solomon, E. I., Hare, J. W., and Gray, H. B., Spectroscopic studies and a structural model for blue copper centers in proteins. *Proc. Natl. Acad. Sci. U.S.A.*, 1976, **73**:1389–1393.
26. Steffens, G. J., and Buse, G., Chemical constitution and subunit function of polypeptide II from cytochromecoxidase. In: Cytochrome Oxidase (T. E. King, Y. Orii, B. Chance, and K. Okunuki, eds.) Developments in Biochemistry, vol. 5, Elsevier, New York, 1979.
27. Tashian, R. E., Evolution and regulation of the carbonic anhydrase isozymes. *Isozymes: Curr. Top. Biol. Med. Res.*, 1977, **2**:21–62.
28. Tracy, R. P., and Chan, S. H. P., Analysis of electrophoretic behaviour of cytochrome *c* oxidase subunits. *Biochim. Biophys. Acta*, 1979, **576**:109–117.

PROTEIN STRUCTURE AND FUNCTION

Glycoprotein Hormones: Similar Molecules with Different Functions

JOHN G. PIERCE AND THOMAS F. PARSONS

Department of Biological Chemistry
UCLA School of Medicine
Los Angeles, California

During the evolution of vertebrates, a number of proteins and peptides originated that exhibit a variety of hormonal responses. The determination of their amino acid sequences has revealed considerable conservation of structure with a remarkable diversity of physiologic function, although some of the basic biochemical mechanisms by which final function is generated may be similar. Groups of hormones with related structures include insulin–relaxin–IGF (insulin-like growth factor); adrenocorticotropic hormone-β-lipoprotein, melanophore-stimulating hormone (ACTH–β-lipotropin–MSH) (all of which arise from a single precursor molecule); growth hormone–prolactin–human placental lactogen; gut hormones and related peptides; and the glycoprotein hormones, namely, pituitary thyrotropin (TSH), lutropin (LH), and follitropin (FSH), together with placental hormones such as human choriogonadotropin and the equine choriogonadotropin (PMSG or pregnant mare serum gonadotropin).

The glycoprotein hormones represent the most complex molecules with hormonal activity; each consists of two peptide chains or subunits designated α and β, with both subunits glycosylated at specific residues and highly cross-linked internally with disulfide bonds (67). The β subunit of human chorionic gonadotropin (hCG) is unusual in that both asparagine- and serine-linked oligosaccharide moieties are present. The subunit structure of these hormones is found in all mammals thus far studied, and β subunits with similar properties have been observed in lower vertebrates down to teleost fishes (25,45). Particularly fascinating is that, within a species, the sequence of the α subunit has been completely conserved (except for truncations at the amino terminus). A summary of α sequences of mammalian hormones is

ISBN 0-12-643150-7

presented in Fig. 1*. The β subunits of the different hormones are also similar, as can be seen from the sequences in Fig. 2, but only when half-cystine residues are aligned. The data strongly suggest that β subunits evolved from a common precursor, and it has been proposed that the first of these hormones to evolve was a single gonadotropin-like molecule (24).

Each subunit by itself does not recognize the receptor on a target cell except possibly with an extremely low affinity (e.g., 82), and no significant hormone activity is expressed until formation of an α-β complex between the two subunits. Three-dimensional structure is also conserved because hormonally active hybrids can be formed between any α and β subunit, not only within a species but from species to species, including mammalian–nonmammalian hybrids (25,45). The particular hormone activity expressed by the complex is always that of the β subunit. Thus the information in the β structure somehow determines the specificity of an α-β complex; how this is accomplished is a major question. The larger question is why, of all the protein and peptide hormones, has a subunit structure evolved for this particular group and why, when they have different physiologic activities, do they share one subunit with a common structure? Possible functions of α subunits follow:

1. To enable correct folding of a single polypeptide containing α and β sequences, i.e., a prohormone exists.
2. To exert control over the amount of hormone secreted by limiting the amount of subunit available for combination.
3. To endow the β subunit, after combination with an α subunit, with the correct conformation to interact with receptor.
4. To interact itself, when in the intact hormone, with receptor either exclusive of or in conjunction with the β subunit.
5. To be internalized, after binding of the hormone to the receptor, with subsequent stimulation of biochemical events leading to expression of hormone activity. This role of an α subunit would be analogous to the mode of action of some toxins that are subunit proteins.
6. To prevent too rapid degradation of circulating hormone.

It is also possible that an entirely unrecognized function of α subunits exists.

With respect to possibility 1, recent results of the translation of both pituitary and placental messenger RNA in cell-free systems demonstrate that α and β subunits are translated from separate messengers as "prehormones" with a small signal or leader prepeptide at their amino termini (e.g., 15,33,41,42). We have found that isolated α subunits can readily refold, *in vitro*, to their native conformation after reduction of their five disulfide bonds (29,30); thus the necessity for a "pro" α has not been demonstrated. As yet, refolding of

*Recent and detailed reviews of the structures of the glycoprotein hormones appear in McKerns (53a).

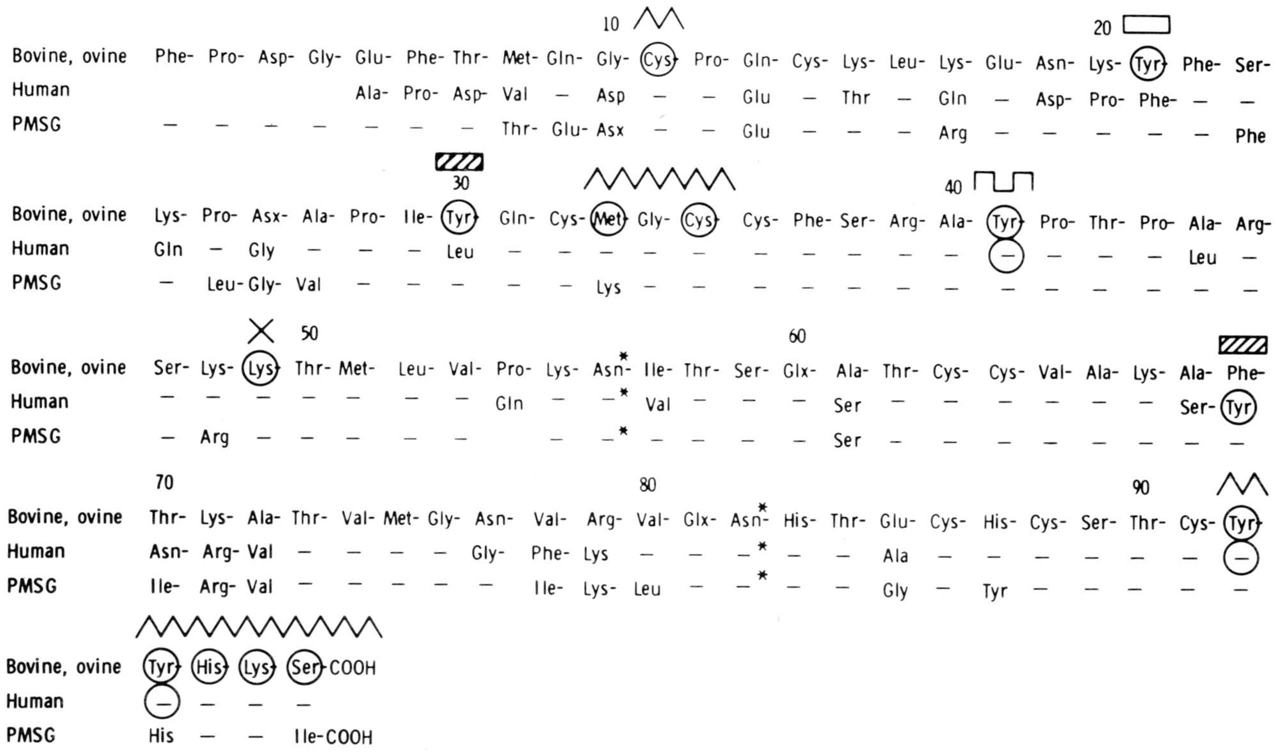

Fig. 1. Amino acid sequences of the common, α, subunit of the glycoprotein hormones. The bovine, ovine sequence is used as the reference; dashes indicate residues identical to this sequence. Asterisks show the points of attachment of carbohydrate. References for sequence determinations are contained in Giudice and Pierce (31); and for PMSG α, Moore *et al.* (55a). Those residues that have been chemically modified are circled, with the degree or effect of modification indicated above each residue: modified in the intact hormone, ▨ ; protected in the intact hormone, ▭ ; involved in an α-β cross-link, ×, modification resulting in a significant decrease in receptor interaction, ⋀⋀⋀ ; modification resulting in a significant decrease in subunit interaction, ⊓⊔⊓⊔⊓ .

	10 · 20
hTSHβ	Phe — Ile — Thr- Glx- Tyr(Met, Thr, His, Val,) — Arg- Arg- Glx — Ala-
bTSHβ	Phe — Ile — Thr- Glu- Tyr- Met- Met- His- Val — Arg- Lys- Glu — Ala-
pTSHβ	— — Ile — Thr- Glu- Tyr- Met- Met- His- Val — Arg- Lys- Glu — Ala-
B,OLHβ	Ser- Arg- Gly- Pro- Leu- Arg- Pro- Leu- Cys- Gln- Pro- Ile- Asn*- Ala- Thr- Leu- Ala- Ala- Glu- Lys- Glu- Ala- Cys- Pro-
pLHβ	— — — — — — — — — Arg — — —* — — — — — — Asp — — — —
hLHβ	— — Glu — — — — Trp — His — — —* — Ile — — Val — — — Gly — —
hCGβ	— Lys- Glu — — — — Arg — Arg — — —* — — — — Val — — — Gly — —
hFSHβ	Asn- Ser — Glu- Leu- Thr- —* Ile — Ile — Ile — — — Glu — Arg-
pFSHβ	— Glu- Leu- Thr- — Ile — Ile- Thr- Val — Val- Lys — Leu- Thr-
eFSHβ	Asx- Ser — Glx- Leu- Thr —* Ile — Ile — Val — — — Gly — Arg-

	30 · 40
hTSHβ	Tyr — Leu — Ile- Asn* — Thr — — — — — — (Met, Thr,) Arg- Asx- Ile- Asx- Gly- Lys- Leu- Phe-
bTSHβ	Tyr — Leu — Ile- Asn* — Thr- Val — — — — — Met- Thr- Arg- Asx- Val- Asx- Gly- Lys- Leu- Phe-
pTSHβ	Tyr — Leu — Ile- Asn*- Ser- Thr — — — — — — Met- Thr- Arg- Asp- Phe- Asp- Gly- Lys- Leu- Phe-
b,oLHβ	Val- Cys- Ile- Thr- Phe- Thr- Thr- Ser- Ile- Cys- Ala- Gly- Tyr- Cys- Pro- Ser- Met- Lys- Arg- Val- Leu- Pro- Val- Ile-
pLHβ	— — — — — — — — — — — — — — — — — Arg — — — — Ala- Ala-
hLHβ	— — — — Val- Asn* — Thr — — — — — — — Thr — Arg (Met) Leu — Glx- Ala- Val-
hCGβ	— — — — Val- Asn* — Thr — — — — — — — Thr — Thr — — — Gln- Gly- Val-
hFSHβ	Phe — Leu — Ile- Asn* — Thr- Trp — — — — — Tyr- Thr- Arg- Asp- Leu — Tyr- Lys- Asn- Pro-
pFSHβ	Phe — — Ser- Ile- Asn — Thr- Trp — — — — — Thr- Thr- Gly- Arg- Asx- Leu- Val- Tyr- Lys- Asx-
eFSHβ	Phe — — — Ile- Asn* — — Trp — — — — — Tyr- Thr- Arg- Asp- Leu — Tyr- Lys- Asp- Pro-

	50 · 60 · 70
hTSHβ	— — Lys- Tyr- Ala- Leu- Ser — Asx — — — — Arg- Asp- Phe- Ile- Tyr- Arg- Thr — Glx- Ile —
bTSHβ	— — Lys- Tyr- Ala- Leu- Ser — Asp — — — — Arg- Asp- Phe- Met- Tyr- Lys- Thr- Ala- Glu- Ile —
pTSHβ	— — Lys- Tyr- Ala- Leu- Ser — Asp — — — — Arg- Asp- Phe- Met- Tyr- Lys- Thr — Glu- Ile —
b,oLHβ	Leu- Pro- Pro Met- Pro Gln- Arg- Val- Cys- Thr- Tyr- His- Glu- Leu- Arg- Phe- Ala- Ser- Val- Arg- Leu- Pro-
pLHβ	— — — Val — — Pro — — — — Arg — — Ile — — — Ser — — —
hLHβ	— — — Val — — Pro — — — — Arg- Asx- Val — — Glx — Ile — — —
hCGβ	— — Ala leu — — Val — — Asn — Arg- Asp- Val — — Glu — Ile — — --
hFSHβ	Ala- Arg- Pro- Lys- Ile — Lys- Thr — — Phe- Lys — — Val- Tyr- Glu- Thr — — Val —
pFSHβ	— Ala- Arg- Pro- Asx- Ile- Glx- Lys- Thr — — — Arg- Glx — Val- Tyr- Glx- Thr — Lys- Val —
eFSHβ	Ala- Arg- Pro- Lys- Ile- — Lys- Thr — Thr- Phe- Lys — — Val- Tyr- Glu- Thr — Lys- Val —

Fig. 2. Amino acid sequences of the β subunit of the glycoprotein hormones. The sequence of bovine, ovine (b,o) LH-β is used as the reference; dashes indicate residues identical to this sequence; parentheses indicate regions where sequence is based on composition only. Asterisks show the points of attachment of carbohydrate. References for sequence determinations are contained in Giudice and Pierce (31); and for eFSH-β, Fujiki *et al.* (25a). Those residues that have been chemically modified are circled, with the degree or effect of modification indicated above each residue: modified in the intact hormone, ▭ ; protected in the intact

(Continued)

reduced β subunits has not been accomplished; thus the possibility that the original translation product of β messenger RNA is a "pre*pro*hormone" with some residues necessary for folding and formation of the six disulfide bonds cannot be dismissed. However, estimates of the size of the original, nonglycosylated, biosynthetic β subunit indicate it to be only about M_r

									80										90					
hTSHβ	—	—	—	Leu-	His	—	(Ala,	—,	Tyr)	Phe	—	Tyr	—	—	—	—	—	—	Lys	—	—	Lys	—	Asx-
bTSHβ	—	—	—	Arg-	His	—	Thr	—	Tyr-	Phe	—	Tyr	—	—	—	Ile	—	—	Lys	—	—	Lys	—	Asx-
pTSHβ	—	—	—	His-	His	—	Thr	—	Tyr-	Phe	—	Tyr	—	—	—	Ile	—	—	Lys	—	—	Lys	—	Asp-
b,oLHβ	Gly-	Cys-	Pro-	Pro-	Gly-	Val-	Asp-	Pro-	Met-	Val-	Ser-	Phe-	Pro-	Val-	Ala-	Leu-	Ser-	Cys-	His-	Cys-	Gly-	Pro-	Cys-	Arg-
pLHβ	—	—	—	—	—	—	—	—	Thr	—	—	—	—	—	—	—	—	—	—	—	—	—	—	—
hLHβ	—	—	—	Arg	—	—	—	—	Val	—	—	—	—	—	—	—	—	—	Arg	—	—	—	—	—
hCGβ	—	—	—	Arg	—	—	Asn	—	Val	—	—	Tyr	Ala	—	—	—	—	—	Gln	—	Ala-	Leu	—	—
hFSHβ	—	—	Ala-	His-	His-	Ala	—	Ser-	Leu-	Tyr-	Thr-	Tyr	—	—	—	Thr-	Gln	—	—	—	—	Lys	—	Asp-
pFSHβ	—	—	Ala-	His-	His-	Ala-	Asx-	Ser-	Leu-	Tyr-	Thr-	Tyr	—	—	—	Thr-	Glu	—	—	—	—	Lys	—	Asx-
eFSHβ	—	—	Ala-	His-	His-	Ala-	—	Ser-	Leu-	Tyr-	Thr-	Tyr	—	—	—	Thr-	Glx	—	—	—	—	Lys	—	Asx-

				100										110										
hTSHβ	Thr-	Asx-	Tyr-	Ser	—	—	Ile-	His	(Glu,	Ala,	Ile)	Lys-	Thr-	Asx-	Tyr	—	Thr-	Lys	—	Glx-	Lys-	Ser-	Tyr-	COOH
bTSHβ	Thr-	Asx-	Tyr-	Ser	—	—	Ile-	His-	Glu-	Ala-	Ile-	Lys-	Thr-	Asn-	Tyr	—	Thr-	Lys	—	Gln-	Lys-	Ser-	Tyr-	Met-COOH
pTSHβ	Thr-	Asp-	Tyr-	Ser	—	—	Ile-	His-	Glu-	Ala-	Ile-	Lys-	Thr-	Asn-	Tyr	—	Thr-	Lys	—	Glu-	Lys-	Ser-	Tyr-	COOH
b,oLHβ	Leu-	Ser-	Ser-	Thr-	Asp-	Cys-	Gly-	Pro-	Gly-	Arg-	Thr-	Glx-	Pro-	Leu-	Ala-	Cys-	Asx-	His-	Pro-	Pro-	Leu-	Pro-	Asp-	Ile-
pLHβ	—	—	—	Ser	—	—	—	—	—	—	Ala	—	—	—	—	—	—	Arg	—	—	—	—	Gly-	Leu–
hLHβ	(Arg)	—	Thr-	Ser	—	—	—	Gly-	Pro-	Lys-	Asx	(His)	—	—	Thr	—	—	(Glx,	Asx,	Ser,	Lys)	Gly-	COOH	
hCGβ	Arg	—	Thr	—	—	—	—	Gly-	Pro-	Lys-	Asp	His	—	—	Thr	—	Asp-	Asp	—	Arg-	Phe-	Gln	—	Ser-
hFSHβ	Ser-	Asp	—	—	—	—	Thr-	Val-	Arg-	Gly-	Leu-	Gly	—	Ser-	Tyr	—	Ser-	Phe-	Gly-	Glu-	Met-	Lys-	Gln-	Tyr-
pFSHβ	Ser-	Asx	—	—	Asx	—	Thr-	Val-	Arg-	Gly-	Leu-	Gly	—	Ser-	Tyr	—	Ser-	Phe-	Gly-	Glu-COOH				
eFSHβ	Ser-	Asp	—	—	Asx	—	Thr-	Val-	Arg-	Gly-	Leu-	Gly	—	Ser-	Tyr	—	Ser-	Phe-	Gly-	Asp-	Met-	Lys-	Glx-	Tyr-

										130										140				
b, o, pLHβ	Leu-COOH																							
hCGβ	Ser-	Ser*-	Ser-	Lys-	Ala-	Pro-	Pro-	Pro-	Ser*-	Leu-	Pro-	Ser-	Pro-	Ser*-	Arg-	Leu-	Pro-	Gly-	Pro-	Ser*-	Asp-	Thr-	Pro-	Ile-
hFSHβ	Pro-	Thr-	Ala-	Leu-	Ser-	Tyr-COOH																		
eFSHβ	Pro-	Val-	Ala-	Leu-	Ser-	Tyr-COOH																		

hCGβ — Pro- Gln- $CONH_2$

Fig. 2. *(Continued)*
hormone, ▨ ; involved in an α-β cross-link, ×; modification resulting in a significant decrease in receptor interaction, /\/\/\/\ ; modification resulting in a significant decrease in subunit interaction, _⊓⊔⊓_.

18,000, which would accommodate a "signal peptide" but not many more residues (41). The second possibility seems unlikely because many laboratories have found excess free α subunits present in the pituitary.

This paper describes the results of chemical modifications directed toward investigations of the third and fourth possibilities, namely, are there recognition sites on the α subunit for hormone receptor present on the target cells? Failing crystallographic information, modification studies also are, despite the caution that must be used in their interpretation, a good way of determining the domains of subunit–subunit contact in the α–β complex. The similarities and differences between the glycoprotein hormones and certain toxins will also be reviewed briefly. A comparison of properties is of considerable interest because some toxins—such as the lectins, ricin, and abrin, as well as bacterial toxins such as cholera toxin—consist of subunits with one type of subunit binding to cells and the other "internalized" for expression of toxin activity. Thus the interesting hypothesis that these toxins and the glycoprotein hormones share common structural features and mechanisms of action has been made and raises the question of whether convergent evolution of subunit hormones and these toxins has occurred.

With respect to chemical modifications, several have been made both on isolated subunits and on intact hormones (for reviews, see 68,69,71,79), with particular interest in the location of possible functional regions on the common α subunit. As mentioned, care must be used in attributing the loss of biological activity of a modified hormone to a particular region or residue, with significant results obtained only with reagents that are suitably selective, and when the modified residue(s) can be located in the polypeptide chain. Figures 1 and 2 also summarize results from this laboratory and others that not only implicate regions of binding to receptors and of subunit–subunit interaction but also locate the modified residue(s). A special problem with these hormones is that a modification may lower the affinity between subunits to a point where the modified hormone dissociates at the low concentrations used for bioassay or binding studies (1 nM–1 pM), and thus activity is lost because of dissociation rather than directly from modification. Two methods of ruling out dissociation at low concentrations as a consequence of modification have been used in this laboratory (32,66,70). One is a radioimmunoassay for the β subunit of bovine lutropin, in which the assay is 40-fold more sensitive to free β subunit than to the complexed β subunit in the intact hormone. If, at concentrations similar to those used in receptor assays, a modified hormone produces competition curves in the radioimmunoassay (displacement of ^{125}I-labeled lutropin) similar to unmodified controls, dissociation cannot have occurred. The second method is to cross-link covalently the subunits by a carbodiimide-mediated coupling (see also 5a,17a,80,81). Cross-linked bovine lutropin retains 50–100% of its binding and biological activity, and thus further modifications can be made without subunit dissociation. Also, if a loss in subunit interaction or receptor binding is to be attributed to modification of a particular residue, significant conformational changes should be ruled out, as changes in tertiary structure might involve domains distant from the modified groups. Circular dichroism (31,32,34,66) and proton nuclear magnetic resonance (4) have been utilized in our laboratory to assess conformational changes as a result of a modification [see also Maghuin-Rogister *et al.* (52)]. Although slight changes in tertiary structure may not be detected by the preceding techniques, they are the best available.

Examples of the types of modifications that have been made in the glycoprotein hormones are acylation of amino groups, nitration and iodination of tyrosine residues, substitution of amides for carboxyl groups, alkylation of methionine and cysteine residues (after specific reduction), oxidation of tryptophan residues, and treatment with carboxypeptidase or CNBr. Two types of results have been obtained: those attributing the loss of biological activity of a modified hormone to a particular region or residue and those indicating whether residues or regions are exposed or protected in the intact hormone as compared to the free subunits.

Some of the modifications of the α subunit are relatively nonspecific, such as those involving lysines or carboxyl groups. No lysines in the α subunit either protected in the intact hormone or with unusual reactivity have been reported (20,22,50,59,74). Most of the 10 α-lysine residues can be modified and most products obtained either do not recombine with native β subunits or, if they do, lose much of the activity. However, methylation of the lysines (1,22), which results in retention of charge, yields a highly active hormone, whereas introduction of a larger group (ethyl) results in an inactive product. Modification of the carboxyl groups by carbodiimide-mediated coupling with glycine methyl ester (23) yields an α subunit with eight residues of glycine methyl ester incorporated. The modified subunit will no longer recombine with native β subunit, but a modified β subunit, with six to seven glycine methyl esters incorporated, will recombine with native α. Thus a general observation is that the α subunit, as compared to the β subunit, has a stronger requirement for the retention of charge in order for recombination to take place.

The site-specific modifications of the α subunit that have been located and implicated in receptor binding or subunit–subunit interaction are shown in Fig. 1. Perhaps the most interesting observation made thus far is that the α subunit is very sensitive to modifications in its COOH-terminal region with either recombination prevented or the biological activity of the recombinants lost. Integrity of the COOH-terminal pentapeptide (Tyr-Tyr-His-Lys-Ser-COOH) apparently is required for receptor interaction by the intact hormone. Iodination or nitration of tyrosine residues 92 and 93 in several different laboratories using α subunits from various sources (5,7,13,19,36,47) results in a modified subunit capable of recombining with native β, yet the recombinant has little biological activity. Removal of this pentapeptide by carboxypeptidase treatment (14) results in a modified α subunit also capable of recombining with native β subunit, but again biological activity is lost. A recombinant with a carboxypeptidase-treated α subunit was covalently cross-linked (66). This derivative also has negligible receptor-binding activity, and its circular dichroism is not significantly different from that of native hormone or non-cross-linked recombinant. Thus the loss in biological activity resulting from the removal of the COOH-terminal pentapeptide does not appear to be a function of dissociation of the modified α and native β subunit or of significant conformational changes.

The integrity of the immediate NH_2-terminal region, however, is not requisite for hormone activity. Bovine–ovine α preparations contain molecules with incomplete NH_2 termini with as many as seven residues missing, and the human subunits have either three or five residues less than the complete bovine–ovine sequence but can form active hybrids with bovine or ovine β chains. In addition, cyanogen bromide treatment of bovine LH-α removes eight residues from the NH_2 terminus without total loss of activity (21).

The region of the disulfide bond formed between cysteine residues 11 and 35, which includes methionine-33, is also implicated as a receptor-binding region of the α subunit. The carboxymethylation of methionine-33 (11,12) or selective reduction of the disulfide bond formed between residues 11 and 35 and subsequent carboxymethylation of these reduced cysteine residues (32), results in an inactive product. The 11,35-bis(*S*-carboxymethyl) derivative will recombine with native β, but the recombinant has negligible binding activity. However, the recombinant formed between native β and an 11,35-bis(*S*-carbamoylmethyl) α derivative binds significantly to gonadotropin receptors. Radioimmunoassay in the homologous lutropin β system showed that the loss of binding by the *S*-carboxymethyl hormone did not occur because of modification-induced dissociation. The circular dichroism of the modified recombinant differs very slightly from that of a recombinant of native subunits; thus the negative charges at positions 11 and 35 must interfere with binding either by prohibiting the direct interaction of that portion of the α subunit with receptor or possibly by preventing the hormone from attaining an extremely stringent conformation needed for interaction with receptor.

Several studies on the topology of the α subunit indicate that tyrosine residues 21, 92, and 93 can all be iodinated or nitrated (5–7,13,19,36,47,49,75,83) in the intact hormone, implying their exposure to the solvent, but Combarnous and Maghuin-Rogister (19) found that iodination of tyrosine-21 has little effect on biological activity. Tyrosine-30 in the bovine, ovine, and porcine subunits is protected not only in the intact hormone but also in the free subunit (5,13). Similarly, tyrosine-41 is protected in the intact hormone, but nitration of this residue in the free subunit (5,7,13) prevents recombination with native β. Nitration studies with human α subunits (7,36,47) indicate tyrosine residue 69 to be at least partially protected in the intact hormone. It is thus likely that tyrosine-41 as well as the tyrosine/phenylalanine at position 69 is in a contact region with the β subunit.

The isolation of the carbodiimide-mediated cross-linked hormones (66,80,81) in good yields indicates an orientation between an activated carboxyl group(s) and a nucleophile(s) in the intact hormone that favors coupling. Recent studies (81) indicate the major amide cross-link in bovine lutropin was formed by covalent coupling of α-lysine-49 and β-aspartic acid-113, indicating that these residues are in regions of subunit–subunit interaction.

The site-specific modifications of the β subunit that not only have been located but also have been implicated in receptor binding or subunit–subunit interaction are shown in Fig. 2. Results on modification of tyrosines in several different β subunits carried out in several laboratories generally agree. Both iodination and nitration studies support the hypothesis that the tyrosines or phenylalanines in the similar regions of the β subunits (i.e., at positions 37,

61, and 84) are in regions of subunit–subunit interaction. These tyrosines are protected against modification in the intact hormone; however, modification in the free β subunit does not affect either recombination with native α subunit or biological activity of the modified recombinant. Combarnous and Hennen (18), however, found that dansylation of free porcine lutropin β subunit results in modification of only tyrosine-37, which prevents recombination with the native α subunit. They also show that dinitrophenylation results in preferential modification of tyrosine-61. The dinitrophenylated β will recombine with native α; however, the reconstituted dimer has very low biological activity. The tryptophan residue in position 33 of the human FSH-β subunit is also protected in the intact hormone (34,72), strengthening the idea that the highly conserved region containing residues 31–38 is in an area of contact between subunits.

The effect of modification of lysines in the β subunit is described in a series of elegant papers by Ward and his colleagues. If isolated ovine LH-β, which contains two lysines, is modified such that the positive charge is either retained or lost, as in guanidination or carbamylation (46,48,50), neither recombination with native α subunit nor biological activity of the recombinant is seriously affected. Introduction of a negative charge in the LH-β lysyl residues by maleylation or citriconylation (50,74) allows recombination in good yield; however, most of the hormonal activity is lost. The ability of these modified β subunits to recombine with native α is not unexpected in that other β subunits have neutral or negatively charged rather than lysine residues at these positions. The loss of biological activity resulting from the maleylation or citriconylation could, however, implicate the regions around these lysine residues as possible regions for receptor recognition.

Other residues implicated as possible receptor recognition sites of the β subunits are tryptophan-8 in the human LH-β chain (34), methionine-53 in bovine LH-β (11, 12), and the COOH-terminal dipeptide (14). An interesting modification has been reported by de la Llosa *et al.* (21) in which the removal of the dodecapeptide lysine-42–methionine-53 of ovine LH-β by cyanogen bromide treatment results in a "mutilated" β subunit capable of recombination with native α subunit. The "mutilated" recombinant retained nearly 50% relative biological potency of a recombinant containing native subunits. However, this apparent nonrequirement of this dodecapeptide for receptor interaction is in contrast to data implicating lysine-42 and methionine-53 as possible receptor interaction regions, and the CNBr treatment needs more investigation. Another interesting modification is the selective reduction of the disulfide bond between cysteine residues 95 and 102 (73) and subsequent carboxymethylation or carbamoylmethylation. Neither the 95,102-bis(*S*-carboxymethyl) nor the 95,102-bis(*S*-carbamoylmethyl) derivatives of bovine LH-β will recombine with native LH-α subunit (G. A. Bloomfield, J. R. Reeve, and J. G. Pierce, unpublished experiments).

The data from the modification studies made thus far provide considerable evidence, though not proof, that portions of the α subunit may indeed be involved directly in recognition of receptors, most probably in concert with domains on the β subunits. However, alternate or conjoint possibilities for α function exist. One that has been suggested is that after hormone–receptor interaction, this subunit is either "internalized" into the cell or "intercalated" within the plasma membrane, where it subsequently elicits a hormone response via an adenylate cyclase system. This mode of action would explain nicely the function of the common subunit, and the concept of a "binding" and an "active" subunit in the glycoprotein hormones parallels in many ways the proposed actions of some bacterial and plant toxins and was first suggested by Olsnes *et al.* in 1974 [63; see also Boquet and Pappenheimer (2)]. The suggestion was based on studies with diphtheria toxin and the toxic plant lectins, abrin, and ricin. Diphtheria toxin is released from the bacterium as a single polypeptide chain with two disulfide bridges. Proteolytic cleavage results in an A and B fragment linked by a disulfide bond. Subsequent reduction yields free A fragment containing NAD glycohydrolase activity *in vitro*. The intact toxin is, however, enzymatically inactive in a cell-free system and proteolytic clips and reduction are needed for activation. The B fragment recognizes specific receptors on many cells, and the A fragment, presumably internalized, is thought to exert its killing effect by ADP ribosylation of elongation factor 2, a component of the cell's protein-synthesizing machinery. The enzymatic reaction has been clearly shown *in vitro* (e.g., 17). Abrin and ricin have a similar type of structure with a B chain that binds to membrane receptors and an A chain that exerts toxic effects, again by interfering with protein synthesis, though not via elongation factor 2 (e.g., 62). Cholera toxin, a third type of toxin, also contains binding and toxic subunits: the B component consists of five identical noncovalently linked polypeptide chains, and the A component after proteolytic cleavage consists of two polypeptide chains, A_1 and A_2, linked by a disulfide bond. The B component can bind to many cell surfaces, presumably providing for the entry of a single A subunit or A_1 fragment causing marked and prolonged stimulation of adenylate cyclases. The A_1 fragment alone is active in broken cell preparations; however, the holotoxin is required in most intact cell systems. NAD is necessary for activity, as in the case of diphtheria toxin, and recent work indicates that cyclase activation is probably via ADP ribosylation of the GTP-binding component of the cyclase system (e.g., 8,27). The structure of cholera toxin and its mechanism of action have been reviewed recently (26,56).

Interest in possible similarities between the action of toxins and the glycoprotein hormones was further stimulated by the observations of Kohn and his colleagues (58), who found that certain gangliosides inhibit the binding of TSH to thyroid membranes and the binding of hCG to membrane components of rat testes. Formation of a complex between hormone and ganglioside was found to be analogous to that between cholera toxin and the ganglioside

G_{M1}, the latter is believed to be an integral component of the cholera toxin receptor (26,56). However, the specificity of glycoprotein hormone–ganglioside interaction is different, with the ganglioside G_{D1b} the most effective in complexing with bovine TSH, mixed gangliosides and G_{D1a} most effective for hCG, and G_{T1} most effective for bovine LH (43,44,58). Membrane preparations from rat thyroid tumors, which bind TSH poorly, were found to contain G_{M3}, which interacts less well with TSH, whereas normal rat thyroid preparations contain some of the more complex gangliosides that have been shown to inhibit TSH binding but that, interestingly enough, do not compete with prebound TSH. A glycoprotein component of the TSH receptor has also been reported (55). In addition, a similarity in sequence was found between the binding subunit of cholera toxin and a highly conserved region of the β subunits of the glycoprotein hormones (37,39,58) (see Fig. 2, residues 31–38). The following are residues involved in the toxin B subunit in comparison to the similar region in bovine LH:

```
                     4                         10     12
cholera toxin B    -Ile-Thr-Asp-Leu-Cys-Ala-Glu-Tyr-His-
                     31                               38
bLH-β              -Thr — Ser- Ile ———  Gly — Cys
```

The region was then suggested as a recognition site between either toxin B or hormone β subunits and receptors (58). The region (often referred to as the C A G Y region, based on the single-letter notation for amino acid sequences) is also in a number of serine proteases (37). A more restricted area of similarity was found between glycoprotein hormone α subunits and the toxin A_1 protein (the active portion of the A subunit) with the sequence -Cys-Pro-Gln-Cys- near the amino terminus of α chains and Cys-Pro-Asx-Ala- in the cholera toxin A_1 subunit (40). A similar sequence (with, however, only a Cys-Pro as an identity) is in oxytocin and vasopressin, hormones of the posterior lobe of the pituitary. This region was proposed by Kohn and his colleagues as a possible site of interaction in all of these proteins with cyclase system after recognition by the target cell (35).

The preceding, then, are the observations concerning the glycoprotein hormones and several of the toxins that have suggested similar modes of action. However, many contradications concerning analogous modes of action between these hormones and the toxins can also be found. Recent studies with hCG under more physiologic conditions than the earlier assays (65) show that rat testes gangliosides do not inhibit the binding of hCG to rat testes preparations, and subsequently, acidic phospholipids were found to be more potent inhibitors of TSH (but not competitors of prebound hormone) in a radioreceptor-ligand assay (64). A requirement for NAD in the stimulation of adenylate cyclase activity by TSH also has not been demonstrated (58), and the NAD glycohydrolase activity present in rather highly purified hCG preparations was found to be a contaminant (57). The evidence from chemical modification experiments also strongly supports the concept that the C A G Y region of the β subunits is a region of subunit–subunit contact in the glycoprotein hormones, because residues in this region of LH-β, hCG-β, and FSH-β are not reactive in the intact hormone but are in the free subunit. The occurrence of analogous regions that are in close proximity to a disulfide bridge or bridges in other proteins such as cholera toxin may be to allow a particular folding pattern to occur. Finally, a recent report (78) shows that the glycoprotein hormones TSH and LH probably do not activate adenylate cyclase by a mechanism identical to cholera toxin, in that cooling distinguished between the cholera toxin and hormonal effects on cyclic AMP accumulation in intact cells.

The small similarities between the α subunit of the posterior lobe hormones and cholera toxin do not seem to us persuasive of similar function. For example, the single disulfide bond in the A protomer of cholera toxin must be reduced to generate enzymatic activity in the A_1 fragment, whereas the single disulfide of the posterior lobe hormones must be intact for activity. In the glycoprotein hormones, the half cystine "analogous" to the half cystine in the posterior lobe hormones can be reduced (see earlier) with retention of activity, and it is linked in the native subunit to a half-cystine 25 residues removed in the sequence rather than 5 as found in the posterior lobe hormones. Thus, although it appears that the evidence does not yet support marked similarities in action between the glycoprotein hormones and toxins such as diphtheria toxin, abrin–ricin, or cholera toxin, the possibility is intriguing and nothing is really known of how the glycoprotein hormones with their subunit structure manifest both their specificity and hormone action. As stated by Kurosky *et al.* (38),

> The curious chemical similarity of the β chain of cholera toxin to the β chain of the glycoprotein hormones, together with the possible structural similarity between cholera toxin and the enterotoxins synthesized by *Escherichia coli* and *Salmonella typhimurium* presents an interesting story of protein evolution. Generally, in the case of chemical similarity between proteins, one immediately questions whether the observed similarity is the result of a convergent or a divergent mechanism of evolution; and certainly this question applies in the case of the glycoprotein hormones and cholera toxin. Considerably more information will be required to establish one or the other possibility. However, there is yet another mechanism by which such structural similarity may have evolved. It is conceivable that by some undefined genetic mechanism the bacteria have somehow entrapped a eukaryotic segment of DNA containing a glycoprotein hormone or hormone-like gene. Primary structures of the remaining chains of cholera toxin (α and γ) as well as of other enterotoxins are now essential to further delineate the evolution of the bacterial enterotoxins.

One must note with interest that reports of material with hCG-like immunoreactivity or properties obtained from various bacteria have appeared (16,51,53).

Interesting experiments are suggested by the need of a binding component to facilitate the entry of an enzymatically active fragment of the toxins and by the subunit structure of the glycoprotein hormones. Recently, Oeltmann and

Heath (60,61) have synthesized a hybrid consisting of the A fragment of ricin covalently linked via a disulfide bond to an amino group on hCG-β. When the effects on protein synthesis in a rat Leydig cell tumor line, which is gonadotropin sensitive, were tested, both intact ricin and the hybrid were inhibitory. In L-cells (a cultured cell line not responsive to hCG) only ricin, not the hybrid, inhibits protein synthesis. Both cell lines were insensitive to either the ricin A or B fragments alone. Whether the recognition of the hybrid by the Leydig cell and the mode of entry of the A (toxic) fragment of ricin are the same as those in the action of the glycoprotein hormones themselves remains to be determined. The amount of hybrid required to inhibit protein synthesis was 50–100 μg/ml of medium, which is 10,000–50,000 times greater than the amount of hCG required for hormonal stimulation of Leydig cells in many *in vitro* assays (9) and is at least 500 times less potent than ricin itself. hCG also blocked the inhibition of protein synthesis by the hybrid, but a concentration of 600 μg/ml was required, whereas about 10 ng/ml will block the binding of ^{125}I-hCG in several assay systems. It should be noted that if there is any residual intrinsic binding of isolated β subunits of glycoprotein hormones, it must be extremely weak. We have recently shown that β subunits of LH from which contaminants have been removed by immunoaffinity chromatography (82) will not compete at a concentration of 200 μg/ml with ^{125}I-LH in a radioligand receptor assay. Reports persist, however, of some intrinsic binding activity of isolated β subunits (for references, see 82), although the amount of contamination with intact hormone was not assessed. Nevertheless, the formation of the ricin A chain–hCG-β hybrid does cause the A chain to affect protein synthesis in hormone-sensitive cells, albeit at high concentrations, whereas the ricin A chain itself does not. Obviously if hybrid molecules can be constructed in which binding of a toxin was specifically directed only against hormone-sensitive cells, selective destruction of tumors of a given cell type could be achieved.

Other interesting hybrids have been described. One, prepared by Collier and his colleagues at the University of California, Los Angeles (28), is a disulfide-linked conjugate of concanavalin A with the A fragment of diphtheria toxin. The hybrid has a low level of toxicity (2–0.2%) toward cells to which ConA binds. Because of the low toxicity of the hybrid, the entry of the A fragment may be by a low-efficiency mechanism, perhaps endocytosis, rather than by a mechanism in which channels are formed by the toxin B fragments (3). Nevertheless, the toxicity of the conjugate is 100–1000 times greater than that of the A fragment itself. The second hybrid is a cross-linked derivative of human placental lactogen and diphtheria toxin A fragment. However, it was not toxic in cultured mammary gland transplants (10). A hybrid toxin between ricin A and a nontoxic mutant protein of diphtheria toxin has also been prepared (76).

One property of cholera toxin and the glycoprotein hormones has recently been found to be similar, namely, that after covalent cross-linking of their subunits, both retain a large portion of their biological activity (77). Therefore dissociation as a prerequisite for activity is very unlikely. Also, recent evidence shows that most probably reduced and nicked but nondissociated cholera toxin retains enzymatic activity (54).

In summary, this volume illustrates well many of the remarkable advances that have been made in our understanding of the structures and functions of many groups of proteins and of the evolutionary relationships between them. In this contribution, we have tried to show the interesting questions both answered and raised by determination of the amino acid sequences of the glycoprotein hormones. Immediate questions that should be answered include conclusive allocation of disulfide bonds and preparation of crystals suitable for X-ray analysis. Both tasks have proven far more difficult than anticipated. Ways to overcome what appear to be the extremely stringent conformational requirements for binding to receptors with high affinity are needed. If this can be accomplished together with precise determination of the roles or role of the common subunit, many interesting and potentially useful derivatives can be made.

ACKNOWLEDGMENTS

The work from the authors' laboratory has been supported for many years by a research grant (AM 18005) from the National Institutes of Health. Appreciation is expressed by JGP to the many students and associates who have contributed greatly to our knowledge of the glycoprotein hormones. The presence of Emil L. Smith and his colleagues at UCLA gave impetus and support to the discovery of the relationships between the subunits of these hormones. His help and the help and collaboration of Alex Glazer and Douglas Brown have been particular pleasures.

REFERENCES

1. Ascoli, M., and Puett, D., Tritium labeling of luteinizing hormone by reductive methylation. *Biochim. Biophys. Acta*, 1974, **371**:203–210.
2. Boquet, P., and Pappenheimer, A. M., Jr., Interaction of diphtheria toxin with mammalian cell membranes. *J. Biol. Chem.*, 1976, **251**:5770–5778.
3. Boquet, P., Silverman, M. S., Pappenheimer, A. M., Jr., and Vernon, W. B., Binding of triton X-100 to diphtheria toxin, crossreacting material 45 and their fragments. *Proc. Natl. Acad. Sci. U.S.A.*, 1976, **73**:4449–4453.
4. Brown, F. F., Parsons, T. F., Sigman, D. S., and Pierce, J. G., Proton nuclear magnetic resonance studies on bovine lutropin, its subunits and on the α subunit of pregnant mare serum gonadotropin: Assignment of histidine resonances in the α subunit. *J. Biol. Chem.*, 1979, **254**:4335–4338.
5. Burleigh, B. D., Liu, W-K., and Ward, D. N., Reaction of tetranitromethane with lutropin, oxytocin, and vasopressin. *J. Biol. Chem.*, 1976, **251**:308–315.

5a. Burleigh, B. D., Liu, W-K., and Ward, D. N., Photocoupling of the subunits of ovine lutropin using a specific aryl azide derivative of the β subunit. *J. Biol. Chem.*, 1978, **253**:7179–7185.

6. Canfield, R. E., Morgan, F. J., Kammerman, S., and Ross, G. T., Studies of the structure and recombination of the subunits of human chorionic gonadotropin.

In: *Structure-Activity Relationships of Protein and polypeptide Hormones* (M. Margoulies and F. C. Greenwood, eds.). Excerpta Medica, Amsterdam, 1972:341–343.

7. Carlsen, R. B., and Bahl, O. P., The reaction of tetranitromethane with human chorionic gonadotropin. *Arch. Biochem. Biophys.*, 1976, **175**:209–220.
8. Cassel, D., and Pfeuffer, T., Mechanism of cholera toxin action: Covalent modification of the guanyl nucleotide-binding protein of the adenylate cyclase system. *Proc. Natl. Acad. Sci. U.S.A.*, 1978, **75**:2669–2673.
9. Catt, K. J., and Dufau, M. L., Interactions of LH and hCG with testicular gonadotropin receptors. In: *Receptors for Reproductive Hormones* (B. W. O'Malley and A. R. Means, eds.). Plenum, New York, 1973:379–418.
10. Chang, T., Dazord, A., and Neville, D. M., Jr., Artificial hybrid protein containing a toxic protein fragment and a cell membrane receptor-binding moiety in a disulfide conjugate. II. Biochemical and biologic properties of diptheria toxin fragment A-S-S human placental lactogen. *J. Biol. Chem.*, 1977, **252**:1515–1522.
11. Cheng, K-W., Carboxymethylation of methionine residues in bovine pituitary luteinizing hormone and its subunits: Effects on the binding activity with receptor sites and interactions between subunits. *Biochem. J.*, 1976, **159**:71–77.
12. Cheng, K-W., Carboxymethylation of methionine residues in bovine pituitary luteinizing hormone and its subunits: Location of specifically modified methionine residues. *Biochem. J.*, 1976, **159**:79–87.
13. Cheng, K-W., and Pierce, J. G., The reaction of tetranitromethane with pituitary luteinizing and thyroid-stimulating hormone. *J. Biol. Chem.*, 1972, **247**:7163–7172.
14. Cheng, K-W., Glazer, A. N., and Pierce, J. G., The effects of modification of the COOH-terminal regions of bovine thyrotropin and its subunits. *J. Biol. Chem.*, 1973, **248**:7930–7937.
15. Chin, W. W., Habener, J. F., Kieffer, J. D., and Maloof, F., Cell-free translation of the messenger RNA coding for the α subunit of thyroid-stimulating hormone. *J. Biol. Chem.*, 1978, **253**:7985–7988.
16. Cohen, H., and Strampp, A., Bacterial synthesis of substance similar to human chorionic gonadotropin. *Proc. Soc. Exp. Biol. Med.*, 1976, **152**:408–410.
17. Collier, R. J., Effect of diptheria toxin on protein synthesis: Inactivation of one of the transfer factors. *J. Mol. Biol.*, 1976, **25**:83–98.

17a. Combarnous, Y., and Hennen, G., Luteinizing hormone derivatives with covalently-linked subunits. *FEBS Lett.*, 1974, **44**:224–228.

18. Combarnous, Y.,and Hennen, G., Chemical study of the topography of porcine lutropin (LH) using dinitrofluorobenzene and dansyl chloride. *Int. J. Pep. Protein Res.*, 1976, 8:491–498.
19. Combarnus, Y., and Maghuin-Rogister, G. Luteinizing hormone: Relative reactivities of tyrosyl residues of the porcine hormone toward iodination. *Eur. J. Biochem.*, 1974, **42**:13–19.
20. de la Llosa, P., Durosay, M., Tetrin-Clary, C., and Jutisz, M., Chemical modification of lysine residues in ovine luteinizing hormone. *Biochim. Biophys. Acta*, 1974, **342**:97–104.
21. de la Llosa, P., Tetrin-Clary, C., and Jutisz, M., Influence on "mutilation" of the polypeptide chain of the luteinizing hormone on its biological activity. In: *Structure-Activity Relationships of Protein and Polypeptide Hormones* (M. Mar-

goulies and F. C. Greenwood, eds.). Excerpta Medica, Amsterdam, 1972:344–347.

22. de la Llosa-Hermier, M. P., de la Llosa, P., and Hermier, C., Studies of the binding activity to different gonadal receptors of ovine luteinizing hormone (LH) after chemical modification of lysine residues. *Gen. Comp. Endocrinol.*, 1977, **31**:302–306.
23. Faith, M. R., and Pierce, J. G., The carboxylic acid groups of bovine luteinizing hormone. *J. Biol. Chem.*, 1975, **250**:6923–6929.
24. Fontaine, Y. A., and Burzawa-Gerard, E., Esquisse de l'évolution des hormones gonadotropes et thyréotropes des vertébrés. *Gen. Comp. Endocrinol.*, 1977, **32**:341–347.
25. Fontaine, Y. A., and Burzawa-Gerard, E., Biochemical and biological properties of fish gonadotropins and their subunits: Comparison with mammalian hormones. In: *Structure and Function of the Gonadotropins* (K. W. McKerns, ed.). Plenum, New York, 1978:361–380.

25a. Fujiki, Y., Rathman, P., and Saxena, B. B., Amino acid sequence of the β-subunit of the follicle-stimulating hormone from equine pituitary glands. *J. Biol. Chem.*, 1978, **253**:5363–5368.

26. Gill, D. M., Mechanism of action of cholera toxin. *Adv. Cyclic Nucleotide Res.*, 1977, **8**:85–118.
27. Gill, D. M., and Meren, R., ADP-ribosylation of membrane proteins catalyzed by cholera toxin: Basis of the activation of adenylate cyclase. *Proc. Natl. Acad. Sci. U.S.A.*, 1978, **75**:3050–3054.
28. Gilliland, D. G., Collier, R. J., Moehring, J. M., and Moehring, T. J., Chimeric toxins: Toxic, disulfide-linked conjugate of concanavalin A with fragment A from diphtheria toxin. *Proc. Natl. Acad. Sci. U.S.A.*, 1978, **75**:5319–5323.
29. Giudice, L. C., and Pierce, J. G., Studies on the disulfide bonds of glycoprotein hormones: Complete reduction and reoxidation of the disulfide bonds of the α subunit of bovine luteinizing hormone. *J. Biol. Chem.*, 1976, **251**:6392–6399.
30. Giudice, L. C., and Pierce, J. G., Studies on the reduction and reoxidation of the disulfide bonds of the α and β subunits of human choriogonadotropin. *Biochim. Biophys. Acta*, 1978, **533**:140.
31. Giudice, L. C., and Pierce, J. G., Glycoprotein hormones: Some aspects of studies of secondary and tertary structure. In: *Structure and Function of the Gonadotropins* (K. W. McKerns, ed.). Plenum, New York, 1978:81–110.
32. Giudice, L. C., and Pierce, J. G., Studies on the disulfide bonds of glycoprotein hormones. Formation and properties of 11,35-bis(S-alkyl) derivatives of the α subunit. *J. Biol. Chem.*, 1979, **254**:1164–1169.
33. Giudice, L. C., and Weintraub, B. D., Cell-free biosynthesis, specific immunoprecipitation, and processing of TSH pre-α and β subunits. In: *61st Annual Meeting of the Endocrine Society*, Anaheim, California. Endocrine Society, Bethesda, Maryland, 1979 (Abstr. 83).
34. Giudice, L. C., Pierce, J. G., Cheng, K-W., Whitley, R., and Ryan, R. J., Circular dichroism of mammalian follitropins and the effects of treatment with N-bromosuccinimide. *Biochem. Biophys. Res. Commun.*, 1978, **81**:725–733.
35. Grollman, E. F., Leed, G., Ambesi-Impiomato, F. S., Meldolesi, M. F., Aloj, S. M., Coon, H. G., Kaback, H. R., and Kohn, L. D., Effects of thyrotropin on the

thyroid cell membrane: Hyperpolarization induced by hormone-receptor interaction. *Proc. Natl. Acad. Sci. U.S.A.*, 1977, **74**:2352–2356.

36. Hum, V. G., Knipfel, J. E., and Mori, K. E., Human chorionic gonadotropin, reaction with tetranitromethane. *Biochemistry*, 1974, **13**:2359–2364.
37. Kurosky, A., Markel, D. E., and Fitch, W. M., Primary structure of cholera toxin β-chain: A glycoprotein hormone analog. *Science*, 1977, **195**:299–300.
38. Kurosky, A., Markel, D. E., and Peterson, J. W., Covalent structure of the β chain of cholera enterotoxin. *J. Biol. Chem.*, 1977, **252**:7257–7264.
39. Lai, C-Y., Determination of the primary structure of cholera toxin B subunit. *J. Biol. Chem.*, 1977, **252**:7249–7256.
40. Lai, C-Y., Cancedda, F., and Chang, D., Primary structure of cholera toxin subunit A_1. Isolation, partial sequences and alignment of the BrCN fragments. *FEBS Lett.*, 1979, **100**:85–89.
41. Landefeld, T. D., Identification of *in vitro* synthesized pituitary glycoprotein α subunit. *J. Biol. Chem.*, 1979, **254**:3685–3688.
42. Landefeld, T. D., McWilliams, D., and Boime, I., The isolation of mRNA encoding the alpha subunit of human chorionic gonadotropin. *Biochem. Biophys. Res. Commun.*, 1976, **72**:381–390.
43. Lee, G., Aloj, S. M., Brady, R. O., and Kohn, L. D., The structure and function of glycoprotein hormone receptors: Ganglioside interactions with human chorionic gonadotropin. *Biochem. Biophys. Res. Commun.*, 1976, **73**:370–377.
44. Lee, G., Aloj, S. M., and Kohn, L. D., The structure and function of glycoprotein hormone receptors: Ganglioside interactions with luteinizing hormone. *Biochem. Biophys. Res. Commun.*, 1977, **77**:434–441.
45. Licht, P., Papkoff, H., Farmer, S. W., Muller, C. H., Tsui, H.-W., and Crews, D., Evolution of gonadotropin structure and function. *Recent Prog. Horm. Res.*, 1977, **33**:169–248.
46. Liu, W-K., and Ward, D. N., preparation of lutropin with acetyl or acetimidinyl substituents on the amino groups of the β subunit. *Biochim. Biophys. Acta*, 1975, **405**:522–526.
47. Liu, W-K., and Ward, D. N., Effect of selective nitration of ovine lutropin on the subunit association and biological activity of the hormone. *J. Biol. Chem.*, 1976, **251**:316–319.
48. Liu, W-K., Esfahani, M., and Ward, D. N., Guanidination of ovine luteinizing hormone and effects on activity. *Endocr. Res. Commun.*, 1975, **2**:47–63.
49. Liu, W-K., Furlong, N. B., and Ward, D. N., Effects of β subunit acylation on lutropin receptor site binding. *J. Biol. Chem.*, 1977, **252**:522–527.
50. Liu, W-K., Yang, K-P., Nakagawa, Y., and Ward, D. N., The role of the amino group in subunit association and receptor site interaction for ovine luteinizing hormone as studied by acylation. *J. Biol. Chem.*, 1974, **249**:5544–5550.
51. Livingston, V. W-C., and Livingston, A. M., Some cultural, immunological and biochemical properties of *Progenitor cryptocides*. *Trans. N.Y. Acad. Sci.* [2], 1974, **36**:569–582.
52. Maghuin-Rogister, G., Degelaen, J., and Roberts, G. C. K., ^{2}H nuclear-magnetic-resonance studies of porcine lutropin and its α and β subunits. *Eur. J. Biochem.*, 1979, **96**:59–68.
53. Maruo, T., Cohen, H., and Koide, S. S., Studies on choriogonadotropin from a

microorganism. In: *61st Annual Meeting of the Endocrine Society*, Anaheim, California. Endocrine Society, Bethesda, Maryland, 1979 (Abstr. 951).

53a. Mckerns, K. W. (ed.), *Structure and Function of the Gonadotropins*. Plenum, New York, 1978.

54. Mekalanos, J. J., Collier, R. J., and Romig, W. R., Enzymic activity of cholera toxin. II. Relationships to proteolytic processing, disulfide bond reduction and subunit composition. *J. Biol. Chem.*, 1979, **254**:5855–5861.

55. Meldolesi, M. F., Fishman, P. H., Aloj, S. M., Kohn, L. D., and Brady, R. O., Relationship of gangliosides to the structure and function of thyrotropin receptors: Their absence on plasma membranes of a thyroid tumor defective in thyrotropin receptor activity. *Proc. Natl. Acad. Sci. U.S.A.*, 1976, **73**:4060–4064.

55a. Moore, W. T., Jr., Ward, D. N., and Burleigh, B. D., Primary structure of pregnant mare serum gonadotropin, *Fed. Pro., Fed. Am. Soc. Exp. Biol.*, 1979, **38**:462 (Abstr. 1231).

56. Moss, J., and Vaughan, M., Activation of adenylate cyclase by choleragen. *Annu. Rev. Biochem.*, 1979, **48**:581–600.

57. Moss, J., Ross, P. S., Agosto, G., Birken, S., Canfield, R. E., and Vaughan, M., Mechanism of action of choleragen and the glycopeptide hormones: Is the nicotinamide adenine dinucleotide glycohydrolase activity observed in purified hormone preparations intrinsic to the hormone? *Endocrinology*, 1978, **102**:415–419.

58. Mullin, B. R., Aloj, S. M., Fishman, P. H., Lee, G., Kohn, L. D., and Brady, R. O., Cholera toxin interactions with thyrotropin receptors on thyroid plasma membranes. *Proc. Natl. Acad. Sci. U.S.A.*, 1976, **73**:1679–1683.

59. Nureddin, A., and Atassi, M. Z., Differentiation of the contribution of the two subunits of lutropin to its *in vivo* activity. *Biochim. Biophys. Acta*, 1978, **533**:257–262.

60. Oeltmann, T. N., and Heath, E. C., A hybrid protein containing the toxic subunit of ricin and the cell-specific subunit of human chorionic gonadotropin. *J. Biol. Chem.*, 1979, **254**:1022–1027.

61. Oeltmann, T. N., and Heath, E. C., A hybrid protein containing the toxic subunit of ricin and the cell-specific subunit of human chorionic gonadotropin. II. Biologic properties. *J. Biol. Chem.*, 1979, **254**:1028–1032.

62. Olsnes, S., Fernandez-Puentes, C., Carrasco, L., and Vazquez, D., Ribosome inactivation by the toxic lectins abrin and ricin. *Eur. J. Biochem.*, 1975, **60**:281–288.

63. Olsnes, S., Pappenheimer, A. M., Jr., and Meren, R., Lectins from Abrus precatorius and Ricinus communis. II. Hybrid toxins and their interaction with chain-specific antibodies. *J. Immunol.*, 1974, **113**:842–847.

64. Omodeo-Sale, F., Brady, R. O., and Fishman, P. H., Effect of thyroid phospholipids on the interaction of thyrotropin with thyroid membranes. *Proc. Natl. Acad. Sci. U.S.A.*, 1978, **75**:5301–5305.

65. Pacuszka, T., Osborne, J. C., Jr., Brady, R. O., and Fishman, P. H., Interaction of human chorionic gonadotropin with membrane components of rat testes. *Proc. Natl. Acad. Sci. U.S.A.*, 1978, **75**:764–768.

66. Parsons, T. F., and Pierce, J. G., Biologically active covalently cross-linked glycoprotein hormones and the effects of modification of the carboxylterminal region of their α subunits. *J. Biol. Chem.*, 1979, **254**:6010–6015.

67. Pierce, J. G., Eli Lilly Lecture: The subunits of pituitary thyrotropin—their relationship to other glycoprotein hormones. *Endocrinology*, 1971, **89**:1331-1344.
68. Pierce, J. G., Structural homologies of glycoprotein hormones. *Pro. Int. Congr. Endocrinol., 5th, 1976* Int. Congr. Ser. No. 403, 1977, **2**:99.
69. Pierce, J. G., Bloomfield, G. A., and Giudice, L. C., The function of the common subunit of glycoprotein hormones. *Trans. Biochem. Soc.*, 1978, **6**:57–60.
70. Pierce, J. G., Bloomfield, G. A., and Parsons, T. F., Purification and receptor binding properties of complexes between lutropin and monovalent antibodies against its α subunit. *Int. J. Pept. Protein Res.*, 1979, **13**:54–61.
71. Pierce, J. G., Faith, M. R., Giudice, L. C., and Reeve, J. R., Structure and structure-function relationships in glycoprotein hormones. *Ciba Found. Symp.*, 1976, **41** (new ser.):225-250.
72. Rathman, P., and Saxena, B. B., Studies on modification of tryptophan, methionine, tyrosine, and arginine residues of human follicle: Stimulating hormone and its subunits. *Biochim. Biophys. Acta*, 1979, **576**:81–87.
73. Reeve, J. R., Cheng, K-W., and Pierce, J. G., Partial reduction of disulfide bonds in the hormone-specific subunits of TSH and LH. *Biochem. Biophys. Res. Commun.*, 1975, **67**:149-155.
74. Sairam, M. R., and Li, C-H., Reaction of ovine interstitial cell stimulating-hormone with citraconic and maleic anhydrides. *Arch. Biochem. Biophys.*, 1975, **167**:534-539.
75. Sairam, M. R., Papkoff, H., and Li, C-H., Reaction of ovine interstitial cell stimulating hormone with tetranitromethane. *Biochim. Biophys. Acta*, 1972, **278**:421–432.
76. Uchida, T., Yamizumi, M., and Okada, Y., Formation of a hybrid toxin from ricin agglutinin and a non-toxic mutant protein of diphtheria toxin. *Biochem. Biophys. Res. Commun.*, 1978, **81**:268-273.
77. Van Heyningen, S., Activity of covalently cross-linked cholera toxin with the adenylate cyclase of intact and lysed pigeon erythrocytes. *Biochem. J.*, 1977, **168**:457–463.
78. Van Sande, J., Pochet, R., and Dumont, J. E., Dissociation by cooling of hormone and cholera toxin activation of adenylate cyclase in intact cells. *Biochim. Biophys. Acta*, 1979, **585**:282-292.
79. Ward, D. N., Chemical approaches to the structure-function relationships of luteinizing hormone (lutropin). In: *Structure and Function of the Gonadotropins* (K. W. McKerns, ed.). Plenum, New York, 1978:31–45.
80. Weare, J. A., and Reichert, L. E., Jr., Studies with carbodiimide cross-linked derivatives of bovine lutropin. *J. Biol. Chem.*, 1979, **254**:6964–6971.
81. Weare, J. A., and Reichert, L. E., Jr., Studies with carbodiimide cross-linked derivatives of bovine lutropin. *J. Biol. Chem.*, 1979, **254**:6972–6979.
82. Williams, J. F., Davies, T. F., Catt, K. J., and Pierce, J. G., Receptor binding of highly purified bovine LH, TSH and subunits. *Endocrinology*, 1980, **106**:1353-1359.
83. Yang, K-P., and Ward, D. N., Iodination of ovine luteinizing hormone and its subunits. *Endocrinology*, 1972, **91**:317–320.

Thymus Hormones: Diverse Molecules Producing Similar Biological Actions

ABRAHAM WHITE

Institute of Biological Sciences
Syntex Research
Palo Alto, California
and
Department of Biochemistry
Stanford University School of Medicine
Stanford, California

The topics to be considered in this chapter are the following:

1. A brief historical account of the development of present knowledge of the functions of the thymus.
2. Criteria for designation of a preparation as a thymus hormone.
3. A brief description of purified molecules that have been described and that exhibit thymic hormone-like activity.
4. Recent studies from our laboratory of a protein fraction from human plasma with thymic hormone-like activity.
5. Concluding comments.

Historical Background

The thymus gland was among the earliest of anatomical structures to be explored. In the first century A.D., Rufus of Ephesus spoke of the thymus as the seat of the soul and of courage (*cf.* 39). In the same century, Galen described the now well-known maximal size of the thymus at or near birth and the decrease that occurs rapidly with age (*cf.* 13). Galen suggested the name *thymus* for the gland that he had dissected; the name was derived from the resemblance of the gland to a bunch of thyme flowers. This early history of the thymus has been described at some length by Haugsted in 1831 and by Simon in 1845 (*cf.* 39).

Some 17 centuries after the first theories the functions of the thymus were

THE EVOLUTION OF PROTEIN STRUCTURE AND FUNCTION

ISBN 0-12-643150-7

sought by the early classical approach of the physiologist, namely, extirpation of the gland with observations of alterations of function in the thymectomized, experimental animal. Initial studies of this type were conducted in 1845 by Restelli (37), who removed the thymus in sheep, dogs, and calves. Although no definitive conclusions were reached by this investigator, he did establish that experimental thymectomy was feasible. Only 6 of 98 animals survived the operation, and all died within 23 days of the operation.

The ensuing 55 years of the nineteenth century recorded many additional studies of the effects of thymectomy in a wide variety of species. The results of a significant number of these early studies indicated a possible role of the thymus in influencing, among several functions, skeletal growth, gonadal development, and the maintenance of muscle tone. In view of present-day knowledge, it is noteworthy that the literature of this earlier period frequently describes postoperative death following thymectomy as being due to severe infections in the experimental animal. Overall, these early studies and those of the first three decades of this century in adult, thymectomized animals did not suggest a significant role for the thymus in mammals.

Thus, in contrast to other endocrine glands, surgical extirpation of the gland in adult animals produced no clues regarding its function, inasmuch as loss of the thymus in adult life appeared to have little physiologic consequence. However, the initial suggestion that thymic extirpation might have serious biological effects was a serendipitous observation made in the course of studies designed to examine the role of the thymus in the development of leukemia in mice. With this latter goal, Miller (28) surgically removed the thymus gland from newborn mice within 24–48 hr of birth. The gross physiologic effects, namely, failure to grow, cachetic appearance, and early death resulting from infections, proved to be the basis for an entirely new era in thymology.

At the same time, Archer and Pierce (1) described reduced immunologic responses in rabbits that were thymectomized shortly following birth. This history and some of the earlier observations have been reviewed in several publications (16,17,29,41,50). The present status of the thymus as an endocrine gland and its hormones has been reviewed recently (46). The information available may be summarized as follows:

1. The thymus functions in producing and exporting to the peripheral structures, notably the lymphoid organs and the spleen, cells that are either the immunologically competent lymphocytes or progenitors of such cells. In this role, the gland also appears to be an essential site in the direction, development, and selection of immunologic properties inherent in mature thymocytes, e.g., cells involved in self- and non-self-recognition and in certain cell–cell interactions within the microenvironment of the gland.
2. The thymus synthesizes and secretes one or more polypeptide hor-

mones that function in accelerating and modulating the conversion of immunologically incompetent precursor cells into mature, immunologically competent cells that can then participate in either humoral or cell-mediated immunologic phenomena.

3. Replacement of thymic function in the neonatally thymectomized mouse has been successfully achieved by several of the diverse isolated and homogeneous products that have been described as putative thymic hormones.

4. Thymic hormone therapy studies are in progress in a variety of clinical dyscrasias that have an immunologic component or an inadequate amount of the hormone that has a role in the maturation of immunologically incompetent, to competent cells.

The preceding brief description is presented to direct attention to the fact that an anatomical structure for which a function has been sought over many centuries may in some cases accidentally be revealed as an endocrine gland having sufficient importance for normal physiological functioning. Perhaps a similar account will be presented in the future regarding other anatomical structures in search of a dynamic function. The more recent experience with the polypeptide hormones of the gastrointestinal tract (51) perhaps supports this conclusion.

The rapidity with which interest in the thymus and thymic hormones has expanded in the present decade is reflected by the publication of two volumes (26,43), several symposia (*cf.* 15), and a number of reviews on the subject of the thymus and its hormones (2–4,17,22,25,42,50).

The growth of interest in the thymus and thymic hormones not only has resulted in a very large number of publications but also has provided evidence that a variety of fractions and preparations from thymic tissue and from blood, as well as other tissues, may synthesize and secrete both polar and nonpolar compounds that may be eligible for designation as thymic hormones. These newer data have been obtained during a period of more precise dissection of specific cell populations whose immunologic responses may be modulated by the secretions of the endocrine thymus.

Criteria for Designation of a Preparation as a Thymic Hormone

The most critical assessment for acceptance of any putative endocrine product as a hormone is its ability to replace specific functions of the extirpated or absent gland in experimental conditions that would, in the lack of activity of the product, fail to ameliorate the deleterious consequences of loss of the organ. For the thymus, past experience indicates that a product with thymic hormone or thymic hormone-like activity should exhibit activity in one or more of the following biological models:

1. Amelioration of immunologic impairment in (a) the neonatally thymectomized animal; (b) the adult thymectomized, immunosuppressed animal; and (c) the nude, athymic mouse.

2. Other biological criteria: Enhancement of immunologic response evaluated in a variety of *in vitro* and *in vivo* assays reflecting the roles of the two major classes of lymphoid cells basic in host immunity, namely, the T and B cells. In addition, a third cell, the macrophage, has been utilized in assays designed to test production of macrophage-inhibitory factor (MIF).

The designation of a product as a thymic hormone logically connotes that the isolated material is obtained from the thymus and exhibits biological effects that have been established as characteristic of this gland. Alternatively, a preparation with thymic hormonelike activity might be present in nonthymic tissue, for example, the blood, as a *de novo* secretory product from the thymus and/or as a fraction derived from chemical alterations that have occurred in the gland of origin or subsequent to entrance into the circulation. In this last case, in the absence of evidence that the putative hormone can be identified as a constituent of the thymus, its biological actions should, as a minimum, resemble those of extracts of the gland itself.

Purified Molecules with Thymic Hormone-like Activity

A number of purified and apparently homogeneous peptides, polypeptides, and proteins with thymic hormone-like activity have been isolated. These are listed in Table 1. It is not possible in the space allotted to this chapter to describe the preparation, properties, and biological activities of each of these products. This has been done in a recent review (46), which can be consulted for additional details. However, it may be noted here that of these substances, 11 have been isolated from calf thymus, one from porcine blood, and one from human plasma. Also of interest is the fact that there is no apparent homology among those molecules whose sequences have been established. However, each exhibits some degree of thymic hormone-like activity in certain selected *in vitro* or *in vivo* assays.

Not listed in Table 1 is a nonpolar, homogeneous molecule isolated by Potop, Milcu, and their colleagues (35,36) from defatted, calf thymic tissue. Chemical studies of this product indicate that it is a steroid; it has been named *thymosterin.* The latter exhibits thymic hormone-like activity in a number of animal assay systems. The relation of thymosterin to the products described in Table I is not clear at present, but serves to emphasize further the diversity of substances that have been related to the functions of the thymus gland.

Immune-enhancing proteins with biological activity resembling that of thymus preparations have also been prepared in purified form from calf parotid glands (31,33). The significance of the occurrence in the parotid gland

TABLE 1
HOMOGENEOUS PEPTIDES, POLYPEPTIDES, AND PROTEIN PREPARATIONS WITH THYMIC HORMONE-LIKE ACTIVITY

Preparation	Molecular weight	Sequence reported	Reference[a]
Thymosin	12,500	No	18,23
Thymosin α_1	3100	Yes	20
Homeostatic thymus hormone (HTH)	~ 2000	No	9
Thymic humoral factor (THF)	3200	No	24,52
Thymoproteins I and II	5500	Yes	22
Ubiquitin	8450	Yes	22
Lymphocyte-stimulating hormone (LSH_h)	~15,000	No	27
Lymphocyte-stimulating hormone (LSH_r)	80,000	No	38
Hypocalcemic factor TP1	66,000	No	30
Hypocalcemic factor TP2	~57,000	No	32
Thymus extract TFX	4200	No	40
Porcine serum thymic factor (FTS)	860	Yes	34
Human blood plasma thymic factor	56,700	Yes	7

[a] Numbers are references presented in the list at the end of the chapter.

of these proteins with immune-enhancing activity is not clear at present. One recalls similar relationships among the thymus, the thyroid, and the parathyroid glands, which are of common embryological origin and from each of which the hormone calcitonin has been isolated (50). Also, immunologic interrelationships obtain between the thyroid and the thymus, as well as among the latter and other endocrine glands (10). The possible significance of these interrelationships for immunologic phenomena (49) and other metabolic parameters (10) has been recognized and studied in detail. However, their possible relevance for the production, secretion, and functions of products elaborated by the thymus remains to be explored.

PROTEIN FRACTION FROM HUMAN PLASMA WITH THYMIC HORMONE-LIKE ACTIVITY

Thymic hormone-like activity has been detected in the blood of several species, including the human (5,6,19). A nonapeptide with such activity has been isolated from porcine blood, its structure established, and synthesis accomplished (12,34). The biological activity of the synthetic nonapeptide was identical to that of the isolated product.

The demonstration of thymic hormone-like activity in human blood led to our current studies with the goal of isolating and characterizing the active component (7,47). We have utilized the convenient *in vitro* assay of Dardenne and Bach (11) to assess the thymic hormone-like activity of fractions obtained in the purification procedure.

Gel filtration of fresh human serum on Sephadex G-150 revealed that approximately 95% of the total azathioprine-sensitive spleen cell rosette assay activity was present in a fraction that coincided with the albumin and prealbumin peak. Only traces of activity (less than 5%) were recovered in the lower-molecular-weight fractions. The peak containing the major portion of the total serum activity corresponded to a fraction with a molecular weight of approximately 52,000 (Fig. 1). On electrophoretic analysis, this fraction appeared to be composed largely of albumin and small amounts of prealbumin.

The purification procedure developed for isolation of the thymic hormone-like activity from pooled human plasma is diagrammed in Fig. 2. Inasmuch as human serum albumin and prealbumin are present in fraction IV-1 in modifications of the procedure described initially by Cohn and his associates (8), this fraction was used as the starting material. Each step was carried out at 5°C unless otherwise specified. Generally, aliquots of 40 g of lyophilized Cohn fraction IV-1 of pooled human plasma were stirred for 3 hr at room temperature with 1 liter of 50 m*M* Tris buffer (pH 8.0) containing 0.02% sodium azide. The suspension was centrifuged at 13,000 *g* for 30 min. Solid $(NH_4)_2SO_4$ was then added to the supernatant to 40% saturation. The precipitate was allowed to settle and then centrifuged at 13,000 *g*. The $(NH_4)_2SO_4$ concentration of the supernatant was then brought to 60% saturation, the precipitate collected, dissolved in cold, distilled H_2O; dialyzed exhaustively against H_2O; and lyophilized (fraction 1). Two successive chromatography steps on a Sephadex G-150 column equilibrated with 50 m*M* Tris/100 m*M* NaCl/0.02% sodium azide (pH 8.0) provided a fraction in the 40,000–68,000 dalton area. This material was collected, desalted by diafiltration through an Amicon UM-10 membrane, and lyophilized. The active fraction was subjected to preparative gel electrophoresis at pH 8.9 in a Canalco apparatus with cooled gel columns. The active fraction was desalted by diafil-

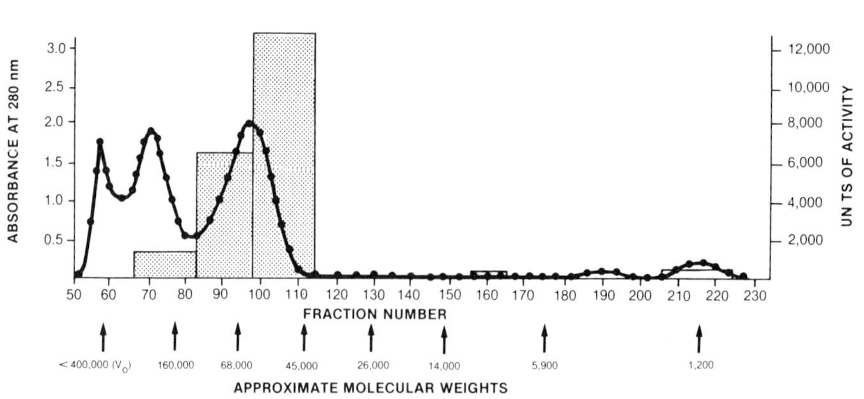

Fig. 1. Gel filtration of 11 ml of freshly drawn human serum obtained from a 30-year-old female donor. Chromatography was carried out on a 5 × 90 cm Sephadex G-150 column equilibrated with 50 m*M* Tris/100 m*M* NaCl (pH 8.0). Fractions containing 9 ml were collected. Shaded bars represent units of activity in the rosette assay. [Adapted from Burton *et al.* (7).]

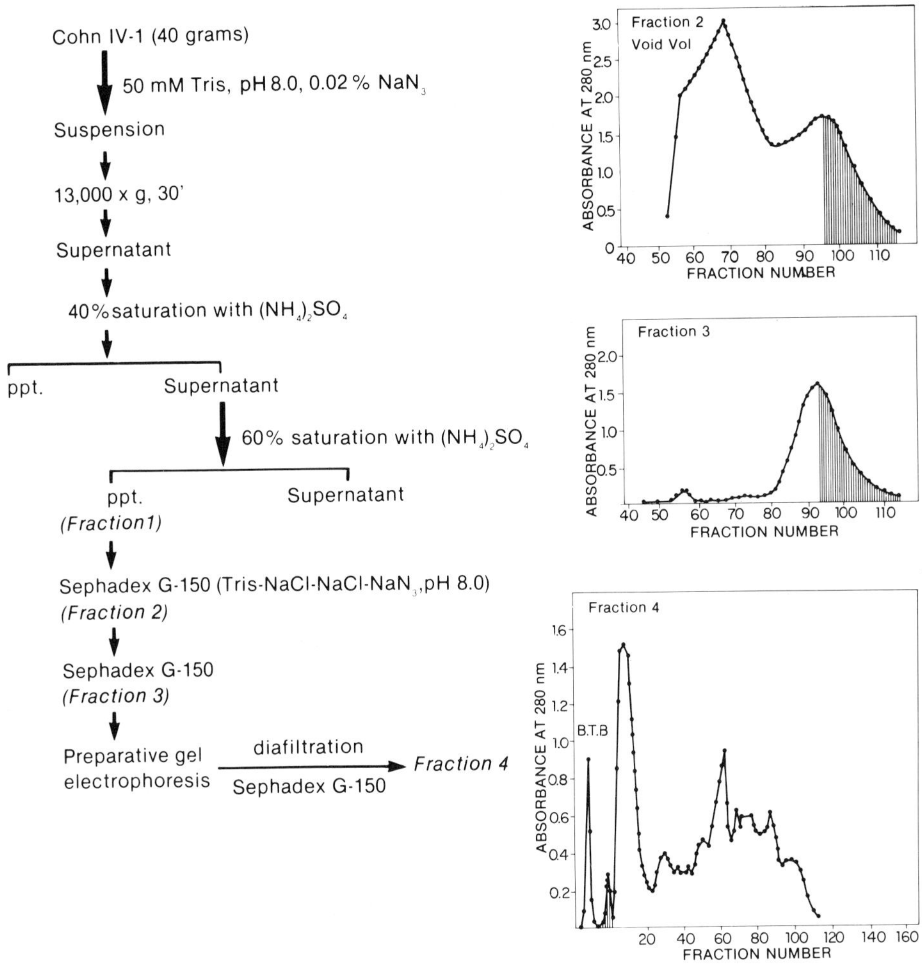

Fig. 2. Flow diagram for the purification of thymic hormone-like activity from Cohn fraction IV-1 obtained from pooled human plasma. On right, Sephadex G-150 chromatographic patterns of fractions 2 and 3 and analytical preparative disc gel electrophoretic pattern of Fraction 4. Shaded areas indicate fractions with highest activity in the rosette assay and are those taken for the next purification step. B.T.B., bromothymol blue dye. [Adapted from Burton *et al.* (7).]

tration, followed by gel filtration on a small Sephadex G-150 column equilibrated with 20 m*M* ammonium bicarbonate (pH 7.8). The peak containing the biological activity (fraction 4) was collected and lyophilized. Yields and relative activities of each fraction are summarized in Table 2. It may be noted that the high degree of enhancement of activity is probably due, in part, to the partial removal by thc 40% ammonium sulfate saturation step and the two subsequent Sephadex G-150 chromatography steps of a protein fraction that is inhibitory in the rosette assay.

Fraction 4, the final product, behaved as a homogeneous protein on analytical polyacrylamide gel electrophoresis at pH 8.9. The mobility of this

TABLE 2
YIELDS AND RELATIVE ACTIVITIES OF FRACTIONS OBTAINED FROM COHN FRACTION IV-1 OF POOLED HUMAN PLASMA[a]

Fraction	Weight (g)	Activity[b] (ng protein)	Enhancement of activity[a]
Cohn IV-1	40	12,000	1
1	17	1000	12
2	4	300	40
3	2	30	400
4	0.02	0.3	40,000

[a] Adapted from Burton *et al.* (7).

[b] Minimum weight of material necessary to reduce by 50% the number of sheep erythrocyte rosettes formed *in vitro* by spleen cells from adult thymectomized mice in the presence of a constant amount of azathioprine.

[c] Relative to that of initial Cohn IV-1 fraction.

fraction and its amino acid composition were identical with preparations of authentic prealbumin (Table 3). The physicochemical properties of fraction 4 and prealbumin are compared in Table 4. Immunodiffusion experiments against prealbumin antiserum with fractions 3 and 4 and with a commercially obtained prealbumin resulted in the formation of single confluent precipitant arcs.

Studies are being conducted to assess the degree to which the thymic hormone-like activity (see the following discussion) of the isolated product is dependent on the intact structure of the prealbumin molecule.

BIOLOGICAL ACTIVITIES OF FRACTION 3

Fraction 3 obtained in our laboratory from pooled human plasma has been utilized for extensive biological assays. This fraction is available in significant quantity, compared to prealbumin. Moreover, in unpublished studies we have established that all the azathioprine-sensitive E-rosette activity in fraction 3, which exhibits approximately eight components on polyacrylamide gel electrophoresis, can be accounted for by the 1% of prealbumin present in fraction 3. However, limited *in vivo* data with fraction 4 (pure prealbumin) have been conducted by injection of the protein into adult, thymectomized mice for three days and determining whether sensitivity of the spleen cells to the inhibitory action of azathioprine *in vitro* had been restored. The data obtained are presented in Table 5. It is evident that a dose–response relationship for thymic hormone-like activity in this assay model was obtained for both fractions 3 and 4. It may be noted that the injection of retinol, which is transported in the blood by the retinol-binding protein, which in turn is bound to prealbumin, was negative in this assay. Also, thyroxine, another

TABLE 3
AMINO ACID ANALYSES OF FRACTION 4 AND AUTHENTIC PREALBUMIN

Amino acid	Residues/mole: Fraction 4	Residues/mole: Authentic prealbumin[a]
Alanine	40.4	42.7
Arginine	13.2	13.9
Aspartic acid	27.6	28.6
Cysteine	3.7	3.4
Glutamic acid	43.0	43.3
Glycine	34.5	34.9
Histidine	12.9	13.6
Isoleucine	16.6	17.6
Leucine	24.4	25.7
Lysine	26.1	27.6
Methionine	2.9	2.9
Phenylalanine	16.1	17.2
Proline	27.5	28.1
Serine	38.6	41.1
Threonine	39.5	41.5
Tyrosine	9.6	12.0
Valine	39.6	42.9
Tryptophan	ND[b]	ND

[a] Obtained from Behring Diagnostics (Somerville, N.J.).
[b] ND = not determined.

molecule that binds to prealbumin, showed no activity when tested in the *in vitro* azathioprine-sensitive E-rosette assay.

A number of additional biological assays have been conducted with fraction 3. These have included utilizing the most critical of the models for establishing thymic hormone-like activity of a product, namely, studies of the ability of the fraction to reconstitute the impaired immunologic competence of neonatally thymectomized mice, notably, enhancement of survival following operation and ability to reject an allogeneic skin graft. Some of the biological

TABLE 4
SOME COMPARATIVE PHYSICAL AND CHEMICAL PROPERTIES OF FRACTION 4 AND PREALBUMIN

Property	Fraction 4	Prealbumin
Sedimentation constant ($s_{20,w}$)	4.07	4.10
Molecular weight	56,000 ± 300	54,980
Isoelectric point	4.5	4.7
Subunits (MW)	4(13,500)	4(13,500)
Amino terminal residue	glycine	glycine

TABLE 5
EFFECT OF INJECTION OF FRACTIONS 3 AND 4 ON THE AZATHIOPRINE SENSITIVITY OF SPONTANEOUS ROSETTE-FORMING SPLEEN CELLS

Experimental group	Daily dose[a] (μg protein)	Azathioprine sensitivity[b] (μg)
Normal mice, 0.9% NaCl	—	0.39
Tx[c] mice, 0.9% NaCl	—	25.0
Tx, fraction 3	10	1.17
Tx, fraction 3	50	0.78
Tx, fraction 3	100	0.20
Tx, fraction 3	250	0.20
Tx, fraction 4	1	1.17
Tx, fraction 4	2	0.39
Tx, fraction 4	4	0.20
Tx, retinol	2.5	25.0
Tx, retinol	5.0	25.0
Tx, retinol	10.0	25.0

[a] Each of two mice received one intraperitoneal injection daily for 3 days and was killed 24 hr after the last injection. The spleens were removed and the azathioprine sensitivity was determined for each spleen.

[b] Amount of azathioprine resulting in 50% inhibition of rosette formation. The values are means of two animals in each group.

[c] Tx, thymectomized.

activities that have resulted from these unpublished studies of fraction 3 are summarized in Table 6.

The increase in the percentage of peripheral blood lymphocytes from cancer patients forming E-rosettes with sheep erythrocytes after incubation with fraction 3, based on extensive similar studies with calf thymosin fraction 5 (16,21), is interpreted to indicate that fraction 3 enhances the maturation of immunologically incompetent blood lymphocytes to immunologically competent cells.

TABLE 6
SUMMARY OF *in Vivo* AND *in Vitro* BIOLOGICAL ACTIVITIES OF FRACTION 3

In Vivo Activity
Restores sensitivity of spleen cells from adult thymectomized C57B1/6J mice to azathioprine
Enhances survival of neonatally thymectomized C_3H/Bi mice
Accelerates allogeneic skin-graft rejection by neonatally thymectomized C_3H/Bi mice
Restores IgM antibody synthesis in neonatally thymectomized C57B1/6 mice
Increases ability of normal DBA/2J mice to restrain growth of a slow growing mammary adenocarcinoma
Increases survival of DBA/2J mice bearing mammary adenocarcinoma
In Vitro Activity
Enhances sensitivity of spleen cells from adult thymectomized C57B1/6J mice to azathioprine
Enhances E-rosette formation by lymphocytes from patients with Hodgkin's disease and several types of malignancies
Enhances mixed lymphocyte reaction between spleen cells of C57B1/6J mice and lethally irradiated P815 cells
Enhances cytotoxicity of spleen cells of C57B1/6 mice

Conclusion

The addition of the thymus to the list of endocrine glands is clearly appropriate, since the several postulates for designation of an organ as an endocrine gland (48) have been satisfactorily documented.

The diverse chemical nature of products reported to exhibit biological activities attributed to the thymus, particularly those products of unequivocal chemical homogeneity, lends support to the suggestion (44,45) that the thymus gland produces a family of several hormones, each of which may alter the extent of expression of one or more, possibly all, of the immunologic parameters reflecting the significant role of the thymus and its hormonal products in the regulation of host immunologic competence. This interpretation is in harmony with the clear evidence that several distinct single chemically pure or physical chemically homogeneous preparations are highly active in the wide diversity of assays in which they have been tested. The biological significance of these data remains to be explored.

Perhaps clarification of the possible relationships to one another of the diverse products with thymic hormone-like activity will be obtained by the application of specific radioimmune assays as well as additional data that may provide answers to the following questions:

1. Are the larger, biologically active preparations precursors or prohormones of the molecules of lower molecular weight? The demonstrations that homogeneous thymosin fraction 8 can be dissociated into smaller, dissimilar subunits and that thymosin α_1 is apparently synthesized in the thymus as a larger precursor (14) begin to provide the type of data needed to answer this question. Additional significant data should derive from current studies of products synthesized and secreted by thymic epithelial cells in culture, including precautionary measures to assure the absence of proteolytic activity.

2. Do the divergent properties reported for the different products reflect, in part, artifacts of the isolation procedures? This possibility has, of course, no significance for explaining the biological activities of synthetic products, e.g., the Bach nonapeptide and calf thymosin α_1. However, in other instances, aggregation phenomena and protein–protein interactions may influence the nature of the products isolated from thymic tissue or blood.

Finally, it may be suggested that critical application of *specific* radioimmune techniques to the several preparations that have been obtained in highly purified form may contribute useful information to possible interrelationships of certain of the described products. Moreover, if each of the described factors is indeed a separate hormone, each should appear in the circulation and be detectable ultimately by a *specific* biological or immunologic procedure.

Acknowledgment

The data from our laboratory have been obtained with the collaborative efforts of Dr. Pamela Burton, Ms. Susan Iden, Ms. Kathleen Mitchell, and Mr. Barry Morgan.

REFERENCES

1. Archer, O. K., and Pierce, J. C., Role of the thymus in development of the immune response. *Fed. Proc., Fed. Am. Soc. Exp. Biol.*, 1961, **20**:26.
2. Bach. J.-F., Thymic hormones: Biochemistry and biological and clinical activities. *Annu. Rev. Pharmacol. Toxicol.*, 1977, **17**:281–291.
3. Bach, J.-F., and Carnaud, C., Thymic factors. *Prog. Allergy*, 1976, **21**:418–484.
4. Bach, J.-F., Bach, M.-A., Blarot, D., Bricas, E. Charreire, J., Dardenne, M., Fournier, C., and Pleau, J.-M., Thymic serum factor. *Bull. Inst. Pasteur, Paris*, 1978, **76**:325–398.
5. Bach, J.-F., Dardenne, M., and Bach, M.-A., Demonstration of a circulating thymic hormone in mouse and in man. *Transplant. Proc.*, 1973, **5**:99–104.
6. Bach, J.-F., Dardenne, M., Pleau, J.-M., and Bach, M.-A., Isolation, biochemical characteristics and biological activity of a circulating thymic hormone in the mouse and in the human. *Ann. N.Y. Acad. Sci.*, 1975, **249**:186–210.
7. Burton, P., Iden, S., Mitchell, K., and White, A., Thymic hormone-like restoration by human prealbumin of azathioprine sensitivity of spleen cells from thymectomized mice. *Proc. Natl. Acad. Sci. U.S.A.*, 1978, **75**:823–827.
8. Cohn, E. J., Strong, L. E., Hughes, W. L., Jr., Mulford, D. J., Ashworth, J. N., Melin, M., and Taylor, H. L., Preparation and properties of serum plasma proteins. IV. A system for the separation into fractions of the protein and lipoprotein components of biological tissues and fluids, *J. Am. Chem. Soc.*, 1946, **68**:459–475.
9. Comsa, J., Thymus substitution and HTH, the homeostatic thymus hormone. In: *Thymic Hormones* (T. D. Luckey, ed.). University Park Press, Baltimore, Maryland, 1973:39–58.
10. Comsa, J., Leonhardt, H., and Schwarz, J. A., Influence of the thymus-corticotropin-growth hormone interaction on the rejection of skin allografts in the rat. *Ann. N.Y. Acad. Sci.*, 1975, **249**:387–401.
11. Dardenne, M., and Bach, J.-F., The sheep cell rosette assay for the evaluation of thymic hormones. In: *The Biological Activity of Thymic Hormones* (D. W. Van Bekkum, ed.). Kooyker Scientific Publications, Rotterdam, 1975:235–243.
12. Dardenne, M., Pleau, J.-M., Man, N. K., and Bach, J.-F., Structural study of circulating thymic factor: A peptide isolated from pig serum. I. Isolation and purification. *J. Biol. Chem.*, 1977, **252**:8040–8044.
13. Duckworth, W. L. H. (trans.), *Galen on Anatomical Procedures. The Later Books* (M. C. Lyons and B. Towers, eds.). Cambridge University Press, London and New York, 1962.
14. Freire, M., Crivellaro, O., Isaacs, C., Moshera, J., and Horecker, B. L., Translation of mRNA from calf thymus in the wheat germ system: Evidence for a precursor of thymosin α_1. *Proc. Natl. Acad. Sci. U.S.A.*, 1978, **75**:6007–6011.
15. Friedman, H. (ed.), Thymic factors in immunity. *Ann. N.Y. Acad. Sci.*, 1975, **249**:1–547.
16. Goldstein, A. L., The history of the development of thymosin: Chemistry, biology and clinical applications. *Trans. Am. Clin. Climatol. Assoc.*, 1976, **88**:79–94.
17. Goldstein, A. L., and White, A., The thymus gland: Experimental and clinical studies of its role in the development and expression of immune functions. *Adv. Metab. Disord.*, 1971, **5**:149–182.

18. Goldstein, A. L., Guha, A., Zatz, M., Hardy, M. A., and White, A., Purification and biological activity of thymosin, a hormone of the thymus gland. *Proc. Natl. Acad. Sci. U.S.A.*, 1972, **69**:1800–1803.
19. Goldstein, A. L., Hooper, J. A., Schulof, R. S., Cohen, G. H., Thurman, G. B., McDaniel, M. C., White, A., and Dardenne, M., Thymosin and the immunopathology of aging. *Fed. Proc., Fed. Am. Soc. Exp. Biol.*, 1974, **33**: 2053–2056.
20. Goldstein, A. L., Low, T. L. K., McAdoo, M., McClure, J., Thurman, G. B., Rossio, J. L., Lai, C-Y., Chang, D., Wang, S-S., Harvey, C., Ramel, A. H., and Meinhofer, J., Thymosin α_1: Isolation and sequence analysis of an immunologically active thymic polypeptide. *Proc. Natl. Acad. Sci. U.S.A.*, 1977, **74**:725–729.
21. Goldstein, A. L., Thurman, G. B., Low, T. L. K., Rossio, J. L., and Trivers, G. E., Hormonal influences on the reticuloendothelial system: Current status of the role of thymosin on the regulation and modulation of immunity. *Res. J. Reticuloendothel. Soc.*, 1978, **23**:253–266.
22. Goldstein, G., Polypeptides regulating lymphocyte differentiation. *Cold Spring Harbor Conf. Cell Proliferation*, 1978, **5**:455–465.
23. Hooper, J. A., McDaniel, M. C., Thurman, D. B., Cohen, G. H., Schulof, R. S., and Goldstein, A. L., Purification and properties of bovine thymosin. *Ann. N.Y. Acad. Sci.*, 1975, **249**:125–144.
24. Kook, A. I., Yakir, Y., and Trainin, N., Isolation and partial characterization of THF, a thymus hormone involved in immune maturation of lymphoid cells. *Cell. Immunol.*, 1975, **19**:151–157.
25. Low, T. L. K., and Goldstein, A. L., Structure and function of thymosin and other factors. In: *The Year in Hematology* (R. Silber, J. LoBuc, and A. S. Gordon, eds.). Plenum, New York, 1978:281–319.
26. Luckey, T. D. (ed.), *Thymic Hormones*. University Park Press, Baltimore, Maryland, 1973.
27. Luckey, T. D., and Venugopal, B., Isolation and quantification of LSH and the evaluation of related serum basic proteins in normal adults and cancer patients. *Ann. N.Y. Acad. Sci.*, 1975, **249**:166–176.
28. Miller, J. F. A. P., Immunological function of the thymus. *Lancet*, 1961, **2**:748–749.
29. Miller, J. F. A. P., and Osoba, D., Current concepts of the immunological function of the thymus. *Physiol. Rev.*, 1967, **47**:437–520.
30. Mizutani, A., Matsushita, Y., and Mizutani, T., Some properties of bovine thymic hypocalcemic factor TP1. *Chem. Pharm. Bull.*, 1977, **25**:3306–3311.
31. Mizutani, A., Mizutani, T., Suzuki, I., and Kuo, P-F., Effect of parotid gland hypocalcemic factor on increasing antibody-producing cells in mice. *Chem. Pharm. Bull.*, 1977, **25**:1601–1606.
32. Mizutani, A., Suzuki, I., Matsushita, Y., and Hayakawa, K., Effect of bovine hypocalcemic factor on increasing antibody-producing cells in mice. *Chem. Pharm. Bull.*, 1977, **25**:2156–2162.
33. Mizutani, A., Suzuki, I., and Yamamoto, H., Purification and characterization of bovine parotid hypocalcemic factor obtained via extraction with glacial acetic acid. *Chem. Pharm. Bull.*, 1978, **26**:1957–1965.

34. Pleau, J.-M., Dardenne, M., Blouquit, Y., and Bach, J.-F., Structural study of circulating thymic factor: a peptide isolated from pig serum. II. Amino acid sequence. *J. Biol. Chem.*, 1977, **252**:8045–8047.
35. Potop, I., Isolation of thymosterine and determination of its structure by spectral analysis. *Rev. Roum. Endocrinol.*, 1972, **9**:385–389.
36. Potop, I., and Milcu, S. M., Isolation, biologic activity and structure of thymic lipids and thymosterin. In: *Thymic Hormones* (T. D. Luckey, ed.). University Park Press, Baltimore, Maryland, 1973:205–273.
37. Restelli, A. D., De Thymo Observationes Anatomico-Physiologico-Pathologicae. *Ticini Regii ex Typog. Fusi et Soci*, 1845.
38. Robey, W. G., Further characterization of an antibody-stimulating factor isolated from bovine thymus. *Ann. N.Y. Acad. Sci.*, 1975, **249**:211–219.
39. Rolleston, H. D., *The Endocrine Glands in Health and Disease with an Historical Review.* Oxford University Press, London and New York, 1936.
40. Skotnicki, A. B., The biological activity and physico-chemical characteristics TFX of thymic extract. *Pol. Tyg. Lek.*, 1978, **33**:1119–1122.
41. Trainin, N., Thymic hormones and the immune response. *Physiol. Rev.*, 1974, **54**:272–315.
42. Trivers, G. E., and Goldstein, A. L., The endocrine thymus: A role for thymosin and other thymic factors in cancer immunity and therapy. *Cancer Biol. Rev.*, 1979 (in press).
43. Van Bekkum, D. W. (ed.), *The Biological Activity of Thymic Hormones.* Kooyker Scientific Publications, Rotterdam, 1975.
44. White, A., Nature and biological activities of thymus hormones: Prospects for the future, *Ann. N.Y. Acad. Sci.*, 1975, **249**:523–530.
45. White, A., Thymus hormones: A new field in endocrinology and its significance for host immunological competence. In: *The Biological Activity of Thymic Hormones* (D. Van Bekkum, ed.). Kooyker Scientific Publications, Rotterdam, 1975:17–23.
46. White, A., Chemistry and biological actions of products with thymic hormone-like activity. In: *Biochemical Actions of Hormones* (G. Litwack, ed.), Vol. 7. Academic Press, New York (in press).
47. White, A., and Burton, P., Isolation from human plasma of a protein fraction with thymic hormone-like activity. *Ann. N.Y. Acad. Sci.*, 1979, **332**:1–8.
48. White, A., and Goldstein, A. L., Is the thymus an endocrine gland? Old problem, new data. *Perspect. Biol. Med.*, 1968, **11**:475–489.
49. White, A., and Goldstein, A. L., Hormonal regulation of host immunity. In: *Immunogenicity: Physical, Chemical and Biological Aspects* (F. Borek, ed.). American Elsevier, New York, 1972:334–364.
50. White, A., and Goldstein, A. L., The endocrine role of the thymus and its hormone, thymosin, in the regulation of the growth and maturation of host immunological competence. *Adv. Metab. Disord.*, 1975, **8**:359–374.
51. White, A., and Levine, R., History of hormones. In: *Biological Regulation and Development* (R. Goldberger, ed.), Vol. 3. Plenum, New York (in press).
52. Yakir, Y., Kook, A. I., and Trainin, N., Enrichment of *in vitro* and *in vivo* immunological activity of purified fractions of calf thymic hormone. *J. Exp. Med.*, 1978, **148**:71–83.

Chromosome Biology and Chemistry

JAMES BONNER

Division of Biology
California Institute of Technology
Pasadena, California

INTRODUCTION

In April 1967, at the annual meeting of the National Academy of Sciences in Washington, I had an opportunity to speak with Emil Smith about protein sequencing. He had talked earlier in the meeting about the sequencing of interesting proteins, although I have forgotten which proteins they were. In any case, I said, "I have some proteins which would really be interesting for you to sequence." He inquired as to their nature and I said, "The histones We have pure histone species." He was dubious about this but agreed to examine our specimens and find out whether they were, in fact, pure, single-protein species.

My interest in the histones was awakened by the discovery in 1962 by my colleague Ru Chi C. Huang and myself (26) that acid-soluble proteins, combined with the DNA of the interphase chromatin of higher organisms cause that DNA to be not transcribable by RNA polymerase. This is not a trivial property of histones. There are many other basic proteins that, when combined with DNA, leave it perfectly transcribable, for example, cytochrome *c*. At that time, there was very little knowledge of the chemistry of the histones. Such information as was available is reviewed in "The Nucleohistones" (6). This appeared in 1964 as a result of the First World Conference on the Biology and Chemistry of the Histones, organized by Paul T'so and myself. At this conference opinions as to the function of the histones were as various as the opinions concerning the number of histone species. Some held that there were literally thousands of histone species. Some, myself included, felt that the evidence tended to indicate that there were only a few species, but the answer was not clear. The problems of these early days are summarized not

ISBN 0-12-643150-7

only in "The Nucleohistones" (6) but also in the book edited by Phillips some years later and entitled simply "Histones and Nucleohistones" (38).

Several fine investigators were involved in early histone purification studies of the chemistry of the histones. One was E. W. Johns of the Chester Beatty Research Institute. Others were D. M. Phillips and J. A. V. Butler of the same institute. In the United States important contributions were made by J. Murray Luck at Stanford and his postdoctoral fellow, Kenneth Murray. However, at Caltech we decided that what was really needed for the total understanding of histone biology and chemistry was the bringing up of a new generation of histone chemists. To this end Dr. Huang packed off a new graduate student, Douglas Fambrough, to Kenneth Murray at Stanford for a brief indoctrination about how to separate the histone classes (lysine rich, arginine rich, and intermediate) by what we now call Biorex P70 chromatography with a guanadinium chloride gradient as a developing agent. Douglas Fambrough learned rapidly, returned to Caltech, and during the ensuing few years showed (1) that acid urea polyacrylamide gel electrophoresis clearly separates histones (Fig. 1); (2) that in peas and cows there are five species of histone molecules and that these same five species occur in all tissues; (3) that each of these five species is very similar (as between peas and cows) as judged by polyacrylamide gel electrophoresis, although the five species themselves are very different from one another; (4) that the tryptic digest maps of the corresponding histones of peas and cows are very similar, if not almost identical (14–16). Fambrough also learned that the controversy about the number of histones present in higher organisms was generated, at least in part, by the fact that although histones traditionally are not supposed to contain cysteine,

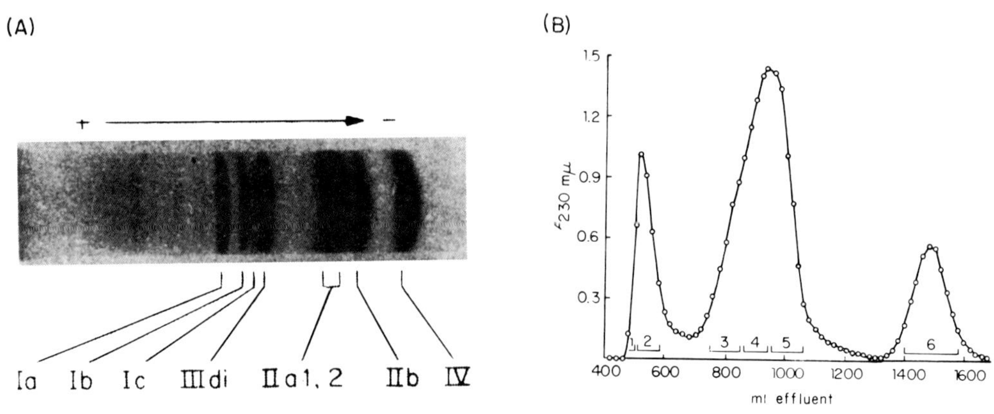

Fig. 1. (A) Polyacrylamide gel electrophoresis of pea histones in Fambrough acid, urea, polyacrylamide gel. The histones are from left to right: H1 (Ia,Ib,Ic), H3 (dimer) (IIIdi), H2a (IIa,1.2.), H2b (IIb), and H4 (IV). (B) Separation, of H4 (the smallest histone) from others by exclusion chromatography or Biogel P60. The developing agent is 0.01 *N* Hcl. H4 is the final included peak.

one histone (H), now known as H3, does contain cysteine. In fact, the H3 of higher mammals contains two cysteines per molecule, and this enables the histone preparation as it oxidizes itself in the refrigerator to produce dimers, trimers, and multimers of many kinds. Polyacrylamide gel electrophoresis of the histone mixture produces an infinitely complicated pattern. These complications are all removed simply by reduction of the histone mixture with β-mercaptoethanol.

It therefore became clear that there are five histone species, at least in peas and cows. (We now know there is another species in chicken erythrocytes and still others in varied sperm.)

Clearly, histone structure is conserved during evolution and it is therefore interesting to determine the exact degree to which histone primary structure is conserved. We had earlier worked simultaneously with peas and cows, and this comparison seemed a reasonably wide one. It was on the basis of this knowledge that I proposed to Emil Smith that we undertake a joint project on the sequencing of H4, the smallest and therefore the most readily obtainable histone in pure form.

Primary Structure of the Histones

Douglas Fambrough provided Emil Smith with electrophoretically purified H4 from each of our two creatures. These samples satisfied his stringent purity criteria and Bob DeLange set to work to determine the primary structures of each. It quickly turned out that each contained an 18 amino acid C-terminal fragment that is separated from the rest of the molecule by a methonine and thus removable by cyanogen bromide cleavage. The C-terminal peptide had an identical sequence in peas and cows. By November 1968 it was possible to announce at the Fall meeting of the National Academy of Sciences that the primary structure of H4 of peas and cows is essentially identical (13). Each is 102 amino acids long. They differ by two conservative substitutions; an arginine for lysine and a valine for an isoleucine, a basic for a basic and a hydrophobic for a hydrophobic.

I should note that in those days large quantities of proteins were required for amino acid sequencing. Fambrough and his colleageus prepared 2 g of H4 of peas and 2 g of H4 of cows. That of cows was prepared from calf thymus. Chromatin was prepared in relatively pure form. The histones were extracted with acid and separated from one another by exclusion chromotography on Biogel P60 using 0.0 *M* HC1 as the developing agent. For peas the situation was more difficult. Pea seedlings were germinated in barrels—25 lb of pea seeds per barrel. The resulting seedlings were put through a mill invented for this purpose and called a "pea seedling disassembling mill," which separated growing seedling axes (shoot and root) from the fleshy cotyledons. Chromatin was prepared from the shoots and roots and H4 was separated as described above for calf thymus. In that year, the summer of 1967, we used 25 tons of dried pea seeds to produce 2 g of H4 and were the largest consumers of the Alaska variety of pea seeds in the United States.

The revelations of November 1968 appeared, bit by bit, in the *Journal of Biological Chemistry* and were completely revealed by early 1969 (12). In the meantime, Emil Smith gave a talk in London about the conservation of primary structure of H4. This resulted in a classic comment by "our cell biology correspondent" in *Nature* to the effect that "this is the first evidence that histones constitute anything other than glue" (1). I may mention, in passing, that our colleagues in both the United States and Britain, and, in particular, Francis Crick, pointed out the uselessness of studying isolated chromatin and the futility of studying histones, because, as anyone could clearly see, histones are only glue. It is interesting how attitudes change.

Histone Sequencing—A Growth Industry

My colleague Jerry Johnson prepared pure pea plant H3. This was sequenced by DeLange and his group at UCLA (37). Histone3 was also prepared by Roger Chalkley from carp and this also was sequenced by the UCLA group (25). Cole and his colleagues at Berkeley (28) began their sequencing with H1, which was completed by Gordon Dixon and his group (private communication), with the H1 of trout. The other two major histones, H2a and H2b, were the last to be sequenced but are also now sequenced in two or more species (23,32,40,48). In general we can conclude that the H4 and H3 are the most conservative sequence over evolution of higher eukaryotes. H2a and H2b are very conservative but less so than H3 and H4. H1 is the least conserved and the most variable. There are, in fact, multiple forms of H1 in, for example, maize. In any case, histone sequencing became an industry that developed not only in the United States but also in England, France, Spain, Japan, South Africa, and probably other places.

Although the histones of higher eukaryotes are similar to one another, the histones of lower eukaryotes are less similar. Thus the histones of *Dictyostelium discoideum*, a slime mold of unknown taxonomic relation, are somewhat different from those of higher eukaryotes (3). The most different thus far studied are those of yeast. Yeast contains H2a, H2b, H3, and H4, but the primary structures of these histones appear to differ more widely from those of higher eukaryotes than histones of the higher eukaryotes do from one another (I. Isenberg, personal communication). If we were to build a genetic tree based on the relatedness of creatures as judged by histone structure, it would look as follows: The aboriginal prokaryote gave rise to the aboriginal eukaryote and from this stem the fungi branch off. At a later point in evolution the higher plants and higher animals branch off.

Histone Biology

A feature of histone chemistry noted in our first revelations of H4 primary structure is that the N-terminal portion of the molecule is highly basic, whereas the C-terminal portion of the molecule contains many hydrophobic amino acids. We pointed out at that time that the H4 molecule appears to be designed for interaction with other proteins at the C-terminal end. The same is true of H3, H2a, and H2b. H1, interestingly enough, was shown by Cole (28) to be just the reverse—to have a highly basic C-terminal end and a hydrophobic N-terminal end.

Early in the study of histone chemistry it became clear that chromatin possesses a histone stoichiometry as shown in Table 1. In chromatin there is one molecule of H1 for each two molecules of each of the other four species of histones. There is, therefore, a 1:2:2:2:2 ratio among the histones. Since chromatin possesses a histone/DNA mass ratio of 1, it follows that the mass of DNA equal to the simplest unit of histone mass (nine molecules) equals approximately 200 base pairs. In the late 1960s also, Jack Griffith, electron microscopist, made use of Douglas Brutlag's (8) discovery that H1 can be covalently bound to DNA by exposure of chromatin to 1% formaldehyde for 1 hr at 0°C. If the resulting chromatin is then banded in cesium chloride, the other four histones all are eluted from the DNA and the H1/DNA complex can be investigated. Griffith studied such H1/DNA complexes by high-resolution electron microscopy (22). It turned out that the H1 molecules are bound to DNA approximately 200 base pairs center to center apart, i.e., the H1 molecules are regularly spaced (see Table 2 and Fig. 2).

This DNA unit of chromatin structures was discovered by an entirely different method by Hewish and Burgoyne (24) and further studied by Kornberg and Noll (29). It appears that nucleases such as micrococcal nuclease attack chromatin, causing clips at regular 200 base pair spacings. This unit is a fundamental element of chromatin structure. Let us stress again that were histones to bear a 1:1 mass ratio to DNA, as they do in chromatin, then two

TABLE 1
Stoichiometry of the Histones in Chromatin in Pea Tissues[a]

		Ratios	
Histone	Moles/100 g	to H1	to nearest integer
H1	8×10^{-4}	1.0	1
H2a	14×10^{-4}	1.8	2
H2b	17×10^{-4}	2.1	2
H3	15×10^{-4}	1.9	2
H4	17×10^{-4}	2.1	2

[a] After Fambrough *et al.* (16).

TABLE 2
CALCULATIONS OF H1 SPACINGS ON CHROMATIN DNA FROM DATA OF FIG. 2

Average distance between H1 molecules	= 580Å/3.4Å = 170 base pairs
Average distance covered by an H1 molecule	= 112Å/3.4Å = 36 base pairs
Total of 1 inter H1 spacing + 1 H1 Spacing	= 170 + 36 = 206 base pairs

molecules each of H2a, H2b, H3, and H4 and one molecule of H1 would provide a mass equal to that of 200 base pairs.

D'Anna and Isenberg (10) showed that the four major species of histones interact strongly with one another, and apparently through their hydrophobic C-terminal peptides. The order and association constants of these interactions are shown in Fig. 3. As a result of a vast amount of work on the nature of the resultant complex (e.g., cross-linking), it is now generally agreed that two molecules each of H3 and H4 first react to form a tetramer and this is then

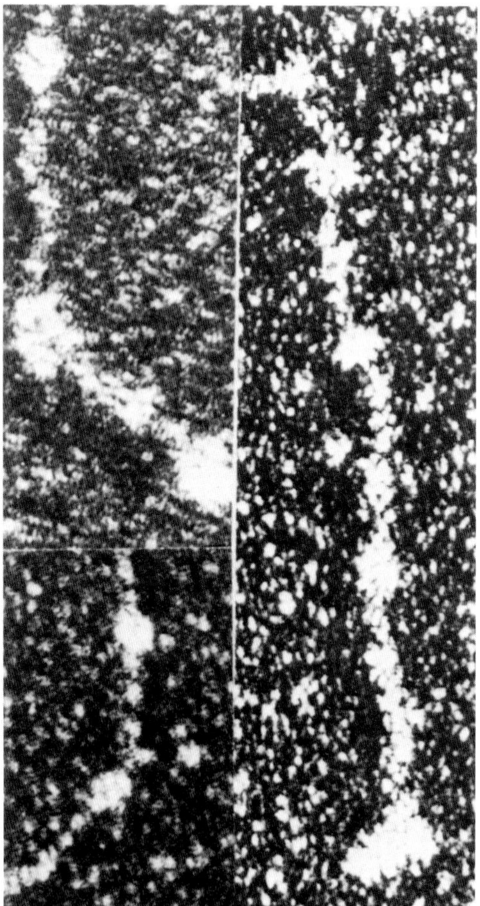

Fig. 2. Pea nucleohistone with all histones other than H1 removed. High-resolution electron microscopy by Griffith (21).

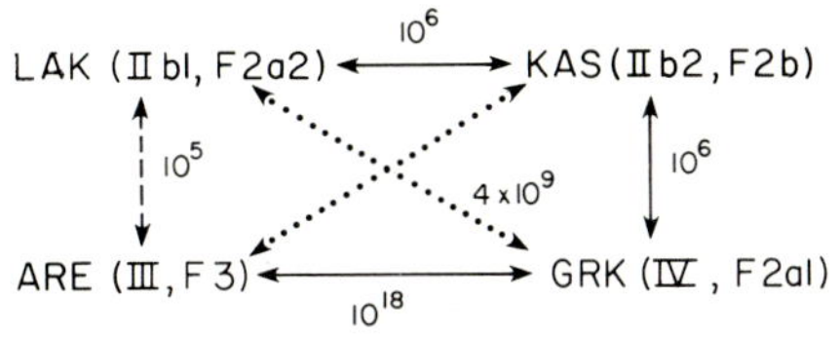

Fig. 3. Histone cross-complexing pattern together with relevant association constants after D'Anna and Isenberg (10).

joined by two H2a and H2b dimers to form the final octamer (Fig. 4). This in time becomes complexed with DNA.

It has been known for some time that interphase chromatin as viewed by high-resolution electron microscopy consists to a large degree of a knobby fiber about 100 A in width (21). That the knobby fiber actually consists of discrete but closely packed beads of diameter about 100 A (now known as nucleosomes) was shown by Olins and Olins (36). Griffith (22) followed this demonstration with useful studies of the SV40 minichromosome. The circular chromosome of SV40 in an appropriate host multiplies and, using the host histones, forms nucleosomes. The SV40 genome consists of approximately 4500 base pairs. It forms approximately 22 nucleosomes. At physiologic ionic strength these are tightly pressed against one another to form a knobby strand about 100 Å in diameter (Fig. 5). In low ionic strength the nucleosomes separate, making the spacer DNA clearly visible. Approximately 145 base pairs of DNA are complexed with the histone core of the nucleosome and approximately 55 base pairs are in the spacer. H1 presumably occupies the

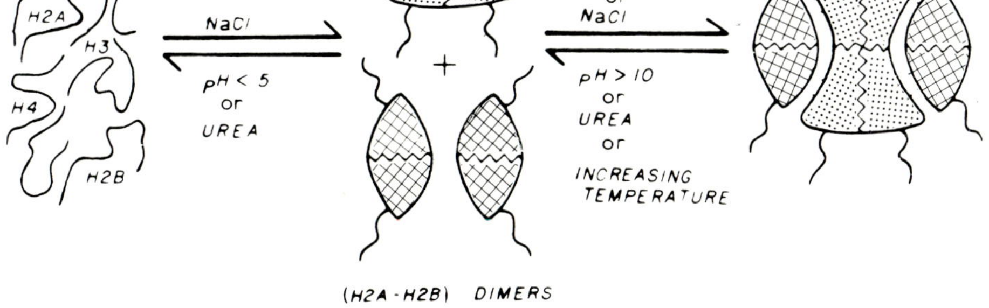

Fig. 4. Speculative possible mode of association of H3–H4 tetramers with 2 H2a–H2b dimers to form histone nucleosomal core.

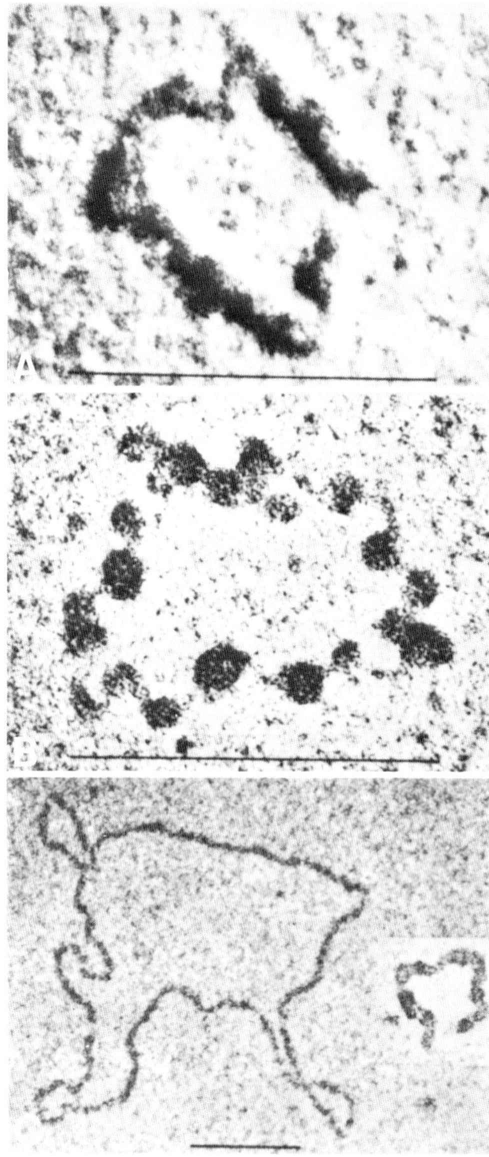

Fig. 5. The SV40 minichromosome (A) at physiologic ionic strength; (B) at very low ionic strength; (C) deproteinized SV40 DNA and SV40 minichromosome (at physiologic ionic strength) compared at identical magnification. The packing ratio of the DNA in the minichromosome is 7.

spacer. From the ratio of the contour length of the minichromosome (at physiologic ionic strength) to the contour length of the DNA contained in it, we can calculate that the packing ratio of the DNA in the "beads-on-a-string" (Fig. 6) structure of chromatin is 7. This requires the the DNA loop twice around the nucleosomal histone core. These findings have led to a tremendous outpouring of work in nucleosomal structure, as reviewed by Kornberg (29).

The structure of interphase chromatins is rather well understood on the level of beads on a string. More interesting today are the problems associated with how it is that the beads-on-a-string structure coils up into higher-order structures and finally into metaphase chromosomes with a packing ratio of 10,000 or more to 1. These questions are unresolved.

The Expressed Portion of the Genome

Since it is known that only a small portion of the total sequence complexity of genomal DNA is expressed in any given kind of specialized eukaryotic cell, it is obviously an interesting problem to see if one can physically separate the transcribed portion of chromatin from the nontranscribed portion of the chromatin (9). A great deal of effort has been put into this question and no convincing result was produced until the findings of Marushige (34). Marushige fractionated rat liver interphase chromatin by means of a nuclease, i.e., DNase II. Mild treatment with DNase II (which makes double-stranded clips in DNA) produces two fractions, a readily pelletable and a soluble one. The soluble chromatin was treated with 0.15 *M* NaCl (in which chromatin is known to be insoluble) and again centrifuged (Fig. 7). The

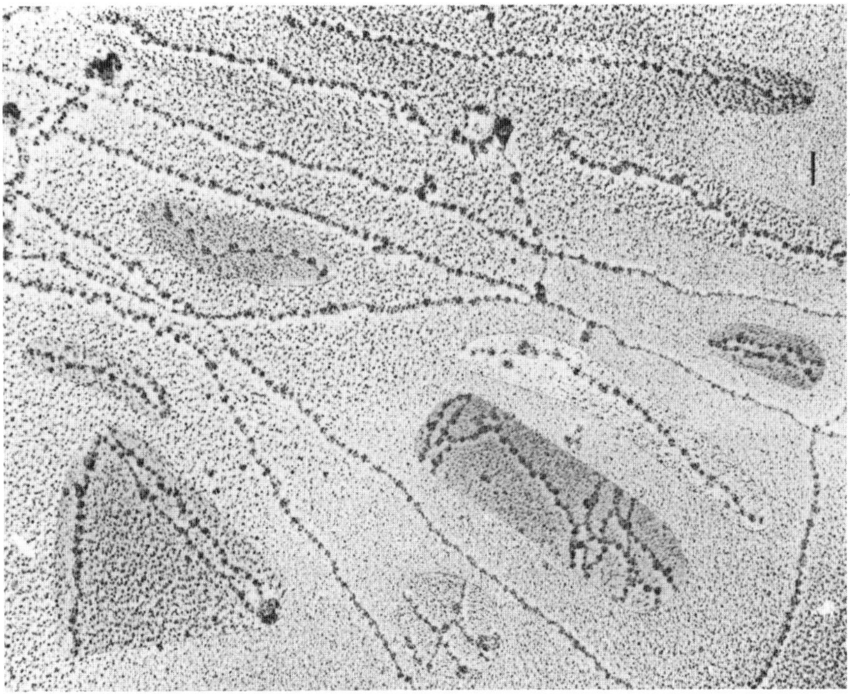

Fig. 6. "Beads-on-a-string" structure of bulk rat liver chromatin at low ionic strength. The diameter of the bead is about 110 Å. (Courtesy of J. R. Wu.)

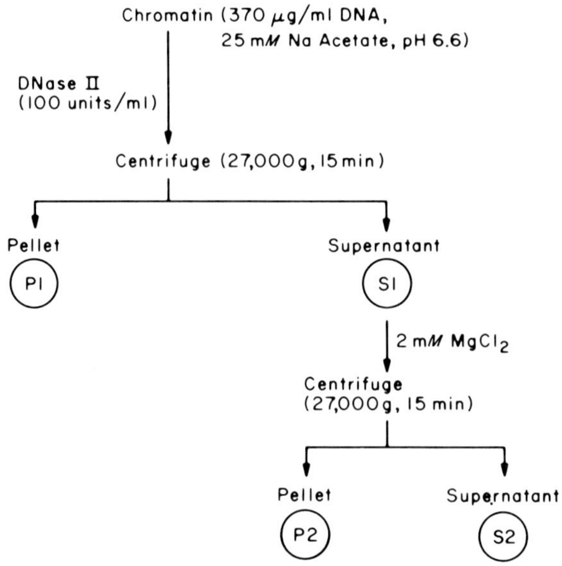

Fig. 7. Protocol for the separation of active (S2) chromatin from inactive (P1 and P2) chromatin by the 1971 DNase II method of Marushige and Bonner. [Adapted from Marushige and Bonner (34).]

portion of chromatin left in solution in 0.15 *M* NaCl (or 2 m*M* $MgCl_2$) is known as S2. The portion that precipitates is known as P2. The two fractions, P1 and S2, have very different properties. In the first place the soluble portion has a high template activity for RNA polymerase, approximately as high as that of deproteinized DNA. The P1 fraction has virtually zero activity in support of such transcription (Fig. 8). In the second place the S2 fraction melts at a low temperature, i.e., its DNA is not stabilized to any considerable degree by the presence of histones. The DNA of bulk chromatin is stabilized, as is the DNA of the P1 fraction. This is true even though the S2 fraction has approximately the same histone-to-DNA ration as does whole chromatin. In

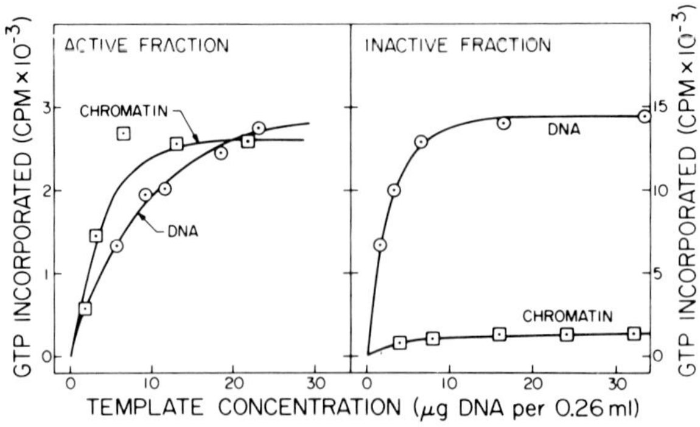

Fig. 8. Template activities in support of RNA transcription of "active" and "inactive" fractions of chromatin.

the third place the studies of Gottesfeld *et al.* (19) have revealed that the S2 fraction contains a subset of the single-copy sequences of whole genomal DNA and a subset of the families of repetitive sequences (Fig. 9). It is clearly a specific subset of genomal complexity and a subset that is specific to the chromatin of each different tissue of the same creature.

One interesting question is, "What is the length of the chromosomal segment that is transcribed by RNA polymerase as a single transcriptional unit?" When we attack chromatin with DNase II for a sufficient length of time to yield about 10% of the DNA as S2, the DNA is reduced in length to about 500 base pairs per fragment. If shorter and shorter times of digestion are used, the yield of S2 becomes less but the lengths of the fragments released become greater. We can extrapolate this to zero time. At zero time the fragments released by DNase II would be about 3000–4000 base pairs in length. Statistical considerations show that the length of a fragment released by two random hits in a susceptible segment of a macromolecule is, on the average, one-third of the length of the segment from which it is excised. Therefore the length of the stretches that are attacked by DNase II in chromatin are of the order of 9000–12,000 thousand base pairs in length. This size is interestingly similar to the length of the hnRNA molecules of the nucleus.

DNase I makes single-stranded nicks on chromatin (as it does on pure DNA) and also works selectively and releases nascent RNA from chromatin (4). However, DNase I rapidly digests the DNA fragments released to acid-soluble material. It has been shown that DNase I and DNase II attack the same sequences of chromatin. Thus chromatin can be attacked by DNase I, the readily digestible material can be digested, the residual chromatin can be pelleted and deproteinized to yield DNA, which has been freed of the readily DNase I digestible material. In a separate portion of the experiment we attack chromatin with DNase II and obtain the S2 fraction, which is the

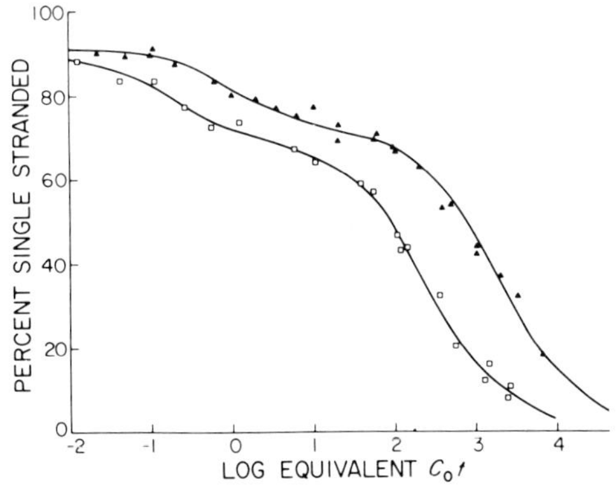

Fig. 9. Reassociation kinetics (C_0t curves) of (right) whole rat liver DNA and (left) DNA of the S2 fraction. Since the reassociation of S2 is approximately 10× more rapid than that of whole DNA, its complexity is 10× less. [After Gottesfeld *et al.* (19).]

readily DNase-II attackable portion of chromatin. We then hybridize the P1 DNA (that which is minus the DNase-I sensitive material of chromatin) as tracer driven by the S2 DNA as driver. No hybridization results; i.e., the sequences contained in S2 have been removed from chromatin by DNase I digestion.

I note in passing that these facts have not gone unnoticed by others in the chromatin community. Axel (17) and Weintraub (46), in particular, have DNase-I treated chromatin and hybridized the resulting DNA with cDNA probes for expressed genes. For example, reticulocyte chromatin, in which the globin gene is expressed, is digested with DNase I. It is then probed with cDNA to the globin mRNA, i.e., a globin probe. DNase I treatment of chromatin reduces or nearly abolishes the ability of the globin probe to hybridize to the chromosomal DNA. However, DNase I treatment of liver chromatin does not remove the globin gene. Active genes only are digested by DNase I.

In our laboratory Bruce Wallace (44) has shown that in induced Friend-cell chromatin (i.e., Friend cells induced by dimethyl sulfoxide to produce hemoglobin), fractionation of the chromatin of such cells by DNase II produces an S2 fraction in which the globin sequences are enriched several fold over whole chromatin. Similar findings have been reported by Gottesfeld (18). Nuclease sensitivity constitutes a sensitive and useful tool for probing the active genes in chromatin. DNase I can destroy active genes. DNase II can extirpate them and release them from the chromatin matrix in which they are embedded.

What Makes Chromatin Template Active?

What is it that causes a portion of chromatin to be template active and to have such other odd properties as low melting? From what we know today it would appear that template-active chromatin differs from nontemplate-active chromatin in two ways. The first way is by acetylation of the N-terminal peptides of the core histones. This has been indicated in a general sense by Allfrey *et al.* (2), who showed that when lymphocytes are caused to become active and start to produce RNA and to divide, they acetylate their histones. Marushige (33) carried this study farther; he acetylated chromatin with acetic anhydride and showed that, first of all, acetic anhydride, in the proper low concentration, acetylates the same ϵ-amino groups of lysines as are acetylated in nature by the acetyltransferase in chromatin. He showed further that this acetylation causes chromatin to become transcribable by RNA polymerase (Fig. 10). We have found (45), in confirmation of Marushige's results, that acetylation of chromatin with acetic anhydride mimics nature in respect to the number of ϵ-amino groups acetylated and in their positions. Acetylation of chromatin or nucleosomes by acetic anhydride causes all the changes that occur *in vivo* as well as all the physical changes associated with the conversion of the template-inactive to template-active chromatin that have been de-

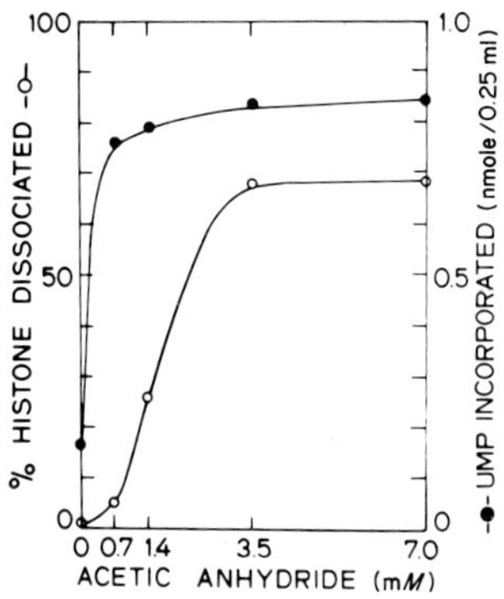

Fig. 10. Increased template activity in support of RNA synthesis of rat liver chromatin as a result of acetylation of the latter with acetic anhydride. [After Marushige (33).]

scribed. These include template transcribability, low melting, nonprecipitability by 0.15 *M* NaCl or m*M* $MgCl_2$, ready attack by nucleases (5,7,45). The circular dichroism spectrum of acetylated chromatin returns to that of the extended B form as contrasted to the nonextended form of DNA in chromatin in which the packing ratio of DNA in the octamers of histone is 7.

The packing ratio of DNA in S2, expressed, chromatin is known with confidence only for ribosomal DNA. In this case the DNA contour length equals that of the transcripts (45S rRNA precursor). No nucleosomes are visible by electron microscopy even though the histones are still present in such chromatin. In regard to nonribosomal active DNA, no direct measurement of the packing ratio is possible because the RNA as it is transcribed is packaged as ribonucleoprotein particles. However, McKnight *et al.* (35) estimate a packing ratio of almost 1 for nonribosomal active genes on the basis of electron microscopy. It should be noted that although nucleosomes are generally not visible in expressed chromatin, such chromatin nonetheless is digested by micrococcal nuclease into about 200 base pair fragments, just as is inactive chromatin (20).

Our findings concerning the acetylation of the histones and its direct relation to gene activation and alteration of the physical state of nucleosomal RNA have been widely confirmed. Thus Levy-Wilson *et al.* (30) fractionated, by means of DNaseII, the chromatin of cultured *Drosophila* cells cultured in labeled acetate. It was found that the histones of the template-active (S2) fraction incorporate acetate 40% more rapidly than do the histones of the template-inactive fraction. We have found the same with induced Friend cells. Most interesting of all are the steady-state compositions of the template-active and inactive fractions. Davie and Candido (11) have found, with trout testes chromatin, that essentially all the multiacetylated H4 is in the template-active fraction of chromatin as judged by polyacrylamide gel electrophoresis (separation of the acetylated from nonacetylated forms of other histones could not be achieved by Davie and Candido).

Riggs *et al.* (39) found that incubating cells in low concentrations of sodium butyrate appears to inhibit histone deacetylation, i.e., reduces acetyl turnover. This in turn results in relatively massive accumulation of multiacetylated histones, as shown by Sealy and Chalkley (41). Vidally *et al.* (42,43) have shown further not only that butyrate causes accumulation of multiacetylated histones in chromatin but also that chromatin whose histones are thus acetylated becomes susceptible to DNase digestion, i.e., becomes more like template-active chromatin with respect to susceptibility to DNase I. Their conclusion parallels ours, namely, acetylation of histone N-terminal peptides causes relaxation of histone binding to DNA.

As a digression, I note that our separation of chromatin into P1 and S2 fractions was carried out in acetate buffer. This is because acetate buffer is a good buffer to use at the pH (about 6) that we wished. It is also true, however, that acetate, like butyrate, is an inhibitor of deacetylase and thus our use of acetate buffer may be presumed to inhibit any deacetylation taking place during the course of incubation of chromatin with DNase II. A few other investigators have been unable to repeat our fractionation with DNase II, notably Gary Felsenfeld, who has used reticulocyte chromatin. Recently it has turned out (personal communication) that if he fractionates chromatin in the presence of butyrate, the frationation works perfectly well. In the absence of inhibitor the fractionation does not work. It appears that in the absence of an inhibitor of histone deacetylase, the fractionation fails to work because there the S2 fraction is deacetylated and becomes identical to the P1 fraction. In order to maintain the S2 fraction in its acetylated form it is necessary to have an inhibitor of deacetylase present in the nuclease reaction mixture, at least with some cell types.

Histone acetylase (acetyl transferase) is a chromosomal enzyme that we have purified in our laboratory. It appears to be totally sequence nonspecific with regard to the histones that it acetylates in chromatin. Clearly some further agent regulates the histones that are acetylated by histone acetylase.

High-Mobility Group and Related Proteins

Johns *et al.* (27) and Levy-Wilson and colleagues (31) have described a group of nonhistone chromosomal proteins extractable from chromatin by 0.35 *M* NaCl and characterized by low molecular weight, high lysine content (about 22 mole %), and high mobility in polyacrylamide gel electrophoresis. To this group Johns has given the name "high-mobility group proteins" (HMG proteins). These are present in chromatin in amounts much smaller than the histones.

Weintraub (47) reports that when erythrocyte chromatin is attacked by DNase I, which preferentially digests the globin gene, HMG proteins are released. These same proteins are released by 0.35 *M* NaCl extraction of erythrocyte chromatin. After such extraction the globin sequences become insensitive to DNase I attack. Reconstitution of the extracted proteins onto chromatin causes the globin gene of erythrocytes, but not of brain chromatin, to become DNase I sensitive. Thus the HMG proteins, according to Weintraub, are associated with an expressed gene but are not specific for this gene in all chromatins, as is shown by the brain-chromatin reconstitution experiment.

Dixon (31) has also found that HMG proteins are associated with the expressed sequences of the genome, although again, without any clear sequence specificity. Gottesfeld *et al.* (20) have shown that the nonhistone proteins associated with the S2, i.e., the template-active chromatin portion, do include members present in lesser amounts in whole chromatin. The relationship of these findings to template activity and to sequence-specific control of gene expression are not clear. Whether HMG proteins selectively bind to polyacetylated nucleosomes or whether the reverse is true we do not know. In any case, neither agency confers, so far as we know now, sequence specificity on the derepression of repressed DNA sequences. The principal problem in this field is to discover the agency that confers sequence specificity on the alteration of nucleosomes to make them transcribable by RNA polymerase.

A General Picture

It appears to be increasingly clear that the final act in the turning on of gene expression is histone acetylation. Thus acetylation of the N-terminal peptides of histones causes the histones to lose their grip on DNA and therefore allows the DNA to relax and extend itself into an extended conformation. In this conformation it is recognized as the B form of DNA. It is transcribable by RNA polymerase. The form in which the histones are present in template-active chromatin is not clear. No "beads-on-a-string" structure is evident in electron micrographs of actively transcribed ribosomal DNA chromatin. The structure of transcribed chromatin other than ribosomal is also reasonably similar to that of rDNA chromatin. The principal remaining question is, "What is it that directs acetylation to the histones of nucleosomes associated with particular DNA sequences?" and "How does sequence specificity get into this act?"

Acknowledgment

Report of work supported in part by U.S. Public Health Service Grant GM13762.

REFERENCES

1. Our cell biology correspondent. *Nature* (*London*), 1969, **223**:892.
2. Allfrey, V., Faulkner, R., and Mirsky, A., Acetylation and methylation of histones and their possible role in the regulation of RNA synthesis. *Proc. Natl. Acad. Sci. U.S.A.*, 1964, **51**:786–794.

3. Bakke, A., and Bonner, J., The histones of *Dictyostelium discoideum*. *Biochemistry*, 1979, **18**:4456–4562.
4 Billing, R., and Bonner, J., Structure of chromatin as revealed by deoxyribonuclease digestion studies. *Biochim. Biophys. Acta*, 1972, **281**:453–462.
5. Bonner, J., Some progress in our understanding of chromatin organization. In: *Molecular Human Cytogenetics* (R. S. Sparkes, D. E. Comings, and C. F. Fox, eds.). Academic Press, New York, 1977:53–64.
6. Bonner, J., and T'so, P. O. P. (eds.), *The Nucleohistones*. Holden-Day, San Francisco, California, 1964.
7. Bonner, J., Wallace, R. B., Sargent, T., Murphy, R., and Dube, S. K., The expressed portion of the eukaryotic chromatin. *Cold Spring Harbor Symp. Quant. Biol.*, 1977, **42**:851–859.
8. Brutlag, D., Schlehuber, C., and Bonner, J., Properties of formaldehyde-treated nucleohistone. *Biochemistry*, 1969, **8**:3214–3218.
9. *Cold Spring Harbor Symposia on Quantitative Biology*, 1977, **42.**
10. D'Anna, J., and Isenberg, I., A histone cross-complexing pattern. *Biochemistry*, 1974, **13**:4992–4997.
11. Davie, J., and Candido, E., Acetylated histone H4 is preferentially associated with template-active chromatin. *Proc. Natl. Acad. Sci. U.S.A.*, 1978, **75**:3574–3578.
12. DeLange, R. J., Fambrough, D. M., Smith, E. L., and Bonner, J., Calf and pea histone H4: Comparison with the homologous calf thymus histone. *J. Biol. Chem.*, 1969, **244**:5669–5679.
13. DeLange, R. J., Smith, E. L., Fambrough, D. M., and Bonner, J., Amino acid sequence of H4. *Proc. Natl. Acad. Sci. U.S.A.*, 1968, **61**:7–8 (Abstr.)
14. Fambrough, D. M., and Bonner, J., On the similarity of plant and animal histones. *Biochemistry*, 1966, **5**:2563–2570.
15. Fambrough, D. M., and Bonner, J., Sequence homology and role of cysteine in plant and animal arginine-rich histones. *J. Biol. Chem.*, 1968, **243**:4434–4439.
16. Fambrough, D. M., Fujimura, F., and Bonner, J., Quantitative distribution of histone components in the pea plant. *Biochemistry*, 1968, **7**:575–585.
17. Garel, A., and Axel, R., Selective digestion of transcriptionally active ovalbumin genes from oviduct chromatin. *Proc. Natl. Acad. Sci. U.S.A.*, 1976, **73**:3966–3970.
18. Gottesfeld, J., and Partington, G., Distribution of messenger RNA-coding sequences in fractionated chromatin. *Cell*, 1977, **12**:953–962.
19. Gottesfeld, J., Bagi, G., Berg, B., and Bonner, J., Sequence composition of the template-active fraction of rat liver chromatin. *Biochemistry*, 1976, **15**:2472–2483.
20. Gottesfeld, J., Murphy, R., and Bonner, J., Structure of transcriptionally active chromatin. *Proc. Natl. Acad. Sci. U.S.A.*, 1975, **72**:4404–4408.
21. Griffith, J., *High Resolution Electron Microscope Studies of Chromosomal Fibers*. PhD. Thesis, California Institute of Technology, Pasadena, 1969.,
22. Griffith, J., Chromatin structure: Deduced from a minichromosome. *Science*, 1975, **187**:1202–1203.
23. Hayashi, H., Iwai, K., Johnson, J., and Bonner, J., Pea histones H2a and H2b. *J. Biochem.*, 1977, **82**:503–510.

24. Hewish, D., and Burgoyne, L., Chromatin substructure. The digestion of chromatin in DNA at regularly spaced sites by a nuclear deoxyribonuclease. *Biochem. Biophys. Res. Commun.*, 1973, **52**:504–510.
25. Hooper, J. A., Smith, E. L., Somer, K. R., and Chalkley, R., Histone H3—amino acid sequence of H3 of the testes of the carp. *J. Biol. Chem.*, 1973, **248**:3275–3279.
26. Huang, R. C. C., and Bonner, J., Histone: A suppressor of chromosomal RNA synthesis. *Proc. Natl. Acad. Sci. U.S.A.*, 1962, **48**:1216–1222.
27. Johns, E., Goodwin, G., Walker, J., and Sanders, C., Chromosomal proteins related to histones. *Ciba Found. Symp., 1975,* **28** (new ser.):95–112.
28. Jones, G. M., Rall, S. C., and Cole, R. D., Extension of the amino acid sequence of a lysine-rich histone. *J. Biol. Chem.*, 1974, **249**:2548–2553.
29. Kornberg, R., Structure of chromatin. *Ann. Rev. Biochem.*, 1977, **46**:931–954.
30. Levy-Wilson, B., Gjerset, R., and McCarthy, B., Acetylation and phosphorylation of *Drosophila* histones. *Biochim. Biophys. Acta*, 1977, **475**:168–175.
31. Levy-Wilson, B., Wong, N. C., Watson, D., Peters, E., and Dixon, G., Structure and function of the low salt extractable chromosomal proteins: Preferential association of trout testes proteins H6 and HMGT with chromatin regions selectively sensitive to nuclease. *Cold Spring Harbor Symp. Quant. Biol.*, 1977, **42**:793–801.
32. Martinage, A., Mangeat, P., Sautierre, P., Marchis-Mouren, G., and Biserte, G., Amino acid sequence of rat thymus histone H2b. *Biochimie*, 1979, **61**:61–69.
33. Marushige, K., Activation of chromatin by acetylation of histone side chains. *Proc. Natl. Acad. Sci. U.S.A.*, 1976, **73**:3937–3941.
34. Marushige, K., and Bonner, J., Fractionation of liver chromatin. *Proc. Natl. Acad. Sci. U.S.A.*, 1971, **68**:2941–2944.
35. McKnight, S., Bustin, M., and Miller, C., Electron microscope analysis of chromosome metabolism in the *Drosophila melanogaster* embryo. *Cold Spring Harbor Symp. Quant. Biol.*, 1977, **42**:741–754.
36. Olins, A., and Olins, D., Spheroid chromatin units (ν bodies). *Science*, 1974, **183**:330–332.
37. Patthy, L., Smith, E., and Johnson, J., H3: The amino acid sequence of pea embryo H3. *J. Biol. Chem.*, 1973, **248**:6834–6840.
38. Phillips, D. M. (ed.), *Histones and Nucleohistones.* Plenum, New York, 1971.
39. Riggs, M., Whittaker, R., Newmann, J., and Ingram, W., n-Butyrate causes histone modification in Hela and Friend cell erythroleukemia cells. *Nature (London)*, 1977, **268**:462–464.
40. Sautiere, P., Structure of histone H2a. *Ciba Found. Symp.*, 1975, **28** (new ser.):77–88.
41. Sealy, L., and Chalkley, R., The effect of sodium butyrate on histone modification. *Cell*, 1978, **14**:115–121.
42. Vidali, G., Baffa, L., Bradbury, E. M., and Allfrey, V. G., Butyrate suppression of histone deacetylation leads to accumulation of multiacetylated histones H3 and H4 and increased DNase I sensitivity of the associated DNA sequences. *Proc. Natl. Acad. Sci. U.S.A.*, 1978, **75**:2239–2245.
43. Vidali, G., Baffa, L., Mann, R., and Allfrey, V. G., Reversible effects of sodium butyrate on histone acetylation. *Biochem. Biophys. Res. Commun.*, 1978, **82**:223–227.

44. Wallace, R. B., Dube, S. K., and Bonner, J., Localization of the globin gene in the template-active fraction of chromatin in Friend Leukemia cells. *Science*, 1977, **198**:1166–1168.
45. Wallace, R. B., Sargent, T., Murphy, R., and Bonner, J., Physical properties of chemically acetylated rat liver chromatin. *Proc. Natl. Acad. Sci. U.S.A.*, 1977, **74**:3244–3248.
46. Weintraub, H., and Groudine, M., Chromosomal subunits in active genes have an altered conformation. *Science*, 1976, **193**:248.
47. Weisbrod, S., and Weintraub, H., Isolation of a subclass of nuclear proteins responsible for conferring DNase I-sensitive structure on globin chromatin. *Proc. Natl. Acad. Sci. U.S.A.*, 1979, **76**:630–634.
48. Yeoman, L. C., Olson, M., Sugano, N., Jordon, J., Taylor, C., Starbuck, W., and Busch, H., Amino acid sequence of the lysine-rich histone from calf thymus: The total sequence. *J. Biol. Chem.*, 1972, **247**:6018.

The Evolution of Histones in Relationship to Recent Advances in Elucidating Chromatin Structure

ROBERT J. DeLANGE
Department of Biological Chemistry
UCLA School of Medicine and Molecular Biology Institute
University of California Los Angeles
Los Angeles, California

A very productive method for studying molecular evolution results from the determination and comparison of amino acid sequences of proteins from various species. Professor Emil L. Smith and his colleagues have made numerous contributions to our understanding of evolutionary processes by using this approach (e.g., 29,30). I have had the distinct privilege of being closely associated with Professor Smith for nearly 14 years (the first two years as a postdoctoral associate), during which time we shared facilities, equipment, and grants and collaborated on several projects, including the structure of subtilisins (4,26,32), the sites of methylation in cytochromes *c* (11,12), and the structure of histones (see later). The last two projects were conducted in collaboration with Professor A. N. Glazer (cytochrome *c*) and Professor James Bonner (histones), both of whom were also participants in this symposium.

Of the proteins that have been studied from various species by sequencing methods, histones 3 (III, F3) and 4 (IV, F2A1) appear to be the most conserved proteins known (for reviews see refs. 5–8,16). This conservation of sequence had previously been observed by the use of peptide mapping (17) and electrophoresis (17,31) and has since been demonstrated for other species by electrophoretic techniques (27) and additional sequence work (see Table 1).

Calf thymus histone 4 was the first histone (H), sequenced completely, as reported by this laboratory 11 years ago (9,14; see also ref. 25). Shortly afterward, we reported the complete sequence of H4 from pea seedling (10), which differed in sequence from the calf thymus histone at only two positions of its 102 residues, as noted in Table 1. Both changes are very conservative

ISBN 0-12-643150-7

TABLE 1
CHANGES IN THE SEQUENCES OF H3 AND H4 FROM VARIOUS ORGANISMS[a]

	Residues in H3				Residues in H4		
	41	53	90	96	60	73	77
Calf, etc.	Tyr	Arg	Met	Cys	Val	Thr	Lys
Other organisms	Phe[b]	Lys[c]	Ser[c]	Ala[c], Ser[d]	Ile[c]	Cys[e]	Arg[c]

[a] Only those residues that have been demonstrated to differ in various organisms are shown. All other residues are identical in all organisms examined, except for H4 in *Tetrahymena*, which has a number of other changes (20). For reviews on histone sequence work, see DeLange and Smith (5–8), Elgin and Weintraub (16), and van Holt *et al.* (35).

[b] In pea and cycad.

[c] In pea.

[d] In pea (some molecules), carp, chicken, cycad, limpet, mouse, sea urchin, and shark.

[e] In sea urchin and starfish.

replacements. In comparison, cytochrome *c* from a plant (wheat) and an animal (bovine) differ (3) at 40% of the residues (including the residues in the extension at the NH_2-terminus of the wheat cytochrome). Until recently only one other change in the sequence of H4 has been found in other organisms, namely, cysteine at residue 73 (threonine in other organisms) in echinoderms (34). However, recent partial sequence studies on H4 from *Tetrahymena* have shown many more changes in sequence (20). Whether this is a specialized exception or is indicative of a greater generalized potential for variability in sequence than was previously thought is not clear at present (see later).

We interpreted the extensive sequence conservation of calf and pea H4's to indicate that the functions of H4 are so completely dependent on its entire structure that little change in the structure of all H4 molecules in chromatin can be tolerated without producing lethal effects (10). The same interpretation appeared to be valid for H3, which we also showed to be highly conserved (13,28; Table 1). It has been of interest to observe the development of our present concepts of chromatin structure in which H3 and H4 apparently play such key roles (see later).

The presently accepted model of chromatin structure (see ref. 18 and 23 for reviews) is based on octomer histone complexes, consisting of two molecules each of H2a, H2b, H3, and H4, around which 140 base pairs of DNA are supercoiled to form nucleosome particles. Between nucleosome particles exist regions of spacer DNA (often about 60 base pairs, but varying in length, depending on the source of chromatin) to which H1 is apparently complexed. We had predicted from our sequence studies that the very basic NH_2-terminal regions of H3 and H4 with net charges of +18 (residues 1–53) and +16 (residues 1–45), respectively, would be primary DNA-binding sites (9,10,13,28). We had also predicted that the COOH-terminal regions of H3 and H4 would be involved in forming complexes with DNA (particularly at sites of basic clusters), histones, and other chromatin proteins. These predictions are being substantiated by other studies, and the currently accepted model of chromatin structure incorporates these structural features (18,23).

Since mild trypsin digestion of chromatin removes 20–30 residues from the NH_2-terminal regions of H3 and H4 (and H2b and H2a?) and leaves the remainders of each histone intact in a particulate complex (36), the models depict a basic "tail" (of 20–30 residues) for each of the eight histone molecules that participate in forming a single nucleosome. These "tails" are not part of the particulate "core" and are assumed to bind to DNA. However, since the proteolytic removal of the "tails" does not destroy the histone–DNA complex, it is obvious that there are other DNA-binding sites in histones. We feel that the basic residues, many of which are in clusters in the remainder of the basic NH_2-terminal regions (up to residue 53 in H3, residue 45 in H4), will also be demonstrated to be primary DNA-binding sites, but certainly basic residues in the remainders of the molecules may also play significant roles. Other residues COOH-terminal to residues 20–30 in each histone would be expected to participate in histone–histone interactions to form the core structure around which the DNA coils.

Many studies have identified specific histone–histone interactions in nucleosomes (e.g., see ref. 2). Strong interactions are known to occur between H3 and H4, H2a and H2b, H2b and H4. Some histone–histone complexes have been isolated directly from chromatin, but cross-linking reagents have been used to produce stable complexes. In collaboration with H. G. Martinson (University of California, Los Angeles, Department of Chemistry) we are investigating specific sites of histone–histone interaction in chromatin by the use of certain zero-length cross-linking techniques. The advantage of these techniques is that spacer molecules are not introduced in forming the cross-links and therefore more direct identification of interacting sites is possible.

Ultraviolet irradiation of chromatin or nuclei produces almost quantitative conversion of H2a and H2b to a cross-linked dimer (24). We have investigated the sites of cross-linking and find that proline residue 26 in H2a is cross-linked to tyrosine residue 37, 40, or 42 in H2b (15). This cross-link (see Fig. 1) is apparently formed by the attack of the tyrosine residue (after activation by ultraviolet irradiation) on the juxtaposed proline residue in nucleosomes. This indicates that near the basic DNA-binding "tails" of these two histones there is a site of strong interaction that is important to the formation and maintenance of nucleosomal structure. Although the sequences of these two histones (particularly H2b) are not as highly conserved as the sequences of H3 and H4, it is of interest that these tyrosine residues in H2b are conserved (35). We are currently determining which of the three tyrosines is involved in the cross-link. Together with other zero-length cross-linking studies in progress, this type of investigation is beginning to give the first real insight into chromatin architecture at the submolecular level.

On the basis of the comparative sequence studies (9,10,13,28) we had

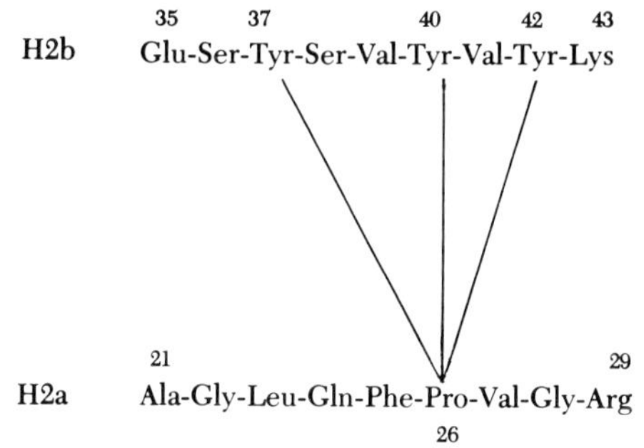

Fig. 1. Sequences of H2a and H2b in the vicinity of the cross-link formed between the two histones by ultraviolet irradiation of chromatin [after DeLange *et al.* (15)]. It is not yet known which of the three tyrosines in H2b is cross-linked to proline residue 26 in H2a.

suggested that the greater sequence conservation of H3 and H4 indicated a more fundamental role for these two histones in chromatin structure and function. Almost any change in any part of the structure (other than the very conservative ones mentioned earlier) would seem to affect a vital site of interaction or reaction (e.g., posttranslational modifications) in an unacceptable manner. It has recently been found that the tetrameric complex of H3 and H4, can induce DNA supercoiling in the absence of H2a and H2b (see ref. 22, and cited references). Thus the fundamental roles of H3 and H4 are beginning to be elucidated. A recent report (20) on the partial sequence of *Tetrahymena* H4 indicates that the structures of H3 and H4 may undergo greater change than previously thought, but the changes in *Tetrahymena* H4 definitely have a profound effect on the ability of this histone to form normal nucleosomal complexes. Perhaps there have been compensatory changes in other histones or chromatin components to allow this modified H4 to function in *Tetrahymena* without serious problems.

Although the sequences of H3 and H4 are highly conserved, the molecules can be extensively modified *in vivo* (5,7,8,16) by posttranslational modification reactions such as acetylation, methylation, phosphorylation, etc. More than 200 subfractions of H4 are possible from the known modifications of this histone (7). This presents a paradox of structure conservation (sequence), on the one hand, with structure variation (posttranslational modification), on the other. The resolution of the paradox seems to be that changes in sequence produce unacceptable changes in chromatin structure throughout the genome, whereas posttranslational modifications are used to modify only certain regions of chromatin where changes in function are required. For example, acetylation of ϵ-NH_2 groups of lysine, discovered simultaneously in our laboratory (9,14) and by Gershey and his colleagues (19), appears to be an important part of the process by which genes are activated for transcription (e.g., see ref. 1), and phosphorylation of histones appears to be important in the processes involved in replication and cell division (e.g., see ref. 21).

We have recently determined the sequence of a histonelike protein from

10

Met-Val-Gly-Ile-Ser-Glu-Leu-Ser-Lys-Glu-Val-Ala-Lys-Lys-Ala

20 30

Asn-Thr-Thr-Gln-Lys-Val-Ala-Arg-Thr-Val-Ile-Lys-Ser-Phe-Leu

40

Asp-Glu-Ile-Val-Ser-Glu-Ala-Asn-Gly-Gly-Gln-Lys-Ile-Asn-Leu

50 60

Ala-Gly-Phe-Gly-Ile-Phe-Glu-Arg-Arg-Thr-Gln-Gly-Pro-Arg-Lys

70

Ala-Arg-Asn-Pro-Gln-Thr-Lys-Lys-Val-Ile-Glu-Val-Pro-Ser-Lys

80 90

Lys-Lys-Phe-Val-Phe-Arg-Ala-Ser-Ser-Lys-Ile-Lys-Tyr-Gln-Gln

Fig. 2. The complete amino acid sequence of a histonelike DNA-binding protein from *Thermoplasma acidophilum*. The protein has also been isolated in a form lacking the NH_2-terminal methionine residue (12a, 15a).

the prokaryote *Thermoplasma acidophilum* (12a, 15a). This basic protein is isolated in two forms containing 89 and 90 residues, respectively. The 90-residue form contains an NH_2-terminal methionine, which is absent from the 89-residue form (see Fig. 2). This protein is complexed to the DNA of this organism and may function to protect the DNA from thermal denaturation at the elevated temperatures (about 59°C) that are optimal for growth (33). This protein (see Fig. 2) resembles eukaryotic histones in several ways: (1) small size (90 residues), (2) basicity (21% basic residues), (3) clustering of basic residues, and (4) complex formation to form nucleosome-like particles with DNA (28a). There are some suggestions of homology with the histones, but this is not striking. It appears that the complex of this protein with DNA can serve in many ways as a simple model of eukaryotic chromatin.

Considerable progress has been made in the last 11 years in understanding the structure and function of chromatin. However, what has been achieved is only the beginning in this exciting area of biology. So many important questions relating to the regulation of chromatin functions remain to be answered, but we are sure that continued investigation of chromatin structure will provide the bases on which many of the answers will be formulated.

Acknowledgments

I especially want to express public appreciation to Professor Emil L. Smith for his guidance, collaboration, scientific inspiration, and friendship over the past 14 years. It has been a genuine pleasure to be associated with one who has such great knowledge in so many areas and has made so many significant contributions to science.

I would also like to express appreciation to those who have been associated with me in this research, particularly Ms. Dorothy McNall, Ms. Lynn Williams, and Mr. Douglas Brown for their excellent technical assistance.
This work was supported in part by U.S. Public Health Service Grant GM 11061.

REFERENCES

1. Allfrey, V. G., Functional and metabolic aspects of DNA-associated proteins. In: *Histones and Nucleohistones* (D. M. P. Phillips, ed.). Plenum, New York, 1971:241–294.
2. D'Anna, J. A., Jr., and Isenberg, I., A histone cross-complexing pattern. *Biochemistry*, 1974, **13**:4992–4997.
3. Dayhoff, M. O. (ed.), *Atlas of Protein Sequence and Structure*. Vol. 5., National Biomedical Research Foundation, Washington, D.C., 1972:D-14, D-23.
4. DeLange, R. J., and Smith, E. L., Subtilisin Carlsberg. II. Sequences of the tryptic peptides. *J. Biol. Chem.*, 1968, **243**:2143–2164.
5. DeLange, R. J., and Smith, E. L., Histones: Structure and function. *Annu. Rev. Biochem.*, 1971, **40**:279–314.
6. DeLange, R. J., and Smith, E. L., Structures of histones. *Acc. Chem. Res.*, 1972, **5**:368–373.
7. DeLange, R. J., and Smith, E. L., Histone function and evolution as viewed by sequence studies. *Ciba Found. Symp.*, 1975, **28** (new ser.):59–76.
8. DeLange, R. J., and Smith, E. L., Chromosomal proteins. In: *The Proteins* (H. Neurath and R. L. Hill, eds.), 3rd ed., Vol. 4. Academic Press, New York, 1979: 119–243.
9. DeLange, R. J., Fambrough, D. M., Smith, E. L., and Bonner, J., Calf and pea histone IV. II. The complete amino acid sequence of calf thymus histone IV; presence of ε-*N*-acetyllysine. *J. Biol. Chem.*, 1969, **244**:319–334.
10. DeLange, R. J., Fambrough, D. M., Smith, E. L., and Bonner, J., Calf and pea histone IV. III. Complete amino acid sequence of pea seedling histone IV. Comparison with the homologous calf thymus histone. *J. Biol. Chem.*, 1969, **244**:5669–5679.
11. DeLange, R. J., Glazer, A. N., and Smith, E. L., Presence and location of an unusual amino acid, ε-*N*-trimethyllysine in cytochrome *c* of wheat germ and *Neurospora*. *J. Biol. Chem.*, 1969, **244**:1385–1388.
12. DeLange, R. J., Glazer, A. N., and Smith, E. L., Identification and location of ε-*N*-trimethyllysine in yeast cytochromes *c*. *J. Biol. Chem.*, 1970, **245**:3325–3327.
12a. DeLange, R. J., Green, G. R., and Searcy, D. G., A histone-like protein (HTa) from *Thermoplasma acidophilum*. I. Purification and properties. Submitted for publication.
13. DeLange, R. J., Hooper, J. A., and Smith, E. L., Histone III. III. Sequence studies on the cyanogen bromide peptides; complete amino acid sequence of calf thymus histone III. *J. Biol. Chem.*, 1973, **248**:3261–3274.
14. DeLange, R. J., Smith, E. L., Fambrough, D. M., and Bonner, J., Amino acid sequence of histone IV: Presence of ε-*N*-acetyllysine. *Proc. Natl. Acad. Sci. U.S.A.*, 1968, **61**:1145–1146.

15. DeLange, R. J., Williams, L. C., and Martinson, H. G., Identification of interacting amino acids at the histone 2A-2B binding site. *Biochemistry*, 1979, **18**:1942-1946.
15a. DeLange, R. J., Williams, L. C., and Searcy, D. G., A histone-like protein (HTa) from *Thermoplasma acidophilum*. II. Complete amino acid sequence. Submitted for publication.
16. Elgin, S. C. R., and Weintraub, H., Chromosomal proteins and chromatin structure. *Annu. Rev. Biochem.*, 1975, **44**:725-774.
17. Fambrough, D. M., and Bonner, J., Sequence homology and role of cysteine in plant and animal histones. *J. Biol. Chem.*, 1968, **243**:4434-4439.
18. Felsenfeld, G., Chromatin. *Nature* (*London*), 1978, **271**:115-122.
19. Gershey, E. L., Vidali, G., and Allfrey, V. G., Chemical studies of histone acetylation. The occurrence of ϵ-*N*-acetyllysine in the f2al histone. *J. Biol. Chem.*, 1968, **243**:5018-5022.
20. Glover, C. V. C., and Gorovsky, M. A., Amino-acid sequence of *Tetrahymena* histone H4 differs from that of higher eukaryotes. *Proc. Natl. Acad. Sci. U.S.A.*, 1979, **76**:585-589.
21. Gurley, L. R., D'Anna, J. A., Barham, S. S., Deaven, L. L., and Tobey, R. A., Histone phosphorylation and chromatin structure during mitosis in Chinese hamster cells. *Eur. J. Biochem.*, 1978, **84**:1-15.
22. Klevan, L., Dattagupta, N., Hogan, M., and Crothers, D. M., Physical studies of nucleosome assembly. *Biochemistry*, 1978, **17**:4533-4540.
23. Kornberg, R. D., Structure of chromatin. *Annu. Rev. Biochem.*, 1977, **46**:931-954.
24. Martinson, H. G., Shetlar, M. D., and McCarthy, B. J., Histone-histone interactions within chromatin. Cross-linking studies using ultraviolet light. *Biochemistry*, 1976, **15**:2002-2007.
25. Ogawa, Y. Quagliarotti, G., Jordan, J., Taylor, C. W., Starbuck, W. C., and Busch, H., Structural analysis of the glycine-rich, arginine-rich histone. III. Sequence of the amino-terminal half of the molecule containing the modified lysine residues and the total sequence. *J. Biol. Chem.*, 1969, **244**:4387-4392.
26. Olaitan, S. A., DeLange, R. J., and Smith, E. L., The structure of subtilisin Novo. *J. Biol. Chem.*, 1968, **243**:5296-5301.
27. Panyim, S., Sommer, K. R., and Chalkley, R., Oxidation of the cysteine-containing histone F3. Detection of an evolutionary mutation in a conservative histone. *Biochemistry*, 1971, **10**:3911-3917.
28. Patthy, L., Smith, E. L., and Johnson, J., Histone III. V. The amino acid sequence of pea embryo histone III. *J. Biol. Chem.*, 1973, **248**:6834-6840.
28a. Searcy, D. G., and Stein, D. B., Nucleoprotein subunit structure in an unusual prokaryotic organism: *Thermoplasma acidophilum*. *Biochim. Biophys. Acta*, in press.
29. Smith, E. L., Evolution of enzymes. In: *The Enzymes* (P. D. Boyer, ed.), 3rd ed., Vol. 1., Academic Press, New York, 1970:267-339.
30. Smith, E. L., The evolution of glutamate dehydrogenase and a hypothesis for the insertion or deletion of multiple residues in the interior of polypeptide chains. *Proc. Am. Philos. Soc.*, 1979, **123**:73-84.
31. Smith, E. L., DeLange, R. J., and Bonner, J., Chemistry and biology of the histones. *Physiol. Rev.*, 1970, **50**:159-170.

32. Smith, E. L., DeLange, R. J., Evans, W. H., Landon, M., and Markland, F. S., Subtilisin Carlsberg. V. The complete sequence; comparison with subtilisin BPN ; evolutionary relationships. *J. Biol. Chem.*, 1968, **243**:2184–2191.

33. Stein, D. B., and Searcy, D. G., Physiologically important stabilization of DNA by a prokaryotic histone-like protein. *Science*, 1978, **202**:219–221.

34. Strickland, M., Strickland, W. N., Brandt, W. F., and von Holt, C., Sequence of the cysteine-containing portion of histone F2al from the sea urchin *Parechinus angulosus*. *FEBS Lett.*, 1974, **40**:346–348.

35. von Holt, C., Strickland, W. N., Brandt, W. F., and Strickland, M. S., More histone sequences. *FEBS Lett.*, 1979, **100**:201–218.

36. Weintraub, H., and Van Lente, F., Dissection of chromosome structure with trypsin and nucleases. *Proc. Natl. Acad. Sci. U.S.A.*, 1974, **71**:4249–4253.

Thermal Stability and Protein Structure

PATRICK ARGOS, MICHAEL G. ROSSMANN, AND ULRICH M. GRAU

Department of Biological Sciences
Purdue University
West Lafayette, Indiana

and

HERBERT ZUBER, GERHARD FRANK, AND JON DURI TRATSCHIN

Institut für Molekularbiologie und Biophysik
ETH-Hönggerberg
Zürich, Switzerland

Introduction

Many amino acid sequences have been determined and a substantial number of three-dimensional protein structures are now known. Nevertheless there are only a few instances where knowledge of both sequence and structure is available for a given protein from a meso- and a thermophilic organism. One such case is for glyceraldehyde-3-phosphate dehydrogenase (GAPDH), where the structure and sequence of both the mesophilic lobster enzyme (6,8) and the thermophilic *Bacillus stearothermophilus* enzyme (4,14) have all been determined. Another case is for T4 phage lysozyme, where both the mesophilic wild-type structure (19) and sequence have been established and where, in addition, the effect of a single mutation that provides thermal stability has been investigated crystallographically (12). In contrast, thermolysin has been investigated intensively (20,31) without the possibility of comparison with the neutral protease. Suffice it to say that there is little supporting evidence for the various mechanisms that have been given as causes of thermal stability. However, Perutz and Raidt (24) carefully compared the amino acid substitutions in ferredoxins and hemoglobins for organisms of different temperature preferences, in light of the known three-dimensional polypeptide backbone structure of the mesophile. They concluded that thermal stability is enhanced by the formation of a few hydrogen bonds and salt bridges, results that were later supported by further sequence comparisons (11). These concepts have been well reviewed by Perutz (23)

ISBN 0-12-643150-7

within the larger framework of electrostatic effects on protein hydration, folding, assembly, and catalysis.

Since the most abundant comparative data are available for GAPDH and lactate dehydrogenase (LDH) and since Emil Smith has had a long-standing interest in the functional adaptation of dehydrogenases, it is appropriate here to attempt an analysis of these stabilizing forces. As some comparative data on the thermal stability of ferredoxins are also available, this information will be included.

It must be said at the outset that all attempts at correlating stability with structure by visual inspection of LDH and GAPDH models around the known amino acid changes failed to produce a satisfactory understanding of thermal stability. Admittedly Biesecker and colleagues (4) proposed some structural alteration between *B. stearothermophilus* and lobster GAPDH and suggested some stabilizing salt bridges, but these proposals still need to be verified with refined structures. It is equally possible that the amino acid alterations produce some very small stabilizing or destabilizing effects, the sum total of which provides the necessary decrease of energy for a thermally stable structure. Such changes would be suitable for the protection of all sides of the molecular structure, whereas it is difficult to envisage how a few single larger changes can have a significant effect on the stability of the molecule tens of Ångstroms away from the altered site. The concept of many very small changes to provide structural stability is by no means new (12). A statistical approach must be used to discern the total effect of many small changes on an enzyme. Such a method will be described here.

Method

Visual inspection of the amino acid sequences showed that thermophiles tended to contain arginines where there were lysines in the mesophiles. Furthermore, the abundance of alanines tended to increase in thermophiles. It had also been shown (33) that pig LDH-isoenzyme H_4 significantly increased its thermal stability when its external lysines had been modified with methyl acetimidate to generate arginine-like amino acids. These observations suggested that certain exchanges might be beneficial to stability provided that they occurred in suitable parts of the molecule. Accordingly, tests were devised to find preferred amino acid exchanges between aligned meso- and thermophilic sequences.

Matrices were formed that represented the frequency with which a given exchange occurred between any two sequences. For instance, in the comparison of lobster and *Thermus aquaticus* GAPDH, there were found five occurrences of a glycine in lobster changing to an alanine in *B. stearothermophilus*, whereas the reverse exchange occurred only once. Similarly, there were four lysine to arginine transitions and none in the reverse direction. Comparison of such matrices for GAPDH and LDH showed similar characteristics. Thus it was necessary to determine whether these observations were statistically significant and, if that were the case, why a given exchange might impart stability.

Let a_{ij} be the elements of such a matrix of exchanges. Here i refers to an amino acid in the mesophile, whereas j refers to an amino acid in the thermophile. For instance, using the preceding example,

$$a_{\mathrm{Lys,Arg}} = 4$$

and

$$a_{\mathrm{Arg,Lys}} = 0$$

These matrices will be square and nonsymmetrical.

The significance of each matrix formed from pairs of aligned sequences will depend on the temperature difference between the optimal temperature of the aligned molecules and the number of homologous amino acid pairs. The greater the temperature difference, the more likely it is that a given exchange will alter the structural stability; and the greater the number of pairs, the better is the statistical base of the results.

Let us now define a new matrix with elements

$$b_{ij} = \frac{a_{ij}}{\Delta T}$$

which normalizes the observed element to a unit of temperature change. Then let us take a weighted average of all the n matrices that can be formed by taking pairs of homologous sequences. Thus

$$c_{ij} = \sum_n \omega_n b_{ij} \Big/ \sum_n \omega_n$$

where ω_n is the weight to be attributed to a given comparison. If we, somewhat arbitrarily but nevertheless physically reasonably, set

$$\omega_n = \frac{1}{(\Delta T_n)^2}$$

then

$$c_{ij} = \sum_n (\Delta T_n) a_{ij} \Big/ \sum_n (\Delta T_n)^2$$

These c_{ij} coefficients are then a weighted average of all our observations.

Before going farther it is necessary to note that, as was the case for the Gly→Ala example given earlier, there are frequently exchanges that may go either way, although there might be a dominance in a given direction. Thus if such exchanges do produce alterations of stability we are observing a preference for stability without excluding the possibility for destabilizing events in the evolutionary process.

An analogy can be made for the migration of students to the Southeastern

U.S. seaboard in early spring of each year. The elements a_{ij} measure the number of cars traveling from a Northern (cold) to a Southern (warm) state on a given day, whereas a_{ji} measure the reverse trend. A 49 × 49 matrix would tabulate all the traffic in the continental United States. Then the gross traffic $g_{ij} = a_{ij} + a_{ji}$, whereas the net traffic $d_{ij} = a_{ij} - a_{ji}$. The size of d_{ij} will then measure any drift from one state to another (as, for instance, from Indiana to Florida on some day in February). In particular, the size of d_{ij} for any North–South exchange is likely to be much bigger than an East–West exchange.

Similarly, the drift between meso- and thermophile exchanges can be defined as

$$d_{ij} = c_{ij} - c_{ji}$$

The significance of a particular drift can be determined by computing the standard deviation $\sigma(d_{ij})$ of all nonzero elements in the d_{ij} matrix. The number of times E_{ij} a given drift is larger than the standard deviation can be computed from

$$E_{ij} = d_{ij} / \sigma(d_{ij})$$

Values of $E_{ij} > 3$ are likely to be significant drifts.

The values found for E_{ij} using the GAPDH, LDH, or ferredoxin data separately (see Table 1 for the data used in these calculations) are shown in Table 2. Shown also are the combined results from all three proteins. There is reasonable agreement between the results from each protein. Of the top 10 substitutions found in the overall statistics, 8 out of 10 with $E_{ij} > 1.9$ appeared in GAPDH, seven out of eight in LDH, and three out of five in ferredoxin. The ferredoxin results do not give equally good agreement because (1) they include only about one-sixth of the residues compared to those in GAPDH, reducing the number of possible exchanges and the reliability of the statistics; (2) differences in optimal temperatures are smaller; and (3) perhaps because ferredoxin is not a tetramer or a dehydrogenase. Nevertheless, there is reasonable agreement between the three different proteins, suggesting that thermal stability may be attained by similar principles in these, and possibly other, proteins.

The results obtained by taking an overall weighted average between all three proteins (Table 2) are shown diagrammatically in Fig. 1. It is immediately apparent that there is a considerable preference for alanine in the thermophiles. Furthermore, the thermal stabilizing process could occur in a series of steps—for example, exchanging a serine first to glycine and then to alanine.

Stabilizing Effects of the Preferred Exchanges

The top 10 preferred exchanges, which might provide stability for thermophiles, are shown in Table 2. These will impart certain altered properties

TABLE 1
AMINO ACID SEQUENCES USED TO REPRESENT THE FERREDOXIN, GAPDH, AND LDH PROTEINS

Protein	Species	Sequence determination (reference no.)	Three-dimensional structure determination (reference no.)	Assumed midpoint of temperature stability (°C)[a]
Ferredoxin	*Clostridium thermosaccharolyticum*	(28)		55
	Clostridium tartarivorum	(27)		46
	Clostridium acidi-urici	(25)		41
	Peptostreptococcus elsdenii	(3)		37
	Clostridium pasteurianum	(29)		37
	Micrococcus aerogenes	(32)	(1)	37
GAPDH	*Thermus aquaticus*	(14)		71
	Bacillus stearothermophilus	(14)	(4)	65
	Yeast	(15)		37
	Pig	(13)		37
	Lobster	(8)	(6)	20
LDH	*Bacillus stearothermophilus*	Zuber, Frank, and Tratschin, unpublished		65
	Chicken M_4	(9)		37
	Pig M_4	(9,16)	(9)	37
	Dogfish M_4	(9,30)	(26)	20

[a] Assumed to correspond to normal environmental temperatures and are taken for bacteria from *Bergey's Manual* (5) or as body temperatures for pig and chicken. While the exact mean thermal stable range for a given enzyme is debatable, the general temperature trend used here is clearly reasonable.

on structures, some of which may be beneficial and others of which may be deleterious. Furthermore, the position of the exchange in the structure is likely to be critical. Thus a given exchange on the molecular exterior may be beneficial, whereas the same exchange in the interior may devastate the structure. Thus in this analysis it will be necessary to bring in a knowledge of structure, whereas in the previous analysis only sequence had been used.

Five characters were selected (Table 3) with an eye to those properties that might have a bearing on the stability of the protein. They were (1) α-helix preference, $P(\alpha)$; (2) β-sheet preference, $P(\beta)$; (3) hydrophobicity, Hy; (4) bulkiness, Bu; and (5) polarity, Po. In addition, the minimum base change per codon was considered, since it is a crude measure of the mutational difficulty in achieving a desirable change. The bulkiness and polarity scales were taken from Zimmerman and colleagues (35), the hydrophobicity scale was derived by Nozaki and Tanford (21), and the $P(\alpha)$ and $P(\beta)$ values were those calculated by Levitt (17) and are similar to those originally published by Chou and Fasman (7) and by Fasman and colleagues (10). The limits of the various scales are $0.52 \leq P(\alpha) \leq 1.47$, $0.64 \leq P(\beta) \leq 1.49$, $0.0 \leq \text{Hy} \leq 3.15$, $3.4 \leq \text{Bu} \leq 21.7$, and $0 \leq \text{Po} \leq 52.0$. The differences Δ in character C, listed in Table 3, are given as $\Delta = \text{C(hot)} - \text{C(cold)}$.

Shown also in Table 3 are the number of occurrences with which a given

TABLE 2
EXCHANGES WITH $E_{ij} \geq 1.9$[a]

Exchange cold → hot	Weighted average of GAPDH, LDH, and ferredoxin	GAPDH (weighted average)	LDH (weighted average)	Ferredoxin (weighted average)
Gly → Ala	1 (4.3)	1 (5.0)		
Ser → Ala	2 (4.1)	2 (3.2)	6 (2.0)	1 (2.2)
Ser → Thr	3 (3.9)	5 (2.6)	4 (2.6)	2 (2.1)
Lys → Arg	4 (3.4)	6 (2.2)	1 (3.1)	
Asp → Glu	5 (3.0)	4 (3.0)		
Ser → Gly	6 (2.9)		3 (2.8)	5 (1.9)
Lys → Ala	7 (2.8)	9 (1.9)	5 (2.1)	
Val → Ala	8 (2.5)	9 (1.9)	7 (1.9)	
Asp → Asn	9 (2.5)	3 (3.1)		
Val → Ile	10 (1.9)		2 (2.9)	

[a] The first number is the rank within a column and the second number, set in parenthesis, is the ratio E_{ij} of the net weighted exchange per °C to the standard error.

exchange has been found in helical (H) or sheet (S) regions and whether the position was on the molecular surface (external, E) or within it (internal, I). Assignments for GAPDH are taken from Olsen and colleagues (22) and for LDH from Eventoff and colleagues (9). Ferredoxin was omitted from these statistics, since there are no explicit published data and, furthermore, it has little secondary structure. Thus for a given exchange to have been beneficial, the products H Δ P(α) (helix formation), S Δ P(β) (sheet formation), (I-E) Δ Hy (internal hydrophobicity), (E-I) Δ Po (external polarity), and (I-E) Δ Bu (internal packing) would have to be large and positive. Presumably, any given exchange need not be beneficial to every physical characteristic, but there should be an overall improvement.

A simplification of the preceding results is shown in Table 4. It is abundantly clear that, for those exchanges with E_{ij} greater than three standard deviations, there is an overall beneficial effect in every case. Furthermore,

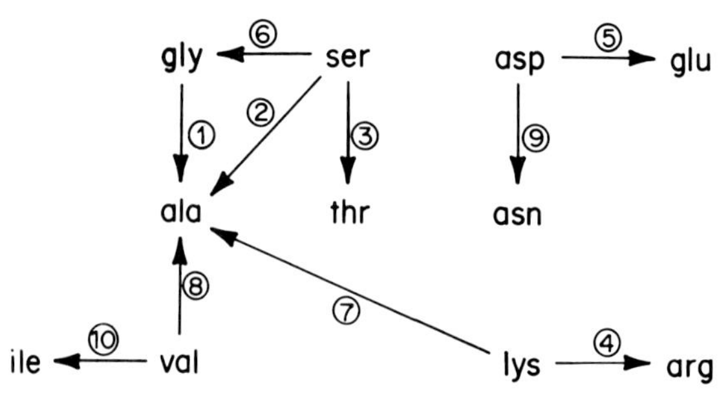

Fig. 1. Direction of observed preferred exchanges. Arrows point from the mesophilic to thermophilic protein. Numbers indicate the ranking of the significance for the given exchange.

TABLE 3
PHYSICAL CHARACTERISTICS OF PREFERRED EXCHANGES WITH $E_{ij} > 3\sigma^a$

Exchange cold → hot	H	ΔP (α)	S	ΔP (β)	E	I	ΔHy	ΔPo	ΔBu	MBC/C
1. Gly → Ala	8	+0.73	8	−0.02	1	19	+0.77	0	+8.1	1
2. Ser → Ala	15	+0.47	3	−0.05	9	15	+0.80	0	+2.0	1
3. Ser → Thr	8	0.00	13	+0.26	12	6	0.00	0	+6.3	1
4. Lys → Arg	6	−0.27	6	+0.22	20	2	−0.75	+3.0	−1.4	1
5. Asp → Glu	6	+0.40	2	+0.03	20	2	0.00	0	+1.9	1

[a] (i) Difference in physical characteristic, C, is calculated as [C(hot) − C(cold)]. (ii) H, S, E, I are the number of times the specific exchanges are found in helix, sheet, externally, and internally for all GAPDH and LDH pair-wise comparisons where ΔT > 0. (iii) ΔP (α), ΔP (β), ΔHy, ΔPo, ΔBu are changes in the α-helix forming ability, β-sheet forming ability, hydrophobicity, polarity and bulkiness, respectively. (iv) MBC/C is the minimum number of base changes per codon required for the exchange. (v) Top five exchanges are those with E_{ij} ‴ 3.0.

although each character imparts some structural benefits, the greatest contributions are provided by helix stability and internal hydrophobicity. The latter character has been generally anticipated as an important property (2,18,34), but its presence could not be recognized (4,24). It can also be remarked that, among the 10 top exchanges (Table 2), nine had a minimum base change of only 1, demonstrating that these exchanges are generally easy evolutionary events.

The preceding results were further verified by computing some controls. These were

TABLE 4
BENEFIT OF EXCHANGE TO STABILITY FOR EXCHANGES WITH $E_{ij} > \ddot{\sigma}^a$

Exchange cold → hot	(i) Helical region	(ii) Sheet region	(iii) Internal hydrophobicity	(iv) External polarity	(v) Internal packing
1. Gly → Ala	++		++		++
2. Ser → Ala	++		+		+
3. Ser → Thr		+			−
4. Lys → Arg	−	+	++	++	+
5. Asp → Glu	+				−

[a] Criteria used are given in the tabulation below.

	(i) H ΔP(α)	(ii) S ΔP(β)	(iii) (I-E) ΔHy	(iv) (E-I) ΔPo	(v) (I-E) ΔBu
++	>5	>5	>10	>50	>50
+	>0.4	>0.4	>0	>0	>0
−	<0.4	<0.4	<0	<0	<0
−−	<5	<5	<10	<50	<50

1. Grouping the proteins simply into two groups of "high" and "low" temperature, thus avoiding detailed examination of the attributed optimal temperatures (Table 1). Results were similar to those already given.

2. Using only the proteins whose temperatures were set at 37°C in Table 1 and arbitrarily allocating these into all possible "high" and "low" temperature groups. The results showed no $E_{ij} > 2.0\sigma$ and none of the eight $E_{ij} > 1.0\sigma$ coincided with those in Tables 2, 3, and 4.

3. Using a series of eukaryotic sequences aligned with prokaryotic sequences but where the organisms are likely to have similar environmental conditions. This experiment examined the dominance of the eukaryotic (cold) to prokaryotic (hot) exchanges in the data used to derive the results given in this paper. That is, a test was necessary to determine whether the present results were truly due to differences in thermal stability or in fact due to the fundamentally different nature of the biological organization. For this experiment all E_{ij} were found to be less than 2.9σ, although 11 were greater than 1.9σ. Nevertheless, seven of the 11 showed a minimum base change per codon equal to 2, and there was no consistent stabilizing effect. The only exchange in common with the most significant five in Tables 4 and 5 was Ser→Ala, but its positioning in the control structures did not enhance stability.

It is possible that the conclusions drawn here are biased by the dominance of data from dehydrogenases. Unfortunately, no data are currently available to perform a suitable control.

Useful Exchanges for Evolving Stability

The preceding results show five fairly significant exchanges that, when suitably positioned in the structure, might frequently impart stability. Other useful exchanges no doubt exist but could not be detected with the limited amount of data available. This section attempts to determine the nature of such exchanges.

Apparently each exchange can impart only a limited effect, except in special circumstances such as those discussed by Grütter and colleagues (12). Hence the change in physical character must be small. If a limit is set on the possible permissible change with the following, somewhat arbitrary rules, then only 23 exchanges remain possible. These rules were

1. Minimum base change per codon = 1
2. $\Delta P(\alpha) < 0.75$, $\Delta P(\beta) < 0.75$
3. $\Delta Hy < 1.5$
4. $\Delta Po < 47.0$
5. $\Delta Bu < 11.0$
6. The only allowed charge changes were to neutrality
7. Cysteine and tryptophan residues are excluded (since their infrequent occurrence implies an inability to find a significant sample of exchanges).

Nine of these 23 exchanges are the same as the top 10 listed in Table 2 and shown in Fig. 1. Four more are found in exchanges with $E_{ij} > 1.9$ for GAPDH, LDH, or ferredoxin taken individually. The remaining substitutions are mainly among the hydrophobic residues Val, Ile, Phe, Met, and Tyr. Quite possibly these exchanges do occur, but their stabilizing effect is only marginal, since they are all rather hydrophobic. Hence their preference would not be easily detected with the limited amount of data available for this study.

Conclusions

Thermal stability is largely achieved in GAPDH, LDH, and ferredoxin by an additive series of very small improvements at many locations within the molecule. The most significant changes increase internal hydrophobicity and the helix-forming ability of residues in helices. In addition, stabilization is achieved by increasing the sheet-forming tendency of residues in β-sheets and by slight increases in bulk for internal residues to obtain better packing organization. These optimal changes can be determined experimentally and proposed from theory.

Acknowledgments

We are grateful to Sharon Wilder and Maria Victoria Manterola for help in the preparation of the manuscript. This work was supported by a National Science Foundation grant (PCM78-16584) and a National Institutes of Health grant (GM 10704) to M.G.R., by a National Science Foundation grant (PCM77-20287) and an American Cancer Society grant (FRA173) to P.A. U.M.G. was the recipient of a Deutsche Forschungsgemeinschaft postdoctoral fellowship.

REFERENCES

1. Adman, E. T., Sieker, L. C., and Jensen, L. H., The structure of a bacterial ferredoxin. *J. Biol. Chem.*, 1973, **248**:3987–3996.
2. Amelunxen, R. E., and Singleton, R., Jr., Thermophilic glyceraldehyde-3-P dehydrogenase. In: *Enzymes and Proteins from Thermophilic Microorganisms* (H. Zuber, ed.). Birkhaeuser Verlag, Basel, 1976:107–120.
3. Azari, P., Glantz, M., Tsunoda, J., and Yasunobu, K. T., unpublished results, cited by Yasunobu, K. T., and Tanaka, M., The evolution of iron-sulfur protein containing organisms. *Syst. Zool.*, 1973, **22**:570–589.
4. Biesecker, G., Harris, J. I., Thierry, J. C., Walker, J. E., and Wonacott, A. J., Sequence and structure of D-glyceraldehyde 3-phosphate dehydrogenase from *Bacillus stearothermophilus*. *Nature* (*London*), 1977, **266**:328–333.
5. Buchanan, R. E., and Gibbons, N. E. (eds.), *Bergey's Manual of Determinative Bacteriology*. 8th ed. Williams & Wilkins, Baltimore, Maryland, 1974.
6. Buehner, M., Ford, G. C., Moras, D., Olsen, K. W., and Rossmann, M. G., Three-dimensional structure of D-glyceraldehyde-3- phosphate dehydrogenase. *J. Mol. Biol.*, 1974, **90**:25–49.
7. Chou, P. Y., and Fasman, G. D., Conformational parameters for amino acids in

helical, β-sheet, and random coil regions calculated from proteins. *Biochemistry*, 1974, **13**:211–222.

8. Davidson, B. E., Sajgò, M., Noller, H. F., and Harris, J. I., Amino-acid sequence of glyceraldehyde 3-phosphate dehydrogenase from lobster muscle. *Nature (London)*, 1967, **216**:1181–1185.
9. Eventoff, W., Rossmann, M. G., Taylor, S. S., Torff, H. J., Meyer, H., Keil, W., and Kiltz, H. H., Structural adaptations of lactate dehydrogenase isozymes. *Proc. Natl. Acad. Sci. U.S.A.*, 1977, **74**:2677–2681.
10. Fasman, G. D., Chou, P. Y., and Adler, A. J., Prediction of the conformation of the histones. *Biophys. J.*, 1976, **16**:1201–1238.
11. Friedman, S. M. (ed.), *Biochemistry of Thermophily*. Academic Press, New York, 1978.
12. Grütter, M. G., Hawkes, R. B., and Matthews, B. W., Molecular basis of thermostability in the lysozyme from bacteriophage T4. *Nature (London)*, 1979, **277**:667–669.
13. Harris, J. I., and Perham, R. N., Glyceraldehyde 3-phosphate dehydrogenase from pig muscle. *Nature (London)*, 1968, **219**:1025–1028.
14. Harris, J. I., and Walker, J. E., Structure and properties of glyceraldehyde 3-phosphate dehydrogenase from thermophilic microorganisms. In: *Pyridine Nucleotide-Dependent Dehydrogenases* (H. Sund, ed.). de Gruyter, Berlin, 1977:43–61.
15. Jones, G. M. T., and Harris, J. I., Glyceraldehyde 3-phosphate dehydrogenase: Amino acid sequence of enzyme from baker's yeast. *FEBS Lett.*, 1972, **22**:185–189.
16. Kiltz, H. H., Keil, W., Griesbach, M., Petry, K., and Meyer, H., The primary structure of porcine lactate dehydrogenase: Isoenzymes M_4 and H_4. *Hoppe-Seyler's Z. Physiol. Chem.*, 1977, **358**:123–127.
17. Levitt, M., Conformational preferences of amino acids in globular proteins. *Biochemistry*, 1978, **17**:4277–4285.
18. Ljungdahl, L. G., Sherod, D. W., Moore, M. R., and Andreesen, J. R., Properties of enzymes from *Clostridium thermoaceticum* and *Clostridium formicoaceticum*. In: *Enzymes and Proteins from Thermophilic Microorganisms* (H. Zuber, ed.). Birkhaeuser Verlag, Basel, 1976:237–248.
19. Matthews, B. W., and Remington, S. J., The three-dimensional structure of the lysozyme from bacteriophage T4. *Proc. Natl. Acad. Sci. U.S.A.*, 1974, **71**:4178–4182.
20. Matthews, B. W., Jansonius, J. N., Colman, P. M., Schoenborn, B. P., and Dupourque, D., Three-dimensional structure of thermolysin. *Nature (London) New Biol.*, 1972, **238**:37–41.
21. Nozaki, Y., and Tanford, C., The solubility of amino acids and two glycine peptides in aqueous ethanol and dioxane solutions. Establishment of a hydrophobicity scale. *J. Biol. Chem.*, 1971, **246**:2211–2217.
22. Olsen, K. W., Moras, D., Rossmann, M. G., and Harris, J. I., Sequence variability and structure of D-glyceraldehyde-3-phosphate dehydrogenase. *J. Biol. Chem.*, 1975, **250**:9313–9321.
23. Perutz, M. F., Electrostatic effects in proteins. *Science*, 1978, **201**:1187–1191.
24. Perutz, M. F., and Raidt, H., Stereochemical basis of heat stability in bacterial ferredoxins and in haemoglobin A2. *Nature (London)*, 1975, **255**:256–259.

25. Rall, S. C., Bolinger, R. E., and Cole, R. D., The amino acid sequence of ferredoxin from *Clostridium acidi-urici*. *Biochemistry*, 1969, **8**:2486–2496.
26. Rossmann, M. G., Adams, M. J., Buehner, M., Ford, G. C., Hackert, M. L., Lentz, P. J., Jr., McPherson, A., Jr., Schevitz, R. W., and Smiley, I. E., Structural constraints of possible mechanisms of lactate dehydrogenase as shown by high resolution studies of the apoenzyme and a variety of enzyme complexes. *Cold Spring Harbor Symp. Quant. Biol.*, 1971, **36**:179–191.
27. Tanaka, M., Haniu, M., Matsueda, G., Yasunobu, K. T., Himes, R. H., Akagi, J. M., Barnes, E. M., and Devanathan, T., The primary structure of the *Clostridium tartarivorum* ferredoxin, a heat-stable ferredoxin. *J. Biol. Chem.*, 1971, **246**:3953–3960.
28. Tanaka, M., Haniu, M., Yasunobu, K. T., Himes, R. H., and Akagi, J. M., The primary structure of the *Clostridium thermosaccharolyticum* ferredoxin, a heat-stable ferredoxin. *J. Biol. Chem.*, 1973, **248**:5215–5217.
29. Tanaka, M., Nakashima, T., Benson, A., Mower, H., and Yasunobu, K. T., The amino acid sequence of *Clostridium pasteurianum* ferredoxin. *Biochemistry*, 1966, **5**:1666–1681.
30. Taylor, S. S., Amino acid sequence of dogfish muscle lactate dehydrogenase. *J. Biol. Chem.*, 1977, **252**:1799–1806.
31. Titani, K., Hermodson, M. A., Ericsson, L. H., Walsh, K. A., and Neurath, H., Amino-acid sequence of thermolysin. *Nature (London) New Biol.*, 1972, **238**: 35–37.
32. Tsunoda, J. N., Yasunobu, K. T., and Whiteley, H. R., Non-heme iron proteins. IX. The amino acid sequence of ferredoxin from *Micrococcus aerogenes*. *J. Biol. Chem.*, 1968, **243**:6262–6272.
33. Tuengler, P., and Pfleiderer, G., Enhanced heat, alkaline and tryptic stability of acetamidinated pig heart lactate dehydrogenase. *Biochim. Biophys. Acta*, 1977, **484**:1–8.
34. Veronese, F. M., Grandi, C., Boccù, E., and Fontana, A., Comparative conformational properties of thermophilic and mesophilic 6-phosphogluconate dehydrogenase. In: *Enzymes and Proteins from Thermophilic Microorganisms* (H. Zuber, ed.). Birkhaeuser Verlag, Basel, 1976:147–155.
35. Zimmerman, J. M., Eliezer, N., and Simha, R., The characterization of amino acid sequences in proteins by statistical methods. *J. Theor. Biol.*, 1968, **21**:170–201.

EVOLUTION OF PROTEINS

The Cytochromes *c*: An Exercise in Scientific Serendipity

RICHARD E. DICKERSON

Division of Chemistry and Chemical Engineering
California Institute of Technology
Pasadena, California

Just 225 years ago, the English writer Horace Walpole invented the word "serendipity," from a possibly hypothetical Persian folk tale entitled "The Three Princes of Serendip" (or Ceylon) (69). Although the princes never managed to complete their long journey, "they were always making discoveries, by accidents and sagacity, of things which they were not in quest of." The entire cytochrome *c* story is an exercise in scientific serendipity. When MacMunn was criticized so harshly in 1890 by Hoppe-Seyler for his work on "modified myohaematin," it was because MacMunn had stumbled onto a new substance, cytochrome *c*, that neither he nor his critics fully appreciated (34). When David Keilin resumed the work in 1925 that a discouraged MacMunn had abandoned, he had access only to a primitive, hand-held spectroscope that contributed significantly to his ability to find cytochrome *c* in living muscle specimens. I cannot say what serendipitous accidents may have led to Emil Smith's widely known contributions to the cytochrome *c* field—perhaps he himself can tell us. But when we at Caltech began investigation of the X-ray structure of cytochrome *c* in 1963 with the direct inspiration of Emil Smith and Emanuel Margoliash, we were searching for something quite different from what we actually found.

Our goal initially was to understand how electrons flowed into and out of the molecule as it functioned in the respiratory electron transport chain of mitochondria and what conformational changes took place in the protein when the heme iron atom was alternately reduced and oxidized. After complete structure analyses of both reduced and oxidized molecules from tuna heart muscle (42) and supporting structures from horse (22) and bonito (9), we

ISBN 0-12-643150-7

can say that little conformational change occurs—at least not on the scale that had originally been expected. Electrons appear to flow in and out through one edge of the heme group, exposed at the surface of the protein molecule. But comparable X-ray structure analyses (1,37,59,67) and amino acid sequence studies (7,8) from cytochromes *c* of bacteria have led to an entirely serendipitous (and possibly even true) conclusion, which in its plainest and most provocative form can be stated as, *We are the metabolic offspring of defective purple photosynthetic bacteria* (1,23). This paper is devoted to making that thesis plausible.

Electron Transport in Bacteria

The most familiar cytochrome *c*, of course, is that of mitochondrial respiration in eukaryotes. But similar cytochromes *c* in bacteria are found almost everywhere that electrons are transferred during respiration or photosynthesis. Most respiring bacteria use O_2 as their terminal oxidant. Some of these are also capable of using nitrate if O_2 is not available, and the change lies only in the final cytochrome oxidase of the respiratory chain. The sulfate-respiring bacteria such as *Desulfovibrio* seem to be more primitive or at least more distantly related to the other bacteria discussed in this paper, and little will be said of them. The three main branches of photosynthetic bacteria—green, purple, and blue-green—are compared in Table 1. Only the cyanobacteria or blue-green algae have the two-photocenter, water-splitting type of photosynthesis that is familiar from higher plants. The green and purple bacteria have just one photocenter and must obtain their H atoms for reducing NAD^+ from some source other than water.

In the last four years it has been realized that a certain symmetry exists between the green and purple photosynthetic bacteria. Each has two families, one of which must use H_2, H_2S, or other reduced sulfur compounds as sources of reducing equivalents (Chlorobiaceae and Chromatiaceae, respectively), and the other of which can utilize organic compounds as reductants (Chloroflexaceae and Rhodospirillaceae). These latter two families also have the ability to respire in the dark using O_2 as an adjunct to their normal photosynthetic existence. Morphologically the two purple families, Chromatiaceae and Rhodospirillaceae, are virtually indistinguishable (65). Some of the former can use organic molecules as reductants, and some of the latter can use H_2S as well as organics. Individual species have on occasion been shifted back and forth between one family and the other, and present practice seems to be to class as Rhodospirillaceae all those purple photosynthetic bacteria that can use organic molecules *but not* elemental sulfur as reductants, regardless of whether they can use H_2S or not (54,68). The most important distinction from our point of view is that the Rhodospirillaceae appear to be simply an adaptation of the Chromatiaceae to an aerobic, sulfide-deficient world, first by the liberation from a dependence on reduced sulfur compounds in photosynthesis and then by the development of the

TABLE 1

METABOLIC COMPARISON OF PHOTOSYNTHETIC BACTERIA[a]

Family	Chlorobiineae (green photosynthetic bacteria)		Rhodospirillineae (purple photosynthetic bacteria)		
	Chlorobiaceae (green and brown sulfur bacteria, GSB)	Chloroflexaceae (green filamentous bacteria, GFB)	Chromatiaceae (purple sulfur bacteria, PSB)	Rhodospirillaceae (purple nonsulfur bacteria, PNSB)	Cyanobacteria (blue-green bacteria or algae)
Light-harvesting photopigments					
Chlorophylls[b]	Bchl *c*, *d* (green), or *e* (brown)	Bchl *c* or *d*	Bchl *a* or *b*	Bchl *a* or *b*	Chlorophyll *a*
Carotenoids	Aromatic	Aromatic	Aliphatic	Aliphatic	Carotene (also phycobilins)
Light absorption maxima (nm)	700–760 and 480–550 (brown)	700–760	>800	>800	<690
Location of primary light absorbers	Chlorobium vesicles beneath outer membrane	Chlorobium vesicles beneath outer membrane	Infoldings of outer membrane into pockets, tubules, or stacked lamellae, resembling the thylakoid grana of green plant chloroplasts		
Types of photocenter	I	I	I	I	I and II
Main source of reducing H for photosynthesis	H_2S (also S, $S_2O_3^{2-}$, H_2)	Organics (also H_2S)	H_2S (also S, $S_2O_3^{2-}$, H_2)	Organics (also H_2S, H_2, $S_2O_3^{2-}$, *not* S)	H_2O
Photoreduced dinucleotide	NADH	NADH?	NADH	NADH	NADPH
Light-generated reductant strong enough to reduce $NAD(P)^+$ directly?	Yes	Yes?	No	No	Yes
CO_2-fixing pathway	Not Calvin–Benson cycle; probably reductive Krebs cycle		Calvin cycle	Calvin cycle	Calvin cycle
Growth by anaerobic fermentation in the dark?	Slow	None	Slow	Moderate	Slow
Capable of O_2 respiration in the dark?	No	Yes	No	Yes	Yes
Photosynthetic and respiratory electron transport chains	—	Close proximity, probably shared	—	Shared	Unclear
Motile? Mechanism?	Nonmotile	Yes, gliding	Yes, flagellated	Yes, flagellated	Nonmotile or gliding

[a] This is a summary of an extensive original literature. The basic form of the table has been adapted from Gromet-Elhanan (28) and Olson (51), supplemented by in Pfennig *et al.* (55, 61, and 68), and other papers in Clayton and Sistrom (14). The reader is referred to these papers for keys to the original literature.

[b] Green photosynthetic bacteria also have smaller amounts of bacteriochlorophyll *a* at their primary reaction centers.

machinery for oxygen respiration. The Chloroflexaceae appear to have evolved from the Chlorobiaceae in a similar manner. These both probably represent an evolutionary response to a gradual change in atmospheric conditions from reducing to oxidizing, produced by the oxygen-releasing photosynthesis of cyanobacteria.

All the photosynthetic bacteria contain electron transport chains with flavoproteins and/or iron–sulfur proteins, quinone pools of reducing equivalents, cytochromes *b*, and cytochromes *c*, in roughly this order of increasing reduction potential. It is these high-potential cytochromes *c* that have received detailed X-ray and amino acid sequence analyses, sufficient to suggest that they are all members of an evolutionary family derived from a common ancestral gene. This in turn suggests a common evolutionary history for all photosynthetic bacteria and for many respiring bacteria as well.

Cytochrome Sequence and Structure

Emil Smith pioneered the first amino acid sequence analysis of a eukaryotic cytochrome *c*, that of horse heart, with 104 residues (44). This was also the first cytochrome whose three-dimensional structure was determined by X-ray diffraction (22), but those of tuna (42) and bonito (9,22) with 103 residues soon followed. Structures have since been determined for the photosynthetic-respiratory c_2 from *Rhodospirillum rubrum*, with 112 residues (59); respiratory c_{550} from *Paracoccus denitrificans*, with 134 residues (67); respiratory c_{551} from *Pseudomonas aeruginosa*, with 82 residues (1); and photosynthetic c_{555} from *Chlorobium thiosulfatophilum*, with 86 residues (37). An analysis is currently under way of the photosynthetic c_{554} from the cyanobacterium *Anacystis nidulans*, with 85 residues (40). Only the Chloroflexaceae and Chromatiaceae are currently unrepresented by three-dimensional cytochrome *c* structures. A small *c* that probably is homologous with the others is known in *Chromatium vinosum* (11,15), although it has not been well characterized, and the Chloroflexaceae are such a recent discovery that little is yet known about their cytochromes (55,57).

The structures of cytochromes c_{551}, c_{555}, *c*, c_2, and c_{550} are compared in Fig. 1a–e. All these molecules have similar chain folding and side group packing around the heme, but they differ in details of extended loops of chain at the surface of the molecules. They fall naturally into three structural subgroups: The short (S) cytochromes of *Pseudomonas* and *Chlorobium* are as shown in Fig. 1a and b. The medium (M) cytochromes *c* all have an extra loop of chain at the bottom, marked I in Fig. 1c, and are the type that are found in all eukaryotic mitochondria. The long (L) cytochromes have this same region I, plus more extended loops of chain at the bottom left (II and III in Fig. 1d and e), and sometimes also at the right (IV in Fig. 1e). But in spite of these extra insertions, the five molecules are all obviously variations on a common theme.

The similarities between the five molecules are even greater than can be

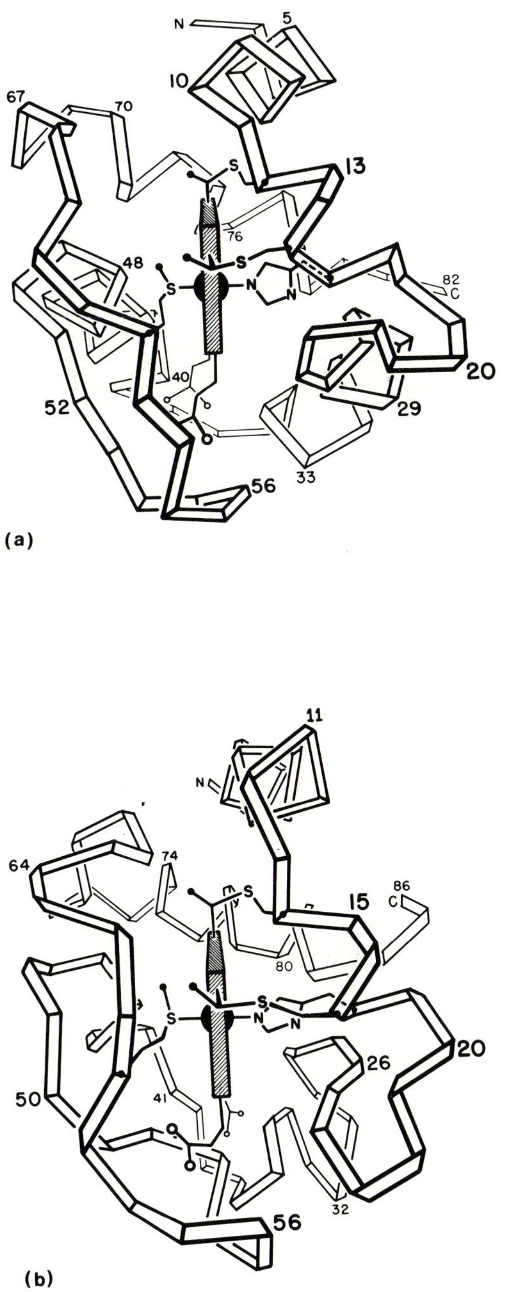

(a)

(b)

Fig. 1. Main-chain folding in five cytochromes *c*. The flat, shaded slab is the heme group viewed on edge, with histidine and methionine ligands to right and left, and propionic acid groups extending downward. (a) and (b) Short cytochromes c_{551} from *Pseudomonas aeruginosa* and c_{555} from *Chlorobium thiosulfatophilum*.

(Continued)

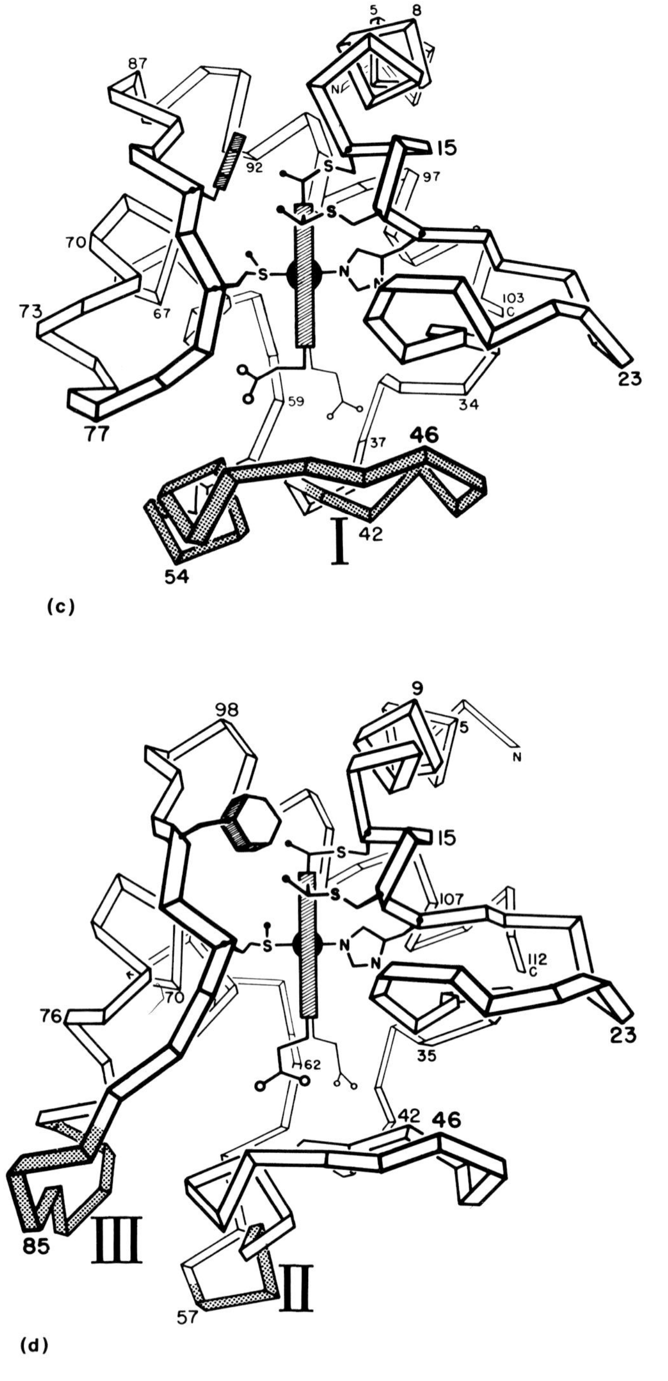

Fig. 1. *(Continued)* (c) Medium cytochrome *c* from tuna, with extra chain as shaded at I. (d) Long cytochromes c_2 from *Rhodospirillum rubrum* and c_{550} from *Paracoccus denitrificans*, with added external loops of chain as shaded at II, III, and IV.

(Continued)

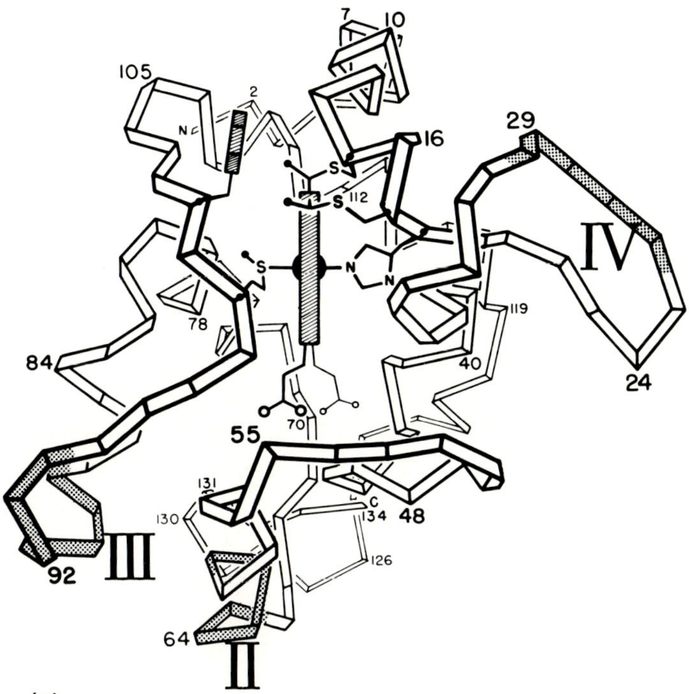

Fig. 1. *(Continued)* (e) Long cytochromes c_2 from *Rhodosprillum rubrum* and c_{550} from *Paracoccus dentrificans*, with added external loops of chain as shaded at II, III, and IV.

shown on these simplified diagrams. The same kinds of hydrophobic side chains—aliphatic or aromatic—tend to be packed around the heme at corresponding positions, both in space and in amino acid sequence. For example, the M and L cytochromes all have a cluster of three aromatic groups at the bottom of the molecule (tyrosines 46 and 48 and tryptophan 59 in tuna), three more to the left of the heme (67, 74, and 82 in tuna numbering), and two more to the upper right (10 and 97 in tuna). Some of these are invariant; others vary between tyrosine and phenylalanine, with tryptophan occasionally substituting. With one or two bacterial exceptions, they all occur at exactly the same positions along their respective amino acid sequences. The pattern of aromatics is more varied in the S class, but the two at the upper right seem to be quite important (tyrosine 10 and 80 in c_{555}, or phenylalanine 7 and tryptophan 77 in C_{551}), as is *something* at the bottom to hydrogen bond to the buried heme propionate. Cytochromes c_{555} and c_{551} solve this latter problem in two different ways: in *Chlorobium* c_{555}, tryptophan 34 occupies the same position along the amino acid chain as tryptophan 59 in tuna, if allowances are made for the previously mentioned chain deletions in c_{555}. But in c_{551} from various pseudomonads this tryptophan is gone, being replaced by tyrosine, phenylalanine, or asparagine. Instead, an entirely unrelated tryptophan 56, corresponding in sequence to the mid-70s in tuna, reaches back from the bottom front of the molecule to make a compensating hydrogen bond with the buried propionic acid.

Table 2 contains most of the presently known amino acid sequences of prokaryotic cytochromes *c* that belong to this evolutionary family. This does not include the high-spin *c*′, multiheme c_3, or flavin cytochromes *c*. The cytochromes in Table 2 have been classed as an interrelated evolutionary family on three lines of evidence: their great similarity in amino acid sequence, their similarity in three-dimensional folding, and their common role near the high-potential ends of photosynthetic and respiratory electron transport chains. (*Desulfovibrio* c_{553} is a low-potential exception.)

One of the more striking aspects of this table is the broad range of variation of cytochromes c_2 of the Rhodospirillaceae (with genera *Rhodospirillum*, *Rhodopseudomonas*, and *Rhodomicrobium*). They are represented in all three size classes: S, M, and L. The cytochromes of respiring bacteria and eukaryotes are thoroughly mixed in with these c_2. *Paracoccus* c_{550} is distinguished only by being the longest of the L class, tuna and all other mitochondrial *c* sit amid the M sequences, and *Pseudomonas* c_{551} make up most of the S class. Although the c_{554} and *f* of the prokaryotic and eukaryotic algae are short cytochromes also, they have been grouped into a separate S* class, since their sequences are quite different from those in class S. In spite of these differences, however, the structure analysis of c_{554} from *Anacystis* (40) has progressed far enough to show that it has the same general chain folding as c_{551} and c_{555}.

The sequence alignments in Table 2 could not be made correctly before the three-dimensional structures of the five cytochromes were known. Although all the information is present in principle in the sequences, since these are sufficient to determine molecule folding, we are not yet clever enough to predict folding successfully from sequence data alone. Prior to the X-ray analyses, four laboratories tried independently to find the correct alignment between cytochrome c_{551}, on the one hand, and *c* and c_2, on the other (17,18,48,50). All arrived at different results, and all were ultimately shown by the X-ray analysis of c_{551} to be incorrect (1). Without the three-dimensional structures, the problem of where to put the massive deletions in the c_{551} sequence was insurmountable, but with the structures in hand it became trivial. This is a good example of the principle that, at our present level of understanding of protein folding, *three-dimensional structures are better guides to similarities between distantly related proteins than are amino acid sequences alone.*

The differences between the sequences of Table 2 are expressed quantitatively in Table 3. For closely related protein sequences, a difference matrix showing the number of amino acid differences in all pairwise comparisons is useful in evaluating the degree of relatedness, and as a starting point in constructing phylogenetic trees of the host organism. This line of approach has been pursued with great success by Fitch and Margoliash (24), Dayhoff

(16), Goodman (49), and others. But for more distantly related sequences an indentities matrix such as Table 3 is better. It avoids the difficult question of whether to treat a long-chain deletion as one evolutionary event or many, and provides a measure of how nearly identical two sequences are in those regions of chain that they have in common. Gross insertions or deletions of chain then can be assessed separately. The upper right half of the matrix in Table 3 shows the number of identical positions between two sequences, and the lower left half shows the number of differences.

It now can be seen that each of the structural classes L, M, S, and S* is a legitimate one, for the sequences within any one class generally are identical in 30 or more positions along the shared chain, whereas the number of identical residues between classes is closer to 20 or less. Only the L and M classes are so similar that one might well consider grouping them into one category. But they differ by having or not having the chain insertions marked II, III, and IV in Fig. 1, which show up only in the difference matrix half of Table 3.

That the S and S* classes differ in sequence in spite of their similar size is not surprising considering their origin. The S cytochromes *c* come from Rhodospirillaceae and from pseudomonads that probably evolved from them; the S* cytochromes come from prokaryotic and eukaryotic algae on the evolutionary branch that developed two-photocenter, water-splitting photosynthesis. Even more distant are the two cytochromes c_{555} from the Chlorobiaceae or green sulfur bacteria, although we know from the X-ray analysis that the chain folding is the same. Least similar of all is cytochrome c_{553} from sulfate-respiring *Desulfovibrio*, which also is anomalous in having a negative reduction potential. Yet the sequence alignment in Table 2 suggest that even this cytochrome might have the common protein folding pattern and belong to this same evolutionary family. In such an event, the family would be an extremely old one. An X-ray analysis of c_{553} from *Desulfovibrio* (not the multiheme c_3) is needed to settle this point.

The cytochromes c_2 of the purple photosynthetic bacteria are a very broad structural class, diverse enough to encompass most of the cytochromes *c* of respiring bacteria. Indeed, there is nothing obvious in either sequences or three-dimensional structures alone to tell us that the respiratory cytochrome *c* from our own mitochondria did not come from a purple photosynthetic bacterium instead. All three of the respiratory *c*'s whose structures are known—*c*, c_{550}, and c_{551}—appear to be only special cases of cytochromes c_2 from the purple photosynthetic bacteria, and therein hangs one of the main themes of this paper: the evolution of respiration from photosynthesis (20).

Confusion from Lateral Gene Transfer?

As Ambler and Kamen have pointed out, the drawing of evolutionary conclusions from protein comparisons in bacteria is not necessarily a straightforward process because of the ease with which genetic material can

TABLE 2

SEQUENCE ALIGNMENTS FOR 37 RELATED BACTERIAL AND ALGAL CYTOCHROMES C[a]

Size Class	Species	Tuna numbering (… 10)
L	*Paracoccus denitrificans* c_{550}	N E G D A A K G E K E F N
	Rhodopseudomonas sphaeroides c_2	Q E G D P E A G A K A F N
	Rhodopseudomonas capsulata c_2	G D A A K G E K E F N
	Rhodospirillum rubrum c_2	E G D A A A G E K V S K
	Rhodospirillum photometricum c_2	A G D A A V G E K I A K
	Rhodopseudomonas palustris c_2	Q D A A K G E A V F K
M	*Rhodopseudomonas acidophila* c_2	A G D P D A G Q K V F L
	Rhodopseudomonas viridis c_2	Q D A A S G E Q V F K
	Tuna *c*	G D V A K G K K T F V
	Rhodomicrobium vannielii c_2	A G D P V K G E Q V F K
	Rhodopseudomonas globiformis c_2	G S A P P G D P V E G K H L F H
	Rhodospirillum molischianum c_2, iso-1	A D A P P — — P A F N
	Rhodospirillum fulvum c_2, iso-1	A D A P — — — T A F N
	Rhodospirillum molischianum c_2, iso-2	A D A P — — — A G F T
	Rhodospirillum fulvum c_2, iso-2	A D A P — — — P A F G
S	*Rhodospirillum tenue* c_{551}	A D E S A L A Q
	Rhodopseudomonas gelatinosa c_{551}	A T P A E L A T
	Pseudomonas aeruginosa c_{551}	E D P E V L F K
	Pseudomonas fluorescens c_{551}	E D G A A L F K
	Pseudomonas stutzeri c_{551}	Q D G E A L F K
	Pseudomonas mendocina c_{551}	A S G E E L F K
	Pseudomonas denitrificans c_{551}	S T G E E L F K
	Azotobacter vinelandii c_{551}	E T G E E L Y K
S*	*Spirulina maxima* c_{554} (cyanobacterium)	G D V A A G A S V F —
	Anacystis nidulans c_{554} (cyanobacterium)	A D L A H G G Q V F —
	Alaria esculenta f (brown alga)	I D I D N G E D I F —
	Porphyra tenera f (red alga)	A D L D N G E K V F —
	Bumilleriopsis filiformis f (yellow-green alga)	A D I E N G E R I F —
	Monochrysis lutheri f (yellow-green alga)	G D I A N G E Q V F —
	Euglena gracilis f (euglenoid alga)	G G A D V F —
	Halotolerant Micrococcus c_{554}	A G D A A A G E D K I —
	Pseudomonas aeruginosa c_4, 1st half	A G D A A A G Q A K A —
	Pseudomonas aeruginosa c_4, 2nd half	L F R G G K I A E G M —
	Pseudomonas mendocina c_5	A A S A G G G A R S A D D I I —
	Chlorobium thiosulfatophilum[b] c_{555}	Y D A A A G K A T Y D
	Prosthecochloris aestuarii[c] c_{555}	A V T K A D V E Q Y D L A N G K T V Y D
	Desulfovibrio vulgaris c_{553}	A D G A A L Y —

(*continued*)

I, II, III and IV indicate chain insertions (as in Fig. 1).

[a] The sequence data (as lettered in Table 3) were taken from the following sources (reference in parentheses): A (66); B–H, J, and L–O, (7); I (38); P,Q, (8); R–W and h–k (4); x, z, a, c, d (6); Y (43); K, b, e, f, g (5). [b] Now listed in *Bergey's Manual* (13) as *Chl. limicola* forma sp. *thiosulfatophilum*. [c] Formerly known as *Chl. limicola* strain 2K (52). [d] Val 55 as listed in Ambler (3, 4) is given as Asn in the earlier reference (2). We have assumed the later determination to be correct in this table.

TABLE 2—*Continued*

Tuna numbering
20 30
K — — C K A C H M I Q A P D G T D I — — K G G K T G P N L Y
Q — — C Q T C H V I V D D S G T T I A G R N A K T G P N L Y
K — — C K T C H S I I A P D G T E I V — K G A K T G P N L Y
K — — C L A C H T F D Q G G — — — — — — — A N K V G P N L F
A K — C T A C H D L N K G G — — — — — — — P I K V G P P L F
Q — — C M T C H R A D — — — — — — — — — — K N M V G P A L G
K — — C A A C H K I G P G A — — — — — — — K N G V G P S L N
Q — — C L V C H S I G P G A — — — — — — — K N K V G P V L N
Q K — C A Q C H T V E N G G — — — — — — — K H K V G P N L W
Q — — C K I C H Q V G P T A — — — — — — — K N G V G P E Q N
T I — C I L C H T D I K G — — — — — — — — R N K V G P S L Y
Q — — C K A C H S I D A G — — — — — — — — K N G V G P S L S
Q — — C K A C H S I E A G — — — — — — — — K N G V G P S L S
L — — C K A C H S V E A G — — — — — — — — K N G V G P S L A
M — — C K A C H S V E A G — — — — — — — — K N G V G P S L A
T K G C L A C H N P E K — — — — — — — — — K — V V G P A Y G
K A G C A V C H Q P T A — — — — — — — — — K — G L G P S Y Q
N K G C V A C H A I D T — — — — — — — — — K — M V G P A Y K
S K P C A A C H T I D S — — — — — — — — — K — M V G P A L K
S K P C A A C H S I D A — — — — — — — — — K — L V G P A F K
S K P C G A C H S V Q A — — — — — — — — — K — L V G P A L K
A K A C V A C H S V D K — — — — — — — — — K — L V G P A F H
T K G C T V C H A I D S — — — — — — — — — K — L V G P S F K
S A N C A A C H M G G — — — — — — — — — — R N V I V A N — K
S A N C A S C H L G G — — — — — — — — — — R N V V N P A — K
T A D C S A C H A G G — — — — — — — — — — N N V I M P E — K
S A N C A A C H A G G — — — — — — — — — — N N A I M P D — K
T A N C A A C H A G G — — — — — — — — — — N N V I M P E — K
T G N C A A C H S V Z — — — — — — — — — — Z Z J T L E L — S
A D N C S T C H V N G — — — — — — — — — — G N V I S A G — K
G T — C V A C H G T D — — — — — — — — — — G Q G L A P I Y P
A V — C G A C H G A B — — — — — — — — — — B B G S A P P F P
P A — C T G C H G S S P — — — — — — — — — V G I A T A G F P
A K H C N A C H G A G V — — — — — — — — — — — L G A P K I G
A S — C A M C H K T G M — — — — — — — — — — — M G A P K V G
A N — C A S C H A A G I — — — — — — — — — — — M G A P K T G
— K S C I G C H S A D G — — — — — — — — — — — G K A M M T N

(*continued*)

TABLE 2—*Continued*

Size Class	Species	Tuna numbering: 40 … 50
L	*Paracoccus denitrificans* c_{550}	G V V G R K I A S E E G F K — Y G E G I
	Rhodopseudomonas sphaeroides c_2	G V V G R T A G T Q A D F K G Y G E G M
	Rhodopseudomonas capsulata c_2	G V V G R T A G T Y P E F K — Y K D S I
	Rhodospirillum rubrum c_2	G V F E N T A A H K D N Y A — Y S E S Y
	Rhodospirillum photometricum c_2	G V F G R T T G T F A G Y S — Y S P G Y
	Rhodopseudomonas palustris c_2	G V V G R K A G T A A G F T — Y S P L N
M	*Rhodopseudomonas acidophila* c_2	G V A N R K A G Q A E G F A — Y S D A N
	Rhodopseudomonas viridis c_2	G L F G R H S G T I E G F S — Y S D A N
	Tuna *c*	G L F G R K T G Q A E G Y S — Y T D A N
	Rhodomicrobium vannielii c_2	D V F G Q K A G A R P G F N — Y S D A M
	Rhodopseudomonas globiformis c_2	G V V G R H S G I E P G Y N — Y S E A N
	Rhodospirillum molischianum c_2, iso-1	G A Y G R K V G L A P N Y K — Y S P A H
	Rhodospirillum fulvum c_2, iso-1	G A Y G R K V G L A P N Y K — Y S A A H
	Rhodospirillum molischianum c_2, iso-2	G V Y G R K A G T I S G F K — F S D P H
	Rhodospirillum fulvum c_2, iso-2	G V Y G R K A G T L A G F K — F S D P H
S	*Rhodospirillum tenue* c_{551}	W V A K K Y A — — — — — — — — — — — —
	Rhodopseudomonas gelatinosa c_{551}	E I A K K Y K — — — — — — — — — — — —
	Pseudomonas aeruginosa c_{551}	D V A A K F A — — — — — — — — — — — —
	Pseudomonas fluorescens c_{551}	E V A A K N A — — — — — — — — — — — —
	Pseudomonas stutzeri c_{551}	E V A A K Y A — — — — — — — — — — — —
	Pseudomonas mendocina c_{551}	D V A A K N A — — — — — — — — — — — —
	Pseudomonas denitrificans c_{551}	D V A A K Y G — — — — — — — — — — — —
	Azotobacter vinelandii c_{551}	E V T A K Y A — — — — — — — — — — — —
S*	*Spirulina maxima* c_{554} (cyanobacterium)	T L — S K S D L A K — — — — — — — — —
	Anacystis nidulans c_{554} (cyanobacterium)	T L — E K A — — — — — — — — — — — — —
	Alaria esculenta f (brown alga)	T L — K K D — — — — — — — — — — — — —
	Porphyra tenera f (red alga)	T L — K K D — — — — — — — — — — — — —
	Bumilleriopsis filiformis f (yellow-green alga	T L — K K D — — — — — — — — — — — — —
	Monochrysis lutheri f (yellow-green alga)	S L — W K A K — — — — — — — — — — — —
	Euglena gracilis f (euglenoid alga)	V L — S K T A I E — — — — — — — — — —
	Halotolerant Micrococcus c_{554}	N L — T G Q S A — — — — — — — — — — —
	Pseudomonas aeruginosa c_4, 1st half	K L — A G Q G E — — — — — — — — — — —
	Pseudomonas aeruginosa c_4, 2nd half	H L — G G Q H A — — — — — — — — — — —
	Pseudomonas mendocina c_5	D T A A — — — — — — — — — — — — — — —
	Chlorobium thiosulfatophilum[b] c_{555}	D K A A — — — — — — — — — — — — — — —
	Prosthecochloris aestuarii[c] c_{555}	T A R K — — — — — — — — — — — — — — —
	Desulfovibrio vulgaris c_{553}	A V K G — — — — — — — — — — — — — — —

(continued)

TABLE 2—*Continued*

Tuna numbering
60 70
L E V A E K N P D L T W T E A N L I E Y V T D P K P L V K K M T
K E A G A — — K G L A W D E E H F V Q Y V Q D P T K F L K E T Y
V A L G A — — S G F A W T E E D I A T Y V K D P G A F L K E K L
T E M K A — — K G L T W T E A N L A A Y V K D P K A F V L E K S
V T M G Q — — K G H T W D D N A L K A Y L L D P K G Y V Q A K S
H N S G E — — A G L V W T Q E N I I A Y L P D P N A Y L K K F L
K N — — — — — S G L T W D E A T F K E Y I T A P Q K K V P — — —
K N — — — — — S G I T W T E E V F R E Y I R D P K A K I P — — —
K S — — — — — K G I V W N N D T L M E Y L E N P K K Y I P — — —
K N — — — — — S G L T W D E A T L D K Y L E N P K A V V P — — —
I K — — — — — S G I V W T P D V L F K Y I E H P Q K I V P — — —
L A — — — — — S G M T I D D A M L T K Y L A N P K E T I P — — —
L A — — — — — S G M T I D E A M L T N Y L A N P K A T I P — — —
I K — — — — — S G L T W D E P T L T K Y L A D P K T V I P — — —
A K — — — — — S G L T W D E P T L T K Y L A D P K G V I P — — —
— — — — — — — — G Q A G A E A K L V A K V M A G G Q G V W A K Q
— — — — — — — — G Q A G A P A L M A E R V R K G S V G I F — — —
— — — — — — — — G Q A G A E A E L A Q R I K N G S Q G V W — — —
— — — — — — — — G V K D A D K T L A G H I K N G T Q G V[d] W — — —
— — — — — — — — G Q D G A A D L L A G H I K N G S Q G V W — — —
— — — — — — — — G V D G A A D V L A G H I K N G S T G V W — — —
— — — — — — — — A Q G D G V A H I T N S I K T G S K G N W — — —
— — — — — — — — G Q A G I A D T L A A K I K A G G S G N W — — —
— — — — — — — Y L K G F D D D A V A — A V A — Y Q V T N — — — —
— — — — — — — D L D E Y G M A S I E — A I T — T Q V T N — — — —
— — — — — — — A L A D N K M V S V N — A I T — Y Q V T N — — — —
— — — — — — — V L E A N S M N T I D — A I T — Y Q V Q N — — — —
— — — — — — — A L E A N G M N A V S — A I T — Y Q V T N — — — —
— — — — — — — S Y L A N F N G D E S — A I V — Y Q V T N — — — —
— — — — — — — E Y L D G G Y — T K E — A I E — Y Q V R N — — — —
— — — — — — — T Y L E S S I — — K A — Y R D G Q R K G G — — — —
— — — — — — — R Y L L K Q M — — H D — I K D G K R T V L — — — —
— — — — — — — T Y V A K Q L — — T D — F R E G T R N D D — — — —
— — — — — — — — W K E R A D — — — H Q G G L D G I L A K I S — —
— — — — — — — — W A P H I A K G M N V M V A N S I K G Y K G — —
— — — — — — — — W N S P L P Q G L A T M I E K S V A G Y E G E Y
— — — — — — — — K Y S D E E L K A L A D Y M K A A M G S A K P V

(*continued*)

TABLE 2—*Continued*

Size Class	Species	Tuna numbering 80
L	*Paracoccus denitrificans* c_{550}	D D K G — — — — — — A K T K M — — — — — T
	Rhodopseudomonas sphaeroides c_2	G D A K — — — — — — A K G K M — — — — — T
	Rhodopseudomonas capsulata c_2	D D K K — — — — — — A K T G M — — — — — A
	Rhodospirillum rubrum c_2	G D P K — — — — — — A K S K M — — — — — T
	Rhodospirillum photometricum c_2	G D P K — — — — — — A N S K M — — — — — I
	Rhodopseudomonas palustris c_2	T D K G Q A D K A T G S T K M — — — — — T
M	*Rhodopseudomonas acidophila* c_2	— — — — — — — — — — G — T K M — — — — — T
	Rhodopseudomonas viridis c_2	— — — — — — — — — — G — T K M — — — — — I
	Tuna *c*	— — — — — — — — — — G — T K M — — — — — I
	Rhodomicrobium vannielii c_2	— — — — — — — — — — G — T K M — — — — — V
	Rhodopseudomonas globiformis c_2	— — — — — — — — — — G — T K M — — — — — G
	Rhodospirillum molischianum c_2, iso-1	— — — — — — — — — — G — N K M — — — G A A
	Rhodospirillum fulvum c_2, iso-1	— — — — — — — — — — G — N K M — — — G A S
	Rhodospirillum molischianum c_2, iso-2	— — — — — — — — — — G — N K M — — — — — V
	Rhodospirillum fulvum c_2, iso-2	— — — — — — — — — — G — N K M — — — — — V
S	*Rhodospirillum tenue* c_{551}	L G — — — — — — — — A E I P M — — — — P A
	Rhodopseudomonas gelatinosa c_{551}	— — — — — — — — — — G K L P M T P T P P
	Pseudomonas aeruginosa c_{551}	— — — — — — — — — — G P I P M — — — — P P
	Pseudomonas fluorescens c_{551}	— — — — — — — — — — G P I P M — — — — P P
	Pseudomonas stutzeri c_{551}	— — — — — — — — — — G P I P M — — — — P P
	Pseudomonas mendocina c_{551}	— — — — — — — — — — G A M P M — — — — P P
	Pseudomonas denitrificans c_{551}	— — — — — — — — — — G P I P M — — — — P P
	Azotobacter vinelandii c_{551}	— — — — — — — — — — G Q I P M — — — — P P
S*	*Spirulina maxima* c_{554} (cyanobacterium)	— — — — — — — — — — G K N A M — — — — P G
	Anacystis nidulans c_{554} (cyanobacterium)	— — — — — — — — — — G K G A M — — — — P A
	Alaria esculenta f (brown alga)	— — — — — — — — — — G K N A M — — — — P A
	Porphyra tenera f (red alga)	— — — — — — — — — — G K N A M — — — — P A
	Bumilleriopsis filiformis f (yellow-green alga)	— — — — — — — — — — G K N A M — — — — P A
	Monochrysis lutheri f (yellow-green alga)	— — — — — — — — — — G K N A M — — — — P A
	Euglena gracilis f (euglenoid alga)	— — — — — — — — — — G K G P M — — — — P A
	Halotolerant Micrococcus c_{554}	— — — — — — — — — — N A A L M — — — — T P
	Pseudomonas aeruginosa c_4, 1st half	— — — — — — — — — — — — E E M — — — — T G
	Pseudomonas aeruginosa c_4, 2nd half	— — — — — — — — — — G T K I M Q — — — S I
	Pseudomonas mendocina c_5	— — — — — — — — — — G I N A M — — — — P P
	Chlorobium thiosulfatophilum[b] c_{555}	— — — — — — — — — — T K G M M — — — — P A
	Prosthecochloris aestuarii[c] c_{555}	R G — — — — — — — — S K T F M — — — — P A
	Desulfovibrio vulgaris c_{553}	K G Q G A E — — — — E L Y K M — — — — K G

(continued)

TABLE 2—*Continued*

Tuna numbering
90 100
F K — — M G K — — — N Q A D V V A F L A Q B B P B A G Z G Z A A G A G S B S Z
F K — — L K K — E A D A H N I W A Y L Q Q V A V R P
F K — — L A K — — — G G E D V A A Y L A S V V K
F K — — L T K — D D E I E N V I A Y L K T L K
F R — — L E K — D D D V A N V I A Y L H T M K
F K — — L A N — D Q Q R K D V A A Y L A T L K
F P G — L P N — E A D R D N I W A Y L S Q F K A D G S K
F A G — I K D — E Q K V S D L I A Y L K Q F N A D G S K K
F A G — I K K — K G E R Q D L V A Y L K S A T S
F V G — L K N — P Q D R A D V I A Y L K Q L S G K
Y P G — Q P D — P Q K R A D I I A Y L E T L K
F G G — L K N — P A D V A A V I A Y L K T V K
F G G — L K K — P E D V K A V I E Y L K T V K
F A G — L K N — P D D V K A V I E Y L K T L K
F A G — L K N — P A D V A A V I A Y L K S L
N N — — V T — — K E E A T R L V K W V L S L K Q I D Y K
A R — — I S — — D A D L K L V I D W I L K T P
N A — — V S — — D D E A Q T L A K W V L S Q K
N Q — — V T — — D A E A L T L A Q W V L S L K
N P — — V T — — E E E A K I L A E W I L S Q K
N P — — V T — — E E E A K T L A E W V L T L K
N A — — V S — — P E E A K T L A E W I V T L K
N P — — V S — — E A E A K T L A E W V L T H K
F N G R L S — — P L Q I E D V A A Y V V D Q A E K G W
F G A K L S — — A D D I E G V A S Y A L D Q S G K E W
F G S R L A — — E T D I E D V A N F V L T B Z B K G W E
F G G R L V — — D E D I E D A A N Y V L S Q S E K G W
F G G R L S — — D S D I E D V A N Y V L S Q S E Q G W D
F G G R L E — — D D E I A B V A S Y V L S K A G
W E G V L S — — E D E I V A V T D Y V Y T Q A G G A W A N V
M A Q G L S — — D E D I A D I A A Y Y S S Q E
L L T B L S — — B Z D I A A L A D Y A S Q K M S V G M A L B B P V A G G E A
A A I K L S — — N K D I A A I S S Y I Q G L H
K G T C A D — — C S D D E L R E A I Q K M S G L
K G G N P K L T D A Q V G N A V A Y M V G Q S K
K G G N P D L T D K Q V G D A V A Y M V N E V L
Y A D G S Y — — G G E R K A M S K L

TABLE 3
PAIRWISE COMPARISONS OF SEQUENCES IN TABLE 2[a]

Size Class	Species		A	B	C	D	E	F	G	H	I	J	K	L	M	N
L	*P. de*	A	**134**	54	64	50	40	47	41	40	37	38	34	34	34	34
	R. sp	B	85	**124**	62	49	43	42	46	37	34	36	32	33	33	33
	R. ca	C	71	62	**116**	51	41	52	34	40	34	36	33	35	36	35
	R. ru	D	86	75	68	**112**	63	46	42	40	37	41	36	37	38	39
	R. ph	E	97	82	79	50	**113**	45	39	41	43	40	37	42	39	42
	R. pa	F	95	88	72	72	74	**114**	44	48	40	48	40	39	37	44
M	*R. ac*	G	96	81	90	76	80	80	**107**	61	46	59	47	45	43	47
	R. vi	H	97	91	84	79	79	76	47	**107**	54	56	47	43	44	47
	Tuna	I	101	92	86	78	72	80	62	54	**103**	47	45	40	41	40
	R. va	J	99	89	85	74	76	73	48	52	58	**104**	47	50	48	55
	R. gl	K	107	97	92	82	81	83	65	66	63	62	**106**	40	37	40
	R. mo	L	105	94	86	78	74	80	64	66	65	56	68	**100**	88	64
	R. fu	M	105	94	85	77	77	82	66	65	64	58	71	12	**99**	66
	R. mo	N	103	92	84	74	72	73	60	60	63	49	66	36	33	**97**
	R. fu	O	101	88	82	76	71	75	59	60	61	47	67	33	36	13
S	*R. te*	P	123	114	107	95	99	105	97	102	97	100	103	101	99	96
	R. ge	Q	128	117	109	102	104	108	96	98	95	96	100	89	89	89
	P. ae	R	124	111	103	91	97	97	90	93	87	87	93	87	87	85
	P. fl	S	125	109	104	90	96	94	87	94	84	88	89	85	88	84
	P. st	T	124	113	102	93	98	96	88	92	86	89	90	86	84	83
	P. me	U	122	113	102	90	95	94	89	89	85	86	86	86	83	79
	P. de	V	127	115	107	96	98	96	93	94	89	89	90	87	84	82
	A. vi	W	128	111	104	94	97	97	89	93	89	91	89	88	88	85
S*	*S. ma*	X	123	111	105	97	104	104	89	93	90	88	95	89	91	91
	A. ni	Y	127	116	108	100	106	103	90	93	95	88	99	90	91	92
	A. es	Z	123	117	109	105	107	105	91	94	99	93	100	92	93	94
	P. te	a	126	114	105	102	105	104	86	92	92	88	99	91	91	91
	B. fi	b	126	115	107	104	104	105	90	93	95	91	99	90	91	92
	M. lu	c	122	112	99	96	99	101	91	92	86	86	94	86	87	86
	E. gr	d	130	115	114	106	112	110	95	98	100	95	103	99	100	95
	H. Mi	e	121	111	103	93	96	100	90	95	90	89	93	90	92	90
	P. ae	f	124	124	123	112	110	114	99	104	107	100	110	104	105	101
	P. ae	g	132	117	115	108	106	111	102	102	96	94	99	96	97	93
	P. me	h	134	122	118	110	106	115	105	106	103	102	104	99	101	100
	C. th	i	127	111	107	102	103	104	95	98	90	95	100	91	94	94
	P. ae	j	134	126	114	110	112	111	110	108	103	105	103	107	107	110
	D. vu	k	131	121	115	107	108	103	110	109	103	106	106	105	104	100

[a] Upper right: identical residues. Lower left: differences between chains. Matrix diagonal: number of amino acids in each sequence. Identification letters along top of table correspond to those at left.

TABLE 3—*Continued*

O	P	Q	R	S	T	U	V	W	X	Y	Z	a	b	c	d	e	f	g	h	i	j	k
36	16	14	15	14	15	17	12	11	18	14	18	15	15	19	11	19	16	9	10	14	14	12
37	15	13	16	18	14	14	12	16	19	14	14	16	16	17	18	17	16	12	10	18	10	10
37	18	15	18	17	19	19	14	17	21	18	18	21	20	24	15	20	14	9	10	16	18	10
37	25	16	24	25	22	25	19	21	24	21	17	19	18	22	18	23	19	9	12	16	16	12
43	21	14	18	19	17	20	17	18	17	15	15	16	18	19	12	20	21	11	16	15	14	11
42	18	13	21	24	22	24	22	21	19	20	19	19	19	19	16	19	20	10	11	17	19	17
48	19	18	21	24	23	22	18	22	23	22	21	26	22	21	19	21	21	10	12	18	15	11
47	14	16	18	17	19	22	17	18	19	19	18	20	19	20	15	17	16	11	12	15	18	12
42	18	14	19	22	20	21	17	17	20	15	12	18	16	21	13	17	13	12	10	18	18	13
57	16	15	21	20	19	22	19	17	23	23	19	23	21	23	19	19	20	15	12	15	17	12
39	16	12	16	20	19	23	19	20	19	15	15	15	16	17	14	16	13	11	8	12	17	13
67	14	18	18	20	19	19	18	17	21	20	19	19	21	21	14	16	16	11	14	17	14	10
63	16	18	18	17	21	22	21	17	19	19	18	19	20	20	13	14	15	10	12	14	14	11
84	18	18	19	20	21	25	22	19	18	17	16	18	18	20	17	15	18	13	12	13	10	14
96	17	17	18	20	18	20	18	16	19	16	15	19	19	21	15	18	17	13	13	14	11	12
97	**92**	30	46	41	43	39	34	40	12	14	14	15	16	16	13	11	11	6	12	14	14	18
90	65	**85**	32	26	31	27	26	32	17	18	15	17	19	16	15	15	12	13	13	15	15	11
86	46	53	**82**	56	56	50	47	50	15	19	17	21	23	19	18	19	12	8	16	14	13	14
84	51	59	26	**82**	60	56	42	48	18	18	16	21	20	18	17	14	13	8	15	17	14	15
86	49	54	26	22	**82**	64	50	54	17	19	18	21	19	18	18	16	14	7	15	15	12	15
84	53	58	32	26	18	**82**	49	51	17	18	20	19	17	20	19	15	14	8	17	14	12	13
86	58	59	35	40	32	33	**82**	47	16	19	16	17	17	17	17	17	16	11	16	14	11	15
88	52	53	32	34	28	31	35	**82**	15	16	20	19	21	17	20	13	12	10	12	13	13	14
90	90	82	81	78	79	79	80	81	**89**	44	43	47	50	37	36	20	19	13	15	20	20	8
93	85	78	74	75	74	75	74	77	45	**85**	43	49	49	37	35	17	18	15	13	18	20	6
95	85	82	77	78	76	74	78	74	47	43	**86**	57	64	38	33	17	17	10	16	14	18	5
90	84	79	72	72	72	74	76	74	42	36	29	**85**	68	42	32	21	21	13	14	19	25	5
91	83	78	71	74	75	77	77	73	40	37	22	18	**86**	46	34	20	20	13	16	18	22	5
85	83	77	71	72	72	70	73	73	52	49	49	44	41	**83**	32	22	19	13	11	16	19	8
97	87	83	77	78	77	76	78	75	56	56	58	59	57	59	**86**	16	17	16	9	14	15	8
87	90	79	72	77	75	76	74	78	72	73	74	69	71	65	77	**83**	38	23	16	18	15	12
102	100	97	94	93	92	92	90	94	84	83	84	80	81	82	84	60	**96**	24	15	17	14	13
93	96	81	84	84	85	84	81	82	81	77	83	79	80	76	79	62	76	**85**	7	6	7	10
99	90	84	78	79	79	77	78	82	86	84	82	83	82	84	93	78	93	88	**86**	21	21	7
93	85	79	77	74	76	77	77	78	79	77	82	76	78	77	87	76	91	89	71	**86**	47	10
109	94	92	91	90	92	92	93	91	92	88	91	83	87	87	99	91	106	100	79	52	**99**	14
101	80	85	79	78	78	80	78	79	96	94	96	95	96	90	95	87	101	90	90	85	90	**82**

be transferred from one bacterium to another (7,8). Plasmids or loops of DNA containing genes conferring antibiotic resistance can sweep through a bacterial population in a hospital, a situation highly advantageous for the host bacteria but less so for us. One would never claim that two strains of bacteria were related simply because both carried genes for tetracycline resistance. Is one any more justified in claiming relatedness because they both contain genes for similar cytochromes *c*? The question cannot be answered in the abstract for all proteins; each must be considered on its own merits.

For a protein that is intrinsically valuable to its host by itself, such as an enzyme that inactivates an antibiotic, the transfer of a single gene could be of great value. But a cytochrome *c* is useless without all the other paraphernalia of photosynthesis or respiration: several cytochromes *b*, *c*, and *a*; dehydrogenase flavoproteins; iron–sulfur proteins; and the enzymes that synthesize quinones, chlorophylls, and carotenoids. One could imagine all these genes being transferred in one long DNA segment if they were genetically linked and contiguous, and recent experiments with Rhodospirillaceae by Marrs suggest that this may be partly true (45,63). The genes for production of the photopigment system—chlorophylls, carotenoids, and associated membrane proteins—are closely linked and can be transferred as a unit by plasmids or by viruslike gene transfer agents. The genes for cytochromes and other electron transport components are not closely linked to these, but one could imagine (without evidence so far) that they were themselves mutually linked in a second DNA "module." The passing on of photosynthesis from one bacterium to another then might only require the simultaneous transfer of the DNA modules for photopigments and electron transport.

These speculations about "jumping genes" may be even more significant in connection with the spread of the ability to respire using O_2. The photosynthetic bacteria, although diverse, are not so much so that one would despair of imagining a common ancestry. But respiration is found everywhere, including bacteria such as *Bacillus* and *Streptomyces*, for which there is little evidence of a photosynthetic ancestor. How can they have adapted a photosynthetic electron transport chain to respiration (the central thesis of this article) if they never had one? Block gene transfer of an entire respiratory system could have been the answer. Close linkage of electron transfer genes might appear to have little advantage for the host bacterium other than possibly one of genetic control. But it would be highly advantageous to the linked genes themselves, by enabling them to spread slowly through a bacterial population, leaving a trail of oxygen respirers behind. There is always the possibility that when we compare cytochromes from different bacteria, we are studying the "phylogeny" of the electron transport genes rather than that of the host bacterium. This in itself would be of great interest if true. Unfortunately, what little hard evidence there is to date concerning linkage of respiratory genes does not support the idea of block transfer (10,30,60). Table 4 shows the positions in *Escherichia coli* and *Sal-*

TABLE 4
LOCATION OF GENES NECESSARY FOR RESPIRATION IN *Escherichia coli* AND *Salmonella typhimurium*

	Position along genome[a]		
Gene	*E. coli*	*Salmonella*	Function
chlA	17	19	Pleiotropic effects on nitrate-chlorate reductase and hydrogen lyase activity
chlB	85	84	F_A factor of nitrate-bound membranous respiratory system
chlC	27	34	Nitrate reductase structural gene
chlD	17	18	Activation of nitrate reductase by molybdate
chlE	18	19	Nitrate reductase activity
chlF	26	—	Formate dehydrogenase structural gene
chlG	0	—	Formate-nitrate reductase activity
entA-F	13	13	Enterochelin synthesis for iron transport
fep	13	—	Iron transport by enterochelin
fesB	13	—	Hydrolysis of iron–enterochelin complex
frdA	93	—	Fumarate reductase structural gene
frdB	28	—	Anaerobic fumarate reduction
glpA	48	47	L-glycerol-3-phosphate dehydrogenase
glpD	74	75	D-glycerol-3-phosphate dehydrogenase
hemA	26	35	Synthesis of aminovlevulinic acid
hemB	8	8	Synthesis of heme for catalase and cytochromes
hemE	89	89	Elimination of uroporphyrin
menA	87	—	Menaquinone synthesis
sdh	16	—	Succinate dehydrogenase gene
ubiA-H	15,48, 62,84, 84,90, 90	—	Steps in synthesis of ubiquinone

[a] Expressed as percent distance along the entire genome.

monella of all those genes that have been located and shown to be necessary for respiration. They are not clustered in one or two places, but are scattered along the entire genome instead. It is difficult to imagine that either of these bacteria could have acquired respiration by block transfer, but the possibility is not ruled out for other bacteria. The linkage maps for *Pseudomonas* (31) and other bacteria are too sketchy for any conclusions to be drawn.

Although block transfer of large segments of the respiratory chain may have occurred several times early in the oxygenated stage of our atmosphere, the evidence is against any rapid, short-term scrambling of cytochrome *c* genes among present-day bacteria, of a type that would invalidate molecular taxonomy completely. Ambler has examined the sequences of cytochromes c_{551} from nine different strains of *Pseudomonas aeruginosa* and six of *P. fluorescens* obtained from such diverse locations as England, France, Denmark, Greenland, Brazil, Puerto Rico, California, and the Philippines (2). Eight of the *P. aeruginosa* sequences are absolutely identical, and the ninth differs only by a single amino acid. Individual *P. fluorescens* sequences from all over the world differ from the type sequence by one to four residues out of

82. By comparison, the type sequences of *P. aeruginosa* and *P. fluorescens* themselves differ at 26 positions. Cytochromes from different species of *Pseudomonas* vary by 18–40 residues, so the variation found by Ambler cannot even represent the lateral transfer of the c_{551} gene from one species to another within the genus *Pseudomonas*. If c_{551} genes were readily transferred, one would expect the species in any one geographic location to be somewhat "homogenized." Instead, the *P. aeruginosa* proteins from England and the Philippines are identical, whereas *P. aeruginosa* and *P. fluorescens* from England differ by 28 positions out of 82. Ambler has explained these findings by assuming that *P. aeruginosa* is a relatively young species that has not had time for extensive gene shuffling with other species, whereas *P. fluorescens* is an older species, in which the shuffling and mixing process is farther advanced. Yet species are not created uniform; they arise by reproductive isolation of one part of a heterogeneous gene pool. If one assumes continual and rapid lateral shuffling of cytochrome c_{551} genes, where would this initial uniformity of *P. aeruginosa* have come from? It seems simpler to interpret the evidence as indicating merely the species integrity of the cytochrome *c* gene in *Pseudomonas*.

Evolution of Photosynthetic Bacteria

Because of this lingering fear of being deceived by gene sharing between bacteria, one cannot build a phylogenetic tree purely on the data from one protein. For prokaryotes even more than for eukaryotes, it is important to assemble all the genetic, biochemical, and metabolic evidence possible, in support of or in opposition to a particular theory of origins. Table 1 is a step in this direction. It shows that there are three clear divisions among the photosynthetic bacteria—green, purple, and blue-green—and that each division has developed a family of respirers. The division of purple photobacteria into the Chromatiaceae or purple sulfur bacteria (H_2S-using, nonrespiring) and Rhodospirillaceae or purple nonsulfur bacteria (organic-using, respiring) is a familiar one, although only in the last dozen years or so has it been realized how metabolically versatile the Rhodospirillaceae are (41,70) and how widespread is their ability to respire under aerobic conditions in the dark (29,35,39,53). But it was only recently that a similar respiring green photosynthetic bacterium was discovered: *Chloroflexus aurantiacus* (57). This and three other species now make up the Chloroflexaceae or family of green filamentous bacteria (55,68), which are oxygen respirers like the Rhodospirillaceae. As Broda was the first to suggest (12), respiration apparently was invented independently more than once in different lines of photosynthetic bacteria.

The way that this invention occurred is best studied in the purple bacteria, about which more is known (36,39,46,47,58). The electron transport chains of eukaryotic mitochondria and purple bacteria are compared in Fig. 2. Light-excited bacteriochlorophyll in these bacteria feeds electrons into a

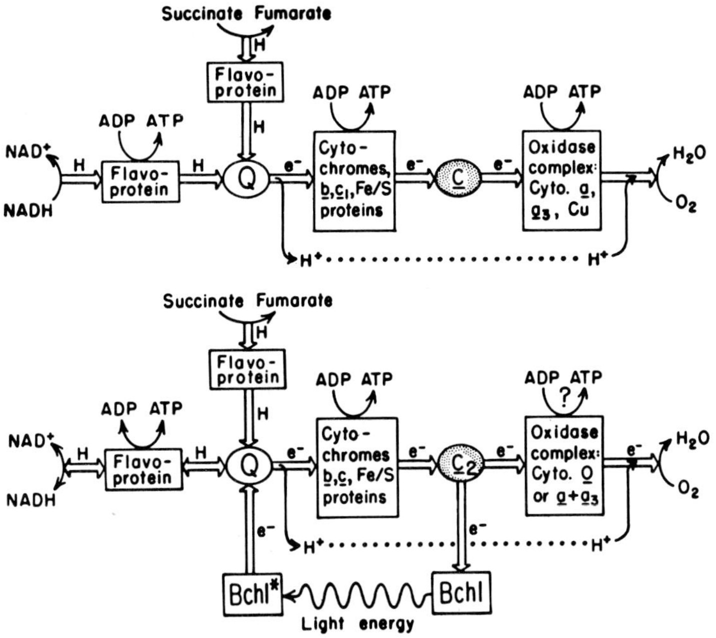

Fig. 2. Respiratory electron transport chain from eukaryotic mitochondria (above), and photosynthetic-respiratory chain from Rhodospirillaceae (below). Both chains contain similar components, except for the bacteriochlorophyll center and for some variation in terminal cytochrome oxidases. Double-headed arrows in the lower diagram represent the generation of NADH from succinate reducing hydrogens in reversed electron flow drived by ATP. Rectangles enclose complexes of several proteins, which are embedded in the bacterial or mitochondrial membrane. Q = ubiquinone; b, c, c_1, c_2, o, a, a_3 = cytochromes; Bchl* = electronically excited bacteriochlorophyll. There is also a c_2-independent pathway from cytochrome b to the oxidase in the lower diagram.

ubiquinone pool, from which they pass to cytochromes b and c_2, and back to chlorophyll again, with the absorbed light energy used to synthesize ATP. Organic molecules such as succinate can feed reducing equivalents into the pool, and these can be used to reduce NAD^+ to NADH by reversed electron flow using the driving energy from ATP. Unlike the green bacteria, the purple photosynthetic bacteria apparently do not generate a low enough potential by photo-absorption to reduce NAD^+ directly to NADH (25,36).

This combination of cyclic photophosphorylation and reversed electron flow is not the entire story. The Rhodospirillaceae possess a complete tricarboxylic acid cycle, and NADH from this can feed reducing equivalents *into* the quinone pool. At the high-potential end of the chain, cytochrome c_2 can pass electrons to a cytochrome oxidase as well as to bacteriochlorophyll, and this oxidase can reduce O_2 to H_2O just as in mitochondrial respiration (39). (An additional pathway also leads directly from cytochrome b, through a different oxidase, to O_2.) If the bacteriochlorophyll electron loop is deleted from the photosynthetic diagram in Fig. 2, then the results are virtually identical with the mitochondrial chain shown above it. It seems probable that some of the Chromatiaceae first developed the ability to obtain reducing equivalents from organic matter, freeing them from a dependence on sulfides and turning them into the ancestors of the Rhodospirillaceae. Some of these then "invented" respiration by adapting one or more heme proteins as bridges to carry electrons from cytochromes to oxygen. This latter step would have come about in response to the gradually increasing availability of free O_2 in the atmosphere, itself a product of water-splitting photosynthesis by the cyanobacteria.

The bacterial cytochrome oxidases are more diverse than the other components of electron transport (11,21,62): cytochrome *a* as in mitochondria (*Rps. sphaeroides, Paracoccus, Bacillus, Mycobacterium* and others), cytochrome *b* of *o* (*Rps. capsulata, R. rubrum,* etc.), or cytochrome *cd* (*Pseudomonas*). Many bacteria, including *Rps. sphaeroides, Halobacterium, Haemophilus, Azotobacter,* and *Thiobacilli,* have both cytochromes *a* and *o*, and others can have both *a* and *cd* (*Paracoccus, Pseudomonas*). Furthermore, as Martin Kamen has pointed out (15a,23a), bacterial cytochromes c_2 generally exhibit good reactivities with mitochondrial reductase but poor reactivities with mitochondrial oxidase, exactly as would be expected if the reductase was a common evolutionary heritage, whereas the oxidases were more diverse. Perhaps this diversity reflects the late, add-on character of the oxidase shunt from cytochrome *b* or *c* to oxygen. All that would be required to turn one of the metabolically versatile Rhodospirillaceae into the ancestors of some of the present-day respiring bacteria would be a mutational lesion in the photosynthetic side of the machinery, at a time when the O_2 concentration in the atmosphere had risen to such a point that the bacterium could survive on respiration alone.

All these relationships between photosynthetic and respiring bacteria are summarized in the metabolic tree shown in Fig. 3. Variants of this tree have appeared at regular intervals over the past four years (1,20,21,23), but it is asymptotically approaching a reasonably final form. The tree is based on what we know about comparative bacterial metabolism. It is *not* merely derived from sequence data, which has been a crippling defect in some earlier published trees. Nevertheless, Fig. 3 is consistent with the cytochrome *c* story and helps to make it understandable. The earliest photosynthesizers to leave modern descendants probably were the green sulfur bacteria, photoreducing NAD^+ directly to NADH with the aid of H_2S and releasing sulfate into the surroundings. Some opportunistic neighbors, not necessarily closely related, then found ways to profit by using this sulfate as an oxidant, leading to today's *Desulfovibrio.* The Calvin–Benson cycle developed by turning gluconeogenesis into a cyclic process in the line that led to the purple and blue-green photosynthetic bacteria. With the energy from two photocenters, the ancestors of cyanobacteria were able to replace H_2S by a weaker but much more plentiful reductant: water. Over perhaps a billion-year time span, the cyanobacteria slowly converted the planetary atmosphere from reducing to

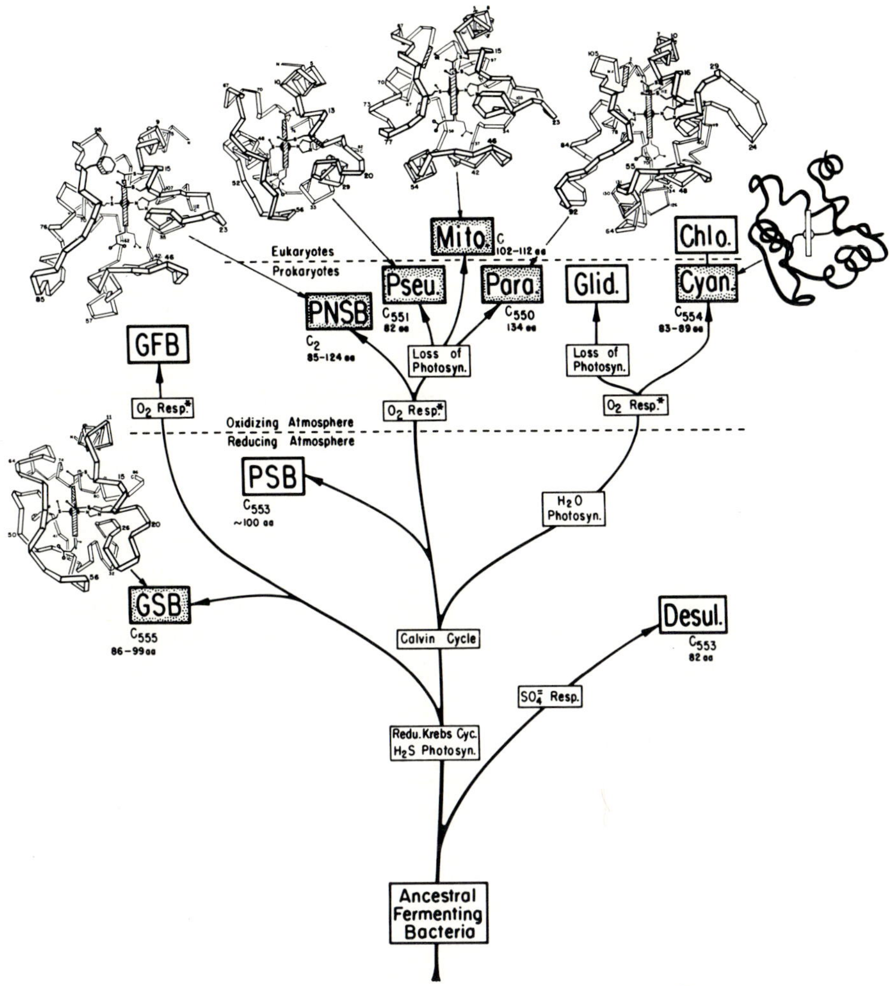

Fig. 3. Phylogenetic tree of the photosynthetic bacteria and their respiring descendants. Sulfate respiration is assumed to have arisen as a response to sulfate-releasing photosynthesis early on the primitive Earth, just as oxygen respiration arose as a consequence of oxygen-releasing photosynthesis at a later era. Oxygen respiration has evolved independently from photosynthesis at least three times: in the green, purple, and blue-green photosynthetic bacteria. The known cytochrome *c* structures are shown, including the nearly completed c_{554} from a cyanobacterium. GSB = green sulfur bacteria (Chlorobiaceae); GFB = green filamentous bacteria (Chloroflexaceae); PSB = purple sulfur bacteria (Chromatiaceae); PNSB = purple nonsulfur bacteria (Rhodospirillaceae); *Pseu.* = *Pseudomonas*, Mito. = eukaryotic mitochondria, *Para.* = *Paracoccus*, Glid. = gliding bacteria related to cyanobacteria, Cyan. = cyanobacteria or blue-green algae, Chlo. = eukaryotic chloroplasts, *Desul.* = *Desulfovibrio*.

oxidizing, forcing all the other previously anaerobic bacteria either to adapt to the presence of O_2 or to retreat to still anaerobic corners of the planet. Oxidation of organic molecules with O_2 is such an efficient source of energy that the general response was to find ways of adapting preexisting photosynthetic electron transport chains to feed electrons to O_2. Respiration was born not once but repeatedly, in different lines of photosynthetic bacteria by adapting a heme protein as an electron shunt from cytochromes to oxygen.

The Chloroflexaceae, Rhodospirillaceae, and gliding bacteria such as *Beggiatoa* and *Leucothrix* most probably are respiring adaptations of the green, purple, and blue-green photosynthetic bacteria, respectively. Just as these gliding bacteria appear to be cyanobacteria that have lost photosynthesis (65), so many of the respiring bacteria probably arose from Rhodospirillaceae by loss of photosynthesis (12,21). Included among these would be the pseudomonads, *Paracoccus*, and the bacterial ancestors of eukaryotic mitochondria. *Paracoccus* has been suggested on many lines of evidence as being most like the probable ancestor of mitochondria (32,33), and Table 3 reveals that *Paracoccus* cytochrome c_{550} most closely resembles the cytochrome c_2 of *Rps. sphaeroides* and *Rps. capsulata*, both of which have respiratory machinery much like that of mitochondria. Perhaps in these two species of purple bacteria we are seeing our own metabolic ancestor.

Which of the three size classes of cytochromes *c* is the oldest: S, M, or L? Has the evolution of cytochrome *c* been a process of trimming down a larger ancestral molecule or building up a smaller one? Chains as short as the S class are widely distributed in the tree of Fig. 3: in *Desulfovibrio*, *Chlorobium*, the cyanobacteria, Rhodospirillaceae, and nonphotosynthetic respirers. In contrast, the larger molecules are found only in the purple bacteria and the respirers that developed from them. Hence one probably should regard the short cytochromes as being more like the common ancestor and the insertions I–IV in Fig. 1 as being late developments that occurred in only one branch. Table 2 shows that many of these insertions are related in different species. *Paracoccus*, *Rps. sphaeroides*, and *Rps. capsulata* all have an insertion in region IV that can be described as

-glycine-threonine-(polar)-isoleucine----(basic)-

Unless we want to postulate a special structural requirement for just this sequence or imagine that evolution repeated itself in some mysterious way, this is evidence for a common ancestral history for these three cytochromes. Similarly, there are enough correspondences within regions I, II, and III to suggest common ancestors for all L cytochromes, and for all M, rather than random, independent enlargements of the chain. Indeed, the fact that these insertions always occur at exactly the same places along the chain also suggests that they had a common origin.

Taxonomy within the Rhodospirillaceae

One serious dilemma is that the foregoing observations cut diametrically across the three morphological genera of the family Rhodospirillaceae as set out in *Bergey's Manual* (13). Long, medium, and short cytochromes c_2 are encountered in species of both *Rhodospirillum* and *Rhodopseudomonas*. Sequence comparisons indicate closer relationships between species of different genera than between some of the species within one genus. This has led to the fear that lateral gene transfer may indeed by extensive (7,8), so much so that nothing can be learned about bacterial phylogeny from molecular criteria. However, the Bergey genera are strictly morphological and determinative, not phylogenetic. The manual is a guide to identification, not a prokaryotic "Almanach de Gotha." If the Bergey genera are set aside and if all the presently available molecular evidence is pooled from cytochrome structures and sequences (19), photosynthetic membrane structure (56), and 16 S ribosomal RNA sequences (26,27), then a different and more orderly picture emerges, as indicated in Table 5, taken from Dickerson (19). Those species that have the small cytochromes c_2 also have Type II ribosomal RNA sequences and photosynthetic membranes with tubular infoldings. Those with M or L cytochromes have Type I RNA and lamellar or vesicular membrane structures. A preliminary assay at constructing a phylogenetic tree of the Rhodospirillaceae based on these molecular criteria is shown in Fig. 4. The major divisions are based on the molecular data indicated at the top of the figure, and the finer divisions, to be regarded more tentatively, come from the identities comparisons of Table 3. Circled crosses indicate where a lengthening of the DNA of the gene has led to insertions of loops of protein

TABLE 5

COMPARISON OF PROPERTIES OF PURPLE PHOTOSYNTHETIC BACTERIA AND THEIR CYTOCHROMES c_2^a

Group	RNA class	Cytochrome c_2 size	Chain insertions (Fig. 1) I	II	III	IV	Type of photosynthetic membrane	Mode of cell divisions	Cell form
Chromatiaceae	III	?	?	?	?	?	Mostly vesicular	Fission	Various
Rhodospirillaceae									
Rps. gelatinosa	II	S	No	No	No	No	Tubular	Fission	Small rod
R. tenue	II	S	No	No	No	No	Tubular	Fission	Thin spiral
Rcy. purpureus	II	?	?	?	?	?	Tubular	Fission	Arc or circle
R. fulvum[a]		M	Yes	No	No	No	Lamellar	Fission	Short spiral
R. molischianum[b]		M	Yes	No	No	No	Lamellar	Fission	Spiral
Rps. globiformis		M	Yes	No	No	No	Vesicular	Fission	Sphere
Rm. vannielii	I	M	Yes	No	No	No	Lamellar	Budding	Stalked ovoid
Rps. viridis	I	M	Yes	No	No	No	Lamellar	Budding	Rod
Rps. acidophila		M	Yes	No	No	No	Lamellar	Budding	Large rod
Rps. palustris	I	L	Yes	Yes	Extra	No	Lamellar	Budding	Rod
R. photometricum		L	Yes	Yes	Yes	No	Lamellar	Fission	Large spiral
R. rubrum	I	L	Yes	Yes	Yes	No	Vesicular	Fission	Spiral
Rps. capsulata	I	L	Yes	Yes	Yes	Yes	Vesicular	Fission	Rod
Rps. sphaeroides	I	L	Yes	Yes	Yes	Yes	Vesicular	Fission	Sphere

[a] Key to abbreviations: R., *Rhodospirillum*; Rps., *Rhodopseudomonas*; Rm., *Rhodomicrobium*; Rcy., *Rhodocyclus*; S, M. L, Short, medium, and long cytochromes c_2 as described in text.

[b] Possess two similar isozymes of c_2.

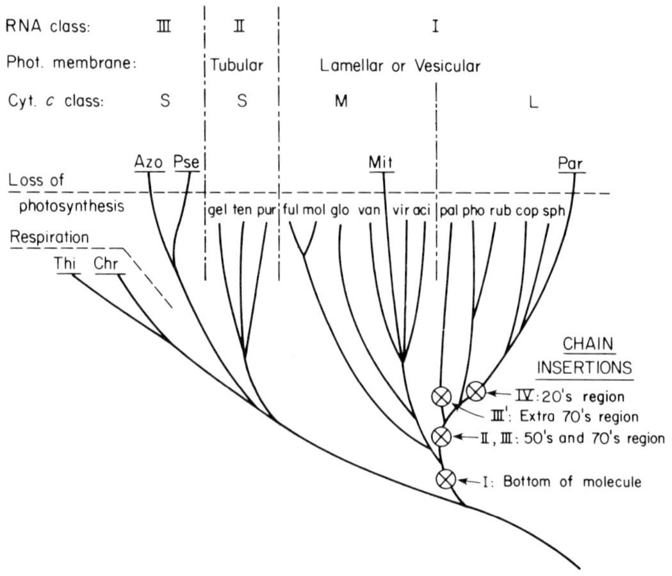

Fig. 4. Tentative phylogenetic tree of the purple photosynthetic bacteria, based on the information in Tables 1–4. The original cytochrome c_2 is assumed to have been small, like *Chlorobium* c_{555} and *Anacystis* c_{554}. Circled crosses mark the four places where specific insertions of polypeptide chain are presmed to have occurred. *Azo* = *Azotobacter; Pse* = *Pseudomonas;* Mit = eukaryotic mitochondria; *Par* = *Paracoccus denitrificans; Thi* = *Thiocapsa; Chr* = *Chromatium;* gel = *Rps. gelatinosa;* ten = *R. tenue;* pur = *Rhodocyclus purpureus;* ful = *R. fulvum;* mol = *R. molischianum;* glo = *Rps. globiformis;* van = *Rm. vannielii;* vir = *Rps. viridis;* aci = *Rps. acidophila;* pal = *Rps. palustris;* pho = *R. photometricum;* rub = *R. rubrum;* cop = *Rps. capsulata;* sph = *Rps. sphaeroides.*

chain. In this phylogeny, each insertion in regions I, II, III, and IV was a unique evolutionary event, making plausible the similarities of inserted sequences. This tree should be regarded only as a starting point for discussion, not the last word. The important point, however, is that nothing in the molecular data so far compels one to invoke a significant lateral shuffling of the cytochrome genes.

In a letter pointing out the fallacy of reading phylogeny into "Bergey's Manual," Norbert Pfennig provided a quotation from Roger Stanier that can serve appropriately as a close to these evolutionary considerations (64):

> It might have happened thus; but we shall surely never know with certainty. Evolutionary speculation constitutes a kind of metascience, which has the same intellectual fascination for some biologists that metaphysical speculation possessed for some medieval scholastics. It can be considered a relatively harmless habit, like eating peanuts, unless it assumes the form of an obsession; then it becomes a vice. The most appropriate response to such speculations (if they are plausible and logically consistent) is an Italian rejoinder, of which the amiable cynicism cannot be adequately translated: "*Se non è vero, è ben trovato.*"

REFERENCES

1. Almassy, R. J., and Dickerson, R. E., Pseudomonas cytochrome c_{551} at 2.0 Å resolution: Enlargement of the cytochrome *c* family. *Proc. Natl. Acad. Sci. U.S.A.*, 1978, **75**:2674–2678.
2. Ambler, R. P., The evolutionary stability of cytochrome c_{551} in *Pseudomonas aeruginosa* and *Pseudomonas fluorescens* biotype C. *Biochem. J.*, 1974, **137**:3–14.
3. Ambler, R. P., Bacterial cytochromes *c* and molecular evolution. *Syst. Zool.*, 1974, **22**:554–565.
4. Ambler, R. P., Amino acid sequences of prokaryotic cytochromes *c*. In: *Handbook of Biochemistry and Molecular Biology*, Proteins (G. D. Fasman, ed.), 3rd ed., Vol. 3. CRC Press, Cleveland, Ohio, 1976:292–307.
5. Ambler, R. P., Personal communition.
6. Ambler, R. P., and Bartsch, R. G., Amino acid sequence similarity between cytochrome *f* from a blue-green bacterium and algal chloroplasts. *Nature (London)*, 1975, **253**:285–288.
7. Ambler, R. P., Daniel, M., Hermoso, J., Meyer, T. E., Bartsch, R. G., and Kamen, M. E., Cytochrome c_2 sequence variation among the recognized species of purple nonsulphur photosynthetic bacteria. *Nature* (*London*), 1979, **278**:659–660.
8. Ambler, R. P., Meyer, T. E., and Kamen, M. D., Anomalies in amino acid sequences of small cytochromes *c* and cytochromes *c′* from two species of purple photosynthetic bacteria. *Nature* (*London*), 1979, **278**:661–662.
9. Ashida, T., Tanaka, N., Yamane, T., Tsukihara, T., and Kakudo, M., The crystal structure of bonito (Katsuo) ferrocytochrome *c* at 2.3 Å resolution. *J. Biochem.* (*Tokyo*), 1973, **73**:463–465.
10. Bachmann, B. J., Low, K. B., and Taylor, A. L., Recalibrated linkage map of *Escherichia coli* K-12. *Bacteriol. Rev.*, 1976, **40**:116–167.
11. Bartsch, R. G., Cytochromes. In: *The Photosynthetic Bacteria* (R. K. Clayton and W. R. Sistrom, eds.). Plenum, New York, 1978:249–279.
12. Broda, E., The origins of bacterial respiration. In: *Chemical Evolution and the Origin of Life* (R. Buvet and C. Ponnamperuma, eds.), Vol. I. North-Holland Publishing Co., Amsterdam, 1971:446–452.
13. Buchanan, R. E., and Gibbons, N., eds., *Bergey's Manual of Determinative Bacteriology*. 8th ed. Williams & Wilkins, Baltimore, Maryland, 1974.
14. Clayton, R. K., and Sistrom, W. R. (eds.), *The Photosynthetic Bacteria*. Plenum, New York, 1978.
15. Cusanovich, M. A., and Bartsch, R. G., A high potential cytochrome *c* from *Chromatium* chromatophores. *Biochim. Biophys. Acta*, 1969, **189**:245–255.

15a. Davis, K. A., Hatefi, Y., Salemme, F. R., and Kamen, M. D., Enzymic redox reactions of cytochromes *c*. *Biochem. Phys. Res. Commun.*, 1972, **49**:1329–1335.

16. Dayhoff, M. O., Computer analysis of protein evolution. *Sci. Am.*, 1969, January:86–95.
17. Dayhoff, M. O. (ed.), *Atlas of Protein Sequence and Structure*, Vol. 5, Suppl. 2. National Biomedical Research Foundation, Washington, D.C., 1976:26–27.

18. Dickerson, R. E., Sequence and structure homologies in bacterial and mammalian-type cytochromes. *J. Mol. Biol.*, 1971, **57**:1–15.
19. Dickerson, R. E., Evolution and gene transfer in purple nonsulfur photosynthetic bacteria. *Nature (London)*, 1980, **283**:136–153.
20. Dickerson, R. E. Cytochrome *c* and the evolution of energy metabolism. *Scientific American*, 1980 (March), **242**:136–153.
21. Dickerson, R. E., and Timkovich, R., Cytochromes *c*. In: *The Enzymes* (P. D. Boyer, ed.), 3rd ed., Vol. 11. Academic Press, New York, 1975:397–547.
22. Dickerson, R. E., Takano, T., Eisenberg, D., Kallai, O. B., Samson, L., Cooper, A., and Margoliash, E., Ferricytochrome *c*. I. General features of the horse and bonito proteins at 2.8 Å resolution. *J. Biol. Chem.*, 1971, **246**:1511–5135.
23. Dickerson, R. E., Timkovich, R., and Almassy, R. J., The cytochrome fold and the evolution of bacterial energy metabolism. *J. Mol. Biol.*, 1976, **100**:473–491.
23a. Errede, B., and Kamen, M. D., Comparative kinetic studies of cytochromes *c* in reactions with mitochondrial cytochrome *c* osidaxe and reductase. *Biochem Biochemistry*, 1978, **17**:1015–1027.
24. Fitch, W. M., and Margoliash, E., Construction of phylogenetic trees. *Science*, 1967, **155**:279–284.
25. Gest, H., Energy conversion and generation of reducing power in bacterial photosynthesis. *Adv. Microb. Physiol.*, 1972, **7**:243–282.
26. Woese, C. R., Gibson, J.,and Fox, G. E., Do genealogical patterns in purple photosynthetic bacteria reflect interspecific gene transfer? *Nature (London)*, 1980, **283**:212–214.
27. Gibson, J., Stackebrandt, E., Fox, G. E., Zablen, L. B., and Woese, C. R., Phylogenetic analysis of the purple photosynthetic bacteria. *Proc. Natl. Acad. Sci. U.S.A.*, 1979 (submitted for publication).
28. Gromet-Elhanan, Z., Electron transport and photophosphorylation in photosynthetic bacteria. *Encycl. Plant Physiol.*, 1977, **5** (new ser.):637–662.
29. Gromet-Elhanan, Z. and Gest, H., A comparison of electron transport and photophosphorylation systems of *Rhodopseudomonas capsulata* and *Rhodospirillum rubrum*. *Arch. Microbiol.*, 1978, **116**:29–34.
30. Haddock, B. A., and Jones, C. W., Bacterial respiration. *Bacteriol. Rev.*, 1977, **41**:47–99.
31. Holloway, B. W., Krishnapillai, V., and Morgan, A. F., Chromosomal genetics of *Pseudomonas*. *Microbiol. Rev.*, 1979, **43**:73–102.
32. John, P., and Whatley, F. R., *Paracoccus denitrificans* and the evolutionary origin of the mitochondrion. *Nature (London)*, 1975, **254**:495–498.
33. John, P., and Whatley, F. R., The bioenergetics of *Paracoccus denitrificans*. *Biochim. Biophys. Acta*, 1977, **463**:129–153.
34. Keilen, D., *The History of Cell Respiration and Cytochrome*. Cambridge University Press, London and New York, 1966:86–116.
35. Keister, D. L., Respiration vs. photosynthesis. In: *The Photosynthetic Bacteria* (R. K. Clayton and W. R. Sistrom, eds.), Plenum, New York, 1978:849–856.
36. Knaff, D. B., Reducing potentials and the pathway of NAD^+ reduction. In: *The Photosynthetic Bacteria* (R. K. Clayton and W. R. Sistrom, eds.), Plenum, New York, 1978:629–640.

37. Korszun, Z. R., and Salemme, F. R., Structure of cytochrome c_{555} of *Chlorobium thiosulfatophilum*. *Proc. Natl. Acad. Sci. U.S.A.*, 1977, **74**: 5244–5247.
38. Kreil, G., Über die Artspezifität von Cytochrom *c:* Vergleich der Aminosäuresequenz des Thunfisch-Cytochroms *c* mit der des Pferde-Cytochroms *c*. *Hoppe-Seyler's Z. Physiol. Chem.*, 1963, **334**:154–166.
39. La Monica, R. F., and Marrs, B. L., The branched respiratory system of photosynthetically grown *Rhodopseudomonas capsulata*. *Biochim. Biophys. Acta*, 1976,**423**:431–439.
40. Ludwig, M. L., Pattridge, K. A., Powers, T. B., Dickerson, R. E., and Takano, T. Structure analysis of a ferricytochrome *c* from the cyanobacterium *Anacystis nidulans*. In: *Interaction Between Iron and Proteins in Oxygen and Electron Transport* (Chien Ho, ed.), Elsevier-North Holland, New York (in press), 1980.
41. Madigan, M. T., and Gest, H., Growth of a photosynthetic bacterium anaerobically in darkness, supported by "oxidant-dependent" sugar fermentation. *Arch. Microbiol.*, 1978, **117**:119–122.
42. Takano, T. and Dickerson, R. E. Redox conformation changes in refined tuna cytochrome *c*. *Proc. Natl. Acad. Sci. U.S.A.*, 1980 (in press).
43. Margoliash, E., Personal communication.
44. Margoliash, E., Smith, E. L., Kreil, G., and Tuppy, H., Amino-acid sequence of horse heart cytochrome *c*. *Nature* (*London*), 1961, **192**:1121–1127.
45. Marrs, B. L., Genetics and bacteriophage. In: *The Photosynthetic Bacteria* (R. K. Clayton and W. R. Sistrom, eds.). Plenum, New York, 1978:873–883.
46. Marrs, B. L., and Gest, H., Genetic mutations affecting the respiratory electron-transport system of the photosynthetic bacterium *Rhodopseudomonas capsulata*. *J. Bacteriol.*, 1973, **114**:1045–1051.
47. Marrs, B. L., and Gest, H., Regulation of bacteriochlorophyll synthesis by oxygen in respiratory mutants of *Rhodopseudomonas capsulata*. *J. Bacteriol.*, 1973, **114**:1052–1057.
48. McLachan, A. D., Cytochrome *c* and cytochrome c_{551}. Chemical similarity as a test of related amino acid sequences. *J. Mol. Biol.*, 1971, **61**:409–424.
49. Moore, G. W., Barnabas, J., and Goodman, M., A method for constructing maximum parsimony ancestral amino acid sequences on a given network. *J. Theor. Biol.*, 1973, **38**:459–485.
50. Needleman, S. B., and Blair, T. T., Homology of pseudomonas cytochrome c_{551} with eukaryotic *c*-cytochromes. *Proc. Natl. Acad. Sci. U.S.A.*, 1969, **63**:1227–1233.
51. Olson, J. M., Precambrian evolution of photosynthetic and respiratory organisms. *Evol. Biol.*, 1978, **11**:1–37.
52. Olson, J. M., Confused history of *Chloropseudomonas ethylica 2K*. *Int. J. Syst. Bacteriol.*, 1978, **28**:128–129.
53. Pfennig, N., Photosynthetic bacteria. *Annu. Rev. Microbiol.*, 1967, **21**:285–324.
54. Pfennig, N., Phototrophic green and purple bacteria: A comparative, systematic survey. *Annu. Rev. Microbiol.*, 1977, **31**:275–290.
55. Pfennig, N., General physiology and ecology of photosynthetic bacteria. In: *The Photosynthetic Bacteria* (R. K. Clayton and W. R. Sistrom, eds.). Plenum, New York, 1978:3–18.

56. Pfennig, N., and Trüper, H. G., The *Rhodospirillales* (phototrophic or photosynthetic bacteria). In: *Handbook of Microbiology* (A. I. Laskin and H. A. Lechevalier, eds.), Vol. I. CRC Press, Cleveland, Ohio, 1973:17–27.
57. Pierson, B. K., and Castenholz, R. E., A photographic gliding filamentous bacterium of hot springs, *Chloroflexus aurantiacus, gen.* and *sp. nov. Arch. Microbiol.*, 1974, **100**:5–24.
58. Prince, R. C., and Dutton, P. L., Physical, chemical and structural descriptions of energetic processes in photosynthetic bacteria: Comparisons with green plants and respiratory systems. In: *Light Transducing Membranes* (D. Deamer, ed.). Academic Press, New York, 1978:167–186.
59. Salemme, F. R., Freer, S. T., Xuong, Ng. H., Alden, R. A., and Kraut, J., The structure of oxidized cytochrome c_2 of *Rhodospirillum rubrum. J. Biol. Chem.*, 1973, **248**:3910–3921.
60. Sanderson, K. E., and Hartman, P. E., Linkage map of *Salmonella typhimurium*, Edition V. *Microbiol. Rev.*, 1978, **42**:471–519.
61. Sirevag, R., Buchanan, B. B., Berry, J. A., and Troughton, J. H., Mechanisms of CO_2 fixation in bacterial photosynthesis studied by the carbon isotope fractionation technique. *Arch. Microbiol.*, 1977, **112**:35–38.
62. Smith, L., and Pinder, P. B., Oxygen-linked electron transport and energy conservation. In: *The Photosynthetic Bacteria* (R. K. Clayton and W. R. Sistrom, eds.). Plenum, New York, 1978:641–654.
63. Solioz, M., and Marrs, B. L., The gene transfer agent of *Rhodopseudomonas capsulata. Arch. Biochem. Biophys.*, 1977, **181**:300–307.
64. Stanier, R. Y., Some aspects of the biology of cells and their possible evolutionary significance. *Symp. Soc. Gen. Microbiol.*, 1970, **20**:1–38.
65. Stanier, R. Y., Doudoroff, M., and Adelberg, E. A., *The Microbial World.* 3rd ed. Prentice-Hall, Englewood Cliffs, New Jersey, 1970.
66. Timkovich, R., and Dickerson, R. E., Amino acid sequence of *Paracoccus denitrificans* cytochrome c_{550}. *J. Biol. Chem.*, 1976, **251**:2197–2206.
67. Timkovich, R., and Dickerson, R. E., The structure of *Paracoccus denitrificans* cytochrome c_{550}. *J. Biol. Chem.*, 1976, **251**:4033–4046.
68. Trüper, H. G., and Pfennig, N., Taxonomy of the Rhodospirillales. In: *The Photosynthetic Bacteria* (R. K. Clayton and W. R. Sistrom, eds.), Plenum, New York, 1978:19–27.
69. Walpole, H., Letter to Horace Mann, In: *Letters of Horace Walpole* (P. Toynbee, ed.), Vol. III. Oxford University Press (Clarendon), London and New York, 1903:204.
70. Yen, H-C., and Marrs, B. L., Growth of *Rhodopseudomonas capsulata* under anaerobic dark conditions with dimethyl sulfoxide. *Arch. Biochem. Biophys.*, 1977, **181**:411–418.

Neutral Changes Revisited

THOMAS H. JUKES

Department of Biophysics and Medical Physics
University of California
Berkeley, California

Comparisons of the amino acid sequences of homologous proteins, such as cytochromes and hemoglobins, first started to become possible about 17 years ago. It was promptly noted that differences in such sequences were roughly proportional to taxonomic or phylogenetic differences. This gave rise to the concept of a molecular evolutionary clock, which says that the difference between homologous proteins, such as α chains of hemoglobins, in two different species is proportional to the length of time that the two species have diverged from a common ancestor. Human and gorilla α globins differed by only one amino acid; human and horse α globins, in contrast, were about 13% different; and so on.

In this same period, Emil Smith and his associates, including Emanuel Margoliash and Hiroshi Matsubara, [references in Jukes (6)] showed evolutionary divergence for the cytochromes *c* of horses, human beings, and other mammals, but the divergence was at a slower rate than that for the hemoglobins. These findings were intellectually very satisfying. They showed that the evolutionary divergence of species took place at the molecular level just as in the case of morphology and other visible phenotypic characteristics. The differences between homologous proteins obviously expressed changes in sequences of DNA molecules, and each protein had a characteristic rate of evolution.

In 1965, I asked the question (6)

> Are the two molecules of cytochrome *c* of dogs and horses splendidly tailored to the different requirements specified by "dogfulness" and "horsefulness," have they evolved to conform to these two different requirements, or have the two cytochromes passively been carried along as dogs and horses evolved separately from a common ancestor? In the latter case, it is probable that separation of the two species would be followed by changes in the genes that in time would result in differences in the two cytochrome *c* molecules.

ISBN 0-12-643150-7

The choice between these two possibilities represents the difference between two approaches to molecular evolution, termed panselectionism and neutralism. This difference was broached in publications by Kimura (14) in 1968 and King and Jukes (17) in 1969. These authors proposed that most amino acid differences between homologous proteins were neutral from the standpoint of natural selection and were fixed by genetic drift.

The opposite viewpoint is that, in protein chemistry, "each amino acid must have a unique survival value in the phenotype of the organism—the phenotype being manifested in the structure of the proteins. This is as true for a single protein as for the whole organism" (27). This viewpoint will undoubtedly be eloquently represented in this volume, so I do not need to elaborate on it.

These two viewpoints have been debated at length over the past decade. Many biochemists find it easy to accept the concept that large portions of protein molecules serve mainly to bring the molecule up to suitable size and shape and have very little specific function as compared with small, specialized active sites. Most of a protein molecule, according to this concept, can evolve freely by random drift. A similar thought, as applied to DNA molecules, was recently expressed by Crick (2), in a discussion of intervening sequences in genes (introns), as follows: "The base sequence of the intron, no longer used for coding, then drifts rather rapidly."

Also, when the divergence rates of various homologous proteins are compared, it is common practice to use the rapid divergence of fibrinopeptides as baselines of unrestrained evolutionary change in proteins. This implies the concept of molecular evolution through genetic drift. It also implies the principle of an evolutionary clock that sets the rate of change in proteins and runs at a specific rate for each protein (29). The evolutionary clock in hemoglobins was one of the key pieces of evidence that led Kimura (14) to propose the neutral theory, because the clock implies a steady accumulation of changes rather than a selective process that would be sporadic.

I shall review some of the recent findings in evolution that have a bearing on the neutral theory, particularly with reference to arguments advanced by King and Jukes in 1969.

Relation of the Genetic Code to Amino Acid Composition of Proteins

We proposed in 1969 that according to the neutral model of evolution, the amino acid composition of proteins should be strongly influenced by the genetic code (Table 1), since the neutral hypothesis states that a significant proportion of the amino acids in proteins has arisen by random mutation and drift (15–17). In contrast, the selectionist or Darwinian model of evolution would state that one particular amino acid will be optimum at a given site in

TABLE 1
The Amino Acid Code

UUU phenylalanine	UCU serine	UAU tyrosine	UGU cysteine
UUC phenylalanine	UCC serine	UAC tyrosine	UGC cysteine
UUA leucine	UCA serine	UAA terminator	UGA terminator
UUG leucine	UCG serine	UAG terminator	UGG tryptophan
CUU leucine	CCU proline	CAU histidine	CGU arginine
CUC leucine	CCC proline	CAC histidine	CGC arginine
CUA leucine	CCA proline	CAA glutamine	CGA arginine
CUG leucine	CCG proline	CAG glutamine	CGG arginine
AUU isoleucine	ACU threonine	AAU asparagine	AGU serine
AUC isoleucine	ACC threonine	AAC asparagine	AGC serine
AUA isoleucine	ACA threonine	AAA lysine	AGA arginine
AUG methionine	ACG threonine	AAG lysine	AGG arginine
GUU valine	GCU alanine	GAU aspartic acid	GGU glycine
GUC valine	GCC alanine	GAC aspartic acid	GGC glycine
GUA valine	GCA alanine	GAA glutamic acid	GGA glycine
GUG valine	GCG alanine	GAG glutamic acid	GGG glycine

each protein, so there should be no relationship between the amino acid frequency and the genetic code. We therefore examined the amino acid frequencies in 53 vertebrate proteins and found that there was a rough correspondence between the percentages found and those predicted from frequencies in the genetic code. A conspicuous exception was arginine, which occurs at less than half the frequency expected, from the fact that it has six codons. In 1969 (17), we summarized the observed frequencies of amino acids in 53 completely sequenced mammalian proteins, and we compared these with the frequencies predicted by the genetic code and random permutations of DNA nucleotides. When arginine was disregarded, the coefficient of correlation between the expected and observed frequencies was 0.89. Since the average composition of mammalian messenger RNA (mRNA) was unknown, we did the best we could by calculating the base composition of mRNA from the first two positions of the codons involved, and from assuming that the composition of the third bases averaged the same as that of the first two bases. Recent findings show that the third bases are much higher in cytosine and guanine and much lower in adenine than are the first two. The actual composition of the presently available sample of mammalian mRNA is 21% uracil (U), 27% cytosine (C), 24% adenine (A), and 28% guanine (G). However, this base composition is the result of the proportions of codons that are present in the sample, so that by this procedure one is employing circular reasoning. Therefore, for measuring randomness, the actual content of amino acids in a sample should be compared with their relative proportions in the genetic code, as shown in Table 2.

Subsequently, we examined the composition of proteins in more detail (12), as shown in Table 2. This included results from 68 completely sequenced

TABLE 2
AMINO ACID DISTRIBUTION PER 61 CODONS[a]

Residue	a	b	c	d	e	Residue	a	b	c	d	e
Ala	4	5.3	5.2	5.2	5.2	Ile	3	3.1	2.8	3.0	2.6
Arg	6	2.6	2.6	2.7	2.8	Leu	6	4.7	4.9	3.9	5.1
Asn	2	3.0	3.3			Lys	2	4.1	4.0	3.9	4.1
Asp	2	3.6	3.4			Met	1	1.1	0.9	1.1	0.9
Asn + Asp	4	6.6	6.7	6.5	6.3	Phe	2	2.3	2.4	2.3	2.4
Cys	2	1.3	1.4	1.4	0.7[b]	Pro	4	2.5	3.0	2.9	3.0
Gln	2	2.4	2.6			Ser	6	4.5	4.6	3.8	4.7
Glu	2	3.3	3.5			Thr	4	3.7	3.8	3.5	3.6
Gln + Glu	4	5.7	6.1	6.5	6.6	Trp	1	0.8	0.9	0.8	0.5
Gly	4	4.8	4.6	4.9	4.7	Tyr	2	2.3	2.1	2.0	2.2
His	2	1.4	1.3	1.3	1.2	Val	4	4.2	4.2	4.1	4.2

[a] (a) From genetic code (see Table 1); (b–e) averages of sequenced proteins. (b) From Jukes *et al.* (12); (c) from Yčas (30); (d) from Reeck and Fisher (25); (e) from Smith (28).
[b] Some "Cys H" values omitted.

proteins containing 12,170 amino acid residues in 47 eukaryotic, 17 prokaryotic, and 4 virus proteins. Only one representative of each family of proteins, such as the globins, was included. The arginine content was 43% of the amount expected from random use of the genetic code. On the average, proline, cysteine, serine, and leucine all occurred at levels lower than one would expect from randomness, in which the nucleotides in DNA are present at equal levels and with equiprobable distribution in DNA sequences. These lower-than-expected frequencies are offset by increased percentages, as compared with expectations, of lysine, aspartic acid, glutamic acid, asparagine, alanine, and glycine. The average amino acid composition of proteins, is therefore nonrandom, but only to a limited extent that modifies but does not contradict our original hypothesis. We conclude that the general properties of protoplasm require that on the average, the acidic side chains in protein molecules should balance the basic side chains, and this is achieved by making the sum of aspartic + glutamic acids approximately equal to the sum of lysine + arginine. However, lysine is definitely preferred over arginine, probably because of the unusual properties of the guanidine group in arginine, which is extremely basic. Lysine ought to have four codons. The need for maintaining a pH near neutrality against basic amino acids calls for use of glutamic and aspartic acids at levels higher than would be anticipated from their representation in the genetic code. Proline, histidine, and cysteine have special functions and properties, and this may diminish their use except for these functions. The needs for arginine, serine, and leucine are not so great as to require six codons apiece. Alanine, asparagine, and glycine have small side chains and may be useful as "fillers" in protein molecules. The other amino acids fall within the ranges expected from their representation in the code and from the neutral model. For example, the proportion between four hydrophobic amino acids, methionine, phenylalanine, isoleucine, and valine (1.1 to 2.3 to 3.1 to 4.1) is close to their representation in the genetic code (1:2:3:4).

The assignment of codons to amino acids depends primarily on the first and second nucleotides in each codon. The nucleotide composition of mRNA molecules reflects this fact, especially when considered in terms of nearest neighbor nucleotide pairs (doublets). The doublets AA and GA in the first two codon positions will be above-average values, for example, and UA, UG, CC, and CG will be below average. This is shown in Table 3.

One Gene, Two Proteins

Two recent findings give important support to the neutral theory. The first of these was the unexpected discovery by Sanger and co-workers (26) that some genes in bacteriophage ΦX174 are translated in two different reading frames, so that two proteins are produced by the same gene. Let us suppose that in the first of the two proteins, the entire sequence of amino acids is related to its function. This would be in accordance with classical selectionist evolutionary theory. The mRNA molecule for this first protein would have a definite nucleotide sequence, except possibly for some permitted variations in the synonymous third positions of codons, although, according to some

TABLE 3
Distribution of "Nearest Neighbor" Nucleotide Doublets in mRNA of Chicken Ovalbumin[a]

	Within codons		Adjoining in codons		Totals	
Pairs	First 2	Second 2	Found	Expected[b]	Found	Expected[c]
UU	27	34	15	19	76	76
UC	23	38	17	17	78	64
UA	11	12	19	32	42	88
UG	9	41	52	35	102	69
CU	25	30	22	18	77	64
CC	14	19	19	16	52	54
CA	22	37	51	31	110	73
CG	2	1	9	34	12	58
AU	42	25	18	19	85	88
AC	15	23	13	17	51	73
AA	37	38	28	32	103	100
AG	28	31	43	35	102	80
GU	31	14	15	15	60	69
GC	35	20	14	13	69	58
GA	47	16	23	25	86	80
GG	19	8	27	27	54	63
Average	24.2	24.2	24.0	24.0	72.4	72.4

[a] Base composition U = 0.2567, C = 0.2153, A = 0.2945, G = 0.2334. Stop codon, but not start codon, is included in mRNA. From McReynolds *et al.* (22).

[b] Expected from the average base composition of the third and first positions of codons.

[c] Expected from the average base composition of the coding regions as given in the first footnote (a).

authors, even the third position bases are fixed by selection. If this nucleotide sequence were to be translated in a second manner by means of a frameshaft, it had always been assumed that the resulting sequence of amino acids would be a biologically meaningless molecule, a so-called nonsense protein. Usually, such nucleotide sequences contain termination codons at average intervals of 20. For example, the first 10 amino acids and codons of rabbit α-globin have the following sequence:

Val- Leu- Ser- Pro- Ala- Asp- Lys- Thr- Asn- Ile-
GUG-CUG-UCU-CCC-GCU-GAC-AAG-ACG-AAC-AUC-

which would be read following frameshift as

Gly- Ala- Val- Ser- Arg- C.T.- Gln- Asp- Glu- His-
GGU-GCU-GUC-UCC-CGC-UGA-CAA-GAC-CAA-CAU-

The demonstration that such nonsense proteins can be biologically functional leads to the conclusion that in certain respects, or in certain regions, it does not matter what a protein is made of. This has profound implications for theories of the origin of life.

In bacteriophage ΦX174 there is a high proportion of codons ending in "U." With the reading frame shifted one space to the right, the end bases become the middle bases of codons for proteins B, E, and K, and codons with "U" in the second position are predominantly those of hydrophobic amino acids. It has been suggested that such proteins rich in hydrophobic residues may be useful for reacting with bacterial membranes (1). In any event, the fact that one nucleotide sequence can code for two different proteins by changes in the reading frame must mean that for one or both of these proteins, much of the amino acid sequence is largely "filler." This supports the neutral model. This is strengthened by the fact that a second virus, G4, is related to ΦX174, and has a set of homologous proteins corresponding to the ΦX174 proteins (3). Once again, three of these proteins are produced by frameshift readings in genes. This takes place even though there are substantial differences in the amino acid sequences of homologous proteins in the two viruses, as will be discussed later.

The Evolutionary Clock

Wilson *et al.* (29) stated, "the discovery of the evolutionary clock stands out as the most significant result of research in molecular evolution."

Kimura (14) drew on comparisons of mammalian hemoglobins to calculate the rate of substitution of nucleotides in the evolutionary history of mammals. His calculation of rate included an estimate of 1% of amino acid substitution in 10^7 years for mammalian hemoglobins, corresponding to 1.2 replacements in 300 base pairs. This would represent a rate of nucleotide substitutions per site per year of 0.4×10^{-9}, uncorrected for multiple hits and back mutations, which showed that, in the evolutionary history of mammals, one nucleotide pair has been substituted in the genome every two years. Kimura concluded that this rate of change was far too great to be adaptive, and therefore many of the substitutions involved must be neutral ones. In 1977 Kimura (15) showed that the rate of substitution is much more rapid in mRNA, because synonymous changes in codons can be detected. Fragments of human and rabbit β-hemoglobin mRNAs showed K_{nuc} (corrected nucleotide substitution rate) $= (2.3 \pm 1.1) \times 10^{-9}$. The mRNAs of histones 4 (IV) from two sea urchins had $K_{nuc} = (3.7 \pm 1.4) \times 10^{-9}$ for third nucleotide positions of codons, which are predominantly synonymous. These higher rates gave additional force to Kimura's (14) original argument.

A contrary argument is that the rapidity of evolutionary changes results from "truncation selection" of large pieces of DNA. This would make it difficult for "fine tuning" of hemoglobin sequences to take place to satisfy the postulated highly specialized requirements of each organism. It would also make it difficult for silent changes to accumulate more rapidly than code-altering changes, unless the former were conceded to be neutral.

According to our proposal in 1969, the evolutionary clock is set in motion by the constant stream of spontaneous alterations in DNA that are continuously fed into the genetic pool. Most of the changes are discarded by natural selection. Among the accepted changes, there are far more that are neutral, or near neutral, than are adaptive. The steady pressure of these changes results in the gradual accumulation of amino acid replacements in protein molecules. The rate at which these replacements accumulate represents the evolutionary clock. Each protein resists change in a characteristic manner, depending on the relation of structure to function. Therefore the evolutionary clock runs at different rates in different proteins.

We noted in 1972 that the evolutionary clock can run at different rates for the same protein in different species (10). This was shown by the fact that rattlesnake cytochrome *c* has evolved more rapidly than turtle cytochrome, as measured by the relative rates at which each has diverged when compared with bird cytochromes *c*. Many other examples have since become available of localized variations in the evolutionary clock, especially among the globins and, recently, the ferredoxins (21). The rate of change of α- and β-globins has greatly slowed down, according to Goodman and co-workers (4), in the mammals as compared with their ancestors. They consider that, in the mammals, the globins are undergoing a process of "microevolution" or "fine tuning."

Against this conclusion, it must be noted that such a slowing process represents only a minor perturbation in the rate of the evolutionary clock for the globins. These molecules are one of the best examples of the relation of time to change. This can be seen when the α and β chains of the shark, bony fish, amphibians, reptiles, birds, marsupial mammals, and placental mammals are compared (Table 4). The divergence is steady and consistent. It is interesting that when the divergences of the α and β chains are averaged, the time-related rate "smooths out."

Moreover, even though the globin molecule is highly specialized with regard to function and tertiary structure, it exhibits a truly astonishing ability to accumulate amino acid replacements at most sites. The average number of different amino acids at each homologous globin site, in sequences of myoglobin and vertebrate hemoglobins that are now available, is 5.8. Obviously this number will increase when more globins are sequenced. It looks as if every possible change that can be squeezed into the globin molecule has been accepted during the course of evolution. Moreover, the favorite amino acid is alanine, which is notable for its lack of distinctive properties and is therefore a suitable candidate for neutral changes (Table 5).

Silent Changes in mRNA Comparisons

However, the story of the evolution of globin genes is only just beginning. Many elaborate phylogenetic trees for globins have been constructed by counting amino acid differences in sequences of globin chains and by using these amino acid differences to calculate evolutionary nucleotide replacements. It now appears that these calculations were wrong. Modern techniques in nucleotide sequencing show that unexpectedly large numbers of synonymous silent changes take place during evolutionary separation and divergence of globin genes. It has been argued that silent replacements are selectively neutral or near neutral, and Kimura, as noted earlier (15), has drawn attention to the preponderance of silent replacements over protein-altering replacements, based on comparisons of 53 nucleotide positions in human and rabbit β-globin mRNAs + 84 positions in histone 4 mRNAs of two

TABLE 4
Average Amino Acid Differences in Hemoglobin Chains

Organisms	Differences per 100 codons		Average
	α chain	β chain	α + β
Placental mammals *inter alia* (except human vs. loris)[a]	17.0	17.5	17.2
Kangaroo vs. placental mammals	22.3	27.7	25.0
Monotremes vs. theria[b]	30.8	24.4	27.5
Chicken vs. mammals	29.8	32.2	31.1
Viper vs. warm blooded	39.1	—	—
Amphibian vs. terrestrial[c]	47.2	43.5	45.4
Bony fish vs. tetrapods[d]	49.8	50.2	50.0
Shark vs. bony vertebrates	56.7	63.8	60.2

[a] Human, loris, mouse, rabbit, dog, cat, pig, bovine, llama.
[b] Echidna, platypus.
[c] Newt, bullfrog.
[d] Carp, goldfish

TABLE 5
EVOLUTIONARY REPLACEMENTS OF AMINO ACID

Residues in Globins			
Sites	No. of residues	Totals of amino acids	
117	13	Alanine	84
9, 22	12	Serine	76
8, 30, 60	11	Glycine	60
21, 24, 66, 74, 77, 113, 116	10	Aspartic Acid	53

sea urchin species. Recently, many more nucleotide sequences of mRNAs have been published. I compared rabbit and human β globin mRNAs (9) and I now present a comparison of a total of 4712 codons in mRNAs and genes for various proteins in Tables 6 and 7. Note that the percentage of silent changes (Table 7) is about 60 in the β comparisons and about 40 in the $\alpha\beta$ comparisons. This is because there are about 2.5 times as many replacements in the α versus β comparisons as in the β comparisons, and this means far more multiple hits in the former set, which decreases the percentage of silent changes.

The question arises, Are these silent changes actually neutral, or have they taken place for adaptive reasons, such as a requirement for a specific secondary structure in mRNA, or a preferential use for certain transfer RNAs in regulating the rate of synthesis of a protein? These two reasons have been advanced to explain the unequal distribution of synonymous codons in the mRNAs of globins and other proteins. However, such argument would not apply to the silent replacements noted when two β-globin messengers (18, 20) are compared, for it scarcely seems conceivable that two closely related mammals would have different requirements for mRNA structure and translation for the same protein. If this were so, a staggering number of species

TABLE 6
COMPARISON OF GENES IN SEQUENCES OF G4 AND ϕX174 CODING REGIONS

Gene	Codon compared	Total nucleotide replacements	Silent changes	Percentage silent
A	507	477	233	52
C	84	83	50	60
D	152	113	81	72
F	427	430	226	53
G	175	263	111	42
H	328	310	186	60
J	25	25	16	64
B	120	87	19	22
E	91	52	11	21
K	56	40	10	25
Totals	1965	1880	943	Average 50.2

TABLE 7
NUCLEOTIDE REPLACEMENTS IN HEMOGLOBIN AND TRYPTOPHAN SYNTHETASE mRNAs

Coding sequences	Nucleotides compared	Total changes	Silent changes	Percentage of silent changes
Rabbit and human β-Hbs[a]	441	49	32	65
Rabbit and mouse β-Hbs[b]	441	84	48	57
Human and mouse β-Hbs	441	77	42	55
Rabbit α- and rabbit β-Hbs[c]	417	177	71	40
Rabbit α- and human β-Hbs	417	172	71	41
Rabbit α- and mouse β-Hbs	417	174	66	38
Escherichia coli and *S. typhimurium* Trp synthetases (initial transcribed regions)[d]	165	16	11	69
E. coli and *Serratia marcescens* TrpG and TrpD portions[e]	138	26	17	65

[a] From Kafatos *et al.* (13).
[b] From Marotta *et al.* (20).
[c] From Heindell *et al.* (5).
[d] From Lee *et al.* (19).
[e] From Miozzarri and Yanofsky (23).

differences would be needed at the molecular level for the thousands of different proteins in each species. We therefore conclude that silent differences between β-globin mRNAs are neutral. This simple explanation probably also applies to comparisons between α- and β-globin mRNAs (Table 7). The results with bacterial mRNAs are similar to those with the globins.

We stated in 1969 that "if DNA divergence in evolution includes the random fixation of neutral mutations, then the third position nucleotides should change more rapidly, because synonymous mutations are more likely to be neutral." This prediction agrees with the findings summarized in Tables 6 and 7.

A great wealth of material for evolutionary comparison has been made available by the sequencing of the DNA molecules of two small bacterial viruses, ΦX174 and G4, whose closely related sequences contain coding regions for the same set of 10 homologous proteins. These are compared in Table 6. Most of the nucleotide differences between ΦX174 and G4 are silent changes. The average percentage of silent changes in the first seven genes listed is roughly 60%, but the percentage is much smaller for genes *B*, *E*, and *K*, about 23%. These three genes are read as frameshifts in genes *A*, *C*, and *D*. Barrell and co-workers (1) and Godson and co-workers (3) concluded that gene E started to function at a later stage of evolution than gene D. This would be consistent with a much lower percentage of silent changes in gene E. Silent changes in gene E would nearly all produce amino acid changes in the predecessor gene D, because third positions of codons in gene E are first positions of codons in gene D. Therefore, silent changes in new genes that have originated from frameshift readings will often be rejected as causing unacceptable amino acid changes in old genes. This conclusion also applies to genes B and K, which are formed by frameshift readings in genes A and C.

The types of evolutionary replacements in comparison of codons are shown in Table 8. There are various types of silent replacements; some are associated with synonymous codons for the same amino acid, and others occur simultaneously with an amino acid replacement. These various types of replacements in the evolution of mRNAs were compiled by Jukes and Holmquist (11). Table 9 summarizes the findings. Obviously, many codons undergo more than one nucleotide replacement during evolutionary divergence. Note also that quite often all three bases in a codon have been replaced, but very seldom does this give rise to a recognizable three-base change. Usually, such events are recorded as two-base changes. The pattern shown by the three viral genes that are read by frameshifts is of unusual interest. Most of the nucleotide replacements are recognizable as code changes.

Table 10 summarizes all the data of this type for the globin comparisons and simultaneously lists the effects that are produced when the sequence of human β-globin mRNA is "read against itself" in a frameshift of two positions to the right, headed "Beta Random." Four terminator codons are thus produced, and these, of course, are not found in actual evolutionary comparisons of mRNAs. Note also that there are 29% silent changes in this comparison.

I turn now to discussion of another question that we raised in 1969: "The relation of the comparative rarity of arginine to possible selection against the nucleotide pair CG" in vertebrate DNA (17). At that time we mentioned four

TABLE 8
TYPES OF EVOLUTIONARY REPLACEMENTS IN CODON COMPARISONS

Type	Example	Frequency[a]
Single-base		
Degenerate	UUU/UUC:Phe/Phe	134
Recognizable	UUU/AUU:Phe/Ile	392
Two-base		
Degenerate	CUC/UUA:Leu/Leu	28
Recorded as single[b]	UUU/AUC:Phe/Ile	1006
Recognizable	UUU/GAU:Phe/Asp	534
Three-base		
Degenerate	UCU/AGC:Ser/Ser	12
Recorded as single[b] (recorded as AGC/GGC)	UCU/GGC:Ser/Gly	308
Recorded as two-base[b] (recorded as GUG/ACG)	GUG/ACA:Val/Thr	1164
Recognizable	UUU/AAG:Phe/Lys	82

[a] In pairwise comparisons of all 61 amino acid codons.
[b] In amino-acid-residue comparisons, using minimum-base-change calculations.

TABLE 9
EVOLUTIONARY REPLACEMENT IN MESSENGER RNAS AND VIRAL GENES

			ϕx174 vs. G4 Genes	
Type	β-Globins	α- vs. β-Globins	In phase	By frameshift
Single-base				
Degenerate	84	64	498	19
Recognizable	27	78	166	74
Two-base				
Degenerate	2		31	
Recorded as single-base	30	95	145	17
Recognizable	13	25	84	14
Three-base				
Degenerate			3	
Recorded as single-base	1	2	23	
Recorded as two-base	2	44	127	8
Recognizable		1	8	

possibilities, two of these were, first, that the CG pair or "doublet" was in some way disadvantageous and was eliminated by natural selection and the second was that this selective process took place so as to lower the arginine content of proteins. It now appears that the second of these two proposals was incorrect; arginine, when needed, is supplied by AGA and AGG codons, so as to avoid the use of CG, which is said to be disadvantageous because it functions as a signal for methylation of C in vertebrate DNA. However, the rarity of CG in vertebrates is not consistent, as shown in Tables 11, 12, and 13.

There are three locations in which CG doublets can occur in coding sequences. The first of these is in the first two codon positions. CG occurs here in four of the six arginine codons. Table 11 shows the comparative use of the two types of arginine codons. There is some selection against CGN codons in some proteins, but not in others.

The second selective process consists of elimination of codons for serine, proline, threonine, and alanine ending in CG. This strategy is easily accomplished if one assumes that third-position changes are neutral. The results are shown in Table 12. There is no selection against CG in mRNA for bovine pituitary hormones.

The third process of selection against a CG pair consists of adjusting the evolutionary ingenuity than the first two methods. It consists of adjusting the base sequences of codons so that C in the third position does not occur next to G in the first position of the next codon. Table 14 shows what has happened in genes of simian virus (SV) 40, which behaves like vertebrate DNA. The base pairs formed by adjoining codons are in most cases what one would predict from randomness, but there has been a pronounced elimination of CG, compensated by increased numbers of doublets such as UG and CA. The

TABLE 10
EVOLUTIONARY REPLACEMENTS IN HUMAN, RABBIT, AND MOUSE GLOBIN MESSENGER RNAs

		β-Globins			α- vs. β-Globins			β Random
Replacements	Example	H/R	H/M	R/M	R/H	R/R	R/M	H/H
Single-base								
Degenerate	UUU/UUC	26	27	31	22	20	22	2
Recognizable	UUU/AUU	5	10	12	26	24	28	10
AA → term	UCA/UAA							
Two-base								
Degenerate	CGC/AGA		1	1				1
Recorded as single-base	UUU/AUC	6	13	11	31	35	29	33
Recognizable	UUU/GAU	3	6	4	7	9	9	36
AA → term	UUU/UAA							2
Three-base								
Degenerate	UCU/AGC							1
Recorded as single-base	UCU/GGC		1		1	1		5
Recorded as two-base	GUG/ACA		2		15	14	15	40
Recognizable	UUU/AAG						1	4
AA → term	AUG/UAA							2
Total nucleotide replacements		49	86	75	172	177	174	312
Silent changes		32	46	44	70	71	66	90
Percentage of silent changes		65	60	59	41	40	38	29

achievement of such results seems to imply a good deal of flexibility in the use of codons. SV40 avoids the juxtaposition of CG in adjoining codons and shows a lowering of total use of C in third positions. Elimination of CG can occur through methylation of the CG pair to 5mCG (5mC=5-methylcytosine) followed by deaminative mutation of 5mC to T (1a). Such neutral or near-neutral changes seem the best explanation for the CG deficiencies described above.

Elimination of CG does not occur in bovine ACTH-LPH precursor (24,

TABLE 11
SELECTION AGAINST CG NUCLEOTIDE PAIRS[a]

Protein	AGR codons	CGN codons
Rabbit β-globin	3	0
Human β-globin	3	0
Mouse β-globin	2	1
Rabbit α-globin	1	2
Rat pre-proinsulin	0	5
Rat pregrowth hormone	2	10
Human C-S	3	6
Bovine ACTH-LPH	3	19
Chicken ovalbumin	13	2
SV40	62	5

[a] In first two codon positions AGR instead of CGN codons used for arginine in β-globins only.

TABLE 12
SELECTION AGAINST CG NUCLEOTIDE PAIRS[a]

	Ending in CU, CC, CA	Ending in CG
Rabbit β-globin	28	1
Human β-globin	28	0
Mouse β-globin	32	0
Rabbit α-globin	36	3
Rat pre-proinsulin	17	2
Rat pregrowth hormone	46	2
Human C-S	23	7
Bovine ACTH-LPH	39	23
Chicken ovalbumin	86	1
SV40	77	0

[a] In second two codon positions of synonymous codons (Ser, Pro, Thr, Ala)

Table 13) and has not been found in prokaryotic sequences, or in the sequences of coliphages (3, 7, 26).

DISCUSSION AND SUMMARY

The neutral approach to molecular evolution is a proposal to prove a negative, which is something like trying to show that a given substance is not a carcinogen. The counterresponse to the publications by Kimura(14) and King and Jukes (17) has been quite strong. Any exceptions to neutrality are usually taken as disproof of it, and many authors have cited such exceptions for this purpose. We have, indeed, developed evidence for such exceptions ourselves (10,12), because a theory should be challenged by those who have postulated it.

For example, the finding that the synonymous codons for each amino acid

TABLE 13
PERCENTAGE OF BASE PAIRS FORMED BY ADJOINING CODONS IN mRNA OF BOVINE CORTICOTROPIN-β-LIPOTROPIN PRECURSOR[a]

Basepair	Percentage Calc.	Percentage Found	Basepair	Percentage Calc.	Percentage Found
UU	1.3	1.9	AU	0.6	0.4
UC	2.6	2.3	AC	1.2	1.1
UA	2.2	1.1	AA	1.0	1.1
UG	3.8	4.6	AG	1.7	1.5
CU	5.4	6.1	GU	5.6	3.4
CC	11.1	12.3	GC	11.5	11.1
CA	9.3	11.1	GA	9.7	8.8
CG	16.3	13.4	GG	16.9	19.5

[a] No significant selection against CG pairs. Data from Nakanishi *et al.* (24).

TABLE 14
PERCENTAGE OF BASE PAIRS FORMED BY ADJOINING CODONS IN SV40 MRNA[a]

Base pair	Percentage Calc.	Percentage Found	Basepair	Percentage Calc.	Percentage Found
UU	7.7	6.6	AU	5.9	3.2
UC	6.1	5.9	AC	4.6	3.5
UA	11.1	9.3	AA	8.9	12.8
UG	11.9	16.1	AG	9.0	8.9
CU	2.8	3.7	GU	4.3	5.0
CC	2.2	2.3	GC	3.4	4.7
CA	4.2	6.8	GA	6.5	3.7
CG	4.2	0.4	GG	6.6	7.0

[a] CG pairs are reduced by pairing U (instead of C) with G by pairing C with A (instead of with G), and by reducing use of C in third position.

are not used in equal amounts in β-hemoglobin mRNA has been cited (13) as disproof of the neutral model, as if such a departure from randomness in a single gene were pertinent. Obviously, many more mRNAs must be examined, and it is already evident that there is great variation in synonymous codon usage in them. In terms of the neutral model, we expect that synonymous codon usage should vary in response to minor pressures. Indeed, we used this as an example, when we noted (17) Cox and Yanofsky's finding that the Treffer's mutator gene would shift the base composition of *Escherichia coli* DNA, apparently, in part, by changing the third bases in synonymous codons, without impairing the viability of the mutated strains.

A possible barrier to acceptance of the neutral theory is the fact that living organisms are engaged in an evolutionary struggle against randomness or against the increase of entropy, and neutral changes tend toward randomness.

Nevertheless, I consider the neutral model of evolution, as expanded by Kimura and Ohta to emphasize near-neutral changes (16), to be alive and well. It has been supported recently by findings with frameshift proteins and by the extensive evidence for rapid accumulation of silent substitutions in the evolution of mRNA sequences.

We said in 1969 (17), "Protein molecules are subjected to incessant probing as a result of point mutations and other DNA alterations. The genome becomes virtually saturated with such changes as are not thrown off through natural selection." I think that this statement is still useful in interpreting molecular evolution.

ACKNOWLEDGMENTS

This work was supported by NASA grant NGR 05-005-460 to the University of California, Berkeley. I thank Amanda Orosco for collaboration in preparation of the manuscript.

REFERENCES

1. Barrell, B. G., Air, G. M., and Hutchinson, C. A., III, Overlapping genes in bacteriophage ΦX-174. *Nature (London)*, **164**:34–41.
1a. Bird, A. P., DNA methylation and the frequency of CpG in mammalian DNA. *Nucleic Acids Research*, 1980, **8**:1499–1504.
2. Crick, F. H. C., Split genes and RNA splicing. *Science*, 1979, **204**:264–270.
3. Godson, G. N., Barrell, B. G., Staden, R., and Fiddes, J. C., Nucleotide sequence of bacteriophage G-4 DNA. *Nature* (*London*), 1978, **276**:236–247.
4. Goodman, M., Moore, G. W., and Matsuda, G., Darwinian evolution in the genealogy of hemoglobin. *Nature* (*London*), 1975, **253**:603–608.
5. Heindell, H. C., Liu, A., Paddock, W. V., Studnicka, G. U., and Salser, W. A., The primary sequence of rabbit α-globin mRNA. *Cell*, 1978, **15**:43–54.
6. Jukes, T. H., *Molecules and Evolution*. Columbia University Press, New York, 1965:9.
7. Jukes, T. H., Nearest neighbor doublets in protein-coding regions of MS2 RNA. *J. Mol. Evol.*, 1977, **9**:299–303.
8. Jukes, T. H., Codons and nearest-neighbor nucleotide pairs in mammalian messenger RNA. *J. Mol. Evol.*, 1978, **11**:121–127.
9. Jukes, T. H., Neutral changes during divergent evolution of hemoglobins. *J. Mol. Evol.*, 1978, **11**:267–269.
10. Jukes, T. H., and Holmquist, W. R., Evolutionary clock: Nonconstancy of rate in different species. *Science*, 1972, **177**:530–532.
11. Jukes, T. H., and Holmquist, W. R., Estimation of evolutionary changes in certain homologous polypeptide chains. *J. Mol. Biol.*, 1972, **64**:173–179.
12. Jukes, T. H., Holmquist, W. R., and Moise, H., Amino acid composition of proteins: Selection against the genetic code. *Science*, 1975, **189**:50–51.
13. Kafatos, F. C., Efstratiadis, A., Forget, B. G., and Weissman, S. M., Molecular evolution of human and rabbit β-globin mRNAs. *Proc. Natl. Acad. Sci. U.S.A.*, 1977, **74**:5618–5622.
14. Kimura, M., Evolutionary rate at the molecular level. *Nature* (*London*), 1968, **217**:624–626.
15. Kimura, M., Preponderance of synonymous changes as evidence for the neutral theory of molecular evolution. *Nature* (*London*), 1977, **267**:275–278.
16. Kimura, M., and Ohta, T., Protein polymorphism as a phase of molecular evolution. *Nature* (*London*), 1971, **229**:467–470.
17. King, J. L., and Jukes, T. H., Non-Darwinian evolution. *Science*, 1969, **164**:788–798.
18. Konkel, D. A., Tilghman, S. M., and Leder, P., The sequence of the chromosomal mouse β-globin major gene: Homologies in capping, splicing and poly-A sites. *Cell*, 1978, **15**:1125–1132.
19. Lee, F., Bertrand, K., Bennett, G., and Yanofsky, C., Comparison of the nucleotide sequences of the initial transcribed regions of the tryptophan operons of *Escherichia coli* and *Salmonella typhimurium*. *J. Mol. Biol.*, 1978, **121**:236–247.
20. Marotta, C. A., Wilson, J. T., Forget, B. J., and Weissman, S. M., Human β-globin messenger RNA. *J. Biol. Chem.*, 1977, **252**:5040–5053.

21. Matsubara, H., Hase, T., Wakabayashi, S., and Wada, K., Structure and evolution of bacterial- and chloroplast-type ferredoxins. *This volume.*
22. McReynolds, L., O'Malley, B. W., Nisbet, A. D., Fothergill, J. E., Givol, D., Fields, S., Robertson, M., and Brownlee, G. G., Sequence of chicken ovalbumin mRNA. *Nature (London)*, 1978, **273**:724–730.
23. Miozzarri, G. F., and Yanofsky, C., Gene fusion during the evolution of the tryptophan operons in enterobacteriaceae. *Nature (London)*, 1979, **277**:486–489.
24. Nakanishi, S., Inoue, A., Kita, T., Nakamura, M., Chang, A. C. Y., Cohen, S. N., and Numa, S., Nucleotide sequence of cloned cDNA for bovine-corticotropin-β-lipotropin precursor. *Nature (London)*, 1979, **278**:423–427.
25. Reeck, G. R., and Fisher, L., A statistical analysis of the amino acid composition of proteins. *Int. J. Pept.Protein Res.*, 1973, **5**:109–117.
26. Sanger, F., Air, G. M., Barrell, B. G., Brown, N. L., Coulson, A. R., Fiddes, J. C., Hutchinson, C. A., III, Slocombe, P. M., and Smith, M. Nucleotide sequence of bacteriophage ΦX174 DNA. *Nature (London)*, 1977, **265**:
27. Smith, E. L., The evolution of proteins. *Harvey Lect.* 1967, **62**:231–246.
28. Smith, M. H. The amino acid composition of proteins. *J. Theor. Biol.*, 1966, **13**:261–269.
29. Wilson, A. G., Carlson, S. S., and White, T. J., Biochemical evolution. *Annu. Rev. Biochem.*, 1977, **46**:573–639.
30. Yčas, M., *The Biological Code.* Elsevier, Amsterdam, 1969:146.

Structure and Evolution of Photosynthetic Accessory Pigment Systems with Special Reference to Phycobiliproteins

ALEXANDER N. GLAZER

Department of Microbiology and Immunology
University of California
Berkeley, California

Everyone who has observed and considered a field of marine algae on the seashore at low tide knows that there are four very conspicuous groups of plants in the ocean. They are the blue-green, red, brown, and green algae. The obvious question is, Why does the sea contain plants of such a variety of colors when the land supports only green vegetation? . . . It is here suggested that the various pigments involved were originally called into existence in response to certain physical environmental forces or variables and that by far the most important of these external factors was illumination, the quantity or quality of which was determined or modified by two different media, (a) the sea water in which the organisms lived and (b) the layer of atmosphere above it.

Josephine E. Tilden, *The Algae and Their Life Relations*, 1937

INTRODUCTION

Photosynthetic organisms living in different environments possess photosynthetic systems that are optimized for the utilization of the available light and that can be modulated in response to changes in its intensity. Light reaching submerged algae progressively changes its spectral composition with depth. Light penetrating deep ocean water is enriched in the green wavelengths and the total photon flux is diminished. The incident radiation is even more severely restricted in river, lake, and coastal waters by the presence of dissolved yellow substances of organic origin (53,54). Such waters transmit light maximally in the 500–650 nm range. In response to these limitations, different accessory light-harvesting systems have evolved. The added effectiveness in the capture of the available solar energy resulting from such systems can be seen in the data presented in Fig. 1. It is clear that relative to a constant amount of chlorophyll *a*, the light-harvesting capabilities of cyanobacteria are superior to those of diatoms, which in turn absorb more light than do green algae. These differences are directly at-

THE EVOLUTION OF PROTEIN STRUCTURE AND FUNCTION

ISBN 0-12-643150-7

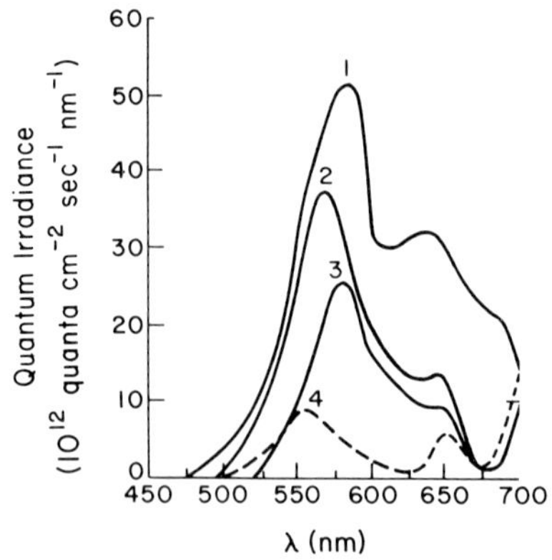

Fig. 1. Calculated spectral distribution of light intensity at a depth of 5 m in model suspension of various algae, present at levels corresponding to 30 mg chlorophyll *a* per cubic meter. Solar altitude is assumed to be 45° and light scattering is ignored. Curve 1 shows the spectral distribution of intensity in the absence of algae in water containing a typical level of yellow substances (53). Curves 2, 3, and 4 show the spectral distribution of intensity in suspensions of green algae, diatoms, and cyanobacteria (which in this example do not contain phycoerythrinoid pigments), respectively. For details of calculation see Kirk (52). [Redrawn from Kirk (54).]

tributable to the spectroscopic characteristics of the photosynthetic accessory pigment–protein complexes (antenna complexes) present in these organisms.

The understanding of the individual solutions to the light-harvesting problem requires an overview of the organization of the photosynthetic apparatus in oxygen-evolving organisms. In the thylakoids of such organisms there are two photosystems operating in series to achieve a light-driven electron transfer from water to ferredoxin, with concomitant evolution of oxygen (Fig. 2). The details of the electron transfer reactions and associated membrane potential changes are discussed extensively in excellent reviews (e.g., 51) and are

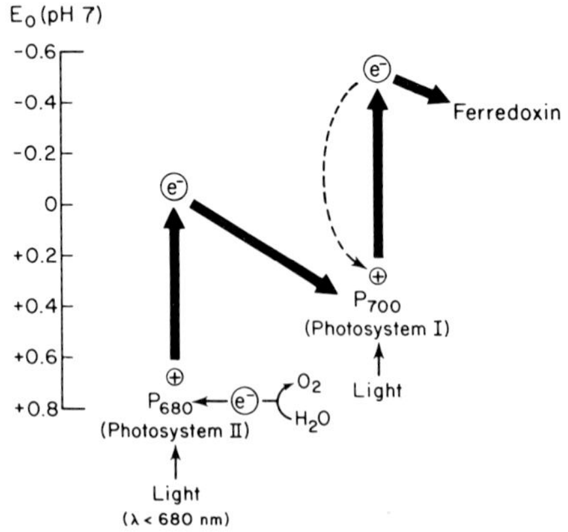

Fig. 2. Schematic representation of light-induced charge separation and electron flow in oxygenic photosynthesis. Cyclic electron flow, which can occur in photosystem I, is indicated by a broken line.

outside the scope of this presentation. Here attention is focused on the light-harvesting components in oxygenic photosynthesis. There is general consensus that these components are present in three distinct structural domains, and several models based on spectroscopic, ultrastructural, and chemical studies have been presented (5,11,79). A diagram based on the available data is shown in Fig. 3. All the photosynthetic accessory pigments appear to exist as protein complexes. The photosystem I domain consists of a reaction center (P_{700}) surrounded by a light-harvesting complex with chlorophyll *a* as the major antenna pigment with some contribution from β-carotene. The photosystem II domain consists of a reaction center (P_{680}) and a smaller antenna complex containing chlorophyll *a*. These two structural domains are intrinsic components of thylakoid membranes and are present in all oxygen-evolving photosynthetic organisms from cyanobacteria to higher plants (5,79). The size of the antenna within each of these two photosystem domains does vary in different organisms. In contrast to these two highly conserved domains, a third domain, designated as "accessory light-harvesting domain" in Fig. 3, shows enormous variation in the chemical nature of the light-harvesting complexes present and in the location and organization of these complexes (Table 1).

Three types of organic compounds are known to serve as chromophores in photosynthetic accessory systems: chlorophylls, carotenoids, and open-chain

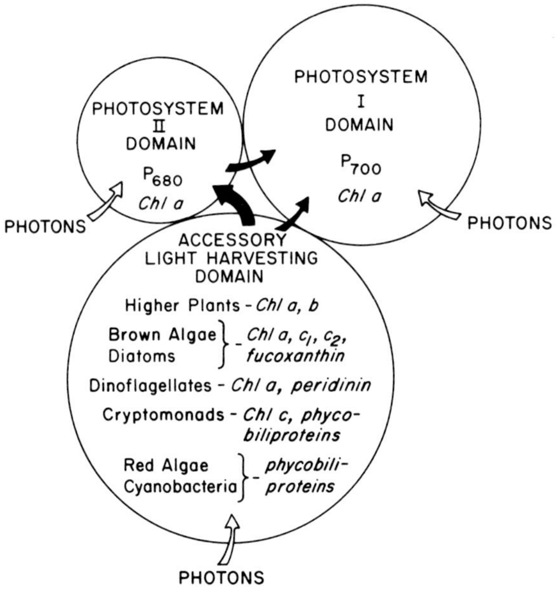

Fig. 3. Energy flow among the light-harvesting domains of the photosynthetic apparatus of cyanobacteria and plants. White arrows indicate direct absorption of light; black arrows show the pathways of migration of excitation energy. P_{680} and P_{700} represent the reaction center chlorophylls of photosystems II and I, respectively. The major light-harvesting components of each domain are shown in italics. Light harvesting range was determined in each instance by the nature of the protein pigment complexes present.

TABLE 1

Major Components of the Accessory Light-Harvesting Domain in Various Classes of Organisms That Perform Oxygenic Photosynthesis

Class of organisms	Light-harvesting complexes	Properties of the complexes
Higher plants, green algae	Chl *a/b* protein	This complex, an intrinsic component of the thylakoid membrane, contains 3 Chl *a*, 3 Chl *b*, and one carotenoid per monomer of 29,500 daltons (79)[a]
Brown algae	Fucoxanthin–chl a/c_2 protein Chl $a/c_1 + c_2$ protein	These complexes are tightly bound to the thylakoid membranes, and their isolation requires use of Triton X-100. The fucoxanthin–chl a/c_2 protein has a ratio of chl a:chl c_2 of 2, whereas the chl $a/c_1 + c_2$ protein has a ratio of chl a:chl $c_1 + c_2$ of 3, and equal amounts of chl c_1 and c_2 (1).
Diatoms	Fucoxanthin–chl *a/c* protein	This complex was isolated from *Phaeodactylum tricornutum* as a component of a much larger particle that exhibited photosystem II activity (50).
Dinoflagellates	Peridinin–chl *a* protein	These proteins are released from the algal cells in water-soluble form after freezing and thawing and grinding (or sonication). The structure of the chromoprotein varies with the organism. The protein from *Glenodinium sp.* contains four molecules of peridinin and 1 of chl *a* per dimer of 35,500 daltons, whereas the protein from *Amphidinium carterae* (Plymouth 450) contains nine molecules of peridinin and two of Chl *a* per 39,200 daltons (49,70).
Cryptomonads	Phycobiliproteins Chlorophyll *a/c* proteins (?)	Phycobiliproteins occupy an intrathylakoidal location (27). They are released as water-soluble aggregates on breakage of the cells. The types of phycobiliproteins present vary with the organism (68).
Red algae, cyanobacteria	Phycobilisomes	Complex particles made up, in the main, of phycobiliproteins. These particles are attached to the outer surface of the thylakoids (22). For details, see text.

[a] It is likely that there are two or more closely related chlorophyll *a/b* proteins.

tetrapyrroles (Fig. 4). Table 1 summarizes the data on the distribution, properties, and intracellular location of the various complexes that contain these pigments. The contributions of such complexes to the intact cell absorption spectra of green algae, brown algae, dinoflagellates, and cyanobacteria are shown in Fig. 5. Measurement of action spectra for oxygen evolution, or for the induction of chlorophyll fluorescence, has shown that, in each instance, the energy absorbed by the accessory light-harvesting complexes is transferred with high efficiency by radiationless processes primarily to the reaction center of photosystem II (e.g., 48). The arrows in Fig. 3 show the flow of absorbed light energy among the domains of the photosynthetic apparatus. Different molecular strategies are employed in the construction of the various light-harvesting complexes described in Table 1. Profound differences exist between the spectroscopic properties of different chromophores (e.g., bilins and carotenoids), for example, and these place specific limitations on the molecular anatomy of complexes that would permit efficient energy transfer (75). The phenotypic expression of the differences in the makeup of accessory light-harvesting domains and their precise juxtaposition relative to the photosystem I and II domains is seen in the gross morphologic differences in the arrangement of the thylakoid lamellae relative to each other in the chloroplasts of various algae (compare Figs. 6 and 7).

The difficulties inherent in the study of integral membrane proteins have hampered studies of the molecular structure of the components within photosynthetic lamellae of higher plants. However, the light-harvesting complexes of certain algae, such as those of cyanobacteria, red algae, and dinoflagellates can be obtained intact in water-soluble form (see Table 1). The remainder of this contribution is devoted to the discussion of one type of complex—the phycobilisome of cyanobacteria and red algae.

The "Accessory Light-Harvesting Domain" in Cyanobacteria and Red Algae

In a series of elegant studies, Gantt and her colleagues (23,24) showed that within the chloroplast of the unicellular red alga *Porphyridium cruentum* arrays of particles (phycobilisomes), some 30–40 mm in diameter and 20–30 nm thick, were regularly disposed on the outer surface of the thylakoids, and that these particles represented the intracellular location of the light-harvesting proteins (phycobiliproteins). Phycobilisomes of varying size and shape have since been demonstrated to be universally present in cyanobacteria and red algae (22,58,85; see Fig. 7). The arrangement of these particles on the surface of the thylakoid membranes correlates with the distribution of particles, believed to represent photosystem II, within the membranes (e.g., 88).

Methods have been developed for the isolation of phycobilisomes in an intact state, as judged by electron microscopy and spectroscopic methods (26,28,47). Phycobiliproteins represent about 85% of the phycobilisome, by

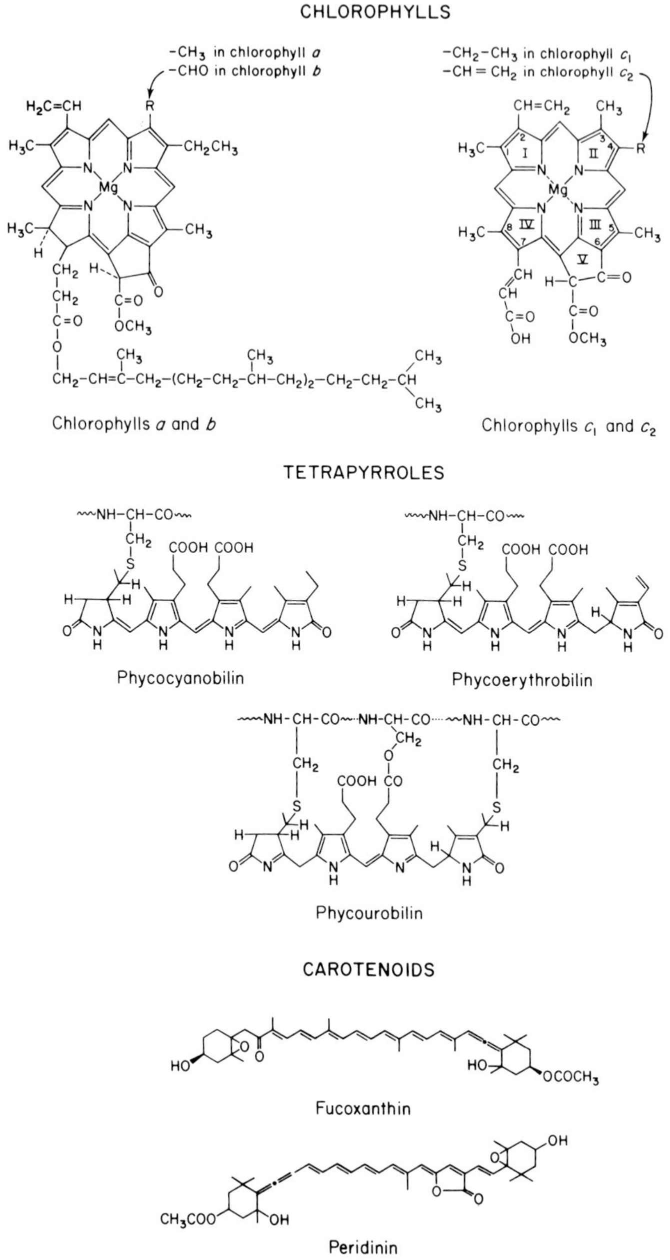

Fig. 4. Examples of the three types of organic compounds that serve as prosthetic groups in photosynthetic accessory systems.

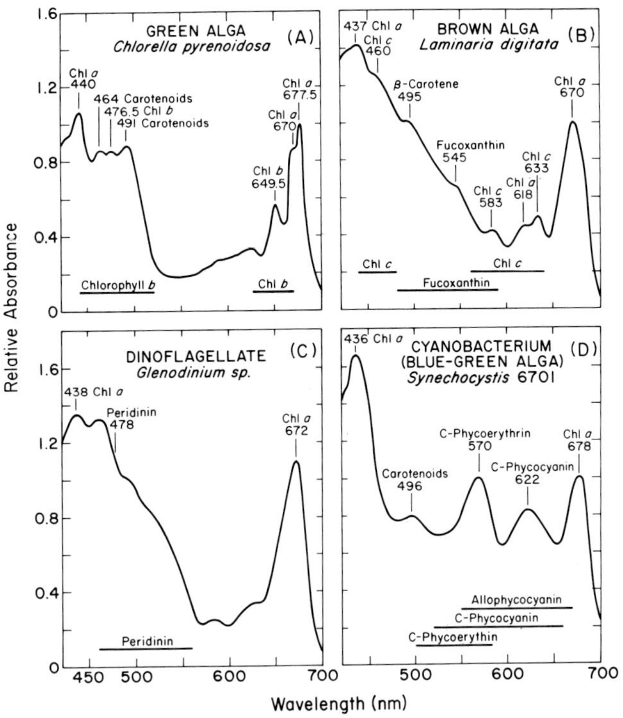

Fig. 5. Absorption spectra of suspensions of algal cells. (A) *Chlorella pyrenoidosa* at 77°K (12). (B) *Laminaria digitata* at 77°K (44). (C) *Glenodinium* sp. (cells cultured at low light intensity) at room temperature [based on Fig. 6 Prézelin *et al.* (71)]. (D) *Synechocystis* 6701 (cells cultured in warm white fluorescent light) at room temperature [based on Fig. 3, Lemasson *et al.* (57), and unpublished data]. For each of the indicated accessory pigments, the solid horizontal lines cover the portion of the spectrum spanned by its major absorption band(s).

weight (77,89), and several nonpigmented polypeptides make up the balance of the structure (77,89). There is evidence that these polypeptides play a role in the attachment of the phycobilisomes to the thylakoid membrane and the assembly of phycobilisomes (77,89). The ability to prepare intact phycobilisomes affords a unique opportunity to define in detail the structure and dynamics of an accessory light-harvesting domain.

Phycobiliproteins—the Light-Harvesting Proteins of Cyanobacteria (Blue-Green Algae) and Red Algae

The distribution and spectroscopic characteristics of common phycobiliproteins are given in Table 2. The color of cyanobacteria ranges from blue-green through violet and red to brown-black. The color in each case results from the simultaneous presence within the cells of these bacteria of the green pigment, chlorophyll *a*, of the blue proteins, allophycocyanin and phycocyanin, and, in some instances, of red proteins such as phycoerythrin. The phycobiliproteins can represent as much as 40% of the soluble protein of the

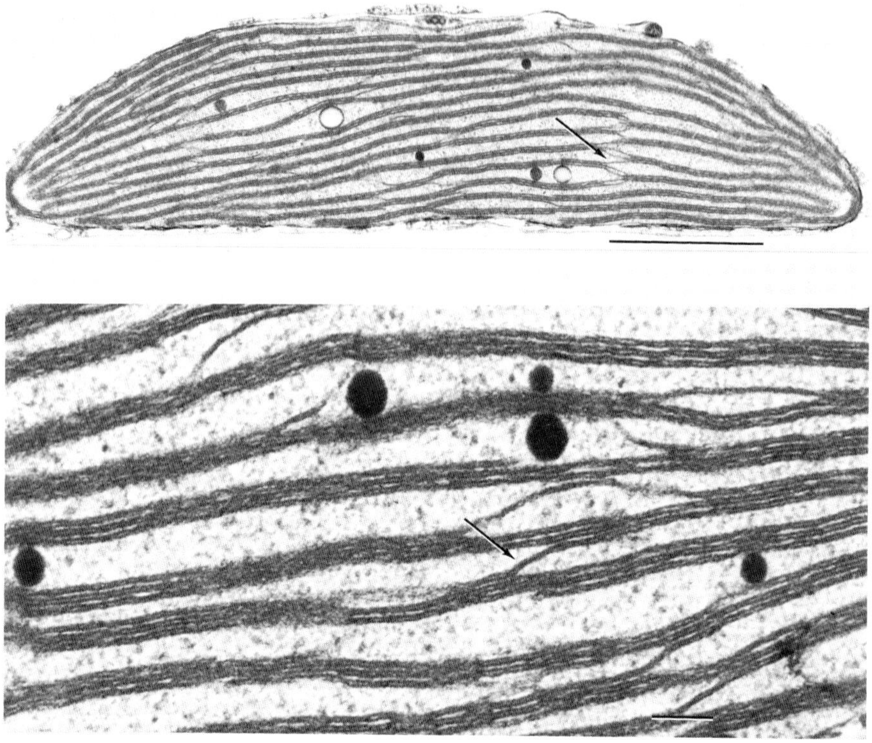

Fig. 6. Sections of the chloroplast of the brown alga *Cutleria hancockii* (order Cutleriales). The thylakoids in brown algal plastids are organized into parallel lamellae of triplets, as seen in section. Individual thylakoids are frequently seen transversing from one lamella to another (arrows), but the triplets are still maintained accurately despite these interchanges. The scale bars represents 1.0 μm in the upper figure and 0.1 μm in the lower one. (Micrographs courtesy of J. W. LaClaire II.)

cell. When phycocyanin predominates, the cells are blue-green, if both phycocyanin and phycoerythrin are present, the cells appear almost black. As shown in Fig. 5, a cyanobacterial cell containing both the red and blue proteins absorbs light strongly over the entire visible portion of the spectrum. Phycobiliproteins are also prominent constituents of red algal chloroplasts, and this fact prompted early suggestions that these chloroplasts represent descendants of cyanobacterial endosymbionts (61,62,76).

For a long time the characterization of phycobiliproteins was impeded by confusion between chemically distinct proteins and different "spectral forms" of a single protein. The spectroscopic properties of the phycobiliproteins show a strong dependence on the state of aggregation of these molecules, largely because of exciton interaction between the bilin chromophores in the higher assembly forms. For example, the prosthetic groups of both allophycocyanin and C-phycocyanin are phycocyanobilins (see Fig. 4). However, the spectra of these two proteins differ markedly from each other when these proteins are in their native aggregation state (Table 2). In contrast, the monomers of allophycocyanin and phycocyanin have very similar absorption spectra (14,42), and the spectra of the fully denatured proteins are qualitatively identical (35).

The diverse phycobiliproteins of cyanobacteria and red algae share in common a number of structural characteristics. Each phycobiliprotein is a higher aggregate of a monomer made up of two dissimilar polypeptide chains, α and β (Table 2). The molecular weights of these polypeptides lie between 16,000 and 22,000 daltons (7,30,67). In every instance the prosthetic groups are open-chain tetrapyrroles and these bilins are covalently attached to the polypeptide through thioether linkages (Fig. 4) (19,43,67,86). The predominant aggregation state of purified phycocyanins and phycoerythrins is either the hexamer $(\alpha\beta)_6$ or the trimer $(\alpha\beta)_3$, which exist in an equilibrium strongly dependent on protein concentration, ionic strength, and pH, with the monomer $(\alpha\beta)$ and the dimer $(\alpha\beta)_2$ (3,30). The spectroscopic differences between the various phycobiliproteins arise from the differences in the nature and number of the tetrapyrrole prosthetic groups (37) as well as from the differences in the primary structure of the polypeptide chains to which these bilins are attached.

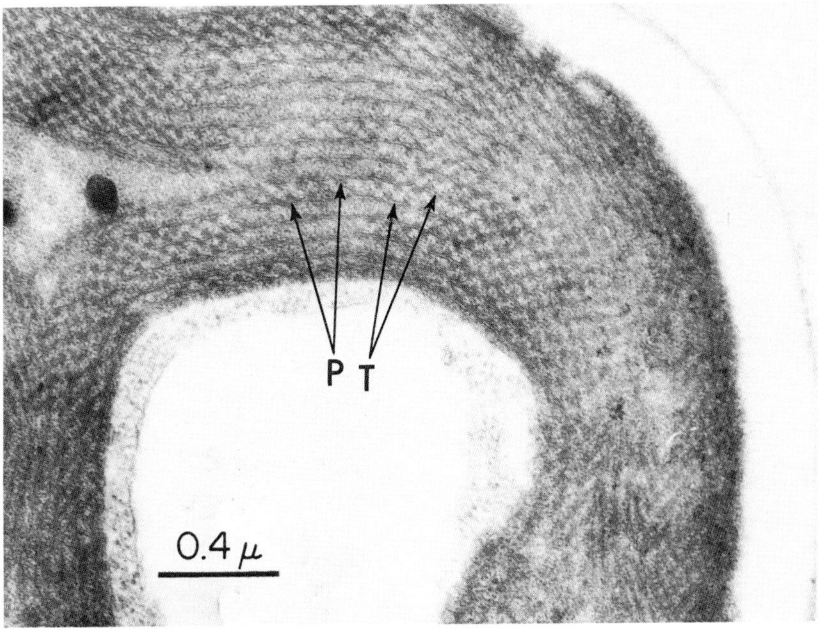

Fig. 7. Section of a portion of the chloroplast of the red alga *Batrachospermum virgatum* (SIRDT) (order Nemalionales). The section is cut perpendicular to the thylakoids (T) and to the phycobilisomes (P), which in this alga appear cylindrical (58). Note that the thylakoids do not come into contact with each other. (Micrograph courtesy of C. Lichtlé.)

TABLE 2

PROPERTIES OF COMMON PHYCOBILIPROTEINS OF CYANOBACTERIA AND RED ALGAE[a]

Biliprotein	Cyanobacteria	Red algae	Subunit structure of stable assembly form	Number and type of bilins attached per subunit[b]			Absorption maxima[c] (nm)	Fluorescence emission maxima (nm)
				α	β	γ		
Allophycocyanin B	+	+	$(\alpha\beta)_3$	1 PCB	1 PCB	–	671 > 618	673
Allophycocyanin	+	+	$(\alpha\beta)_3$	1 PCB	1 PCB	–	650	660
C-Phycocyanin	+	+	$(\alpha\beta)_3$, $(\alpha\beta)_6$	1 PCB	2 PCB	–	620	648
R-Phycocyanin	–	+	$(\alpha\beta)_3$	1 PCB	1 PCB; 1 PEB	–	617 > 555	636
Phycoerythrocyanin	+	–	$(\alpha\beta)_3$	1 PXB	2 PCB	–	568 > 590(s)	607
C-Phycoerythrin	+	–	$(\alpha\beta)_6$	2 PEB	4 PEB[d]	–	565	577
b-Phycoerythrin	–	+	$(\alpha\beta)_n$	2 PEB	4 PEB	–	545 > 563(s)	570
B-Phycoerythrin	–	+	$(\alpha\beta)_6\gamma$	2 PEB	4 PEB	2 PEB; 2(?)PUB	545 > 563 > 498(s)	575
R-Phycoerythrin	–	+	$(\alpha\beta)_6\gamma$			2 PEB; 2(?)PUB	567 > 538 > 498	578

[a] For original references, see Glazer (31).

[b] The abbreviations for the tetrapyrrole prosthetic groups are: PCB, phycocyanobilin; PEB, phycoerythrobilin; PUB, phycourobilin; PXB, a bilin of as yet undetermined structure.

[c] (s) denotes a shoulder.

[d] The β subunits of certain cyanobacterial phycoerythrins have been reported to have only 3 PEB prosthetic groups (65,66).

Phycobiliproteins Are Descended from a Common Ancestral Gene

Immunologic studies of the phycobiliproteins belonging to each of the major spectroscopically distinct classes (allophycocyanins, phycocyanins, phycoerythrins) demonstrated remarkable conservation of determinants for proteins belonging to a given spectroscopic class, independent of organismic origin (2,6,17,40,82). Among heterologous antigens, little difference was seen in the degree of cross-reactivity with proteins from cyanobacteria and those from the chloroplasts of red algae. No cross-reactivity was detected between proteins belonging to different spectroscopic types (e.g., allophycocyanins and phycocyanins) (40).

The immunologic results predicted an extraordinary degree of conservation of the tertiary structure of these proteins over an evolutionary history of over a billion years. This prediction was further strengthened by experiments that demonstrated that stable hybrid monomers could be obtained by hybridization of subunits of phycocyanins from unicellular and filamentous cyanobacteria of very different genera (36). More recently, such hybridization experiments on allophycocyanins resulted in the formation of stable hybrids between subunits derived from cyanobacterial and red algal proteins (14).

The preceding findings led to the expectation that the primary structures of individual phycobiliproteins would be highly conserved. The sequence comparisons in Figs. 8 and 9 show that this is indeed the case. Figure 8 compares the sequence of the α subunit of the C-phycocyanin of a filamentous cyanobacterium with that of an α subunit from the C-phycocyanin of a unicellular rhodophyte. Figure 9 compares the sequence of the β subunit of the C-phycocyanin from the same filamentous cyanobacterium with that derived from the C-phycocyanin of an unrelated unicellular cyanobacterium. In each instance, the extraordinary conservation of primary structure is self-evident. Moreover, there is clear homology between the α and β subunits.

The finding that the monomer of each of the phycobiliproteins was made up of two dissimilar subunits each of which carried bilin chromophores (33), suggested a common evolutionary origin for all of these proteins (30). The amino-terminal sequences of numerous phycobiliproteins, of almost every spectroscopic class, of both cyanobacterial and red algal origin, have now been determined, and it is clear beyond all reasonable doubt that all phycobiliproteins are descended from a common ancestral gene (Fig. 10). Moreover, analysis of the amino-terminal sequences hints that similarities in tertiary structure will be found among all the phycobiliproteins, as indicated by the total conservation of the hydrophobic character of the residues at certain positions in the composite sequence (Fig. 11).

Cryptomonad Phycobiliproteins—A Case of Interspecific Gene Transfer?

The phycobiliproteins occur in three classes of organisms: the cyanobacteria, the red algae, and the cryptomonads. The pigmented cryptomonads are

1									10						
Val	Lys	Thr	Pro	Ile	Thr	Asp	Ala	Ile	Ala	Ala	Ala	Asp	Thr	Gln	Gly
Met						Glu							Asn		
			20										30		
Arg	Phe	Leu	Ser	Asn	Thr	Glu	Leu	Gln	Ala	Val	Asn	Gly	Arg	Tyr	Gln
							40								
Arg	Ala	Ala	Ala	Ser	Leu	Glu	Ala	Ala	Arg	Ala	Leu	Thr	Ala	Asn	Ala
										Ser			Ser		
	50										60				
Gln	Arg	Leu	Ile	Asp	Gly	Ala	Ala	Gln	Ala	Val	Tyr	Gln	Lys	Phe	Pro
					70										80
Tyr	Leu	Ile	Gln	Thr	Ser	Gly	Pro	Asn	Tyr	Ala	Ala	Asp	Ala	Arg	Gly
	Thr	Ser		Met	Pro			*Gln*			Ser	*Ser*		*Val*	
									90						
Lys	Ser	Lys	Cys*	Ala	Arg	Asp	Ile	Gly	His	Tyr	Leu	Arg	Ile	Ile	Thr
	Ala								Tyr				Met		
			100										110		
Tyr	Ser	Leu	Val	Ala	Gly	Gly	Thr	Gly	Pro	Leu	Asp	Glu	Tyr	Leu	Ile
	Cys									Met					
							120								
Ala	Gly	Leu	Asn	Glu	Ile	Asn	Asp	Ala	Phe	Glu	Leu	Ser	Pro	Ser	Trp
			Glu				*Arg*	Thr		Asp					
	130										140				
Tyr	Ile	Glu	Ala	Leu	Lys	Tyr	Ile	Lys	Ala	Asn	His	Gly	Leu	Ser	Gly
	Val				Asn										
					150										160
Gln	Ala	Ala	Asn	Glu	Ala	Asn	Thr	Tyr	Ile	Asp	Tyr	Val	Ile	Asn	Ala
												Ala			
Leu	Ser														

Fig. 8. The continuous sequence is that of the C-phycocyanin α subunit of the filamentous cyanobacterium *Mastigocladus laminosus* (19). The residues indicated below this sequence represent replacements in the sequence of the corresponding subunit of the C-phycocyanin of the unicellular rhodophyte *Cyanidium caldarium* (80). Replacements that require more than a single base change in the codon are indicated in italics. The bilin-linked cysteinyl residue is indicated by an asterisk.

1									10						
Ala	Tyr	Asp	Val	Phe	Thr	Lys	Val	Val	Ser	Gln	Ala	Asp	Ser	Arg	Gly
Thr	Phe		Ala						Ala				Ala		

			20										30		
Glu	Phe	Leu	Ser	Asn	Glu	Gln	Leu	Asp	Ala	Leu	Ala	Asn	Val	Val	Lys
				Asp	Ala					Ser		*Arg*	Leu		*Ala*

							40								
Glu	Gly	Asn	Lys	Arg	Leu	Asp	Val	Val	Asn	Arg	Ile	Thr	Ser	Asn	Ala
					Ile		*Thr*						Gly		

	50										60				
Ser	Thr	Ile	Val	Thr	Asn	Ala	Ala	Arg	Ala	Leu	Phe	Glu	Glu	Gln	Pro
	Ser			Ala								Ala			

					70										80
Gln	Leu	Ile	Ala	Pro	Gly	Gly	Ser	Ala	Thr	Arg	Asn	Gly	Thr	Met	Ala
							Asn		*Tyr*	Thr		[−]	Arg		

									90						
Ala	Cys*	Leu	Arg	Asp	Met	Glu	Ile	Ile	Leu	Arg	Tyr	Ile	Thr	Tyr	Ala
												Val			

			100										110		
Ile	Leu	Ala	Gly	Asp	Ala	Ser	Ile	Leu	Asp	Asp	Arg	Cys	Leu	Asn	Gly
Val	Phe	Thr												Asp	

							120								
Leu	Arg	Glu	Thr	Tyr	Gln	Ala	Leu	Gly	Thr	Pro	Gly	Ser	Ser	Val	Ala
					Leu		Ser		Val			Ala	Leu		

	130										140				
Val	Gly	Ile	Gln	Lys	Met	Lys	Glu	Ala	Ala	Ile	Asn	Ile	Ala	Asn	Asp
Glu		Val	Arg				Asp			Val	*Ala*		Val	Ser	

															160
Pro	Asn	Gly	Ile	Thr	Lys	Gly	Asp	Cys*	Ser	Ala	Leu	Ile	Ser	Glu	Val
Arg					Gln						[−]				Leu

									170		
Ala	Ser	Tyr	Phe	Asp	Arg	Ala	Ala	Ala	Ala	Val	Ala
Gly					Lys						

Fig. 9. The continuous sequence is that of the C-phycocyanin β subunit of the filamentous cyanobacterium *Mastigocladus laminosus* (19). The residues indicated below this sequence represent replacements in the corresponding subunit of the C-phycocyanin of the unicellular cyanobacterium *Synechococcus sp.* 6301 (20). Replacements that require more than a single base change in the codon are indicated in italics. Apparent deletions are indicated by [–]. The bilin-linked cysteinyl residues are indicated by asterisks.

Composite of the amino-terminal sequence of α-type subunits

1									10									
Met	Lys	Thr	Pro	Ile	[Thr]	Glu	[Ala]	[Ile]	Ala	Ala	[Ala	Asp]	Ala	Gln	[Gly]	Arg	[Phe]	[Leu]
Val		Ser	Ile	Val		Thr	[Val]	[Val]	Val	Thr			Asn	Glu	[Ala]		[Tyr]	
Ser			Val	Leu		Asp			Gly	Asn			Thr	Ala				
						Lys			Ser				Ser	Arg				
									Thr				Asp					
													Val					

Composite of the amino-terminal sequences of β-type subunits

Met	Leu	Asp	Thr	Phe	[Thr]	Lys	[Val]	[Val]	Ala	Gln	[Ala	Asp]	Ala	Arg	[Gly]	Glu	[Phe]	[Leu]
Ala	Phe		Ala	Ile	[Ala]	Arg	[Ala]	[Ile]	Val	Ala			Val	Lys	[Ala]	Ala	[Tyr]	[Val]
Thr	Tyr		Val			Ala			Asn	Asn			Ser	Gln		Lys		
Gly	Glu								Gln				Thr			Asn		
									Ser				Arg					

Source of the proteins whose α- and β-subunit amino-terminal sequences were used for the composite sequences shown above:

Unicellular cyanobacteria	
Allophycocyanin	*Synechocystis sp* (ATCC 22663)[a] (39); *Synechococcus sp* 6301 (ATCC 27144) (39)
C-Phycocyanin	*Synechococcus sp* 6301 (20,87)
C-Phycoerythrin	*Synechocystis sp* 6701 (ATCC 27170) (9)
Filamentous cyanobacteria	
Allophycocyanin	*Anabaena variabilis* (ATCC 27892) (39)
C-Phycocyanin	*Anabaena variabilis* (39); *Mastigocladus laminosus* (19); *Oscillatoria agardhii* (84)
Phycoerythrocyanin	*Anabaena variabilis* (39)
Unicellular red algae	
Allophycocyanin	*Porphyridium cruentum* (39)
C-Phycocyanin	*Cyanidium caldarium* (81); *Porphyridium aerugineum* (9)
R-Phycocyanin	
b-phycoerythrin	
B-phycoerythrin	*Porphyridium cruentum* (38,39)

[a] Abbreviation used: ATCC, American Type Culture Collection.
[b] Highly conserved homologous residues in the composite α and β sequences are "boxed."

Fig. 10. Amino-terminal sequences of cyanobacterial and red algal phycobiliproteins.

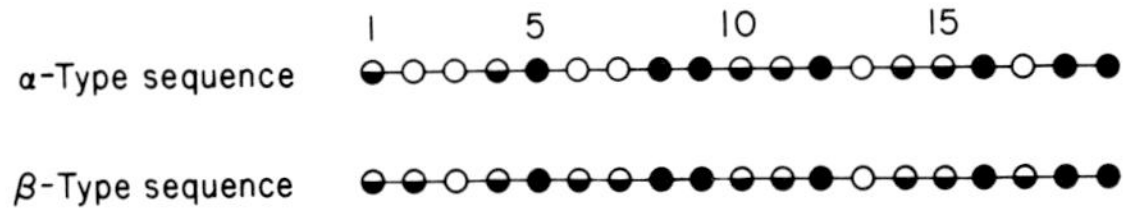

Fig. 11. Composite sequences for the amino-terminal 19 residues of the α-type and β-type subunits of the phycobiliproteins, derived from the data in Fig. 10, compared with respect to the nature of the residues occurring at each position. ○ indicates that only hydrophilic residues are found in this position; ◒ indicates that both hydrophobic and hydrophilic residues occur; and ● means that only hydrophobic residues are found.

biflagellate photoautotrophic unicellular algae that show little taxonomic relationship to the red algae (16). Comparison of the chloroplast of the cryptomonads with that of red algae reveals major differences in chlorophyll and carotenoid composition and ultrastructure (16,76). In contrast to the biliproteins of cyanobacteria and red algae, cryptomonad biliproteins form only low-molecular-weight aggregates *in vitro*, and *in vivo* have been shown to be localized within the thylakoids (27). Whereas the cryptomonad phycobiliproteins do have an αβ monomer structure, the α subunit is small (9,000–10,000 daltons) (8,41,60,63), and certain of the cryptomonad biliproteins carry bilin chromophores chemically different from those observed thus far among cyanobacterial and red algal phycobiliproteins (34).

These considerations have led to the conclusion that red algal and cryptomonad chloroplasts have distinct origins (76) and that the biliproteins of these two algal groups have an independent evolutionary history (72). This conclusion is in apparent harmony with paleontological evidence, which places the earliest cyanobacteria and red algae in the Precambrian and Lower Cambrian, respectively, whereas the cryptomonads first appear in the geologic record in the Holocene (59).

Determination of the amino-terminal sequence of the β subunit of a cryptomonad phycocyanin led to the remarkable finding that 15 of the first 19 residues were identical to those of the β subunit of a red algal phycoerythrin (32). Two alternative explanations may be offered for this surprising finding. The first explanation would postulate the transfer of a biliprotein gene from an ancestor of the cyanobacterium to an unrelated photosynthetic prokaryote. The cyanobacteria are undoubtedly the evolutionary precursors of the red algal chloroplast (76); the unrelated photosynthetic prokaryote would be presumed to be the ancestor of the cryptomonad chloroplast. A more plausible explanation is that the cryptomonad chloroplast arose from a photosynthetic prokaryote different from the ancestor of the cyanobacteria, but that at a late stage in evolution it acquired additional characteristics from a red algal andosymbiont (78). Whatever the explanation, the sequence data strongly support the view that phycobiliproteins arose but once in evolution.

Energy Transfer

The phycobiliproteins in cyanobacteria and red algae are organized in particles called phycobilisomes (see the preceding discussion). These complex structures are stable in concentrated solutions of certain salts (e.g., 0.65–0.75 *M* Na-K phosphate or 0.8 *M* Na sulfate) at pH 7–8 but dissociate on decrease in the salt concentration. Phycobilisome dissociation is accompanied by a very large increase in the fluorescence emission at wavelengths characteristic of the individual biliproteins (Table 3, column 2), whereas in intact phycobilisomes, excitation with green light (λ ~530 nm) produces maximum emission at ~675 nm. Hence fluorescence spectra provide an excellent index of phycobilisome integrity.

Migration of energy from the biliproteins to chlorophyll *a* proceeds by radiationless energy transfer processes. The pathway for this transfer is

```
                                              allophycocyanin
                                              (λmax 650 nm)
Phycoerythrin    →   phycocyanin    →                              →   chlorophyll a
(λmax ~560 nm)       (λmax ~620 nm)                                    (λmax 680 nm)
                                              allophycocyanin B
                                              (λmax 671 nm)
```

This scheme is based on evidence of the type presented in Table 3. Although in the intact system the efficiency of energy transfer from one chromoprotein to the next is 99%, the path of energy transfer may be traced by following the appearance of the small amount of fluorescence emitted by each chromoprotein after excitation by a very short pulse of monochromatic light of appropriate wavelength. For example, the time required for energy to be transferred from phycoerythrin to phycocyanin is observed as a risetime in the fluorescence of the latter pigment to a maximum value. Inspection of column 4 in Table 3 shows unambiguously that, in intact cells, energy transfer follows the path shown earlier. Other important points are illustrated by the data in Table 3. The energy transfer process is very rapid relative to the

TABLE 3

THE FLUORESCENCE CHARACTERISTICS OF THE LIGHT-HARVESTING PIGMENTS OF THE UNICELLULAR RED ALGA *Porphyridium cruentum*[a]

				Mean lifetime of fluorescence (psec)	
Pigment	Emission wavelength (nm)	Fluorescence lifetime (nsec) Isolated phycobiliprotein	Rise time of fluorescence (psec) In whole cells	In whole cells	In isolated phycobilisomes
B-Phycoerythrin	578	7.1	0	70	70
R-Phycocyanin	640	1.8	12	90	
Allophycocyanin	660	2.7[b]	22	118	4 nsec
Chlorophyll *a*	685	5.7	52	175	

[a] Excitation was performed with 6 psec 530 nm light pulses of low photon densities. The absorption at 530 nm is 94–96% due to phycoerythrin and less than 1% due to allophycocyanin. Data from Porter *et al.* (69, 73).

[b] Data from Grabowski and Gantt (45).

lifetime of the excited state of each individual chromoprotein (compare columns 3 and 5). Further, the rate of energy transfer between biliproteins in whole cells and in phycobilisomes appears the same, as indicated by a mean fluorescence lifetime of 70 psec for B-phycoerythrin in both instances (Table 3, columns 5 and 6). By this criterion the phycobilisomes are not damaged during isolation. Since in the phycobilisomes the terminal acceptor, chlorophyll *a*, is not present, the mean lifetime of fluorescence for allophycocyanin increases to the value anticipated for the isolated biliprotein (Table 3, columns 3 and 6).

Phycobilisome Morphology and Its Relation to Function

A model for the structure of phycobilisomes is shown in Fig. 12. This model is derived from electron microscope studies of cyanobacterial (10) and red algal (22,26,64) phycobilisomes. The ultrastructure of these particles varies in detail from one organism to another, but the overall design appears conserved in every instance. From a variety of studies, including cross-linking (13), immunoelectron microscopy (25), partial dissociation (29), isola-

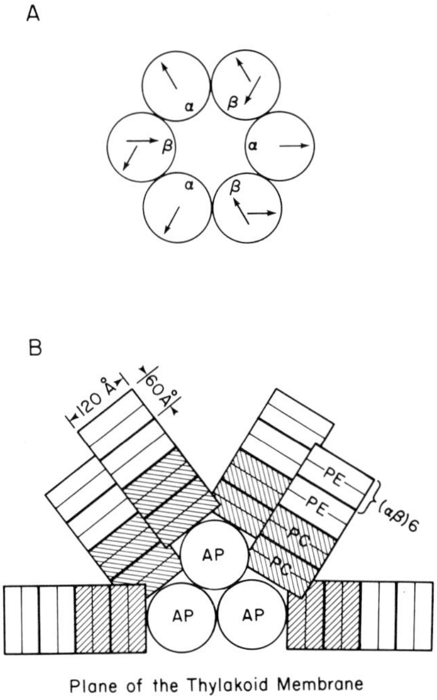

Fig. 12. (A) Top view of a phycocyanin double disk. The arrows indicate schematically orientations of the transition moments of the nine phycocyanobilin groups in the upper $(\alpha\beta)_3$ disk (see text). (B) Side view of a phycobilisome from a cyanobacterium, which produces phycoerythrin (PE), phycocyanin (PC), and allophycocyanin (AP), indicating the locations of these chromoproteins in the structure. [Based on the model presented in Bryant *et al.* (10).]

tion of complexes from partially dissociated phycobilisomes (56), examination of the effect of chromatic adaptation (10), etc., it has been shown that phycoerythrin occupies the peripheral position in the structure and that allophycocyanin is at the core (Fig. 12B). The physical arrangement of the biliproteins parallels the sequence of donors and acceptors in the energy transfer pathway discussed earlier. This arrangement is to be anticipated from the observed very high efficiency of energy transfer and the known r^{-3} to r^{-6} dependence of the rate of radiationless energy transfer processes on the distance (r) between donor and acceptor dipoles (55).

The structure of phycobilisomes is optimized for their light-harvesting function in other respects as well. Electron microscopy of purified phycoerythrin (21) and phycocyanin (4,18) shows that these molecules form double disk assembly forms morphologically indistinguishable from the repeating elements of the rods in the phycobilisome structure, designated as $(\alpha\beta)_6$ in Fig. 12B. On excitation with polarized light, the fluorescence emission of the hexameric aggregates is seen to be largely depolarized (15,83). This indicates that energy migration within these aggregates is very rapid and that the orientation of the bilin dipoles relative to each other is nearly random. A diagrammatic representation of such an arrangement is shown in the top view of a phycocyanin hexamer (Fig. 12A), in which the arrows represent the direction of transition dipole moments of the nine bilins contained in an $(\alpha\beta)_3$ disc (35). The fluorescence emission of intact phycobilisomes is also nearly completely depolarized, indicating that the mutual orientation of the transition moments of the bilin dipoles in the allophycocyanin- and allophycocyanin B-containing core is nearly random (46).

In the cell the arrays of light-harvesting biliproteins rods are held at widely varying orientation with respect to the plane of the thylakoid membrane. It can readily be demonstrated that for such a situation, a maximum absorption of unpolarized light is obtained with a random orientation of absorbing dipoles. An analogous arrangement of photoreceptor dipoles in the unfused rhabdoms of flies and in the outer segment discs of the rods of the vertebrate retina is utilized to ensure maximum absorption of unpolarized light (74).

A striking feature of the organization of the phycobilisome is that this is a structure designed for *polar energy transfer* and that its design minimizes the number of excitation energy jumps between molecules of the same biliprotein by organizing the antenna molecules into individual rods. In a structure in which all the rods were fused, the number of energy transfer paths within each type of light-harvesting layer would be significantly increased with concomitant decrease in the efficiency of the overall transfer process. This inference is supported by the studies of Searle and colleagues (73), who showed that only a few jumps are necessary to transfer energy from one pigment to the next. In this context it should be noted that a C-phycoerythrin double-disk, $(\alpha\beta)_6$, contains 36 bilin chromophores, and that of B-phycoerythrin, with the structure $(\alpha\beta)_6\,\gamma$, contains 40 such chromophores (38).

Acknowledgments

The subject of photosynthetic accessory pigments is particularly appropriate to this occasion, since it was Emil Smith's pioneering work in 1941 that first indicated that chlorophyll in photosynthetic membranes was bound to proteins and that such complexes were of manageable size. I offer this contribution to Professor Emil L. Smith as a token of appreciation for many years of stimulating discussion and encouragement in this and many other areas of research.

The research in the author's laboratory is supported in part by grant PCM 76-15243A02 from the National Science Foundation.

REFERENCES

1. Anderson, J. M., and Barrett, J., Chlorophyll-protein complexes of brown algae: P700 reaction centre and light-harvesting complexes. *Ciba Found. Symp.*, 1979, **61** (new ser.):81–104.
2. Berns, D. S., Immunochemistry of biliproteins. *Plant Physiol.*, 1967, **42:**1569–1586.
3. Berns, D. S., Phycocyanins and deuterated proteins. *Biol. Macromol.*, 1971, **5:**105–148.
4. Berns, D. S., and Edwards, M. R., Electron micrographic investigations of C-phycocyanin. *Arch. Biochem. Biophys.*, 1965, **110:**511–516.
5. Boardman, N. K., Anderson, J. M., and Goodchild, D. J., Chlorophyll-protein complexes and structure of mature and developing chloroplasts. *Curr. Top. Bioenerg.*, 1978, **8:**35–109.
6. Bogorad, L., Studies of phycobiliproteins. *Rec. Chem. Prog.*, 1965, **26:**1–12.
7. Bogorad, L., Phycobiliproteins and complementary chromatic adaptation. *Annu. Rev. Plant Physiol.*, 1975, **26:**369–401.
8. Brooks, C., and Gantt, E., Comparison of phycoerythrins (542, 566 nm) from cryptophycean algae. *Arch. Mikrobiol.*, 1973, **88:**193–204.
9. Bryant, D. A., *Comparative Studies on Cyanobacterial and Rhodophytan Biliproteins.* Ph.D. Thesis, University of California, Los Angeles, 1977.
10. Bryant, D. A., Guglielmi, G., Tandeau de Marsac, N., Castets, A.-M. and Cohen-Bazire, G., The structure of cyanobacterial phycobilisomes. A model. *Arch. Microbiol.*, 1979, **123:**113–127.
11. Butler, W. L., Energy distribution in the photosynthetic apparatus of plants. *Brookhaven Symp. Biol.*, 1977, **28:**338–346.
12. Cho, F., and Govindjee, Low temperature (4–77°K) spectroscopy of *Chlorella;* temperature dependence of energy transfer efficiency. *Biochim. Biophys. Acta*, 1970, **216:**139–150.
13. Cohen-Bazire, G., and Lefort-Tran, M., Fixation of phycobiliproteins to photosynthetic membranes by glutaraldehyde. *Arch. Mikrobiol.*, 1970, **71:**245–257.
14. Cohen-Bazire, G., Béguin, S., Rimon, S., Glazer, A. N., and Brown, D. M., Physico-chemical and immunological properties of allophycocyanins. *Arch. Microbiol.*, 1977, **111:**225–238.
15. Dale, R. E., and Teale, F. W. J., Number and distribution of chromophore types in native phycobiliproteins. *Photochem. Photobiol.*, 1970, **12:**99–117.
16. Dodge, J. D., *The Fine Structure of Algal Cells.* Academic Press, New York, 1973.

17. Eder, J., Wagenmann, R., and Rüdiger, W., Immunological relationship between phycoerythrins from various blue-green algae. *Immunochemistry*, 1978, **15**:315–321.
18. Eiserling, F. A., and Glazer, A. N., Blue-green algal proteins: Assembly forms of C-phycocyanin from *Synechococcus sp. J. Ultrastruct. Res.*, 1974, **47**:16–25.
19. Frank, G., Sidler, W., Widmer, H., and Zuber, H., Complete amino acid sequences of both subunits of C-phycocyanin from the cyanobacterium *Mastigocladus laminosus. Hoppe-Seyler's Z. Physiol. Chem.*, 1978, **359**:1491–1507.
20. Freidenreich, P., Apell, G. S., and Glazer, A. N., Structural studies on phycobiliproteins. II. C-Phycocyanin: Amino acid sequence of the β subunit. Specific cleavage of the α subunit. *J. Biol. Chem.*, 1978, **253**:212–219.
21. Gantt, E., Properties and ultrastructure of phycoerythrin from *Porphyridium cruentum. Plant Physiol.*, 1969, **44**:1629–1638.
22. Gantt, E., Phycobilisomes: Light-harvesting pigment complexes. *BioScience*, 1975, **25**:781–788.
23. Gantt, E., and Conti, S. F., Granules associated with the chloroplast lamellae of *Porphyridium cruentum. J. Cell Biol.*, 1966, **29**:423–434.
24. Gantt, E., and Conti, S. F., Phycobiliprotein localization in algae. *Brookhaven Symp. Biol.*, 1967, **19**:393–405.
25. Gantt, E., and Lipschultz, C. A., Probing phycobilisome structure by immunoelectron microscopy. *J. Phycol.*, 1977, **13**:185–192.
26. Gantt, E., and Lipschultz, C. A., Phycobilisomes of *Porphyridium cruentum*. I. Isolation. *J. Cell Biol.*, 1972, **54**:313–324.
27. Gantt, E., Edwards, M. R., and Provasoli, L., Chloroplast structure of the *Cryptophyceae*. Evidence for phycobiliproteins within intrathylakoidal spaces. *J. Cell Biol.*, 1971, **48**:280–290.
28. Gantt, E., Lipschultz, C. A., Grabowski, J., and Zimmerman, B. K., Phycobilisomes from blue-green and red algae. Isolation criteria and dissociation characteristics. *Plant Physiol.*, 1979, **63**:615–620.
29. Gantt, E., Lipschultz, C. A., and Zilinskas, E., Further evidence for a phycobilisome model from selective dissociation, fluorescence emission, immunoprecipitation, and electron microscopy. *Biochim. Biophys. Acta*, 1976, **430**:375–388.
30. Glazer, A. N., Phycocyanins: Structure and function. *Photochem. Photobiol. Rev.*, 1976, **1**:71–115.
31. Glazer, A. N., Structure and molecular organization of the photosynthetic accessory pigments of cyanobacteria and red algae. *Mol. Cell. Biochem.*, 1977, **18**:125–140.
32. Glazer, A. N., and Apell, G. S., A common evolutionary origin for the biliproteins of cyanobacteria, Rhodophyta and Cryptophyta. *FEMS Lett.*, 1977, **1**:113–116.
33. Glazer, A. N., and Cohen-Bazire, G., Subunit structure of the phycobiliproteins of blue-green algae. *Proc. Natl. Acad. Sci. U.S.A.*, 1971, **68**:1398–1401.
34. Glazer, A. N., and Cohen-Bazire, G., A comparison of cryptophytan phycocyanins. *Arch. Microbiol.*, 1975, **104**:29–32.
35. Glazer, A. N., and Fang, S., Chromophore content of blue-green algal phycobiliproteins. *J. Biol. Chem.*, 1973, **248**:659–662.

36. Glazer, A. N., and Fang, S., Formation of hybrid proteins from the α and β subunits of phycocyanins of unicellular and filamentous blue-green algae. *J. Biol. Chem.*, 1973, **248:**663–671.
37. Glazer, A. N., and Hixson, C. S., Characterization of R-phycocyanin. Chromophore content of R-phycocyanin and C-phycoerythrin. *J. Biol. Chem.*, 1975, **250:**5487–5495.
38. Glazer, A. N., and Hixson, C. S., Subunit structure and chromophore composition of rhodophytan phycoerythrins. *J. Biol. Chem.*, 1977, **252:**32–42.
39. Glazer, A. N., Apell, G. S., Hixson, C. S., Bryant, D. A., Rimon, S., and Brown, D. M., Biliproteins of cyanobacteria and Rhodophyta: Homologous family of photosynthetic accessory pigments. *Proc. Natl. Acad. Sci. U.S.A.*, 1976, **73:**428–431.
40. Glazer, A. N., Cohen-Bazire, G., and Stanier, R. Y., Comparative immunology of algal biliproteins. *Proc. Natl. Acad. Sci. U.S.A.*, 1971, **68:**3005–3008.
41. Glazer, A. N., Cohen-Bazire, G., and Stanier, R. Y. Characterization of phycoerythrin from a *Cryptomonas sp. Arch. Mikrobiol.*, 1971, **80:**1–18.
42. Glazer, A. N., Fang, S., and Brown, D. M., 1973, Spectroscopic properties of C-phycocyanin and of its α and β subunits. *J. Biol. Chem.*, 1973, **248:**5679–5685.
43. Glazer, A. N., Hixson, C. S., and DeLange, R. J., Determination of the number of thioether-linked cysteine residues in cytochromes *c* and phycobiliproteins. *Anal. Biochem.*, 1979, **92:**489–496.
44. Goedheer, J. C., On the pigment system of brown algae. *Photosynthetica*, 1970, **4:**97–106.
45. Grabowski, J., and Gantt, E. Photophysical properties of phycobiliproteins from phycobilisomes: Fluorescence lifetimes, quantum yields, and polarization spectra. *Photochem Photobiol.*, 1978, **28:**39–45.
46. Grabowski, J., and Gantt, E. Excitation migration in phycobilisomes: Comparison of experimental results and theoretical predictions. *Photochem. Photobiol.*, 1978, **28:**47–54.
47. Gray, B. H., Lipschultz, C. A., and Gantt, E., Phycobilisomes from a blue-green alga *Nostoc* species. *J. Bacteriol.*, 1973, **116:**471–478.
48. Halldal, P., The photosynthetic apparatus of microalgae and its adaptation to environmental factors. In: *Photobiology of Microorganisms* (P. Halldal, ed.). Wiley, New York, 1970:17–55.
49. Haxo, F. T., Kycia, J. H., Somers, G. F., Bennett, A., and Siegelman, H. W., Peridinin-chlorophyll *a* proteins of the dinoflagellate *Amphidinium carterae* (Plymouth 450). *Plant Physiol.*, 1976, **57:**297–303.
50. Holdsworth, E. S., and Arshad, J. H., A manganese-copper-pigment-protein complex isolated from the photosystem II of *Phaeodactylum tricornutum. Arch. Biochem. Biophys.*, 1977, **183:**361–373.
51. Junge, W., Physical aspects of light harvesting, electron transport and electrochemical potential generation in photosynthesis of green plants. *Encycl. Plant Physiol.*, 1977, **5** (new ser.):59–93.
52. Kirk, J. T. O., A theoretical analysis of the contribution of algal cells to the attenuation of light within natural waters. III. Cylindrical and spheroidal cells. *New Phytol.*, 1976, **77:**341–358.
53. Kirk, J. T. O., Yellow substance (Gelbstoff) and its contribution to the attenuation

of photosynthetically active radiation in some inland and coastal south-eastern Australian Waters. *Aust. J. Mar. Freshwater Res.*, 1976, **27**:61–71.

54. Kirk, J. T. O., Attenuation of light in natural waters. *Aust. J. Mar. Freshwater Res.*, 1977, **28**:497–508.
55. Knox, R. S., Excitation energy transfer and migration: Theoretical considerations. In: *Bioenergetics of Photosynthesis* (Govindjee, ed.). Academic Press, New York, 1975:183–221.
56. Koller, K.-P., Wehrmeyer, W., and Mörschel, E., Biliprotein assembly in the disc-shaped phycobilisomes of *Rhodella violacea.* On the molecular composition of energy-transfering complexes (tripartite units) forming the periphery of the phycobilisome. *Eur. J. Biochem.*, 1978, **91**:57–63.
57. Lemasson, C., Tandeau de Marsac, N., and Cohen-Bazire, G., Role of allophycocyanin as a light-harvesting pigment in cyanobacteria. *Proc. Natl. Acad. Sci. U.S.A.*, 1973, **70**:3130–3133.
58. Lichtlé, C., and Giraud, G., Aspets ultrastructuraux particuliers au plaste du *Batrachospermum virgatum* (Sirdt) Rhodophycée Némalionale. *J. Phycol.*, **6**:281–293.
59. Loeblich, A. R., Jr., Protistan phylogeny as indicated by the fossil record. *Taxon*, 1974, **23**:227–290.
60. MacColl, R., Habig, W., and Berns, D. S., Characterization of phycocyanin from *Chroomonas* species. *J. Biol. Chem.*, 1973, **248**:7080–7086.
61. Mereschkovsky, C., Über Natur und Ursprung der Chromatophoren in Pflanzenreiche. *Biol. Zentralbl.*, 1905, **25**:593–604.
62. Mereschkovsky, C., La plante considérée comme un complexe symbiotique. *Bull. Soc. Sci. Nat. Ouest Fr.*, 1920, **6**:17–21.
63. Mörschel, E., and Wehrmeyer, W., Cryptomonad biliprotein: Phycocyanin-645 from a *Chroomonas* species. *Arch. Microbiol.*, 1975, **105**:153–158.
64. Mörschel, E., Koller, K. P., Wehrmeyer, W., and Schneider, H., Biliprotein assembly in the disc-shaped phycobilisomes of *Rhodella violacea.* I. Electron microscopy of phycobilisomes *in situ* and analysis of their architecture after isolation and negative staining. *Cytobiologie*, 1977, **16**:118–129.
65. Muckle, G., and Rüdiger, W., Chromophore content of C-phycoerythrin from various cyanobacteria. *Z. Naturforsch.* [*C*], 1977, **32**:957–962.
66. Muckle, G., Otto, J., and Rüdiger, W., On the linkage between chromophore and protein in biliproteins. VII. Amino acid sequence in the chromophore regions of C-phycoerythrin from *Pseudanabaena* W1173 and *Phormidium persicinum. Hoppe-Seyler's Z. Physiol. Chem.*, 1978, **359**:345–355.
67. O'Carra, P., and O'hEocha, C., Algal biliproteins and phycobilins. In: *Chemistry and Biochemistry of Plant Pigments* (T. W. Goodwin, ed.), 2nd ed. Academic Press, New York, 1976:328–376.
68. O'hEocha, C., O'Carra, P., and Mitchell, D., Biliproteins of cryptomonad algae. *Proc. R. Ir. Acad.* [B], 1964, **63**:191–200.
69. Porter, G., Tredwell, C. J., Searle, G. F. W., and Barber, J., Picosecond time-resolved energy transfer in *Porphyridium cruentum.* Part I. In the intact alga. *Biochim. Biophys. Acta*, 1978, **501**:232–245.
70. Prézelin, B. B., and Haxo, F. T., Purification and characterization of peridinin-

chlorophyll *a*-proteins from the marine dinoflagellates *Glenodinium sp.* and *Gonyaulax polyedra. Planta,* 1976, **128:**133–141.
71. Prézelin, B. B., Ley, A. C., and Haxo, F. T., Effects of growth irradiance on the photosynthetic action spectra of the marine dinoflagellate, *Glenodinium* sp. *Planta,* 1976, **130**:251–256.
72. Scagel, R. F., Bandoni, R. J., Rouse, G. E., Schofield, W. B., Stein, J. R., and Taylor, T. M. C., *An Evolutionary Survey of the Plant Kingdom.* Wadsworth Publishing Co., Inc., Belmont, California, 1965:176–177.
73. Searle, G. F. W., Barber, J., Porter, G., and Tredwell, C. J., Picosecond time-resolved energy transfer in *Porphyridium cruentum.* Part II. In the isolated light harvesting complex (phycobilisomes). *Biochim. Biophys. Acta,* 1978, **501**:246–256.
74. Snyder, A. W., and Laughlin, S. B., Dichroism and absorption by photoreceptors. *J. Comp. Physiol.,* 1975, **100:**101–116.
75. Song, P-S., Molecular topography of solar energy harvesting pigments in marine algae. *Trends Biochem. Sci.,* 1978, **3:**25–27.
76. Stanier, R. Y., The origins of photosynthesis in eukaryotes. *Symp. Soc. Gen. Microbiol.,* 1974, **24:**219–240.
77. Tandeau de Marsac, N., and Cohen-Bazire, G., Molecular composition of cyanobacterial phycobilisomes. *Proc. Natl. Acad. Sci. U.S.A.,* 1977, **74:**1635–1639.
78. Taylor, F. J. R., Symbionticism revisited: A discussion of the evolutionary impact of intracellular symbioses. *Proc. R. Soc. London* [*B*], 1979, *204:*267–286.
79. Thornber, J. P., Alberte, R. S., Hunter, F. A., Shiozawa, J. A., and Kan, K-S., The organization of chlorophyll in the plant photosynthetic unit. *Brookhaven Symp. Biol.,* 1977, **28:**132–148.
80. Troxler, R. F., and Brown, A. S., Amino acid sequence of the phycocyanin α subunit from the alga, *Cyanidium caldarium. Fed. Proc., Fed. Am. Soc. Exp. Biol.,* 1979, **38:**510 Abs.
81. Troxler, R. F., Foster, J. A., Brown, A. S., and Franzblau, C., 1975, The α and β subunits of *Cyanidium caldarium* phycocyanin: Properties and amino acid sequences at the amino terminus. *Biochemistry,* 1975, **14**:268–274.
82. Vaughan, M. H., Jr., *Structural and Comparative Studies of the Algal Protein Phycoerythrin.* Ph.D. Thesis, Massachusetts Institute of Technology, Cambridge, 1964.
83. Vernotte, C., Separation et caractérisation spectroscopique du monomère et des polymères de la C-phycocyanine. *Photochem. Photobiol.,* 1971, **14:**1614:163–173.
84. Wallin, R., Selset, R., and Sletten, K., Characterization of chromophoric peptides from C-phycocyanin. *Biochem. Biophys. Res. Commun.,* 1978, **81:**1319–1328.
85. Wildman, R. B., and Bowen, C. C., Phycobilisomes in blue-green algae. *J. Bacteriol.,* 1974, **117:**866–881.
86. Williams, V. P., and Glazer, A. N., Structural studies on phycobiliproteins. I. Bilin-containing peptides of C-phycocyanin. *J. Biol. Chem.,* 1978, **253:**202–211.
87. Williams, V. P., Freidenreich, P., and Glazer, A. N., Homology of amino-

terminal regions of C-phycocyanins from a prokaryote and a eukaryote. *Biochem. Biophys. Res. Commun.*, 1974, **59:**462–466.
88. Wollman, F.-A., Ultrastructural comparison of *Cyanidium caldarium* wild type and III-C mutant lacking phycobilisomes. *Plant Physiol.*, 1979, **63:**375–381.
89. Yamanaka, G., Glazer, A. N., and Williams, R. C., Cyanobacterial phycobilisomes. Characterization of the phycobilisomes of *Synechococcus* sp. 6301. *J. Biol. Chem.*, 1978, **253:**8303–8310.

Structure and Evolution of Chloroplast- and Bacterial-Type Ferredoxins

H. MATSUBARA, T. HASE, S. WAKABAYASHI, AND K. WADA

Department of Biology
Faculty of Science
Osaka University
Toyonaka, Osaka, Japan

We have been studying the molecular evolution of ferredoxin, which has a wide distribution, ranging from nonphotosynthetic, anaerobic bacteria to higher plants and animals. The size of the molecule is convenient and source materials are easily available. Ferredoxins are a group of acidic proteins with equimolar amounts of nonheme iron (Fe) and inorganic sulfur (S) as exemplified by the three different classes with, respectively, $(4Fe\text{-}4S)_2$, 4Fe-4S, and 2Fe-2S clusters; very low redox potentials; characteristic absorption and electron paramagnetic resonance (EPR) spectra; and unique amino acid sequences. They function in a wide variety of biological electron transfer systems (34,51).*

Several nonphotosynthetic, fermentative bacterial ferredoxins have simple amino acids as the only constituents (35), high amounts of Fe (13,31,48,57) and S (13,35), and very low redox potentials (54), suggesting that ferredoxin is one of the proteins preserving the characteristics present in ancient organisms that appeared on earth soon after the appearance of life. The ancestral nature of the clostridial ferredoxins gains support from the fact that several of their amino acids have been detected in carbonaceous chondrites and have been synthesized under simulated prebiotic conditions (16).

Both the internal repetition of the ferredoxin sequence in *Clostridium pasteurianum* (62) and a reconstitution experiment using half-molecules of *C. acidi-urici* ferredoxin (50) suggest that a primordial ferredoxin composed of about 30 residues was synthesized and folded early in evolutionary time and in an anaerobic atmosphere, when iron and sulfur (37) were present probably

THE EVOLUTION OF PROTEIN STRUCTURE AND FUNCTION

ISBN 0-12-643150-7

in abundance on the primitive earth, to produce an active electron carrier with less efficiency and stability than present-day ferredoxins (11,39,41,43). When two nonidentical half-molecules were mixed with iron and sulfur, the reconstituted protein showed an activity comparable with that of the present-day ferredoxin (50). This observation suggests that a later primitive organism might contain a primordial ferredoxin with two nonidentical half-molecules derived by a gene duplication process having fairly high biological activity. The ancestral form of ferredoxin must have been internally duplicated to produce an active and stable electron carrier similar to what we see at present.

The four cysteine residues constituting each of the two Fe-S centers were found to be unaligned in each of the two halves of a bacterial ferredoxin from *Peptococcus aerogenes* (1). Yasunobu and Tanaka (69) thus threw doubt on the idea of the gene duplication mechanism. It is, however, not unreasonable to assume that after elongation of the peptide chain by gene duplication, the chelate structure became more stable and effective by rearrangement of the positions of cysteines coordinating to Fe (42,47). In fact, internal gene duplication is common in the evolution of biologically active proteins (4; see also 16,40,51).

Bacterial and chloroplast ferredoxins are mutually exchangeable in a photoreduction system of $NADP^+$ by chloroplasts (57). Comparison of the amino acid sequence of these ferredoxins led to the conclusion that they had probably developed from a common ancestor present in early evolutionary time (8,32,44).

Although it has been suggested that adrenal ferredoxin and its related proteins (5,15) and rubredoxins (7,67) have a common ancestor with bacterial and plant ferredoxins, we have focused our attention on the molecular evolution of the latter two groups of proteins. We have established several interesting ferredoxin sequences from plants, algae, and bacteria and compared them with the three-dimensional structure of a chloroplast-type ferredoxin from *Spirulina platensis*. In this paper we compare the relationship of these ferredoxins to the three-dimensional structure, present a phylogenetic tree, and discuss the molecular evolution of ferredoxins. Several review articles on evolutionary studies of ferredoxins have appeared (6,9,15–17,30,40,42,43,69).

Chloroplast-Type Ferredoxins: Amino Acid Sequences and Three-Dimensional Structure

Chloroplast-type ferredoxins having a 2Fe-2S cluster are found in many photosynthetic eukaryotes, ranging from red algae to higher plants, and in prokaryotes such as blue-green algae and extreme halophilic bateria. Figure 1 shows the amino acid sequences of 26 chloroplast-type ferredoxins. We have a total of 27 sequences: 14 from 10 vascular plants (Koa, spinach, alfalfa, taro, pokeweed, i.e., *Phytolacca americana* and *esculenta;* fern, *Gleichenia japonica*; wheat, *Triticum aestivum*; and horsetail, *Equisetum telmateia* and *arvense)*; three from two green algae (*Scenedesmus quadricauda* and *Dunaliella salina*); two from two red algae (*Porphyra umbilicalis* and *Cyanidium caldarium*); six from five blue-green algae (*Spirulina platensis* and *maxima*, *Nostoc muscorum*, *Mastigocladus laminosus*, and *Aphanothece sacrum*); and two from two halophilic bacteris (*Halobacterium halobium* and Dead Sea). Several gaps are inserted in this figure to make sequences highly homologous.

Halobacterial ferredoxins have an active-site cluster similar to that of chloroplast-type ferredoxins (33), although *halobacteria* do not live by photosynthesis and their ferredoxins do not function in the photoreduction system of $NADP^+$. Their sequences are strikingly homologous with plant ferredoxin sequences (Fig. 1). They have extra residues, 22 at the amino (N)-terminal and five at the carboxyl (C)-terminal region, respectively, in comparison with other chloroplast-type ferredoxins. For example, there are 38 identical and 16 similar residues between the sequences of *Nostoc* and halobacterial ferredoxins. *Nostoc* ferredoxin has only the four cysteine residues minimally required for the chelate structure, and their positions correspond to those of halobacterial ferredoxin at positions 63, 68, 71, and 102. Thus the spectroscopic evidence is structurally verified.

In the following description the structural characteristics are described on the basis of comparison of 24 ferredoxins, excluding wheat, which has very recently been reported (58), and halobacterial ferredoxins.

The distribution of cysteine residues in various ferredoxins made certain the assignment of four cysteine residues among six possible positions chelating the two Fe, i.e., residues 42, 46, 49, and 79 corresponding to those in *S. platensis* ferredoxin, as proposed by the study of horsetail ferredoxin (2). This has recently been confirmed by the X-ray structural analysis of *S. platensis* ferredoxin in which Cys-41 and -46 chelate one of the two Fe and Cys-49 and -79 chelate the other Fe (14,64), as shown in Fig. 2.

The chain folding of *S. platensis* ferredoxin suggested that there was no α-helix in this molecule but 3_{10} turn-like structures at several places and the parallel and antiparallel β sheets. There was a hydrogen bond between the CO of residue 9 and NH of residue 13.

The Fe-S cluster is near the surface of the upper area, about 8 Å from the surface of the molecule with an approximate dimension of $40 \times 35 \times 25$ Å and is covered only by several residues from Cys-41 to -46 at the top-front side of the cluster. It seems that the cluster may well be in contact with the outer environment. The immediate lower part of the cluster is densely folded by peptide chains and the lower part of the molecule is loosely folded by the chains.

Figure 3 shows the various amino acid residues appearing at each site of the sequences in 24 ferredoxins. The sites can be classified into three groups: (1) invariant sites, where only one amino acid occurs; (2) semi-invariant sites, where substitution at each site occurs only among conservative amino acids

```
                                   10        20        30        40 *
(1)  Koa                     -AFKVKLLT-PDG-PKEFECPDDVYILDQAEELGIDLPYSCRAGS
(2)  Spinach                 AAYKVTLVT-PTG-NVEFQCPDDVYILDAAEEEGIDLPYSCRAGS
(3)  Alfalfa                 ASYKVKLVT-PEG-TQEFECPDDVYILDHAEEEGIVLPYSCRAGS
(4)  Taro                    ATYKVKLVT-PSG-QQEFQCPDDVYILDQAEEVGIDLPYSCRAGS
(5)  P. americana I          ATYKVTLVT-PSG-TQTIDCPDDTYVLDAAEEAGLDLPYSCRAGS
(6)  P. americana II        AASYKVTFVT-PSG-TNTITCPADTYVLDAAEESGLDLPYSCRAGA
(7)  P. esculenta I          ATYKVTLVT-PSG-TQTIDCPDDTYVLDAAEEAGLDLPYSCRAGS
(8)  P. esculenta II        AASYKVTFVT-PSG-TKTITCPADTYVLDAAEDTGLDLPYSCRAGA
(9)  G. japonica             AIFKVKFLT-PDG-ERTIEVPDDKFILDAGEEAGLDLPYSCRAGA
(10) E. telmateia I          -AYKTVLKT-PSG-EFTLDVPEGTTILDAAEEAGYDLPFSCRAGA
(11) E. telmateia II         -AYKVTLKT-PDG-DITFDVEPGERLIDIASEKA-DLPLSCQAGA
(12) E. arvense I            -AYKTVLKT-PSG-EFTLDVPEGTTILDAAEEAGYDLPFSCRAGA
(13) E. arvense II           -AYKVTLKT-PDG-DITFDVEPGERLIDIGSEKA-DLPLSCQAGA
(14) S. quadricauda          ATYKVTLKT-PSG-DQTIECPDDTYILDAAEEAGLDLPYSCRAGA
(15) D. salina I             -SYMVTLKT-PSG-EEKVEVSPDSYILDAAEEAGVDLPYSCRAGS
(16) D. salina II            -AYKVTLKT-PSG-DQTIEVSPDAYILDAAEEAGLDLPYSCRAGA
(17) P. umbilicalis          ADYKIHLVSKEEGIDVTFDCSEDTYILDAAEEEGIELPYSCRAGA
(18) C. caldarium            ASYKIHLVNKDQGIDETIECPDDQYILDAAEEQGLDLPYSCRAGA
(19) S. platensis            ATYKVTLINEAEGINETIDCDDDTYILDAAEEAGLDLPYSCRAGA
(20) S. maxima               ATYKVTLISEAEGINETIDCDDDTYILDAAEEAGLDLPYSCRAGA
(21) N. muscorum I           ATFKVTLINEAEGTKHEIEVPDDEYILDAAEEEGYDLPFSCRAGA
(22) M. laminosus            ATYKVTLINEAEGLNKTIEVPDDQYILDAAEEAGIDLPYSCRAGA
(23) A. sacrum I             ASYKVTLKT-PDG-DNVITVPDDEYILDVAEEEGLDLPYSCRAGA
(24) A. sacrum II            ATYKVTLINEEEGINAILEVADDQTILDAGEEAGLDLPSSCRAGS
(25) H. halobium             LFEKAADAGLDGEDYGTMEVAEGEYILEAAEAQGYDWPFSCRAGA
(26) H. Dead Sea             VFGEASDMDLDDEDYGSLEVNEGEYILEAAEAQGYDWPFSCRAGA

     *  *                          *
(1)  CSSCAGKLVEGDL-DQSDQSFLDD`EQIEEGWV-LTCAAYPRSDVVIETHKEEELTG
(2)  CSSCAGKLKTGSL-NQDDQSFLDDDQIDEGWV-LTCAAYPVSDVTIETHKEEELTA
(3)  CSSCAGKVAAGEV-NQSDGSFLDDDQIEEGWV-LTCVAYAKSDVTIETHKEEELTA
(4)  CSSCAGKVKVGDV-DQSDGSFLDDEQIGEGWV-LTCVAYPVSDGTIETHKEEELTA
(5)  CSSCTGKVTAGTV-DQEDQSFLDDDQIEAGFV-LTCVAFPKGDVTIETHKEEDIV-
(6)  CSSCAGKVTAGAV-NQEDGSFLEEEQMEAGWV-LTCVAYPTSDVTIETHKEEDLTA
(7)  CSSCTGKVTAGTV-DQEDQSFLDDDQIEAGFV-LTCVAYPKGDVTIETHKEEDIA-
(8)  CSSCAGKVTAGSV-NQEDGSFLDEEQMEAGWV-LTCVAYPTSDVTIETHKEEDLSA
(9)  CSSCTGKLLDGRV-DQSEQSFLDDDQMAEGFV-LTCVAYPAGDITIETHAEEKL--
(10) CSSCLGKVVSGSV-DQSEGSFLDDGQMEEGFV-LTCIAIPESDLVIETHKEEELF-
(11) CSTCLGKIVSGTV-DQSEGSFLDDEQIEQGYV-LTCIAIPESDVVIETHKEDEL--
(12) CSSCLGKVVSGSV-DESEGSFLDDGQMEEGFV-LTCIAIPESDLVIETHKEEELF-
(13) CSTCLGKIVSGTV-DQSEGSFLDDEQIEQGYV-LTCIAIPESDVVIETHKEDEL--
(14) CSSCAGKVEAGTV-DQSDQSFLDDSQMDGGFV-LTCVAYPTSDCTIATHKEEDLF-
```

```
(15) C S S C A G K V E S G T V - D Q S D Q S F L D D D Q M(D,S,G,F)V - L T C V A Y A T S D C T I V T H Q E E N L Y -
(16) C S S C A G K V E A G T I - D Q S D Q S F L D D D Q Q G R G F V - L T C V A Y A T S D C T I S T H Q E E S L Y -
(17) C S T C A G K V T E G T V - D Q S D Q S F L D D E Q M L K G Y V - L T C I A Y P E S D C T I L T H V E Q E L Y -
(18) C S T C A G K L L E G E V - D Q S D Q S F L D D D Q V K A G F V - L T C V A Y P T S N A T I L T H Q E E S L Y -
(19) C S T C A G T I T S G T I - D Q S D Q S F L D D D Q I E A G Y V - L T C V A Y P T S D C T I K T H Q E E G L Y -
(20) C S T C A G K I T S G S I - D Q S D Q S F L D D D Q I E A G Y V - L T C V A Y P T S D C T I Q T H Q E E G L Y -
(21) C S T C A G K L V S G T V - D Q S D Q S F L D D D Q I E A G Y V - L T C V A Y P T S D V V I Q T H K E E D L Y -
(22) C S T C A G K L I S G T V - N Q S D Q S F L D D D Q I E A G Y V - L T C V A Y P T S D C V I E T H K E E E L Y -
(23) C S T C A G K L V S G P A P D - E D Q S F L D D D Q I Q A G Y I - L T C V A Y P T G D C V I E T H K E E A L Y -
(24) C S T C A G K L V S G A A P N Q D D Q A F L D D D Q L A A G W V - M T C V A Y P T G D C T I M T H Q E S E V L -
(25) C A N C A S I V K E G E I - D M D M Q Q I L S D E E V E E K D V R L T C I G S P A A D E V K I V Y N A K H L D Y
(26) C A N C A A I V L E G D I - D M D M Q Q I L S D E E V E D K N V R L T C I G S P D A D E V K I V Y N A K H L D Y
           50                   60                   70                   80                  90                   100
```

Fig. 1. Amino acid sequences of chloroplast-type ferredoxins. ε-Amino group of Lys-97 (118 in the whole sequence) in halobacterial ferredoxin, (25) and (26), is acetylated. Their amino- and carboxyl-terminal sequences are deleted from this figure and are as shown below.

```
(25) H. halobium:  P T V E Y L N Y E T L D D Q G W D M
                   1                 10              18
(26) H. Dead Sea:  P T V E Y L N Y E V V D D N G W D M
-----------------------------------------------------
(25) D D D D ---- L Q N R V I
           22      123        128
(26)  Y D D D ---- L Q N R V I
```

This numbering is as in the original paper for *H. halobium* ferredoxin (22).

The references for various sequences, (1)–(6), (10)–(14), (19)–(21), (23), and (24) are listed in Hase *et al.* (24). The others are Takruri *et al.* (59) for sequence (17), Hase *et al.* (25) for (18), Hase *et al.* (22) for (25), and unpublished results of T. Hase and co-workers for (7)–(9), (15), (16), and (26). *Triticum aestivum* ferredoxin (58) is not listed in this figure. A star indicates the position of the cysteine residue chelated to Fe.

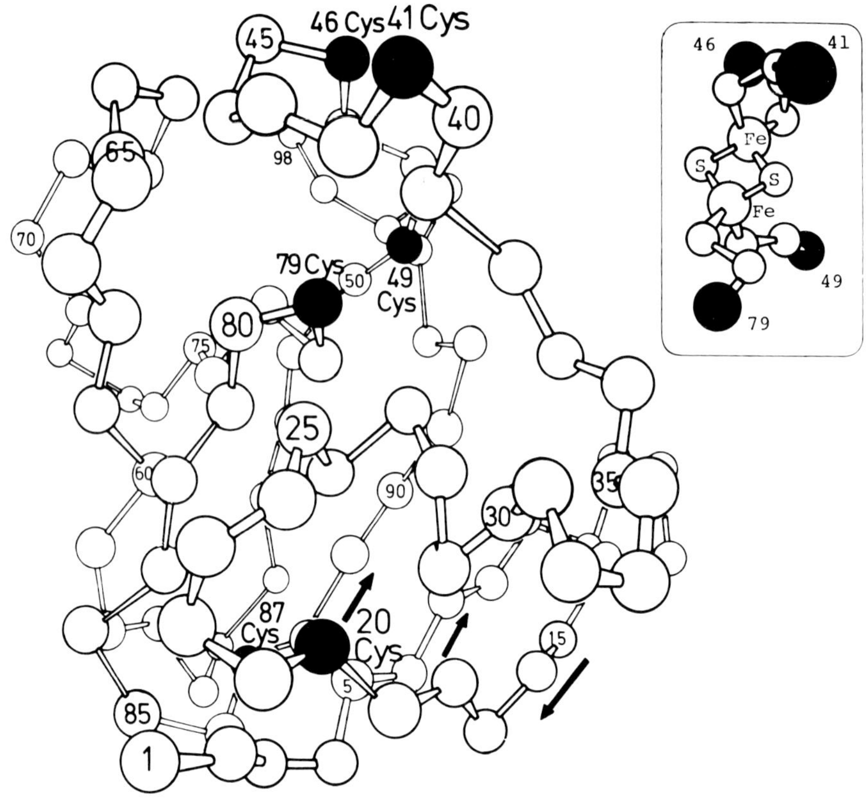

Fig. 2. Drawing of a three-dimensional model of the main chain of *S. platensis* ferredoxin, based on the X-ray structural analysis at 2.5 Å resolution (14). The α-carbon atoms of cysteine residues are shown by black circles. The arrows show the parallel and antiparallel β-sheets. The inset shows the Fe-S cluster deleted from the main chain drawing.

(e.g., Asp and Glu, Ile and Val/Phe/Leu, etc.); and (3) variant sites, where more than six different amino acids occur. On this basis we can calculate 24 invariant sites, including the four most important sites with cysteines chelating the Fe, 22 semi-invariant sites, and 14 variant sites. Semi-invariant sites are at positions 3, 7, 9, 18, 26, 27, 30, 32, 34, 45, 48, 53, 62, 64, 67, 68, 75–77, 82, 85, and 97, numbered as in *S. platensis* ferredoxin.

Their distribution on the primary structure reveals that the regions near the four chelating cysteines are occupied always by invariant or semi-invariant residues and variant sites scatter in various regions of the molecule rather than as clusters.

The distribution of invariant and variant sites on the three-dimensional structure of the molecule is shown in Fig. 4. Almost all the 24 invariant sites are located around the Fe-S cluster, and four cysteines are of course included in this region. The hydrophobic residues with bulky side chains such as Leu-37, Phe-65, and Leu-66 are at the outer boundary of the cluster. Four glycine residues out of five, excluding Gly-44, distribute at the exterior of the molecule and are far from the cluster. Gly-56 and Gly-13, in particular, are located in the middle of variable regions, suggesting that these areas may be preserved for the space required for chain folding.

If the semi-invariant sites are included in this figure, the characteristic feature just described becomes more evident. The surrounding area of the cluster is occupied by invariant residues with small side chains, such as Pro-38, Ser-40, Ala-43, Gly-44, Ser-47, Thr-78, and Ala-81, and by three semi-invariant sites with bulky hydrophobic content, that is, Phe-65, Leu-66, Val, Ile, or Leu-26, Val, Ile or Leu-27, Leu or Ile-37, Tyr, Phe, or Trp-75, Val or Ile-76, Met or Leu-77, Ile, Tyr, or Phe-82 and Val, Ile, or Leu-97.

This feature is derived from the comparison of sequences of many ferredoxins isolated from a wide variety of organisms, including blue-green algae and higher plants, that diverged at least 1.5 billion years ago. The cluster region plays a key role in transferring electrons in various metabolic systems and has been preserved during the long process of evolution.

Another interesting feature is that several hydrophobic residues, such as Phe or Try-3, Val, Leu, or Phe-17, Ile-89, Val, Ile, Leu, or Phe-18, and Cys or Val-20 come together on the lower front of the molecule. The fact that Cys-20, found in ferredoxins from spinach and *S. platensis*, was unreactive to sulfhydryl reagents in native form but reactive to those in the presence

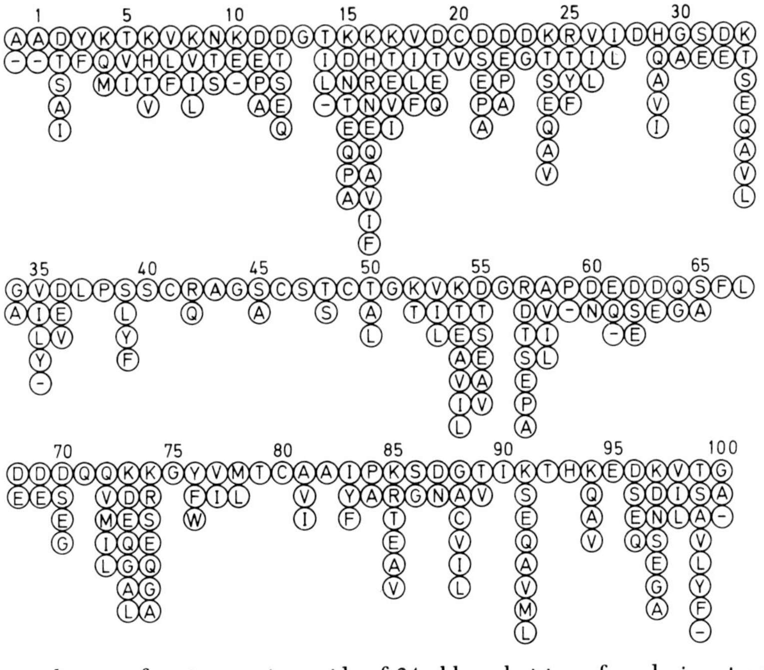

Fig. 3. Distribution of various amino acids of 24 chloroplast-type ferredoxins at each site. Wheat and halobacterial ferredoxins are omitted. ⊖ represents the presence of a deletion. Residue No. 1 corresponds to the N-terminus of spinach ferredoxin.

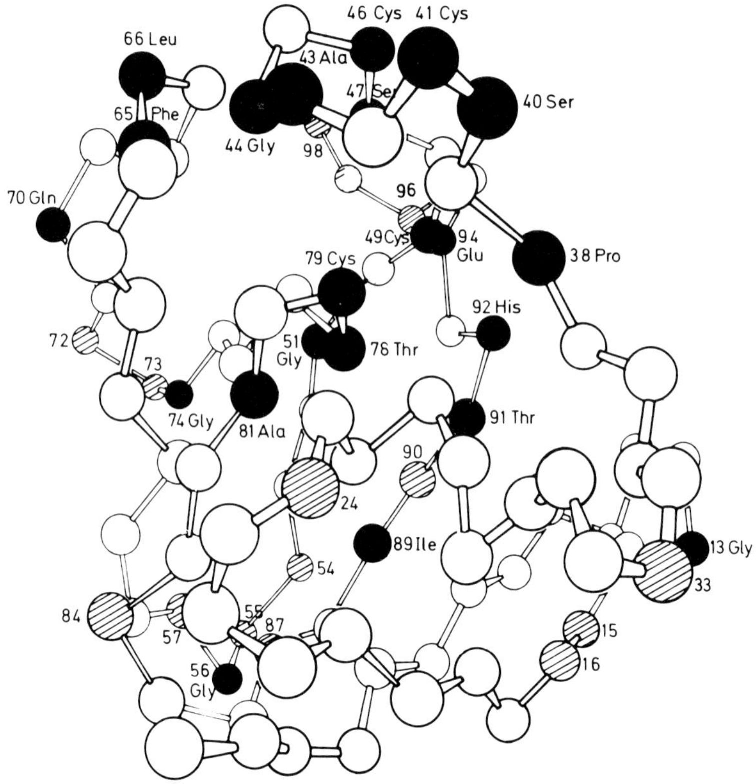

Fig. 4. Distribution of invariant and variant sites on *S. platensis* ferredoxin molecule. The Fe-S cluster is deleted from this figure to avoid confusion. Filled circle, invariant sites; cross-hatched circles, variant sites. Only 22 invariant sites are present in this figure referring to the two additional sequences reported later.

of denaturants (46,52) is compatible with the present observation that Cys-20 is in the hydrophobic region and is replaced only by valine, which is probably buried inside the molecule. This region preserves such a hydrophobicity and may serve a structurally or functionally important role in ferredoxin.

Trp-73 in spinach ferredoxin, corresponding to Tyr-75 in *S. platensis* ferredoxin, is in a position where quenching of fluorescence of Trp-73 occurs because of the presence of a quenching center, the Fe-S cluster (38,46). If the indole ring of Trp-73 points to the cluster, the two chromophore groups are reasonably close to each other to explain the quenching effect.

Hydrophobic residues in invariant and semi-invariant sites are located at the exterior of the molecule, that is, Asp-28, His-92, Glu-94, and Asp or Glu-32, -62, -67, and -68. However, no characteristic feature is observed on the surface area, although the possibility that these sites may play an unknown important role cannot be excluded. Variant sites are also shown in Fig. 4 at positions 15, 16, 24, 33, 54, 55, 57, 72, 73, 84, 87, 90, 96, and 98, corresponding to those in *S. platensis* ferredoxin. These sites are located in the lower half and surface region of the molecule and far from the Fe-S cluster, and a very clear compartmentation can be recognized in locations of variant and invariant sites on the molecule. The variant residues may be excluded from the functional role, for instance, participation in the complex formation with ferredoxin-$NADP^+$ reductase.

Molecular Evolution and Gene Duplication

The chloroplast-type ferredoxin sequences are all homologous, indicating that these ferredoxins are derived from a common ancestor. The comparison of these sequences on a quantitative basis gives an evolutionary relationship of these ferredoxins. A matrix is shown in Table 1 in which 26 sequences of chloroplast-type ferredoxin, including two halobacterial ferredoxins, are compared. Since there is a good correlation between the values calculated by amino acid differences and base differences per codon, the following discussion is based on the matrix calculated only by amino acid differences.

As previously pointed out (43), a linear relationship does not seem to exist between the number of amino acid differences and the elapsed time after divergence of organisms, particularly in the precambrian era. A phylogenetic tree of chloroplast-type ferredoxins has been constructed according to a matrix method (12) as shown in Fig. 5 on the basis of data in Table 1. Notable information deduced from this tree is as follows.

1. Algal ferredoxins are evolutionarily in a very diverse group (3,65) and diverged in early evolutionary time. The relative positions of blue-green algal ferredoxins in this tree are in accordance with their cellular organizations. Ferredoxin of a red alga, *P. umbilicalis*, diverged from the line of blue-green algal ferredoxins and not from that of any others. This relationship agrees with the facts that both the blue-green algae and red algae have phycobiliproteins and also two extra residues at positions 10 and 14, except for *A. sacrum* ferredoxin I (Fig. 1). Other ferredoxins have no such extra residues.

2. *C. caldarium* is a unicellular eukaryotic alga in an uncertain class. The sequence shows a closer similarity to that of another red alga, *P. umbilicalis*, than to the others. A unique amino acid distribution is commonly found only in the ferredoxins of these two algae, namely, Ile-5, His-6, and Lys-10 (Fig. 1), supporting the closer relationship between these two. Therefore *C. caldarium* is considered to be one of the red algae.

3. Halobacterial and other chloroplast-type ferredoxins must have diverged independently from a common ancestor long before the divergence of higher plants and blue-green algae. Halobacteria are strictly aerobic; therefore they must have appeared after ancestral blue-green algae. Recent comparative studies of 16S rRNA (36) have suggested that halobacteria did not arise from the typical bacteria as a halophilic adaptation and are rather similar to methanogenic bacteria belonging to "archaebacteria," which are no more

TABLE 1

MATRIX OF MUTATION DISTANCE AMONG CHLOROPLAST-TYPE FERREDOXINS[a]

	1	2	3	4	5	6	7	8	9	10	11	12	13	14	15	16	17	18	19	20	21	22	23	24	25	26
1. Koa		0.31	0.26	0.25	0.46	0.46	0.43	0.48	0.46	0.61	0.59	0.62	0.60	0.46	0.55	0.55	0.58	0.49	0.49	0.51	0.40	0.40	0.49	0.62	0.99	0.96
2. Spinach	20		0.28	0.26	0.38	0.38	0.35	0.38	0.49	0.60	0.65	0.60	0.66	0.41	0.47	0.49	0.51	0.51	0.45	0.42	0.42	0.42	0.47	0.51	0.98	1.04
3. Alfalfa	23	19		0.21	0.36	0.35	0.35	0.36	0.48	0.59	0.61	0.60	0.62	0.42	0.47	0.49	0.52	0.47	0.46	0.45	0.41	0.44	0.51	0.56	1.03	1.05
4. Taro	20	18	16		0.40	0.38	0.38	0.39	0.47	0.58	0.65	0.59	0.66	0.38	0.47	0.48	0.52	0.52	0.49	0.48	0.47	0.44	0.56	0.61	1.01	1.04
5. *P. americana* I	35	30	27	28		0.27	0.02	0.26	0.40	0.51	0.63	0.52	0.65	0.26	0.41	0.41	0.51	0.46	0.32	0.32	0.41	0.35	0.46	0.52	0.96	1.00
6. *P. americana* II	37	31	27	29	23		0.25	0.06	0.43	0.56	0.67	0.57	0.67	0.35	0.58	0.48	0.56	0.49	0.43	0.42	0.47	0.44	0.47	0.58	0.97	1.02
7. *P. esculenta* I	34	29	26	27	2	22		0.24	0.39	0.53	0.65	0.54	0.66	0.25	0.40	0.40	0.50	0.45	0.31	0.31	0.40	0.34	0.43	0.53	0.95	1.00
8. *P. esculenta* II	37	31	28	30	23	6	22		0.47	0.55	0.67	0.56	0.68	0.33	0.45	0.46	0.55	0.48	0.41	0.39	0.47	0.42	0.46	0.58	0.98	1.03
9. *G. japonica*	34	37	37	36	31	38	30	38		0.50	0.65	0.51	0.63	0.37	0.45	0.45	0.53	0.45	0.46	0.45	0.42	0.37	0.46	0.53	0.94	0.94
10. *E. telmateia* I	36	39	39	36	35	39	35	38	35		0.39	0.01	0.40	0.44	0.52	0.53	0.56	0.65	0.59	0.57	0.48	0.49	0.59	0.68	0.91	0.95
11. *E. telmateia* II	39	45	45	43	43	47	43	47	46	29		0.40	0.01	0.62	0.63	0.60	0.65	0.73	0.60	0.60	0.52	0.53	0.55	0.73	0.99	0.98
12. *E. arvense* I	37	40	40	37	36	40	36	39	36	1	30		0.41	0.45	0.49	0.55	0.57	0.66	0.60	0.58	0.49	0.51	0.57	0.71	0.91	0.94
13. *E. arvense* II	40	46	46	44	44	48	44	48	45	30	1	31		0.63	0.65	0.61	0.66	0.74	0.62	0.61	0.53	0.54	0.57	0.72	1.00	1.00
14. *S. quadricauda*	33	30	29	26	19	25	18	25	28	30	42	31	43		0.24	0.21	0.44	0.40	0.29	0.29	0.38	0.30	0.41	0.51	0.96	1.00
15. *D. salina* I	37	35	33	34	32	37	31	37	34	34	44	35	45	21		0.19	0.51	0.45	0.39	0.39	0.42	0.37	0.44	0.53	0.97	1.04
16. *D. salina* II	36	35	33	33	28	34	27	34	32	34	41	35	42	16	16		0.47	0.42	0.35	0.35	0.44	0.37	0.43	0.55	0.93	0.98
17. *P. umbilicalis*	40	37	37	36	38	40	37	40	44	39	43	40	44	31	37	35		0.39	0.39	0.37	0.53	0.44	0.58	0.61	0.93	0.96
18. *C. caldarium*	39	37	36	37	34	37	33	37	34	44	51	45	52	27	33	29	28		0.35	0.34	0.40	0.35	0.45	0.50	0.89	0.90
19. *S. platensis*	40	36	38	37	28	36	27	36	38	42	44	43	45	25	31	27	29	24		0.04	0.31	0.19	0.41	0.40	0.95	0.96
20. *S. maxima*	39	34	37	36	28	35	27	34	37	40	44	41	45	25	31	27	28	25	4		0.31	0.21	0.40	0.41	0.95	0.98
21. *N. muscorum* I	32	34	34	36	32	37	31	36	36	38	40	39	41	28	34	33	36	29	21	21		0.18	0.31	0.45	0.88	0.86
22. *M. laminosus*	31	31	32	32	29	36	28	35	33	36	39	37	40	24	30	28	31	25	15	16	13		0.26	0.37	0.92	0.92
23. *A. sacrum* I	36	34	37	38	32	35	31	36	35	42	41	42	42	30	34	31	41	33	32	31	27	25		0.48	0.93	0.92
24. *A. sacrum* II	47	40	45	46	42	48	41	49	42	50	56	51	55	40	41	41	45	38	31	31	34	29	37		1.03	1.02
25. *H. halobium*	91	94	96	94	96	95	95	95	93	87	96	87	97	92	95	91	90	89	92	92	87	90	95	98		0.20
26. *H.* Dead Sea	91	98	99	96	98	97	98	97	95	90	97	89	98	94	96	93	92	91	94	94	88	92	95	99	20	

[a] Values in the lower left half of the table show the amino acid differences and those in the upper right half of the table the minimum base difference per codon among ferredoxins. Calculations are based on the alignment of sequences as shown in Fig. 1. A deletion or insertion is counted as one difference for amino acid differences.

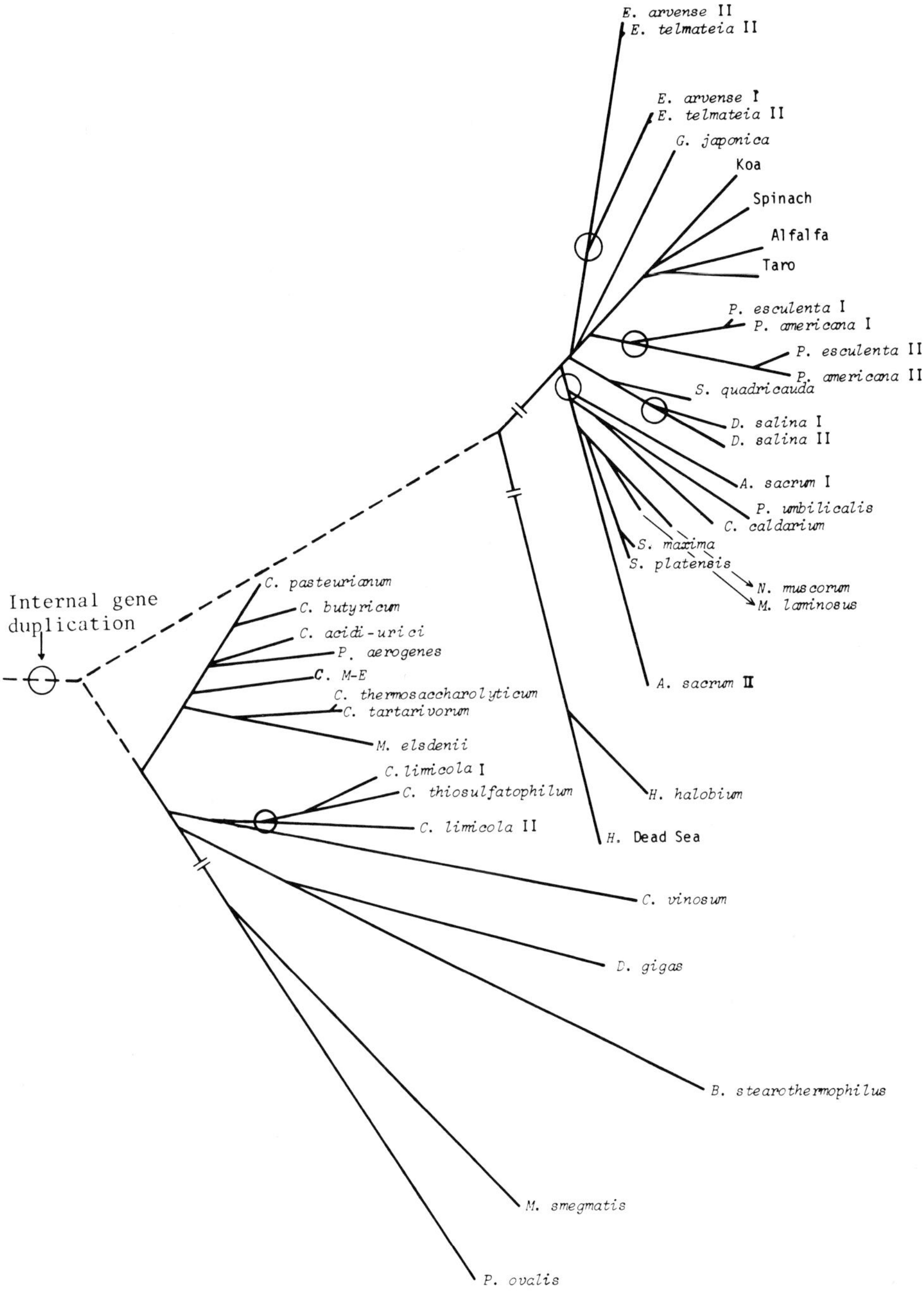

Fig. 5. A phylogenetic tree of ferredoxins of both the chloroplast type and the bacterial type. The tree was constructed from the amino acid differences shown in Tables 1 and 2 according to a matrix method (12). Circles show the points where gene duplications occurred. A break in the solid line represents the distance on this line cut appropriately short to fit in a space. A dashed line shows the relation between the ferredoxins of chloroplast type and bacterial type without quantitative estimation in distance. Refer to the text for full names of organisms.

related to "Eubacteria" than they are to eukaryotic cytoplasm (68). Structural comparisons of 5S rRNA (26) and ribosomal proteins (39) have indicated that halobacteria are closer to eukaryotes than to prokaryotes. These studies, together with the present study, strongly suggest that halobacteria and eukaryotes are close relatives but that they diverged a long time ago on the evolutionary scale.

So far, more than 15 organisms, including higher plants, algae, and bacteria, have been reported to have two, or in very rare cases three, molecular species of ferredoxin (55,61,66,71,72) and the sequence comparison of these isozymes has revealed that, long ago, gene duplication frequently occurred and the divergence of species was preceded by gene duplication (42). Two molecules found in each organism usually showed large substitution numbers of amino acid; e.g., 37 residues between *A. sacrum* ferredoxins I and II, 16 between *D. salina*'s, 29 between *E. telmateia*'s, 31 between *E. arvense*'s, 23 between *P. americana*'s, and 22 between *P. esculenta*'s, as given in Table 1. Similar cases probably occur in each of the two ferredoxins of *N.* MAC (29), *N. verrucosum* (53), and *Petunia* species (28). However, the substitution numbers between the corresponding two ferredoxins over two species are small, 1–6 so far studied. In the phylogenetic tree the points where gene duplications occurred are shown by circles. The relationship between gene duplication and speciation is shown diagrammatically in Fig. 6.

The duplication of ferredoxin genes of *Equisetum*, *Dunaliella*, and *Phytolacca* occurred within their own phyletic lines after divergence from other plant phyla. In other words, the organisms descended with two ferredoxin genes and in relatively recent years they diverged into two independent species (43).

The gene duplication of *A. sacrum* ferredoxin took place within blue-green algal and red algal lines before *A. sacrum* diverged from other algae. As shown in Fig. 1, *A. sacrum* ferredoxin I, higher plant ferredoxins, and green algal ferredoxins have no extra residues at positions 10 and 14, and in this sense they are similar to one another. On this basis we propose the following development.

An ancestral organism had a single gene corresponding to that of *A. sacrum* ferredoxin I. Higher plants and green algae inherited the gene without insertion of the extra residues even after divergence from one another. Gene duplication occurred in some plants and green algae, giving two ferredoxins as described earlier, but some failed to produce doubling of the gene or translation to protein even after doubling.

After gene duplication, one of the two genes acquired insertions at the two positions and these have been inherited by *A. sacrum*. Other blue-green algae and red algae might have inherited these, but in spite of receiving insertions, only one of them is translated to protein. The fact that the N-terminal sequence of one of the two ferredoxins of *N.* MAC (29) has no gap and corresponds to ferredoxin II of *A. sacrum* and that the other has a gap at

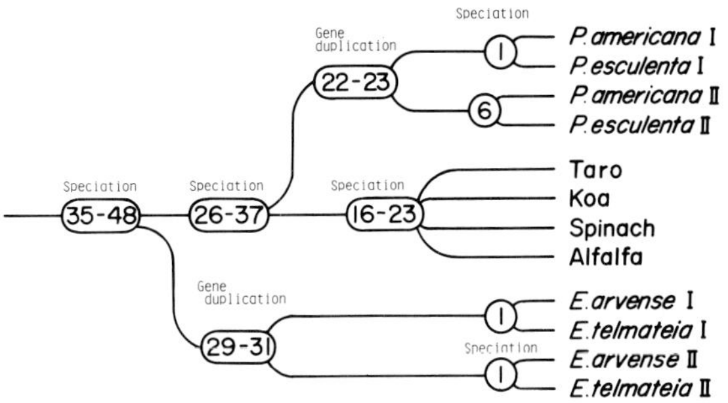

Fig. 6. A diagram showing the relationship between gene duplication and speciation in phyletic relationships of ferredoxins. The numbers show amino acid differences between ferredoxins.

position 10 and an insertion at 2 or 3 indicates that the latter is rather similar to (but not exactly the same as) *A. sacrum* ferredoxin I and seems to represent a transitional form between *A. sacrum* ferredoxin II and eukaryotic ferredoxins.

Bacterial-Type Ferredoxins: Amino Acid Sequences, Characteristics, and Molecular Evolution

Ferrodoxins of bacterial orgins except for those of *Halobacteria* have usually one or two 4Fe-4S clusters and several types of sequence and spectroscopic properties. So far 16 sequences are known, as shown in Fig. 7, with several gaps. They are divided into four groups as follows:

CLOSTRIDIAL TYPE

Ferredoxins belonging to this group have 54–55 amino acids with eight cysteines. They have two 4Fe-4S clusters with redox potentials at around -400 mV. They are isolated from *Clostridium pasteurianum*, *C. butyricum*, *C. acidi-urici*, *C.* M-E, *C. tartarivorum*, *C. thermosaccharolyticum*, *Peptococcus aerogenes*, and *Megasphera elsdenii* [(A)-(H) in Fig. 7]. There are 19 invariant and 13 variant sites among them. Their symmetrical structure suggested the occurrence of internal gene duplication, as described earlier. This characteristic is also preserved in other bacterial ferredoxins to some extent.

PHOTOSYNTHETIC BACTERIAL TYPE

Only four ferredoxin sequences are known. Three of them are of *Chlorobium*, photosynthetic green sulfur bacteria, *C. limicola* I and II, and *C. thiosulfatophilum* Tassajara. These sequences have 60–61 residues with nine cysteines [(I)-(K)]. One of the four sequences is of *Chromatium vinosum*, a photosynthetic purple sulfur bacterium, and is composed of 82 residues with nine cysteines (L). These four sequences have six to seven extra residues

```
     1           10           20            30                40                      50
                *  *      *   *                           *   *                 *   *
(A)  A-YKI--ADSCVSCG--ACASECPVNAISQGDSIFVI----DADTCI-DC----------GNCANVCPVGAPVQE
(B)  A-FVI--NDSCVSCG--ACAGECPVSAITQGDTQFVI----DADTCI-DC----------GNCANVCPVGAPNQE
(C)  A-YVI--NEACISCG--ACDPECPVDAISQGDSRYVI----DADTCI-DC----------GACAGVCPVDAPVQA
(D)  A-YKI--TDGCINCG--ACEPECPVEAISESDAVRVI----DADKCI-DC----------GACANTCPVDAIVEG
(E)  A-YVI--NDSCIACG--ACKPECPVN-IQQG-SIYAI----DADSCI-DC----------GSCASVCPVGAPNPED
(F)  AHI-I--TDECISCG--ACAAECPVEAIHEGTGKYQV----DADTCI-DC----------GACQAVCPTGAVKAE
(G)  AHI-I--TDECISCG--ACAAECPVEAIHEGTGKYEV----DADTCI-DC----------GACEAVCPTGAVKAE
(H)  MH-VI--SDECVKCG--ACASTCPTGAIEEGETKYVV-----TDSCI-DC----------GACEAVCPTGAISAE

     1           10           20            30                 40                   50          60
                *  *      *   *                           *   *                  *   *    *
(I)  ALY-I--TEECTYCG--ACEPECPVTAISAGDDIYVI----DANTCN-EC--AGLDEQ---ACVAVCPAECIVQG
(J)  AH-RI--TEECTYCA--ACEPECPVNAISAGDEIYIV----DESVCT-DC--EGYYDEP--ACVAVCPVDCIIKV
(K)  ALY-I--TEECTYCG--ACEPECPTNAISAGSEIYVI----DAAGCT-EC--VGFADAP--ACAAVCPAECIVQG
(L)  ALM-I--TDQCINCN--VCQPECPNGAISQGDETYVI----EPSLCT-EC--VGHYETS--QCVEVCPVDCIIKDPSHEETEDELRAKYERITGEG
     1          *  *      *   *   *       30              *   *                 *   *             60                    78
(M)  T-FVV--TDNCIKCKYTDCVEVCPVDCFYEGPNFLVI----HPDECI-DC--------A--LCEPECPAQAIFSEDEVPSGMENFIELNAELAEIW
     PNITERKDALPDAEEWDGKPGKIADLER
                *         *   *   *                       *   *                 *   *
(N)  T-YVI--AEPCVDVKDKACIEECPVDCIYEGARMLYI----HPDECV-DC--------G--ACEPVCPVEAIYYEDDVPDQWSSYAQANADFFAEL
     GSPGGASKVGQTDNDPQAIKDLPPQGED

     1           10           20             30                        40         50
                *  *      *   *                                         *    *
(O)  PI-QV---DNCMACQ--ACINECPVDVF-QMDEQGDKAVNI---------PNSNLDDQC-VEAIQSCPA-AIRS
     1          *    *     * 20         30          40           50              *            70       80
(P)  PKYTIVDKETCIACG--ACGAAAP-DIY-DYDEDGIAYVTLDDNQGIVEVPD-ILIDDM-MDAFEGCPTESIKVA---DEPFDGDPNKFD
```

Fig. 7. Amino acid sequences of bacterial-type ferredoxins. (A), *C. pasteurianum;* (B), *C. butyricum;* (C), *C. acidi-urici;* (D), *C.* M-E; (E), *P. aerogenes;* (F), *C. tartarivorum;* (G), *C. thermosaccharolyticum;* (H), *M. elsdnii;* (I), *C. limicola* I; (J), *C. limicola* II; (K), *C. thiosulfatophilum;* (L), *C. vinosum;* (M), *P. ovalis;* (N), *M. smegmatis;* (O), *D. gigas;* (P), *B. stearothermophilus*. The references for sequences (A)–(H) are listed in Yasunobu and Tanaka (69). The others are Tanaka *et al.* (60) for (I), Tanaka *et al.* (61) for (J), Hase *et al.* (20) for (K), Hase *et al.* (18) and Matsubara *et al.* (45) for (L), Hase *et al.* (23) for (M), Hase *et al.* (21) for (N), Travis *et al.* (63) for (O), and Hase *et al.* (19) for (P). Stars above the sequences emphasize the positions of cysteine. Gaps are inserted in various places to make all alignments most probable.

at residue 39 or 40 in clostridial ferredoxins, and *Chromatium* ferredoxin has a long C-terminal of 21 extra residues. They have a common extra cysteine at their C-terminal regions that probably does not participate in iron chelation.

Eleven invariant residues are found among clostridial and photosynthetic bacterial ferredoxins, with invariant and semi-invariant residues scattered all over the three-dimensional structure (1).

DESULFOVIBRIO GIGAS AND *BACILLUS STEAROTHERMOPHILUS* FERREDOXINS

D. gigas (63) and *B. stearothermophilus* (19) ferredoxins contain one 4Fe-4S cluster with 56 and 81 residues, respectively. The former has six cysteines, but the latter has four, fulfilling the minimum requirement for iron chelation. The N-terminal halves of these ferredoxins closely resemble those of the clostridial type of ferredoxin, particularly in the distribution of cysteines [(O) and (P) in Fig. 7]. In *B. stearothermophilus* ferredoxin, the formation of one 4Fe-4S cluster is composed of a -Cys-X-X-Cys-X-X-Cys- segment and a -Cys-Pro- sequence far removed from this segment. This agrees with the result obtained by X-ray structural analysis (1). Therefore the four cysteines chelating Fe in *D. gigas* ferredoxin are probably those at positions 7, 10, 13, and 50.

PSEUDOMONAS OVALIS AND *MYCOBACTERIUM SMEGMATIS* FERREDOXINS

Azotobacter vinelandii ferredoxin I (56), *Rhodospirillum rubrum* ferredoxin IV (73), and *Mycobacterium flavum* ferredoxin I (70) have similar molecular weights, amino acid compositions, absorption spectra, and EPR signals. They have two 4Fe-4S clusters with widely different midpoint potentials, about 800 mV apart (e.g., -420 mV for one cluster and +350 mV for the other in *A. vinelandii* ferredoxin).* These properties seem to be formally equivalent to that represented by a mixture of ferredoxin and *Chromatium* high-potential-iron-protein (10).

P. ovalis ferredoxin I (49) and *M. smegmatis* ferredoxin (T. Imai *et al.*, personal communication) show properties similar to those of *A. vinelandii* ferredoxin. The former two ferredoxins have 106 residues with nine and eight cysteines, respectively. They are homologous with clostridial ferredoxins and have long C-terminal tails. The N-terminal sequences of *A. vinelandii* ferredoxin (27) and *P. ovalis* ferredoxin are very similar to each other, namely, there are at most three amino acid differences out of the 40 residues compared (23). An identical segment also occurs between *P. ovalis*

*An X-ray structural analysis of *Azotobacter* ferredoxin suggests the presence of one 4Fe–4S and one 2Fe–2S cluster (55a). Stont, C.D., Structure of the iron–sulfur clusters in *Azotobacter* ferredoxin or 4 Å resolution. *Nature*, 1979, **279**, 83–84.

ferredoxin, residues 16–24, and *C. vinosum* ferredoxin, residues 49–57, as reported (23).

P. ovalis ferredoxin and *M. smegmatis* ferredoxin have 37 identical residues that are located mostly in their N-terminal halves. The eight cysteines out of nine excluding Cys-24 in *P. ovalis* ferredoxin are at corresponding positions of eight cysteines in clostridial ferredoxins and, therefore, Cys-24 was assumed not to be necessary for Fe chelation (21). However, a cysteine residue corresponding to Cys-11 of *P. ovalis* ferredoxin is replaced by valine in *M. smegmatis* ferredoxin, which has eight cysteines. Therefore Cys-24 must be necessary for the chelation together with others, except one at position 11 in *P. ovalis* ferredoxin. If the four cysteines at positions 20, 39, 42, and 45 construct the low-potential cluster as those in clostridial ferredoxins, the other four at positions 8, 16, 24, and 49 may construct the high-potential cluster. To clarify this important distribution of cysteines further studies are needed.

All bacterial ferredoxins have N-terminal regions homologous with the half-regions of clostridial ferredoxins, which suggests that they are derived by duplication of an ancestral ferredoxin gene followed by insertion, deletion, and recombination to produce various types of ferredoxin. Table 2 gives a matrix of amino acid differences of bacterial ferredoxins aligned as shown in Fig. 7 and a phylogenetic tree is constructed on the basis of this matrix (Fig. 5).

Four major lines are evident. Nonphotosynthetic fermenter ferredoxins are usually in a most primitive form and preserve the traces of internal gene duplication. Ferredoxins from photosynthetic bacteria living on H_2S also show traces of gene duplication to less extent and have extra segments. *C. limicola* has two ferredoxins probably derived by gene duplication, as found in horsetail ferredoxins. The event probably occurred before speciation to *C. limicola* and *C. thiosulfatophilum*, which may have the second ferredoxin. Photosynthetic purple nonsulfur bacterial ferredoxins have not been sequenced yet and can be expected to give additional interesting results.

B. stearothermophilus, *P. ovalis*, and *M. smegmatis* are aerobic bacteria having respiratory chains, and *D. gigas* is an anaerobic sulfate reducer. Their ferredoxins changed more rapidly than clostridial ferredoxins, and at present it is difficult to conclude whether the sulfate reducer originated before or after the appearance of *Clostridia* and photosynthetic bacteria.

P. ovalis and *M. smegmatis* ferredoxins have very special features in terms of cysteine distribution and probably in redox potentials, which are assumed to be similar to those of *Azotobacter* ferredoxin. The distributions of cysteine in these ferredoxins are diagrammatically compared with those of clostridial ferredoxins in Fig. 8.

The distribution of eight cysteine residues in clostridial ferredoxins is a fundamental form. The insertion of Cys-24 occurred to produce an ancestral

TABLE 2

MATRIX OF AMINO ACID DIFFERENCES AMONG BACTERIAL-TYPE FERREDOXINS[a]

	1	2	3	4	5	6	7	8	9	10	11	12	13	14	15	16
1. *C. pasteurianum*		9	14	19	18	23	23	27	31	37	33	60	87	82	55	70
2. *C. butyricum*	9		16	23	18	23	23	26	35	40	38	61	86	82	55	71
3. *C. acidi-urici*	14	16		18	18	25	24	30	28	34	31	57	86	81	55	67
4. *C.* M-E	19	23	18		26	25	25	31	30	35	32	57	85	84	55	68
5. *P. aerogenes*	18	18	18	26		26	26	30	37	40	38	61	85	82	56	70
6. *C. tartarivorum*	23	23	25	25	26		2	21	33	35	36	62	86	86	57	69
7. *C. thermosaccharolyticum*	23	23	24	25	26	2		19	33	35	36	62	86	86	57	68
8. *M. elsdenii*	27	26	30	31	30	21	19		39	38	39	64	85	84	56	71
9. *C. limicola* I	31	35	28	30	37	33	33	39		24	14	51	96	92	53	66
10. *C. limicola* II	37	40	34	35	40	35	35	38	24		22	49	98	92	54	69
11. *C. thiosulfatophilum*	33	38	31	32	38	36	36	39	14	22		51	97	93	56	67
12. *C. vinosum*	60	61	57	57	61	62	62	64	51	49	51		95	95	77	77
13. *P. ovalis*	87	86	86	85	85	86	86	85	96	98	97	95		69	103	112
14. *M. smegmatis*	82	82	81	84	82	86	86	84	92	92	93	95	69		107	110
15. *D. gigas*	55	55	55	55	56	57	57	56	53	54	56	77	103	107		64
16. *B. stearothermophilus*	70	71	67	69	70	68	68	71	66	69	67	77	112	110	64	

[a] Calculations are based on the alignment of sequences as shown in Fig. 7. A deletion or insertion is counted as one difference.

form of *P. ovalis* and *M. smegmatis* ferredoxins. Both of the two clusters in this ancestral form might still have low redox potentials. The accumulation of amino acid substitution in this ancestral form probably made the rearrangement of cysteine residues chelating to Fe unavoidable and produced *P. ovalis* ferredoxin with clusters having very different redox potentials. The Cys-11 was deleted later to produce *M. smegmatis* ferredoxin. There is no apparent sequence homology between high-potential-iron-protein and any one of the ferredoxins described earlier; therefore their functional similarity of high redox potentials results in a convergent evolution.

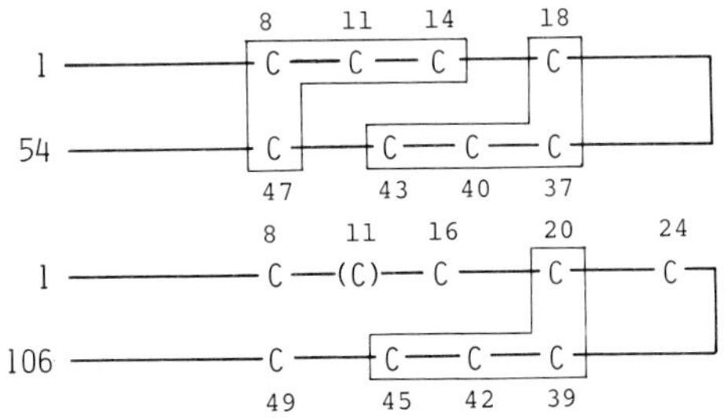

Fig. 8. Diagram of cysteine distribution in clostridial ferredoxins (upper) and *P. ovalis* and *M. smegnatis* ferredoxins (lower). Cysteine groups surrounded by lines probably construct a low-potential Fe-S cluster. C, cysteine residue; (C), cysteine residue present in *P. ovalis* ferredoxin but absent in *M. smegmatis* ferredoxin.

Summary

Comparisons have been made between amino acid sequences of 26 chloroplast-type ferredoxins and 16 bacterial-type ferredoxins. Their structural characteristics are described and related to a three-dimensional structure of a chloroplast-type ferredoxin. Aspects of molecular evolution of these ferredoxins are presented together with a phylogenetic tree including both chloroplast- and bacterial-type ferredoxins.

Acknowledgments

This work was supported in part by grants (Nos. 343028, 312107) from the Ministry of Education, Science and Culture of Japan. We thank Drs. Y. Katsube, T. Tsukihara, K. Fukuyama, N. Tanaka, M. Kakudo, and Mr. Y. Matsuura for their permission to use the data of the three-dimensional structure of *S. platensis* ferredoxin. Some of the data are taken from a dissertation to be submitted by T. Hase in 1979 to Osaka University in partial fulfillment of the requirements of a degree of Doctor of Philosophy.

REFERENCES

1. Adman, E.T., Sieker, L.C., and Jensen, L.H., The structure of a bacterial ferredoxin. *J. Biol. Chem.*, 1973, **284**:3987–3996.
2. Aggarwal, S.J., Rao, K.K., and Matsubara, H., Horsetail ferredoxin: Isolation and some chemical studies. *J. Biochem. (Tokyo)*, 1971, **69**:601–603.
3. Andrew, P.W., Rogers, L.J., Boulter, D., and Haslett, B.G., Ferredoxin from a red alga, *Porphyra umbilicalis. Eur. J. Biochem.*, 1976, **69**:243–248.
4. Barker, W.C., Ketcham, L.K., and Dayhoff, M.O., A comprehensive examination of protein sequences for evidence of internal gene duplication. *J. Mol. Evol.*, 1978, **10**:265–281.
5. Barker, W.C., McLaughlin, P.C., and Dayhoff, M.O., Evolutionary relationship between adrenodoxin and ferredoxin. *Fed. Proc., Fed. Am. Soc. Exp. Biol.*, 1972, **31**:837.
6. Barker, W.C., Schwartz, R.M., and Dayhoff, M.O., Other electron transport proteins. In: *Atlas of Protein Sequence and Structure* (M.O. Dayhoff, ed.) Vol. 5. National Biomedical Research Foundation, Washington, D.C., 1976,:51–65.
7. Bruschi, M., and LeGall, J., Phylogenetic studies of three families of electron transfer proteins from sulfur reducing bacteria: Cytochromes c_3, ferredoxins, and rubredoxins. In: *Evolution of Protein Molecules* (H. Matsubara and T. Yamanaka, eds.). Japan Scientific Societies Press, Center for Academic Publications. Tokyo, 1978:221–232.
8. Dayhoff, M.O. (ed.), *Atlas of Protein Sequence and Structure*. Vol. 4. National Biomedical Research Foundation, Washington, D.C., 1969.
9. Dayhoff, M.O., and Schwartz, R.M., Evolution of prokaryotes and eukaryotes inferred from sequences. In: *Evolution of Protein Molecules* (H. Matsubara and T. Yamanaka, eds.). Japan Scientific Societies Press, Center for Academic Publications. Tokyo, 1978:323–342.
10. Dus, K., De Klerk, H., Sletten, K., and Bartsch, R.G., Chemical characterization of high potential iron proteins from *Chromatium* and *Rhodopseudomonas gelatinosa. Biochim. Biophys. Acta*, 1967, **140**:291–311.
11. Fitch, W.M., Evidence suggesting a partial, internal duplication in the ancestral

gene for heme-containing globins. *J. Mol. Biol.*, 1966, **16**:17–27.

12. Fitch, W.M., and Margoliash, E., Construction of phylogenetic trees. *Science*, 1967, **155**:279–284.
13. Fry, K.T., and San Pietro, A., Studies on photosynthetic pyridine nucleotide reductase. *Biochem. Biophys. Res. Commun.*, 1962, **9**:218–221.
14. Fukuyama, K., *Structure Analysis of Chloroplast-type Ferredoxin at 2.8 Å Resolution.* Ph.D. Thesis, submitted to Osaka University, 1979, December. (Also personal communication after revision based on the analysis at 2.5 Å resolution in 1980.)
15. Hall, D.O., Cammack, R., and Rao, K.K., The plant ferredoxins and their relationship to the evolution of ferredoxins from primitive life. *Pure Appl. Chem.*, 1973, **34**:553–577.
16. Hall, D.O., Cammack, R., and Rao, K.K., Ferredoxin in the evolution of photosynthetic systems from anaerobic bacteria to higher plants. *Space Life Sci.*, 1973, **4**:455–468.
17. Hall, D.O., Rao, K.K., and Cammack, R., The iron-sulphur proteins: Structure, function and evolution of a ubiquitous group of proteins. *Sci. Prog. (Oxford)*, 1975, **62**:285–317.
18. Hase, T., Matsubara, H., and Evans, M.C.W., Amino acid sequence of *Chromatium vinosum* ferredoxin: Revisions. *J. Biochem. (Tokyo)*, 1977, **81**:1745–1749.
19. Hase, T., Ohmiya, M., Matsubara, H., Mullinger, R.N., Rao, K.K., and Hall, D.O., Amino acid sequence of a four-iron-four-sulfur ferredoxin isolated from *B. stearothermophilus. Biochem. J. (Tokyo)*, 1976, **159**:55–63.
20. Hase, T., Wakabayashi, S., Matsubara, H., Evans, M.C.W., and Jennings, J.V., Amino acid sequence of a ferredoxin from *Chlorobium thiosulfatophilum* strain Tassajara, a photosynthetic green sulfur bacterium. *J. Biochem. (Tokyo)*, 1978, **83**:1321–1325.
21. Hase, T., Wakabayashi, S., Matsubara, H., Imai, T., Matsumoto, T., and Tobari, J., *Mycobacterium smegmatis* ferredoxin: A unique distribution of cysteine residues constructing iron-sulfur clusters. *FEBS Lett.*, 1979 **103**:224–228.
22. Hase, T., Wakabayashi, S., Matsubara, H., Kerscher, L., Oesterhelt, D., Rao, K.K., and Hall, D.O., Complete amino acid sequence of *Halobacterium halobium* ferredoxin containing an *N*ε-acetyllysine residue. *J. Biochem. (Toyko)*, 1978, **83**:1657–1670.
23. Hase, T., Wakabayashi, S., Matsubara, H., Ohmori, D., and Suzuki, K., *Pseudomonas ovalis* ferredoxin: Similarity to *Azotobacter* and *Chromatium* ferredoxins. *FEBS Lett.*, 1978, **91**:315–319.
24. Hase, T., Wakabayashi, S., Matsubara, H., Rao, K.K., Hall, D.O., Widmer, H., Gysi, J., and Zuber, H., The amino acid sequence of ferredoxin from the alga, *Mastigocladus laminosus. Phytochemistry*, 1978, **17**:1863–1867.
25. Hase, T., Wakabayashi, S., Wada, K., Matsubara, H., Jüttner, F., Rao, K.K., Fry, I., and Hall, D.O., *Cyanidium caldarium* ferredoxin: A red algal type? *FEBS Lett.*, 1978, **96**:41–44.
26. Hori, H., and Osawa, S., Evolutionary change in 5S RNA secondary structure and a phylogenetic tree of 54 5S RNA species. *Proc. Natl. Acad. Sci. U.S.A.*, 1979, **76**:381–385.
27. Howard, J.B., Lorsbach, T., and Que, L., Iron-sulfur clusters and cysteine

distribution in a ferredoxin from *Azotobacter vinelandii*. *Biochem. Biophys. Res. Commun.*, 1976, **70**:582–588.
28. Huisman, J.G., Stapel, S., and Muijsers, A.O., Two different plant-type ferredoxins in each of two *Petunia* species. *FEBS Lett.*, 1978, **85**:198–202.
29. Hutson, K.G., Rogers, L.J., Haslett, B.G., Boulter, D., and Cammack, R., Comparative studies of two ferredoxins from the cyanobacterium *Nostoc* strain MAC. *Biochem. J.*, 1978, **172**:465–477.
30. Jukes, T.H., *Molecules and Evolution*. Columbia University Press, New York, 1966.
31. Katoh, S., and Takamiya, A., The presence of iron in photosynthetic pyridine nucleotide reductase. *Biochem. Biophys. Res. Commun.*, 1962, **8**:310–313.
32. Keresztes-Nagy, S., Perini, F., and Margoliash, E., Primary structure of alfalfa ferredoxin. *J. Biol. Chem.*, 1969, **244**:981–955.
33. Kerscher, L., Oesterhelt, D., Cammack, R., and Hall, D.O., A new plant-type ferredoxin from *Halobacteria*. *Eur. J. Biochem.*, 1976, **71**:101–107.
34. Lovenberg, W. (ed.), *Iron-sulfur Proteins*. Vols. 1 and 2, Academic Press, New York, 1973; Vol. 3, 1977.
35. Lovenberg, W., Buchanan, B.B., and Rabinowitz, J.C., Studies on the chemical nature of clostridial ferredoxin. *J. Biol. Chem.*, 1963, **238**:3899–3913.
36. Magrum, L.J., Luehrsen, K.R., and Woese, C.R., Are extreme halophiles actually "Bacteria"? *J. Mol. Evol.*, 1978, **11**:1–8.
37. Malkin, R., and Rabinowitz, J.C., The reconstitution of clostridial ferredoxin. *Biochem. Biophys. Res. Commun.*, 1966, **23**:822–827.
38. Masaki, R., Wada, K., and Matsubara, H., Chemical modification of spinach ferredoxin: Properties of acetylated spinach ferredoxin. *J. Biochem. (Tokyo).*, 1977, **81**:1–9.
39. Matheson, A.T., Yaguchi, M., Nagar, R.N., Visentin, L.P., and Willick, G.E., The structure of ribosomes from moderate and extreme halophilic bacteria. In: *Energetics and Structure of Halophilic Microorganisms* (S.R. Caplan and M. Ginzburg, eds.), Elsevier, Amsterdam, 1978:481–501.
40. Matsubara, H., Chemistry and evolution of ferredoxin molecules, In: *Aspects of Cellular and Molecular Physiology* (K. Hamaguchi, ed.). University of Tokyo Press, Tokyo, 1972:31–55.
41. Matsubara, H. and Wada, K., Molecular evolution. In: *Chemistry of Evolution in Modern Chemistry Series* (F. Egami, ed.). Iwanami Publisher, Tokyo, 1979, **19**:141–212 (in Japanese).
42. Matsubara, H., Hase, T., Wakabayashi, S., and Wada, K., Gene duplications during evolution of chloroplast-type ferredoxins. In: *Evolution of Protein Molecules* (H. Matsubara and T. Yamanaka, eds.), Japan Scientific Societies Press, Center for Academic Publications, Tokyo, 1978:209–219.
43. Matsubara, H., Jukes, T.H., and Cantor, C.R., Structural and evolutionary relationship of ferredoxins. *Brookhaven Symp. Biol.*, 1968, **21**:201–211.
44. Matsubara, H., Sasaki, R.M., and Chain, R.K., The amino acid sequence of spinach ferredoxin. *Proc. Natl. Acad. Sci. U.S.A.*, 1967, **57**:439–445.
45. Matsubara, H., Sasaki, R.M., Tsuchiya, D.K., and Evans, M.C.W., The amino acid sequence of *Chromatium* ferredoxin. *J. Biol. Chem.*, 1970, **245**:2121–2131.
46. Matsubara, H., Wada, K., and Masaki, R., Structure and function of

chloroplast-type ferredoxins. In: *Iron and Copper Proteins* (K.T. Yasunobu, H.F., Mower, and O. Hayaishi, eds.). Plenum, New York, 1976:1–15.

47. McLachlan, A.D., Gene duplications in the structural evolution of chymotrypsin. *J. Mol. Biol.*, 1979, **128**:49–79.
48. Morteson, L.E., Valentine, R.C., and Carnahan, J.E., An electron transport factor from *Clostridium pasteurianum. Biochem. Biophys. Res. Commun.*, 1962, **7**:448–452.
49. Ohmori, D., Properties of iron-sulfur protein isolated from *Pseudomonas ovalis. Biochem. Biophys. Res. Commun.*, 1976, **72**:566–574.
50. Orme-Johnson, W.H., Tryptic cleavage of *Clostridium acidicurici* apoferredoxin, and reconstiution of the separated fragments. *Biochem. Soc. Trans.*, 1973, **1**:30–31.
51. Orme-Johnson, W.H., Iron-sulfur proteins: structure and function. *Annu. Rev. Biochem.*, 1973, **42**:159–204.
52. Petering, D.H., and Palmer, G., Properties of spinach ferredoxin in anaerobic urea solution: A comparison with the native protein. *Arch. Biochem. Biophys.*, 1970, **141**:456–464.
53. Shin, M., Sukenobu, M., Oshino, R., and Kitazume, Y., Two plant-type ferredoxins from a blue green alga, *Nostoc verrucosum. Biochim. Biophys. Acta*, 1977, **460**:85–93.
54. Sobel, B.E., and Lovenberg, W., Characteristics of *Clostridium pasteurianum* ferredoxin in oxidation–reduction reactions. *Biochemistry*, 1966, **5**:6–13.
55. Stombaugh, N.A., Burris, R.H., and Orme-Johnson, W.H., Ferredoxins from *Bacillus subtilis*: Low potential iron-sulfur proteins which appear to contain single four iron, four sulfur centers accepting a single electron on reduction. *J. Biol. Chem.*, 1973, **248**:7951–7956.

55a. Stout, C.D., Structure of the iron-sulfur clusters in *Azotobacter* ferredoxin at 4 Å resolution. *Nature (London)*, 1979, **279**:83–84.

56. Sweeney, W.V., Rabinowitz, J.C., and Yoch, D.C., High and low reduction potential (4Fe–4S*) cluster in *Axotobacter vinelandii* $(4Fe\text{-}4s^*)_2$ ferredoxin I: Influence of the polypeptide on the reduction potentials. *J. Biol. Chem.*, 1975, **250**:7842–7847.
57. Tagawa, K., and Arnon, D.I., Ferredoxins as electron carriers in photosynthesis and in the biological reduction and consumption of hydrogen gas. *Nature (London)*, 1962, **195**:537–543.
58. Takruri, I., and Boulter, D., The amino acid sequence of ferredoxin from *Triticum aestivum* (wheat). *Biochem. J.*, 1979, **179**:373–378.
59. Takruri, I., Haslett, B.G., Boulter, D., Andrew, P.W., and Rogers, L.J., The amino acid sequence of ferredoxin from the red alga, *Porphyra umbilicalis Biochem. J.*, 1978, **173**:459–466.
60. Tanaka, M. Haniu, M., Yasunobu, K.T., Evans, M.C.W., and Rao, K.K., Amino acid sequence of ferredoxin from a protosynthetic green bacterium, *Chlorobium limicola. Biochemistry*, 1974, **13**:2953–2959.
61. Tanaka, M., Haniu, M., Yasunobu, K.T., Evans, M.C.W., and Rao, K.K., The amino acid sequence of ferredoxin II from *Chlorobium limicola*, a photosynthetic green bacterium. *Biochemistry*, 1975, **14**:1939–1943.
62. Tanaka, M., Nakashima, T., Benson, A., Mower, H.F., and Yasunobu, K.T.,

The amino acid sequence of *Clostridium pasteurianum* ferredoxin. *Biochem. Biophys. Res. Commun.*, 1964, **16**:422–427; *Biochemistry*, 1966, **5**:1666–1681.

63. Travis, J., Newman, D.J., LeGall, J., and Peck, H.D., Jr., The amino acid sequence of ferredoxin from the sulfur-reducing bacterium, *Desufovibrio gigas. Biochem. Biophys. Res. Commun.*, 1971, **45**:452–458.
64. Tsukihara, T., Fukuyama, K., Tahara, H., Katsube, Y., Matsuura, Y., Tanaka, N., Kakudo, M., Wada, K., and Matsubara, H., X-ray analysis of ferredoxin from *Spirulina platensis* II. Chelate structure of active center. *J. Biochem. (Toyko).*, 1978, **84**:1645–1647.
65. Wada, K., Hase, T., Tokunaga, H., and Matsubara, H., Amino acid sequence of *Spirulina platensis* ferredoxin: A far divergency of blue-green algal ferredoxins. *FEBS Lett.*, 1975, **55**:102–104.
66. Wakabayashi, S., Hase, T., Wada, K., Matsubara, H., Suzuki, K., and Takaichi, S., Amino acid sequences of two ferredoxins from pokeweed, *Phytolacca americana. J. Biochem. (Toyko)*, 1978, **83**:1305–1319.
67. Weinstein, B., An archetype correlation between bacterial rubredoxin and both bacterial and plant ferredoxins. *Biochem. Biophys. Res. Commun.*, 1969, **35**:109–114.
68. Woese, C.R., and Fox, G.E., Phylogenetic structure of the prokaryotic domain: The primary kingdoms. *Proc. Natl. Acad. Sci. U.S.A.*, 1977, **74**:5088–5090.
69. Yasunobu, K.T., and Tanaka, M., Primary structures of iron-sulfur proteins. In: *Iron-sulfur Proteins* (W. Lovenberg, ed.), Vol. 2. Academic Press, New York, 1973:27–130.
70. Yates, M.G., O'Donnell, M.K., Lowe, D.J., and Bothe H., Ferredoxins from nitrogen-fixing bacteria: Physical and chemical characterization of two ferredoxins from *Mycobacterium flavum* 301. *Eur. J. Biochem.*, 1978, **85**:291–299.
71. Yoch, D.C., Purification and properties of two ferredoxins from the nitrogen-fixing bacterium *Bacillus polymyxa. Arch. Biochem. Biophys.*, 1973, **158**:633–640.
72. Yoch, D.C., and Arnon, D.I., Two biologically active ferreredoxins from the aerobic nitrogen-fixing bacterium, *Azotobacter vinelandii J. Biol. Chem.*, 1972, **247**:4514–4520.
73. Yoch, D.C., Carithers, R.P., and Arnon, D.I., Isolation and characterization of bound iron-sulfur proteins from bacterial photosynthetic membranes: I. Ferredoxins III and IV from *Rhodospirillum rubrum* chromatophores. *J. Biol. Chem.*, 1977, **252**:7453–7460.

Amino Acid Sequences of Globin Chains and Their Use in Phylogenetic Divergence Point Estimations

E. O. P. THOMPSON

School of Biochemistry
University of New South Wales
Kensington, Australia.

INTRODUCTION

For the past 10 years my co-workers and I have been studying the amino acid sequences of globin chains from distinctive Australian animals. In some cases, such as the globins of the spiny anteater and the platypus, the animals are unique to Australia and are the sole surviving members of the subclass Prototheria. The inadequacy of the fossil record for dating these animals suggested the possibility of using the changes in amino acid sequences of globin chains as a basis for calculating times since divergence from other mammals.

Zuckerkandl and Pauling (87) were among the first to point out the value of amino acid sequences in proteins as evidence of evolutionary history for the species from which they had been isolated, and they initiated efforts to calculate the rates of change in amino acid sequences in globin chains. An early exploration of the possibilities of this idea of molecular evolution in relation to structure–function relationships from species covering a wide expanse of the taxonomic scale flowed from the studies on cytochrome *c* by Margoliash and Smith (49) and their co-workers. The examination of a group of homologous proteins is still best considered in the way adopted by these workers, namely, constant features of the structure, variable features of the structure and phylogenetic relationships, and evolutionary aspects of structure–function relationships.

The mechanisms of protein evolution are now generally accepted and have been reviewed (21,23).

ISBN 0-12-643150-7

Myoglobins and Hemoglobins

The globin chains of myoglobins and hemoglobins have occupied a rather special place in comparative studies, being among the earliest proteins to have both covalent (14,27) and three-dimensional structure (41,61,78) determined. From an evolutionary viewpoint they have provided clear examples of protein chains that have arisen by gene duplication (39) and divergent evolution leading to the development of new properties (Fig. 1). In particular, the ability to aggregate dissimilar globin chains to form heterotetramers (e.g., $\alpha_2\gamma_2,\alpha_2\beta_2$) with improved oxygen loading–unloading characteristics was clearly an evolutionary development of prime importance for large-bodied animals and was probably vital for animals to leave the water. In Fig. 1 an approximate time scale is placed on these gene duplications (21). The rate of fixation of mutations in globin chains (86) is given as 3.7 million years for α-globin, intermediate between those for histones with a unit evolutionary period of 400 million years (my), lactate dehydrogenase (19 my); cytochrome *c* (15 my), and albumin (3 my) or fibrinopeptides (1.1 my).

The duplication of globin genes has not been restricted to those with different functions. There is abundant evidence for multiple genes coding for α- and γ-chains in humans (68,69), as well as genes for globin chains that appear at different times during ontogeny (79), and knowledge of these is growing rapidly with the developments in DNA sequencing. The red cell in adult humans is remarkable for the high concentration, over 90%, of a major hemoglobin $\alpha_2\beta_2$. In most animals multiple hemoglobins are present in the blood (36,43), arising from both allelic and nonallelic genes.

The concept of molecular disease introduced by Pauling and co-workers (87) with reference to sickle-cell hemoglobin, coupled with the ease of screening and isolation of abnormal human hemoglobins (46), has greatly increased our knowledge of protein variability due to mutations of a single nucleotide in the genetic DNA. These variants have a frequency of occurrence of approxi-

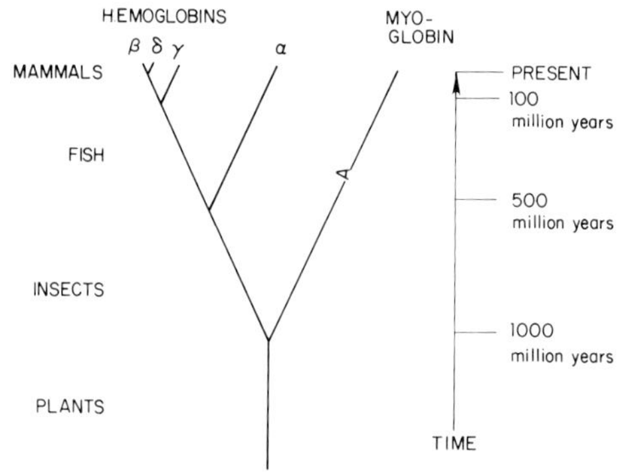

Fig. 1. Simplified representation of evolution of human globin genes. The branches are the result of gene duplications and an approximate time scale is given (21,39). The duplication leading to α and γ genes resulted in the development of tetrameric hemoglobins.

mately 1 in 200 population, with a single amino acid difference in either an α or β chain (45).

Because of the change in only one amino acid residue they have been particularly important in understanding the molecular pathology of hemoglobin (50,59), which is usually insensitive to replacements of most amino acid residues on its surface but very sensitive to even small alterations of internal nonpolar residues, especially those near the heme groups, which affect stability, and those at contacts between α and β subunits that affect respiratory function.

The evolution of our knowledge of the many functional aspects of hemoglobin has been slow (28) but has grown so much more rapidly since the molecular structure became available (57,58) that we now have clear ideas for interpreting cooperative interactions that affect oxygen binding, and involve protons, carbon dioxide, 2,3-diphosphoglycerate (DPG), or other phosphate esters like ATP, GTP, or inositol di-, penta-, or hexaphosphate in nucleated red cells (42,63,64).

The Myoglobin Fold

For myoglobin, which lacks cooperative effects because of its monomeric nature, there are features unique to its structure compared with hemoglobin, even though they have a common tertiary structure. Since sperm whale myoglobin was the first protein to have the folding of its polypeptide chain solved (41), this tertiary structure is usually referred to as the myoglobin fold (Fig. 2).

The common features of this myoglobin fold are eight helical and six nonhelical segments forming a hydrophobic basket for the heme. The helices are lettered A to H and the nonhelical segments NA, AB, . . . , GH, HC from the amino to the carboxyl end. Residues along any of the segments are numbered from the amino end, and by leaving gaps to accommodate deletions or additions, since globin chains vary in length (21), structurally equivalent residues have the same notation in myoglobins and hemoglobins. Perutz (58) has listed those single and tetrameric hemoglobins whose structures have been determined. The list includes leghemoglobin from the root nodules of leguminous plants (77) the structure of which indicates that the myoglobin fold was present in respiratory proteins of organisms having origins 1500 million years ago. In discussions of the functional morphology of both myoglobins and hemoglobins from species other than those whose structure has been solved, it is assumed that there is an overall resemblance in the three-dimensional structure. It is then possible to compare the amino acid residues in the different helical or interhelical regions that have undergone changes between animals of different species and try to predict or explain their differences in function.

For all globin chains, whether monomeric myoglobins and hemoglobins or tetrameric hemoglobins, the conservation of the essential hydrophobic resi-

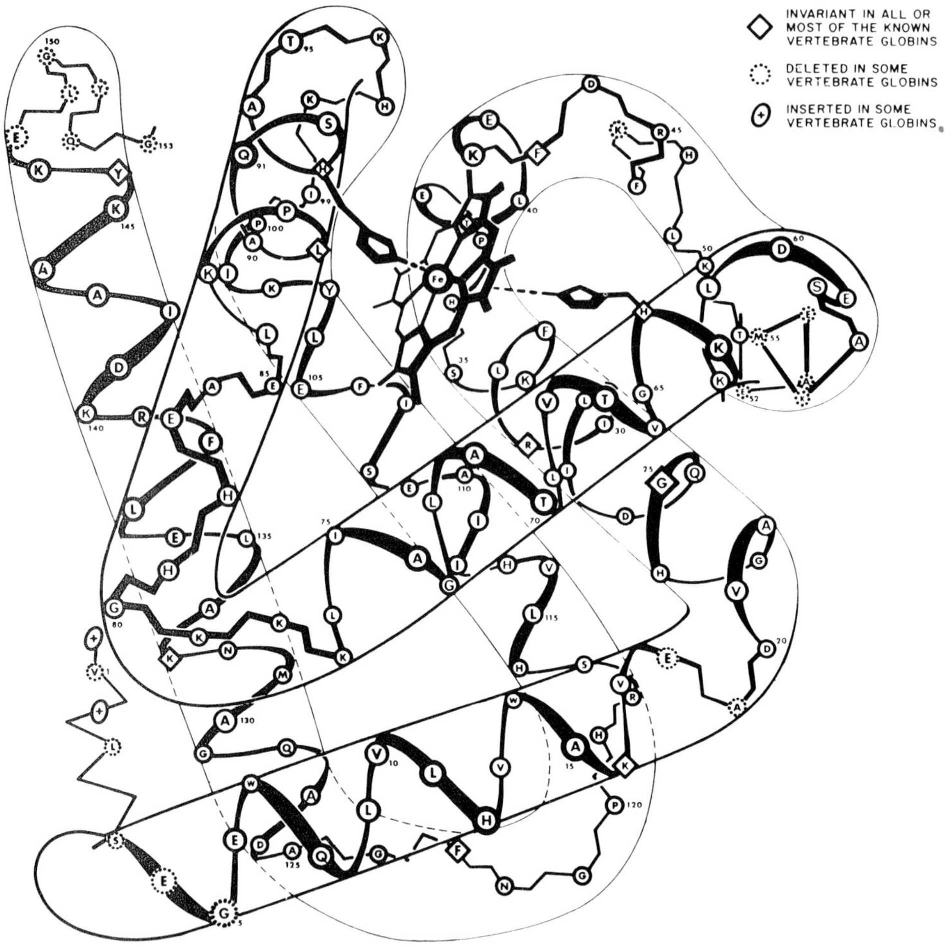

Fig. 2. The myoglobin fold determined for sperm whale myoglobin (41) showing the amino acid sequence (27). Helical regions are A 3–18; B 20–35; C 36–42; D 51–57; E 58–77; F 86–94; G 100–118; H 124–149. The heme group is situated in a hydrophobic basket bounded by helices C, E, F, and G. ◇, Invariant in all or most of the known vertebrate globins. ⁘ , Deleted in some vertebrate globins. ⊕, Inserted in some vertebrate globins. [Adapted from Dickerson (22) and reproduced with permission (21).]

dues that are contact sites with the heme group is essential. These heme contact residues are not identical for myoglobin, α- and β-globins. Perutz (58) has listed the most recent findings regarding interatomic contacts and discussed the differences in terms of the angle of the heme group in the pocket. Eleven sites are common of the 14–16 different sites listed for myoglobin and α and β chains, and in seven of these (E10, F7, FG3, FG5, G4, G5, G8) myoglobin has a different amino acid residue from those present and common to α- and β-globins. Obviously during evolution selective mutations have occurred to conserve the four sites common to them all (CD1, E7, E11, F8) while permitting changes to accumulate at other sites, provided they maintained the essential hydrophobic characteristics.

For myoglobin there are inter- and intrasegmental hydrogen bonds and salk links (78) that have no counterpart in hemoglobin chains. They have evolved to stabilize the monomeric protein, and there are obviously residues on the surface that must maintain charged amino acid side chains to inhibit aggregation of the monomer. It was mutation of these residues to nonpolar contacts that must have lead to the dimeric and tetrameric molecules (67).

Tetrameric Hemoglobins

The length of polypeptide chains is not conserved over very long periods of evolution, but in mammalian globins the myoglobin (153 residues), α-globin (141 residues), and β-globin (146 residues) chains are consistent in length. The additional residues in myoglobin are mostly in the HC terminal region, but like α chain it differs from the β chains by the deletion of one residue, NA2. The β chain differs further from α chains and myoglobin by the deletion of two residues in the AB corner. The shorter length of α chain compared with β chains is due to deletion of six residues in the short D-helix, so that this helix is not present in α chain.

As the number of different globin sequences is compared, the number of constant homologous residues decreases. Only the F8 histidine that binds to the iron of the heme group and CD1 phenylalanine that is a heme contact residue are invariant in all known globin chains. Within myoglobins, α chains, and β chains, the number of invariant residues is greater. For 30 myoglobins, 19 α chains, and 18 β chains whose sequences are known (Tables 1 and 2) the constant residues comprise 28%, 16%, and 22%, respectively.

Contact Residues and Sites of Cooperative Interactions

Because of the many different interactions that involve α and β globin chains there are very few areas of their sequence that are not involved in some defined functional role (35,57,58). There is an axis of symmetry in the tetramer with two different kinds of contacts, $\alpha_1\beta_1$ (equivalent to $\alpha_2\beta_2$) and $\alpha_1\beta_2$ (equivalent to $\alpha_2\beta_1$), which involve 34 residues and 19 residues, respectively (9,61), with the majority of the interactions being nonpolar. The larger $\alpha_1\beta_1$-type contact is stronger and does not break during dissociation of hemoglobin to dimers. It does not vary significantly between the oxy and deoxy conformations of the chains. It is the $\alpha_1\beta_2$-type contact sites that change during oxygenation–deoxygenation with a movement corresponding to a rotation of one pair of subunits with respect to the other pair. If one dimer is held fixed, the other turns by 15° about an off-center axis and shifts slightly along it (58). The axis of symmetry is preserved but it is rotated by 7.5°.

Contact between the two dimers has two stable conformations, the deoxy

TABLE 1

Amino Acid Sequences of Myoglobins[a]

			NA	A															A		B															B	C						C	CD		CD						CD	D
Myoglobin	1	2	3	1	2	3	4	5	6	7	8	9	10	11	12	13	14	15	16		1	2	3	4	5	6	7	8	9	10	11	12	13	14	15	16	1	2	3	4	5	6	7	1	2	3	4	5	6	7	8	9	1
Human	G	—	L	S	D	G	E	W	Q	L	V	L	N	V	W	G	K	V	E	A	D	I	P	G	H	G	Q	E	V	L	I	R	L	F	K	G	H	P	E	T	L	E	K	F	D	K	F	K	H	L	K	S	E
Chimpanzee	.	.	.	.	.	.	.	.	.	.	.	.	.	.	.	.	.	.	.	.	.	.	.	.	.	.	.	.	.	.	.	.	.	.	.	.	.	.	.	.	.	.	.	.	.	.	.	.	.	.	.	.	.
Gibbon	.	.	.	.	.	.	.	.	.	.	.	.	.	.	.	.	.	.	.	.	.	.	.	S	.	.	.	.	.	.	.	.	.	.	.	.	.	.	.	.	.	.	.	.	.	.	.	.	.	.	.	.	.
Baboon	.	.	.	.	.	.	.	.	.	.	.	.	.	.	.	.	.	.	.	.	.	.	.	S	.	.	.	.	.	.	.	.	.	.	.	.	.	.	.	.	.	.	.	.	.	.	.	.	.	.	.	.	.
Macaque	.	.	.	.	.	.	.	.	.	.	.	.	.	.	.	.	.	.	.	.	.	.	.	S	.	.	.	.	.	.	.	.	.	.	.	.	.	.	.	.	.	.	.	.	.	.	.	.	.	.	.	.	.
Tree shrew	.	.	.	.	.	.	.	.	.	.	.	.	.	.	.	.	.	.	.	.	.	V	A	.	.	.	.	.	.	.	.	.	.	.	.	.	.	.	.	.	.	.	.	.	.	.	.	.	.	.	.	T	.
Marmoset	.	.	.	.	.	.	.	.	.	S	.	.	.	.	.	.	.	.	.	.	.	L	A	.	.	.	.	.	I	.	.	.	.	.	T	A	.	.	.	.	.	.	.	.	.	.	.	.	N	.	.	T	P
Woolly monkey	.	.	.	.	.	.	.	.	.	.	.	.	.	I	.	.	.	.	.	.	.	.	.	S	.	.	.	.	.	.	.	S	.	.	.	.	.	.	.	.	.	.	.	.	.	.	.	.	.	.	.	.	.
Potto	.	.	.	.	.	.	.	.	.	S	.	.	.	.	.	.	.	.	.	.	.	L	A	.	.	.	.	.	I	.	.	.	.	.	T	A	.	.	.	.	.	.	.	.	.	.	.	.	N	.	.	T	P
Squirrel monkey	.	.	.	.	.	.	.	.	.	.	.	.	.	I	.	.	.	.	.	.	.	.	.	S	.	.	.	.	.	.	.	S	.	.	.	.	.	.	.	.	.	.	.	.	.	.	.	.	.	.	.	.	.
Lemur	.	.	.	.	.	.	.	.	.	.	.	.	.	.	.	.	.	.	.	.	.	V	G	.	.	.	.	.	.	.	.	.	.	.	T	.	.	.	.	.	.	.	.	.	.	.	.	.	.	.	.	T	A
Fruit bat	.	.	.	.	.	.	.	.	.	.	.	.	.	.	.	.	.	.	.	.	.	.	.	.	.	.	.	.	.	.	.	.	.	.	.	.	.	.	.	.	.	.	.	.	.	.	.	.	.	.	.	.	.
Galago	.	.	.	.	.	.	.	.	.	.	.	.	K	I	.	.	.	.	.	.	.	L	A	.	.	.	.	D	.	.	.	.	.	.	T	A	.	.	.	.	.	.	.	.	.	.	.	.	N	.	.	T	A
Hedgehog	.	.	.	.	.	.	.	.	.	.	.	.	.	.	.	.	.	.	.	.	.	.	.	.	.	.	.	.	.	.	.	.	.	.	.	D	.	.	.	.	.	.	.	.	.	.	.	.	.	.	.	.	.
Harbor seal	.	.	.	.	.	.	.	.	H	.	.	.	.	.	.	.	.	.	.	T	.	L	A	.	.	.	.	.	.	.	.	.	.	.	.	S	.	.	.	.	.	.	.	.	.	.	.	.	.	.	.	.	.
Badger	.	.	.	.	.	.	.	.	.	.	.	.	.	.	.	.	.	.	.	.	.	L	A	.	.	.	.	.	.	.	.	.	.	.	.	.	.	.	.	.	.	.	.	.	.	.	.	.	.	.	.	.	.
Sea lion	.	.	.	.	.	.	.	.	.	.	.	.	.	I	.	.	.	.	.	.	.	L	V	.	.	.	.	.	.	.	.	.	.	.	.	.	.	.	.	.	.	.	.	.	.	.	.	.	.	.	.	.	.
Horse	.	.	.	.	.	.	.	.	.	Q	.	.	.	.	.	.	.	.	.	.	.	.	A	.	.	.	.	.	.	.	.	.	.	.	T	.	.	.	.	.	.	.	.	.	.	.	.	.	.	.	.	T	.
Sheep	.	.	.	.	.	.	.	.	.	.	.	.	.	A	.	.	.	.	.	.	.	V	A	.	.	.	.	.	.	.	.	.	.	.	T	.	.	.	.	.	.	.	.	.	.	.	.	.	.	.	.	T	.
Ox	.	.	.	.	.	.	.	.	.	A	.	.	.	A	.	.	.	.	.	.	.	V	A	.	.	.	.	.	.	.	.	.	.	.	T	.	.	.	.	.	.	.	.	.	.	.	.	.	.	.	.	T	.
Dolphin	.	.	.	.	.	.	.	.	.	.	.	.	.	.	.	.	.	.	.	.	.	V	A	.	.	.	.	D	I	.	.	.	.	.	.	.	.	.	.	.	.	.	.	.	.	.	.	.	.	.	.	T	.
Harbor porpoise	.	.	.	.	E	.	.	.	.	.	.	.	.	.	.	.	.	.	.	.	.	L	A	.	.	.	.	D	.	.	.	.	.	.	.	.	.	.	.	.	.	.	.	.	.	.	.	.	.	.	.	T	.
Killer whale	.	.	.	.	.	.	.	.	.	.	.	.	.	.	.	.	.	.	.	.	.	L	A	.	.	.	.	D	I	.	.	.	.	.	.	.	.	.	.	.	.	.	.	.	.	.	.	.	.	.	.	T	.
Sperm whale	V	.	.	.	E	.	.	.	.	.	.	.	H	.	.	A	.	.	.	.	.	V	A	.	.	.	.	D	I	.	.	.	.	.	.	S	.	.	.	.	.	.	.	.	.	R	.	.	.	.	.	T	.
Kangaroo	.	.	.	.	.	.	.	.	.	.	.	.	.	I	.	.	.	.	.	T	.	V	G	.	.	.	.	D	.	.	.	.	.	.	.	.	.	.	.	.	.	.	.	.	.	.	.	.	.	.	.	.	.
Opossum	.	.	.	.	.	.	.	.	.	.	.	.	.	A	.	.	.	.	.	.	.	.	.	.	.	.	.	.	.	.	.	.	.	.	.	.	.	.	.	.	.	.	.	.	.	.	.	.	.	.	.	.	.
Echidna	.	.	.	.	.	.	.	.	.	.	.	.	K	.	.	.	.	.	.	T	.	.	T	.	.	.	.	D	.	.	.	.	.	.	.	T	.	.	.	.	.	.	.	.	.	.	.	.	.	.	.	T	.
Platypus	.	.	.	.	.	.	.	.	.	.	.	.	K	.	.	.	.	.	.	G	.	L	.	.	.	.	.	.	.	.	.	.	.	.	.	T	.	.	.	.	.	.	.	.	.	.	.	.	G	.	.	T	.
Chicken	.	.	.	.	.	Q	.	.	.	Q	.	.	T	I	.	.	.	.	.	.	.	.	A	.	.	.	H	.	.	M	.	.	.	.	H	D	.	.	.	.	.	D	R	.	.	.	.	.	G	.	.	T	.
Shark	.	.	.	.	Ac	T	.	.	E	H	.	N	K	.	.	A	V	.	.	P	.	.	.	A	V	.	L	A	I	L	.	.	.	.	.	E	.	K	.	.	K	D	L	.	P	.	.	.	E	—	I	P	V

[a] The helical regions A-H and interhelical regions NA, AB, CD . . . etc. are numbered. The one letter code and alignment follows the *Atlas of Protein Sequence and Structure* (21). The sequence of human myoglobin is given in full, dashes indicate deletions or insertions in longer globins. Dots indicate identity with the residue in human myoglobin. Ac represents an acetyl group. The sequences have been published (16–18,21,29,67).

TABLE 1—*Continued*

					D	E																				E	EF								EF	F								F	FG				FG	G			
Myoglobin	3	4	5	6	7	1	2	3	4	5	6	7	8	9	10	11	12		13	14	15	16	17	18	19	20	1	2	3	4	—	5	6	7	8	1	2	3	4	5	6	7	8	9	1	2	3	4	5	1	2	3	4
Human	D	E	M	K	A	S	E	D	L	K	K	H	G	A	T	V	L	—	T	A	L	G	G	I	L	K	K	K	G	H	—	H	E	A	E	I	K	P	L	A	Q	S	H	A	T	K	H	K	I	P	V	K	Y
Chimpanzee	.	.	.	.	.	.	.	.	.	.	.	.	.	.	.	.	.		.	.	.	.	.	.	.	.	.	.	.	.		.	.	.	.	.	.	.	.	.	.	.	.	.	.	.	.	.	.	.	.	.	.
Gibbon	.	.	.	.	.	.	.	.	.	.	.	.	.	.	.	.	.		.	.	.	.	.	.	.	.	.	.	.	.		.	.	.	.	.	.	.	.	.	.	.	.	.	.	.	.	.	.	.	.	.	.
Baboon	.	.	.	.	.	.	.	.	.	.	.	.	.	.	.	.	.		.	.	.	.	.	.	.	.	.	.	.	.		.	.	.	.	.	.	.	.	.	.	.	.	.	.	.	.	.	.	.	.	.	.
Macaque	.	.	.	.	.	.	.	.	.	.	.	.	.	V	.	.	.		.	.	.	.	.	.	.	.	.	.	.	.		.	.	.	.	.	.	.	.	.	.	.	.	.	.	.	.	.	.	.	.	.	.
Tree shrew	.	.	.	.	.	.	.	.	.	.	.	.	.	N	.	.	.		S	.	.	.	.	.	.	.	.	.	.	Q		.	.	.	.	.	.	.	.	.	.	.	.	.	.	.	.	.	.	.	.	.	.
Marmoset	.	.	.	.	.	.	.	.	.	.	.	.	.	V	.	.	.		.	.	.	.	.	.	.	.	.	.	.	.		.	.	.	.	.	.	.	.	.	.	.	.	.	.	.	.	.	.	.	.	.	.
Woolly monkey	.	.	.	.	.	.	.	E	.	.	.	.	.	V	.	.	.		.	.	.	.	.	.	.	.	.	.	.	Q		.	.	.	.	L	.	.	.	.	.	.	.	.	.	.	.	.	.	.	.	.	.
Potto	.	.	.	.	.	.	.	.	.	.	.	.	.	V	.	.	.		.	.	.	.	.	.	.	.	.	.	.	.		.	.	.	.	.	.	.	.	.	.	.	.	.	.	.	.	.	.	.	.	.	.
Squirrel monkey	.	.	.	.	.	.	.	E	.	.	.	.	.	T	.	.	.		.	.	.	.	.	.	.	.	.	.	.	Q		.	.	.	.	L	.	.	.	.	.	.	.	.	.	.	.	.	.	.	.	.	.
Lemur	.	.	.	.	.	.	.	.	.	.	.	.	.	T	.	.	.		.	.	.	.	.	.	.	.	.	.	.	Q		.	.	.	.	L	.	.	.	.	.	.	.	.	.	.	.	.	.	.	I	.	.
Fruit bat	.	.	.	.	.	.	.	.	.	.	.	.	.	.	.	.	.		.	.	.	.	.	.	.	.	.	.	.	Q		.	.	.	Q	L	.	.	.	.	.	.	.	.	.	.	.	.	.	.	.	.	.
Galago	.	.	.	.	.	.	.	.	.	.	.	.	.	V	.	.	.		.	.	.	.	.	.	.	.	.	.	.	Q		.	.	.	.	.	.	.	.	.	.	.	.	.	.	.	.	.	.	.	.	.	.
Hedgehog	.	.	.	.	S	.	.	.	.	.	.	.	.	T	.	.	.		.	.	.	.	.	.	.	.	.	.	.	Q		.	.	.	Q	L	A	.	.	.	.	.	.	.	N	.	.	.	.	.	.	.	.
Harbor seal	.	D	.	R	R	.	.	.	.	R	.	.	.	N	.	.	.		.	.	.	.	.	.	.	.	.	.	.	.		.	.	.	.	L	.	.	.	.	.	.	.	.	.	.	.	.	.	.	I	.	.
Badger	.	.	.	.	G	.	.	.	.	.	.	.	.	N	.	.	.		.	.	.	.	.	.	.	.	.	.	.	.		Q	.	.	.	L	.	.	.	.	.	.	.	.	.	.	.	.	.	.	.	.	.
Sea lion	.	.	.	.	R	.	.	.	.	.	.	.	.	K	.	.	.		.	.	.	.	.	.	.	.	.	.	.	.		.	D	.	.	L	.	.	.	.	.	.	.	.	.	.	.	.	.	.	I	.	.
Horse	A	.	.	.	.	.	.	.	.	.	.	.	.	T	V	.	.		.	.	.	.	.	.	.	.	.	.	.	.		.	.	.	.	L	.	.	.	.	.	.	.	.	.	.	.	.	.	.	I	.	.
Sheep	A	.	.	.	.	.	.	.	.	.	.	.	.	N	.	.	.		.	.	.	.	.	.	.	.	.	.	.	.		.	.	.	.	V	.	H	.	.	E	.	.	.	N	.	.	.	.	.	.	.	.
Ox	A	.	.	.	.	.	.	.	.	.	.	.	.	N	.	.	.		.	.	.	.	.	.	.	.	.	.	.	.		.	.	.	.	V	.	H	.	.	E	.	.	.	N	.	.	.	V	.	I	.	.
Dolphin	A	D	.	.	.	.	.	.	.	.	.	.	.	D	.	.	.		.	.	.	.	A	.	.	.	.	.	.	.		.	D	.	.	L	.	.	.	.	.	.	.	.	.	.	.	.	.	.	I	.	.
Harbor porpoise	A	.	.	.	.	.	.	.	.	.	.	.	.	N	.	.	.		.	.	.	.	.	.	.	.	.	.	.	.		.	.	.	N	L	.	.	.	.	.	.	.	.	.	.	.	.	.	.	I	.	.
Killer whale	A	D	.	.	.	.	.	.	.	.	.	.	.	N	.	.	.		.	.	.	.	A	.	.	.	.	.	.	.		.	D	.	.	L	.	.	.	.	.	.	.	.	.	.	.	.	.	.	I	.	.
Sperm whale	A	.	.	.	.	.	.	.	.	.	.	.	.	V	.	.	.		.	.	.	.	A	.	.	.	.	.	.	.		.	.	.	.	L	.	.	.	.	.	.	.	.	.	.	.	.	.	.	I	.	.
Kangaroo	.	.	.	.	.	.	.	.	.	.	.	.	.	I	.	.	.		.	.	.	.	N	.	.	.	.	.	.	.		.	.	.	.	L	.	.	.	.	.	.	.	.	.	.	.	.	.	.	.	Q	F
Opossum	.	.	.	.	.	.	.	.	.	.	.	.	.	.	.	.	.		.	.	.	.	N	.	.	.	.	.	.	N		.	.	.	.	L	.	.	.	.	.	.	.	.	.	.	.	.	.	S	.	Q	F
Echidna	.	.	.	.	.	.	A	.	.	.	.	.	.	G	V	.	.		.	.	.	.	S	.	.	.	.	.	.	Q		.	.	.	.	L	.	.	.	.	.	.	.	.	.	.	.	.	.	S	I	.	F
Platypus	.	.	.	.	.	.	A	.	.	.	.	.	.	G	.	.	.		.	.	.	.	N	.	.	.	.	.	.	Q		.	.	.	.	L	.	.	.	.	.	.	.	.	.	.	.	.	.	S	I	.	F
Chicken	P	D	.	.	G	.	.	.	.	.	.	.	.	Q	.	.	.		.	.	.	.	A	Q	.	.	.	.	.	.		.	.	.	D	L	.	.	.	.	.	T	.	.	.	.	.	.	.	.	.	.	.
Shark	Q	Q	L	G	N	N	.	.	.	R	.	.	.	V	.	.	.		R	.	.	.	N	.	.	.	Q	.	.	K		.	S	T	N	V	.	E	.	.	D	T	.	I	N	.	.	.	.	.	P	.	N

(*continued*)

TABLE 1—*Continued*

															G	GH				GH	H																									H	HC			
Myoglobin	5	6	7	8	9	10	11	12	13	14	15	16	17	18	19	1	2	3	4	5	1	2	3	4	5	6	7	8	9	10	11	12	13	14	15	16	17	18	19	20	21	22	23	24	25	26	1	2	3	4
Human	L	E	F	I	S	E	C	I	I	Q	V	L	Q	S	K	H	P	G	D	F	G	A	D	A	Q	G	A	M	N	K	A	L	E	L	F	R	K	D	M	A	S	N	Y	K	E	L	G	F	Q	G
Chimpanzee	.	.	.	.	.	.	.	.	.	.	.	.	H	.	.	.	.	.	.	.	.	.	.	.	.	.	.	.	.	.	.	.	.	.	.	.	.	.	.	.	.	.	.	.	.	.	.	.	.	.
Gibbon	.	.	.	.	.	.	.	.	.	.	.	.	.	.	.	.	.	.	.	.	.	.	.	.	.	.	.	.	.	.	.	.	.	.	.	.	.	.	.	.	.	.	.	.	.	.	.	.	.	.
Baboon	.	.	L	.	.	.	S	.	.	.	.	.	.	.	.	.	.	.	.	.	.	.	.	.	.	.	.	.	.	.	.	.	.	.	.	.	N	.	.	.	A	K	.	.	.	.	.	.	.	.
Macaque	.	.	L	.	.	.	S	.	.	.	.	.	.	.	.	.	.	.	.	.	.	.	.	.	.	.	.	.	.	.	.	.	.	.	.	.	N	.	.	.	A	K	.	.	.	.	.	.	.	.
Tree shrew	.	.	.	.	.	.	A	.	.	.	.	.	.	.	.	.	.	.	.	.	.	.	.	.	.	A	.	.	S	.	.	.	.	.	.	.	N	.	I	.	A	K	.	.	.	.	.	.	.	.
Marmoset	.	.	.	.	.	.	A	.	.	H	.	.	.	.	.	.	.	.	.	.	.	.	.	.	.	.	.	.	.	.	.	.	.	.	.	.	N	.	I	.	A	K	.	.	.	.	.	.	.	.
Woolly monkey	.	.	.	.	.	D	A	.	.	H	A	.	.	K	.	.	.	.	.	.	.	.	.	.	.	.	.	.	K	.	.	.	.	.	.	.	N	.	.	.	A	K	.	.	.	.	.	.	.	.
Potto	.	.	.	.	.	.	A	.	.	H	.	.	.	.	.	.	.	.	.	.	.	.	.	.	.	.	.	.	.	.	.	.	.	.	.	.	N	.	I	.	A	K	.	.	.	.	.	.	.	.
Squirrel monkey	.	.	L	.	.	D	A	.	V	H	.	.	.	K	.	.	.	.	.	.	.	.	.	.	.	.	.	.	K	.	.	.	.	.	.	.	N	.	.	.	A	K	.	.	.	.	.	.	.	.
Lemur	.	.	.	.	.	D	A	.	V	H	.	.	H	.	.	.	.	A	E	.	.	.	.	.	.	A	.	.	K	.	.	.	.	.	.	.	N	.	I	.	A	K	.	.	.	.	.	.	.	.
Fruit bat	.	.	.	.	.	.	V	.	.	.	.	.	.	.	.	.	.	.	.	.	.	.	.	.	.	.	.	.	G	.	.	.	.	.	.	.	N	.	I	.	A	K	.	.	.	.	.	.	.	.
Galago	.	.	.	.	.	.	A	.	.	H	.	.	.	N	.	.	S	.	.	.	.	T	.	V	.	.	.	.	S	.	.	.	.	.	.	.	N	.	I	.	A	K	.	.	.	.	.	.	.	.
Hedgehog	.	.	.	.	.	.	A	.	.	.	.	.	K	.	.	.	A	.	.	.	.	.	.	.	.	.	.	.	S	.	.	.	.	.	.	.	N	.	I	.	A	K	.	.	.	.	.	.	.	.
Harbor seal	.	.	.	.	.	.	A	.	.	H	.	.	H	.	.	.	.	A	E	.	.	.	.	.	.	A	.	.	K	.	.	.	.	.	.	.	N	.	I	.	A	K	.	.	.	.	.	.	H	.
Badger	.	.	.	.	.	D	A	.	A	.	.	.	.	.	.	.	.	.	.	.	A	.	E	.	.	.	.	.	K	.	.	.	.	.	.	.	N	.	I	.	A	K	.	.	.	.	.	.	.	.
Sea lion	.	.	.	.	.	.	A	.	.	H	.	.	.	.	.	.	.	.	.	.	.	.	.	T	H	A	.	.	K	.	.	.	.	.	.	.	N	.	I	.	A	K	.	R	.	.	.	.	.	.
Horse	.	.	.	.	.	D	A	.	.	H	.	.	H	.	.	.	.	.	.	.	.	.	.	.	.	.	.	.	T	.	.	.	.	.	.	.	N	.	I	.	A	K	.	.	.	.	.	.	.	.
Sheep	.	.	.	.	.	D	A	.	.	H	.	.	H	A	.	.	.	S	N	.	.	.	.	.	.	.	.	.	S	.	.	.	.	.	.	.	N	.	.	.	A	E	.	.	V	.	.	.	.	.
Ox	.	.	.	.	.	D	A	.	.	H	.	.	H	A	.	.	.	S	N	.	A	.	.	.	.	.	.	.	S	.	.	.	.	.	.	.	N	.	A	.	E	K	.	.	V	.	.	.	H	.
Dolphin	.	.	.	.	.	.	A	.	.	H	.	.	H	.	R	.	.	A	Q	.	.	.	.	.	.	.	.	.	.	.	.	.	.	.	.	.	.	.	I	.	A	K	.	.	.	.	.	.	H	.
Harbor porpoise	.	.	.	.	.	.	A	.	.	H	.	.	H	.	R	.	.	A	E	.	.	.	.	.	.	.	.	.	.	.	.	.	.	.	.	.	.	.	I	.	T	K	.	.	.	.	.	.	H	.
Killer whale	.	.	.	.	.	.	A	.	.	H	.	.	H	.	R	.	.	A	Q	.	.	.	.	.	.	.	.	.	.	.	.	.	.	.	.	.	.	.	I	.	A	K	.	.	.	.	.	.	H	.
Sperm whale	.	.	.	.	.	.	A	.	.	H	.	.	H	.	R	.	.	.	.	.	.	.	.	.	.	.	.	.	.	.	.	.	.	.	.	.	.	.	I	.	A	K	.	.	.	.	.	Y	.	.
Kangaroo	.	.	.	.	.	D	A	.	.	.	.	I	.	.	.	.	A	.	.	.	.	.	.	.	.	A	.	.	K	.	.	.	.	.	.	.	H	.	.	.	A	K	.	.	.	F	.	.	.	.
Opossum	.	.	.	.	.	.	A	.	.	.	.	I	.	.	.	.	.	.	.	.	.	G	.	.	.	A	.	.	G	.	.	.	.	.	.	.	N	.	.	.	A	K	.	.	.	.	.	.	.	.
Echidna	.	.	.	.	.	.	A	.	.	H	.	.	.	.	.	.	S	A	.	.	.	.	.	.	.	A	.	.	G	.	.	.	.	.	.	.	N	.	.	.	T	K	.	.	.	F	.	.	.	.
Platypus	.	.	Y	.	.	.	A	.	.	H	.	.	.	.	.	.	S	A	.	.	.	.	.	.	.	A	.	.	G	.	.	.	.	.	.	.	N	.	.	.	A	K	.	.	.	F	.	.	.	.
Chicken	.	.	.	.	.	.	V	.	.	K	.	I	A	E	.	.	A	A	.	.	.	.	.	S	.	A	.	.	K	.	.	.	.	.	.	.	D	.	.	.	.	K	.	.	.	F	.	.	.	.
Shark	F	V	L	.	T	N	I	A	V	K	.	.	T	E	M	Y	.	S	.	M	T	G	P	M	.	E	S	F	S	.	V	F	T	V	I	C	S	.	L	E	T	L	.	.	.	A	N	.	.	.

TABLE 2

AMINO ACID SEQUENCES OF SOME α- AND β-GLOBINS[a]

			NA	A															A		B															B	C						C	CD						
Globin	1	2	3	1	2	3	4	5	6	7	8	9	10	11	12	13	14	15	16		1	2	3	4	5	6	7	8	9	10	11	12	13	14	15	16	1	2	3	4	5	6	7	1	2	3	4	5	6	7
Human α	V	—	L	S	P	A	D	K	T	N	V	K	A	A	W	G	K	V	G	A	H	A	G	E	Y	G	A	E	A	L	E	R	M	F	L	S	F	P	T	T	K	T	Y	F	P	H	F	—	D	L
Rhesus monkey α	.	.	.	.	.	.	.	.	S	.	.	.	.	.	.	.	.	.	.	G	.	.	.	.	.	.	.	.	.	.	.	.	.	.	.	.	.	.	.	.	.	.	.	.	.	.	.	.	.	.
Loris α	.	.	.	.	.	.	.	.	.	.	.	.	.	.	.	E	.	.	.	S	.	.	.	D	.	.	.	.	.	.	.	.	.	.	.	.	.	.	.	.	.	.	.	.	.	.	.	.	.	.
Rabbit α	.	.	.	.	.	.	.	.	.	.	I	.	T	.	.	E	.	I	.	S	.	G	.	.	.	.	.	.	.	V	.	.	.	.	.	G	.	.	.	.	.	.	.	.	.	.	.	.	.	F
Mouse α	.	.	.	.	G	E	.	.	S	.	I	.	.	.	.	.	.	I	.	G	.	G	A	.	.	.	.	.	.	.	.	.	.	.	A	.	.	.	.	.	.	.	.	.	.	.	.	.	.	V
Dog α	.	.	.	.	.	.	.	.	.	.	I	.	S	T	.	D	.	I	.	G	.	.	.	D	.	.	G	.	.	.	D	.	T	.	Q	.	.	.	.	.	.	.	.	.	.	.	.	.	.	.
Horse α	.	.	.	.	A	.	.	.	.	.	.	.	.	.	.	S	.	.	.	G	.	.	.	.	.	.	.	.	.	.	.	.	.	.	.	G	.	.	.	.	.	.	.	.	.	.	.	.	.	.
Ox α	.	.	.	.	A	.	.	.	G	.	.	.	.	.	.	.	.	.	.	G	.	.	A	.	.	.	.	.	.	.	.	.	.	.	.	.	.	.	.	.	.	.	.	.	.	.	.	.	.	.
Opossum α	.	.	.	.	A	N	.	.	.	.	.	.	G	.	.	S	.	.	.	G	N	S	.	A	.	M	G	.	.	.	Y	.	T	.	.	.	.	.	.	.	.	.	.	.	.	N	Y	.	.	F
Gray kangaroo α	.	.	.	.	A	.	.	.	G	H	.	.	.	I	.	.	.	.	.	G	.	.	.	.	.	A	.	.	G	.	.	.	T	.	H	.	.	.	.	.	.	.	.	.	.	.	.	.	.	.
Echidna 1 α	.	.	.	T	D	.	E	.	K	E	.	T	S	L	.	.	.	A	S	G	.	.	E	.	.	.	.	.	.	.	.	.	L	.	.	.	.	.	.	.	.	.	.	.	S	.	H	.	.	.
Echidna 2 α	.	.	.	T	D	.	E	R	K	E	.	T	S	L	.	.	.	A	S	G	.	.	E	D	.	.	.	.	.	.	.	.	L	.	.	.	.	.	.	.	.	.	.	.	S	.	M	.	.	.
Platypus α	M	.	.	T	D	.	E	.	K	E	.	T	.	L	.	.	.	A	A	G	.	G	E	.	.	.	.	.	.	.	.	.	L	.	Q	A	.	.	.	.	.	.	.	.	S	.	.	.	.	.
Chicken α	.	.	.	.	N	.	.	.	N	.	.	.	G	I	F	T	.	I	A	G	.	.	E	.	.	.	.	.	T	.	.	.	.	.	I	G	.	.	.	.	.	.	.	.	.	.	.	.	.	.
Viper α	.	.	.	.	E	D	.	.	N	R	.	R	T	S	V	.	.	N	P	E	L	P	.	.	.	.	S	.	T	.	T	.	.	.	A	A	H	.	.	.	.	.	.	.	.	.	.	.	.	.
Newt α	M	K	.	.	A	E	.	.	H	.	.	.	T	T	.	D	H	I	K	G	.	E	E	A	L	.	.	.	.	.	F	.	.	.	T	.	L	.	A	.	R	.	.	.	.	A	K	.	.	.
Suckerfish α	S	.	.	.	D	K	.	.	A	D	.	.	I	.	.	A	.	I	S	P	R	.	D	.	I	.	.	.	.	.	G	.	.	L	T	V	Y	.	Q	.	.	.	.	.	A	.	W	A	.	.
Carp α	S	.	.	.	D	K	.	.	A	A	.	.	I	.	.	A	.	I	S	P	K	.	D	D	I	.	.	.	.	.	G	.	.	L	T	V	Y	.	Q	.	.	.	.	.	A	.	W	A	.	.
Shark α	D	.	Y	.	A	.	.	R	A	E	L	A	.	L	S	K	V	L	A	Q	N	.	E	A	F	.	.	.	.	.	A	.	.	.	T	V	Y	A	A	.	.	S	.	.	K	D	Y	K	.	F
Human β	V	H	L	T	P	E	E	K	S	A	V	T	A	L	W	G	K	V	—	—	N	V	D	E	V	G	G	E	A	L	G	R	L	L	V	V	Y	P	W	T	Q	R	F	F	E	S	F	G	D	L
Rhesus monkey β	.	.	.	.	.	.	.	.	N	.	.	.	T	.	.	.	.	.	.	.	.	.	.	.	.	.	.	.	.	.	.	.	.	.	L	.	.	.	.	.	.	.	.	.	.	.	.	.	.	.
Spider monkey β	.	.	.	.	G	.	.	.	.	.	.	A	.	.	.	.	.	.	.	.	.	.	.	.	.	.	.	.	.	.	.	.	.	.	.	.	.	.	.	.	.	.	.	.	.	.	.	.	.	.
Loris β	.	.	.	.	G	.	.	.	.	.	.	.	.	.	.	.	.	.	.	.	.	.	.	D	.	.	.	.	.	.	.	.	.	.	.	.	.	.	.	.	.	.	.	.	.	.	.	.	.	.
Rabbit β	.	.	.	S	S	.	.	.	.	.	.	.	.	.	.	.	.	.	.	.	.	.	E	.	.	.	.	.	.	.	.	.	.	.	.	.	.	.	.	.	.	.	.	.	.	.	.	.	.	.
Mouse β	.	.	.	.	D	A	.	.	A	.	.	S	G	.	.	.	.	.	.	.	.	A	.	.	.	.	.	.	.	.	.	.	.	.	.	.	.	.	.	.	.	.	Y	.	S	.	.	.	.	.
Dog β	.	.	.	.	A	.	.	.	.	L	.	S	G	.	.	.	.	.	.	.	.	.	.	.	.	.	.	.	.	.	.	.	.	.	I	.	.	.	.	.	.	.	.	.	D	.	.	.	.	.
Horse β	.	Q	.	S	G	.	.	.	A	.	.	L	.	.	.	D	.	.	.	.	.	E	E	.	.	.	.	.	.	.	.	.	.	.	.	.	.	.	.	.	.	.	.	.	D	.	.	.	.	.
Sheep β	M	—	.	.	A	.	.	.	A	.	.	.	G	F	.	.	.	.	.	.	K	.	.	.	.	.	A	.	.	.	.	.	.	.	.	.	.	.	.	.	.	.	.	.	.	H	.	.	.	.
Ox β	M	—	.	.	A	.	.	.	A	.	.	.	.	A	.	S	.	.	.	.	H	.	.	.	.	.	.	.	.	.	.	.	.	.	.	.	.	.	.	.	.	.	.	.	.	.	.	.	.	.
Gray kangaroo β	.	.	.	.	A	.	.	.	N	.	I	.	S	.	.	.	.	.	.	.	A	I	E	Q	T	.	.	.	.	.	.	.	.	.	I	.	.	.	.	.	S	.	.	.	D	H	.	.	.	.
Red kangaroo β	.	.	.	.	A	.	.	.	N	.	I	.	S	.	.	.	.	.	.	.	A	I	E	Q	T	.	.	.	.	.	.	.	.	.	I	.	.	.	.	.	S	.	.	.	D	H	.	.	.	.
Potoroo β	.	.	.	S	S	.	.	.	G	L	I	.	S	.	.	.	.	I	.	.	D	I	E	Q	T	.	.	.	.	.	.	.	.	.	I	.	.	.	.	.	S	.	.	.	D	H	.	.	.	.
Echidna β	.	.	.	S	G	S	.	.	T	.	.	.	N	.	.	.	H	.	.	.	.	.	N	.	L	.	.	.	.	.	.	.	.	.	.	.	.	.	.	.	.	.	.	.	.	.	.	.	.	.
Platypus β	.	.	.	S	G	G	.	.	.	.	.	.	N	.	.	.	.	.	.	.	.	I	N	.	L	.	.	.	.	.	.	.	.	.	.	.	.	.	.	.	.	.	.	.	.	A	.	.	.	.
Chicken β	.	.	W	.	A	.	.	.	Q	L	I	.	G	.	.	.	.	.	.	.	.	.	A	.	C	.	A	.	.	.	A	.	.	.	I	.	.	.	.	.	.	.	.	.	A	.	.	.	N	.
Frog β	—	—	—	—	—	—	G	S	D	L	.	S	G	F	.	.	.	.	.	.	D	A	H	K	I	.	.	.	.	.	A	.	.	.	.	.	.	.	.	.	.	.	Y	.	T	T	.	.	N	.
Shark β	.	.	W	S	E	V	.	L	H	E	I	.	T	T	.	K	S	I	.	.	D	K	H	S	L	.	A	K	.	.	A	.	M	F	I	.	.	.	.	.	T	.	Y	.	G	N	L	K	E	F

[a] The helical regions A–H and interhelical regions NA, AB, CD . . ., etc., are numbered. The one letter code and alignment follows the *Atlas of Protein and Sequence and Structure* (21). The sequence of human α- and β-globins are given in full; dashes indicate deletions or insertions in longer chains. The extra residue in fish α-chains in the E helix has no helical number. Dots indicate identity with the human globin residue. The α chains of suckerfish, carp, and shark are acetylated. The extended NA region of shark Ac-Ser (Thr Ser)$_3$- is not shown here, but can be seen in Table 6. The sequences have been published (21,31,54,72,80,81).

(continued)

TABLE 2—*Continued*

Globin	CD 8	D 1	2	3	4	5	6	D 7	E 1	2	3	4	5	6	7	8	9	10	11	12		13	14	15	16	17	18	19	E 20	EF 1	2	3	4	—	5	6	7	EF 8	F 1	2	3	4	5	6	7	8	F 9	FG 1	2	3
Human α	S	H	—	—	—	—	—	G	S	A	Q	V	K	G	H	G	K	K	V	A	—	D	A	L	T	N	A	V	A	H	V	D	D	—	M	P	N	A	L	S	A	L	S	D	L	H	A	H	K	L
Rhesus monkey α	.	.	.	.	.	.	.	.	.	.	.	.	.	.	.	.	.	.	.	.	.	.	.	.	.	L	.	.	G	.	.	.	.	.	.	.	.	.	.	.	.	.	.	.	.	.	.	.	.	.
Loris α	.	.	.	.	.	.	.	.	.	.	.	.	.	A	.	.	.	.	.	.	.	.	.	.	.	.	.	.	S	.	.	.	.	.	.	.	S	.	.	.	.	.	.	.	.	.	.	.	.	.
Rabbit α	T	.	.	.	.	.	.	.	.	E	.	I	.	A	.	.	.	.	.	S	.	Q	.	.	.	K	.	.	G	.	L	.	.	.	L	.	G	.	.	.	T	.	.	.	.	.	.	.	.	.
Mouse α	.	.	.	.	.	.	.	.	.	.	.	.	.	.	.	.	.	.	.	.	.	.	.	.	A	.	.	G	.	.	L	.	.	.	L	.	G	.	.	.	.	.	.	.	.	.	.	.	.	.
Dog α	.	P	.	.	.	.	.	.	.	.	.	.	.	A	.	.	.	.	.	.	.	.	.	.	.	T	.	.	.	.	L	.	.	.	L	.	G	.	.	.	.	.	.	.	.	.	.	Y	.	.
Horse α	.	.	.	.	.	.	.	.	.	.	.	.	.	A	.	.	.	.	.	G	.	.	.	.	.	L	.	.	G	.	L	.	.	.	L	.	G	.	.	.	D	.	.	N	.	.	.	.	.	.
Ox α	.	.	.	.	.	.	.	.	.	.	.	.	.	.	.	.	A	.	.	.	.	A	.	.	.	K	.	.	E	.	L	.	.	.	L	.	G	.	.	.	E	.	.	.	.	.	.	.	.	.
Opossum α	.	A	.	.	.	.	.	.	.	.	.	I	.	T	Q	.	Q	.	I	.	.	.	.	V	G	L	.	.	.	.	L	.	.	.	.	.	T	.	.	.	S	.	.	.	.	.	.	.	E	.
Gray kangaroo α	.	.	.	.	.	.	.	.	.	.	.	I	Q	A	.	.	.	.	I	.	.	.	.	.	G	Q	.	.	E	.	I	.	.	.	L	.	G	T	.	.	K	.	.	.	.	.	.	.	.	.
Echidna 1 α	.	K	.	.	.	.	.	.	.	.	.	.	.	A	.	.	.	R	.	.	.	.	.	.	.	T	.	A	G	.	F	N	.	.	.	D	S	.	.	.	.	.	.	.	.	.	.	.	.	.
Echidna 2 α	.	K	.	.	.	.	.	.	.	.	H	.	R	A	.	.	.	.	.	.	.	.	.	.	.	T	.	.	G	.	F	N	.	.	.	D	G	.	.	.	D	.	.	.	.	.	.	.	.	.
Platypus α	.	.	.	.	.	.	.	.	.	.	.	I	.	A	.	.	.	.	.	.	.	.	.	.	S	T	.	A	G	.	F	.	.	.	.	D	S	.	.	.	.	.	.	.	.	.	.	.	.	.
Chicken α	.	.	.	.	.	.	.	.	.	.	.	I	.	.	.	.	.	.	.	.	.	L	.	I	.	.	.	I	E	.	A	.	.	.	I	S	G	.	.	.	K	.	.	.	.	.	.	.	.	.
Viper α	.	S	.	.	.	.	.	.	.	P	N	L	.	A	.	.	.	.	.	I	.	.	.	.	D	.	.	.	E	G	L	.	.	.	A	V	A	T	.	.	K	.	.	.	.	.	.	Q	.	.
Newt α	.	E	.	.	.	.	.	.	.	S	F	L	H	S	.	.	.	.	.	M	.	G	.	.	S	.	.	.	.	.	I	.	.	.	I	D	A	.	.	C	K	.	.	.	K	.	.	Q	D	.
Suckerfish α	.	P	.	.	.	.	.	.	.	G	P	.	.	—	.	.	.	.	.	I	M	G	.	I	G	D	.	.	T	K	F	.	.	.	L	L	G	G	.	A	S	.	.	E	.	.	.	S	.	.
Carp α	.	P	.	.	.	.	.	.	.	G	P	.	.	—	.	.	.	.	.	I	M	G	.	V	G	D	.	.	S	K	I	.	.	.	L	V	G	G	.	A	S	.	.	E	.	.	.	S	.	.
Shark α	T	A	.	.	.	.	.	A	A	P	S	I	.	A	.	.	A	.	.	V	.	T	.	.	A	K	.	C	D	.	L	.	.	.	L	K	T	H	.	H	K	.	A	T	F	.	G	S	E	.
Human β	S	T	P	D	A	V	M	G	N	P	K	V	K	A	H	G	K	K	V	L	—	G	A	F	S	D	G	L	A	H	L	D	N	—	L	K	G	T	F	A	T	L	S	E	L	H	C	D	K	L
Rhesus monkey β	.	S	.	.	.	.	.	.	.	.	.	.	.	.	.	.	.	.	.	.	.	.	.	.	.	.	.	.	N	.	.	.	.	.	.	.	.	.	.	.	Q	.	.	.	.	.	.	.	.	.
Spider monkey β	.	.	.	.	.	.	.	S	.	.	.	.	.	.	.	.	.	.	.	.	.	.	.	.	.	.	.	.	.	.	.	.	.	.	.	.	.	.	.	.	Q	.	.	.	.	.	.	.	.	.
Loris β	.	S	.	S	.	.	.	.	.	.	.	.	.	.	.	.	.	.	.	.	.	S	.	.	.	.	.	.	N	.	.	.	.	.	.	.	.	.	.	.	K	.	.	.	.	.	.	.	.	.
Rabbit β	.	S	A	N	.	.	.	N	.	.	.	.	.	.	.	.	.	.	.	.	.	A	.	.	.	E	.	.	S	.	.	.	.	.	.	.	.	.	.	.	K	.	.	.	.	.	.	.	.	.
Mouse β	.	S	A	S	.	I	.	.	.	A	.	.	.	.	.	.	.	.	.	I	.	T	.	.	.	.	.	.	N	.	.	.	.	.	.	.	.	.	.	.	S	.	.	.	.	.	.	.	.	.
Dog β	.	.	.	.	.	.	.	S	.	A	.	.	.	.	.	.	.	.	.	.	.	N	S	.	.	.	.	.	K	N	.	.	.	.	.	.	.	.	.	.	K	.	.	.	.	.	.	.	.	.
Horse β	.	N	.	G	.	.	.	.	.	.	.	.	.	.	.	.	.	.	.	.	.	H	S	.	G	E	.	V	H	.	.	.	.	.	.	.	.	.	.	.	A	.	.	.	.	.	.	.	.	.
Sheep β	.	N	A	.	.	.	.	N	.	.	.	.	.	.	.	.	.	.	.	.	.	D	S	.	.	.	.	M	K	.	.	.	D	.	.	.	.	.	.	.	Q	.	.	.	.	.	.	.	.	.
Ox β	.	.	A	.	.	.	.	D	.	.	.	.	.	.	.	.	.	.	.	.	.	D	S	.	.	.	.	M	K	.	.	.	D	.	.	.	.	.	.	.	A	.	.	.	.	.	.	.	.	.
Gray kangaroo β	.	N	A	K	.	.	.	A	.	.	.	.	L	.	.	.	A	.	.	.	.	V	.	.	G	.	A	I	K	N	.	.	.	.	.	.	.	.	.	.	K	.	.	.	.	.	.	.	.	.
Red kangaroo β	.	N	A	K	.	.	.	.	.	.	.	.	L	.	.	.	A	.	.	.	.	V	.	.	G	.	A	I	K	N	.	.	.	.	.	.	.	.	.	.	K	.	.	.	.	.	.	.	.	.
Potoroo β	.	S	A	K	.	.	L	.	.	A	.	.	L	.	.	.	A	.	.	.	.	V	S	.	G	.	A	I	K	N	.	.	.	.	.	.	.	.	.	.	K	.	.	.	.	.	.	.	.	.
Echidna β	.	S	A	.	.	.	.	.	.	A	.	.	.	.	.	.	A	.	.	.	.	T	S	.	G	.	A	.	K	N	.	.	.	.	.	.	.	.	.	.	K	.	.	.	.	.	.	.	.	.
Platypus β	.	S	A	G	.	.	.	.	.	.	.	.	.	.	.	.	A	.	.	.	.	T	S	.	G	.	A	.	K	N	.	.	D	.	.	.	.	.	.	.	K	.	.	.	.	.	.	.	.	.
Chicken β	.	S	.	T	.	I	L	.	.	.	M	.	R	.	.	.	.	.	.	.	.	T	S	.	G	.	A	V	K	N	.	.	.	.	I	.	N	.	.	S	Q	.	.	.	.	.	.	.	.	.
Frog β	G	S	A	.	.	I	C	H	.	A	.	.	L	.	.	.	E	.	.	.	.	A	.	I	G	E	.	.	K	.	P	E	.	.	.	.	A	H	Y	.	K	.	.	.	Y	.	S	N	.	.
Shark β	T	A	—	—	—	—	—	C	S	Y	G	.	.	E	.	A	.	.	.	T	.	.	.	L	G	V	A	V	T	.	.	G	D	.	V	.	S	Q	.	T	D	.	.	K	K	.	A	E	E	.

TABLE 2—*Continued*

Globin	4	FG 5	G 1	2	3	4	5	6	7	8	9	10	11	12	13	14	15	16	17	18	G 19	GH 1	2	3	4	GH 5	H 1	2	3	4	5	6	7	8	9	10	11	12	13	14	15	16	17	18	19	20	H 21	HC 1	2	3
Human α	R	V	D	P	V	N	F	K	L	L	S	H	C	L	L	V	T	L	A	A	H	L	P	A	E	F	T	P	A	V	H	A	S	L	D	K	F	L	A	S	V	S	T	V	L	T	S	K	Y	R
Rhesus monkey α	.	.	.	.	.	.	.	.	.	.	.	.	.	.	.	.	.	.	.	.	.	.	.	.	.	.	.	.	.	.	.	.	.	.	.	.	.	.	.	.	.	.	.	.	.	.	.	.	.	.
Loris α	.	.	.	.	.	.	.	.	.	.	.	.	.	.	.	.	.	.	.	C	.	H	.	.	D	.	.	.	.	.	.	.	.	.	.	.	.	.	.	.	.	.	.	.	.	.	.	.	.	.
Rabbit α	.	.	.	.	.	.	.	.	.	.	.	.	.	.	.	.	.	.	.	N	.	H	.	S	.	.	.	.	.	.	.	.	.	.	.	.	.	.	.	N	.	.	.	.	.	.	.	.	.	.
Mouse α	.	.	.	.	.	.	.	.	.	.	.	.	.	.	.	.	.	.	.	S	.	H	.	.	D	.	.	.	.	.	.	.	.	.	.	.	.	.	.	.	.	.	.	.	.	.	.	.	.	.
Dog α	.	.	.	.	.	.	.	.	.	.	.	.	.	.	.	.	.	.	.	C	.	H	.	T	.	.	.	.	.	.	.	.	.	.	.	.	.	F	.	A	.	.	.	.	.	.	.	.	.	.
Horse α	.	.	.	.	.	.	.	.	.	.	.	.	.	.	.	S	.	.	.	V	.	.	.	N	D	.	.	.	.	.	.	.	.	.	.	.	.	.	S	.	.	.	.	.	.	.	.	.	.	.
Ox α	.	.	.	.	.	.	.	.	.	.	.	.	S	.	.	.	.	.	.	S	.	.	.	S	D	.	.	.	.	.	.	.	.	.	.	.	.	.	.	N	.	.	.	.	.	.	.	.	.	.
Opossum α	K	.	.	.	.	.	.	.	F	.	C	.	N	V	.	.	.	M	.	.	.	.	G	K	D	.	.	.	E	I	.	.	.	M	.	.	.	.	.	.	.	.	.	.	.	.	.	.	.	.
Gray kangaroo α	.	.	.	.	.	.	.	.	.	.	.	.	.	.	.	.	.	F	.	.	.	.	G	D	A	.	.	.	E	.	.	.	.	.	.	.	.	.	.	A	.	.	.	.	.	.	.	.	.	.
Echidna 1 α	.	.	.	.	.	.	.	.	.	.	A	.	.	F	.	.	V	.	.	R	.	H	.	.	.	.	.	.	S	A	.	.	A	M	.	.	.	.	S	R	.	A	.	.	.	.	.	.	.	.
Echidna 2 α	.	.	.	.	.	.	.	.	.	.	A	.	.	F	.	.	V	.	.	R	.	H	.	E	.	.	.	.	S	A	.	.	A	M	.	.	.	.	S	R	.	A	.	.	.	.	.	.	.	.
Platypus α	.	.	.	.	.	.	.	.	.	.	A	.	.	I	.	.	V	.	.	R	.	C	.	G	.	.	.	.	S	A	.	.	A	M	.	.	.	.	S	K	.	A	.	.	.	.	.	.	.	.
Chicken α	.	.	.	.	.	.	.	.	.	.	G	Q	.	F	.	.	V	.	V	.	.	.	.	.	.	L	A	.	K	.	.	.	.	.	.	.	.	.	C	A	.	G	.	.	.	.	A	.	.	.
Viper α	.	.	.	.	A	.	.	.	I	.	.	Q	.	.	.	S	.	.	.	N	.	R	N	P	.	.	G	.	.	.	L	.	.	V	.	.	.	.	C	N	.	.	E	.	.	E	.	.	.	.
Newt α	M	.	.	.	A	.	.	P	K	.	A	.	N	I	.	.	V	M	G	I	.	.	K	.	H	L	.	Y	P	.	.	C	.	V	.	.	.	.	D	V	.	G	H	.	.	.	.	.	.	.
Suckerfish α	.	.	.	.	S	.	.	.	I	.	A	N	.	I	T	.	V	I	M	F	Y	.	.	G	D	.	P	.	E	.	.	.	.	V	.	.	.	F	Q	N	L	A	L	A	.	G	Q	.	.	.
Carp α	.	.	.	.	A	.	.	.	I	.	A	N	H	I	V	.	G	I	M	F	Y	.	.	G	D	.	P	.	E	.	.	M	.	V	.	.	.	F	Q	N	L	A	L	A	.	S	E	.	.	.
Shark α	K	.	.	.	A	.	.	Q	Y	.	.	Y	.	.	E	.	A	.	.	V	.	.	—	T	.	.	S	.	E	T	.	C	A	.	.	.	.	.	T	N	.	C	H	E	.	S	.	R	.	.
Human β	H	V	D	P	E	N	F	R	L	L	G	N	V	L	V	C	V	L	A	H	H	F	G	K	E	F	T	P	P	V	Q	A	A	Y	Q	K	V	V	A	G	V	A	N	A	L	A	H	K	Y	H
Rhesus monkey β	.	.	.	.	.	.	.	K	.	.	.	.	.	.	.	.	.	.	.	.	.	.	.	.	.	.	.	.	Q	.	.	.	.	.	.	.	.	.	.	.	.	.	.	.	.	.	.	.	.	.
Spider monkey β	.	.	.	.	.	.	.	.	.	.	.	.	.	.	.	.	.	.	.	.	.	.	.	.	.	.	.	.	Q	L	.	.	.	.	.	.	.	.	.	.	.	.	.	.	.	.	.	.	.	.
Loris β	.	.	.	.	.	.	.	.	.	.	.	.	.	.	.	V	.	.	.	.	.	.	.	.	D	.	.	.	Q	.	.	S	.	.	.	.	.	.	.	.	.	.	.	.	.	.	.	.	.	.
Rabbit β	.	.	.	.	.	.	.	.	.	.	.	.	.	.	.	I	.	.	S	.	.	.	.	.	.	.	.	.	Q	.	.	.	.	.	.	.	.	.	.	.	.	.	.	.	.	.	.	.	.	.
Mouse β	.	.	.	.	.	.	.	.	.	.	.	.	M	I	.	I	.	.	G	.	.	L	.	.	D	.	.	.	A	A	.	.	.	F	.	.	.	.	.	.	.	.	A	.	.	.	.	.	.	.
Dog β	.	.	.	.	.	.	.	K	.	.	.	.	.	.	.	.	.	.	.	.	.	.	.	.	.	.	.	.	Q	.	.	.	.	.	.	.	.	.	.	.	.	.	.	.	.	.	.	.	.	.
Horse β	.	.	.	.	.	.	.	.	.	.	.	.	.	.	A	V	.	.	.	R	.	.	.	.	D	.	.	.	E	L	.	.	S	.	.	.	.	.	.	.	.	.	.	.	.	.	.	.	.	.
Sheep β	.	.	.	.	.	.	.	.	.	.	.	.	.	.	.	V	.	.	.	R	.	H	.	N	.	.	.	.	V	L	.	.	D	F	.	.	.	.	.	.	.	.	.	.	.	.	.	.	.	.
Ox β	.	.	.	.	.	.	.	K	.	.	.	.	.	.	.	V	.	.	.	R	N	.	.	N	.	.	.	.	V	L	.	.	D	F	.	.	.	.	.	.	.	.	.	.	.	.	.	R	.	.
Gray kangaroo β	.	.	.	.	.	.	.	K	.	.	.	.	I	I	.	I	C	.	.	E	.	.	.	.	.	.	.	I	D	T	.	V	.	W	.	.	L	.	.	.	.	.	.	.	.	.	.	.	.	.
Red kangaroo β	.	.	.	.	.	.	.	K	.	.	.	.	I	I	.	I	C	.	.	E	.	.	.	.	.	.	.	I	D	T	.	V	.	W	.	.	L	.	.	.	.	.	.	.	.	.	.	.	.	.
Potoroo β	.	.	.	.	.	.	.	K	.	.	.	.	.	.	.	I	C	.	.	E	.	.	.	.	D	.	.	I	D	A	.	V	.	W	.	.	.	.	.	.	.	.	.	.	.	.	.	.	.	.
Echidna β	.	.	.	.	.	.	.	N	R	.	.	.	.	.	.	V	.	.	.	R	.	.	S	.	.	.	.	.	E	A	.	.	.	W	.	.	L	.	S	.	.	S	H	.	.	.	.	.	.	.
Platypus β	.	.	.	.	.	.	.	N	R	.	.	.	.	.	I	V	.	.	.	R	.	.	S	.	D	.	S	.	E	.	.	.	.	W	.	.	L	.	S	.	.	.	H	.	.	G	.	.	.	.
Chicken β	.	.	.	.	.	.	.	.	.	.	.	D	I	.	I	I	.	.	.	A	.	.	S	.	D	.	.	.	E	C	.	.	.	W	.	.	L	.	R	V	.	.	H	.	.	.	R	.	.	.
Frog β	.	.	.	.	A	.	.	.	.	.	.	.	.	F	I	T	.	.	.	R	.	.	Q	H	.	.	.	.	E	L	.	H	.	L	E	A	N	F	C	A	.	G	D	.	.	.	K	A	.	.
Shark β	.	.	.	V	.	S	.	K	.	.	A	K	C	F	.	V	E	.	G	I	L	L	K	D	K	.	A	.	Q	T			I	W	E	.	Y	F	G	V	.	V	D	.	I	S	K	E	.	.

quaternary structure or T structure, and the oxy or R structure corresponding to a higher affinity for oxygen. On transition between oxy and deoxy states the dimers click over from one position to the other. Stabilization of the R and T structures (50) is aided by alternative sets of hydrogen bonds formed between amino acid side chains attached to the opposing faces of the dimers (Fig. 3).

Abnormal human hemoglobins that involve change of any of the residues involved in these hydrogen bonds, G1 Asp α to G4 Asn β in oxy or C7 Tyr α to G1 Asp β in deoxy, lead to preference for the T or R structures, respectively, which retains its hydrogen bonding. This results in a corresponding low or high affinity for oxygen in the abnormal hemoglobin.

There are additional bonds present in the deoxy or T structure that contribute to its stabilization and that are broken in the transition to the R state. Salt linkages involving the amino- and carboxyl terminal residues of the α chain including the basic side chain of the C-terminal arginine, as well as the C-terminal histidine residue of the β chain including its basic side chain, are shown diagrammatically in Fig. 4. Chloride ions and carbamate -NH-COO$^-$ groups formed by reaction of carbon dioxide with α-amino groups position

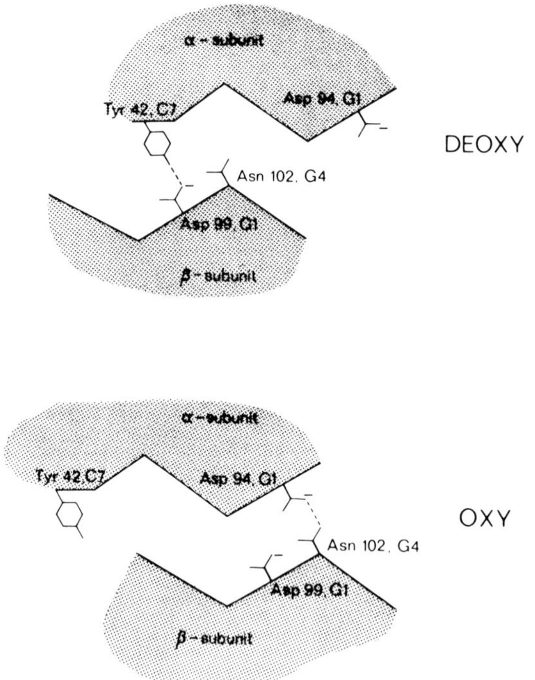

Fig. 3. Change in $\alpha_1\beta_2$ contact on oxygenation. The contact clicks from one quaternary conformation to another involving the switch of hydrogen bonding shown. Recent X-ray studies at higher resolution (M.F. Perutz, private communication) have revealed several other hydrogen bonds between the α_1 and β_2 subunits in the two alternative structures. In some of these, water molecules act as bridges between the subunits. [From Morimoto *et al.* (50). By permission.]

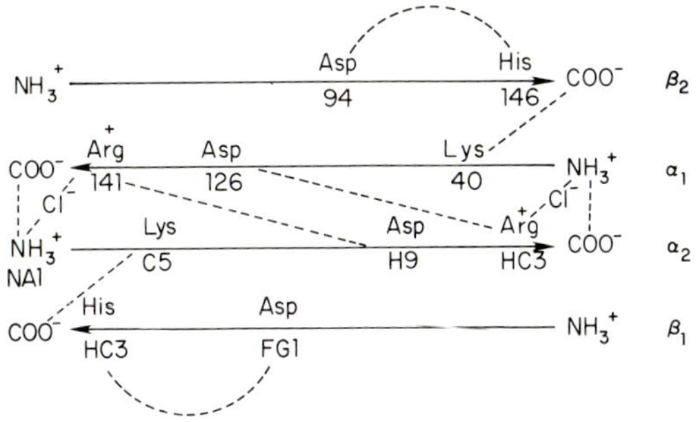

Fig. 4. Diagrammatic representation of the salt links in deoxyhemoglobin broken on oxygenation. The helical positions are indicated as well as the number of the residue in human α- and β-globins.

themselves between the positively charged guanidinium group of the C-terminal arginine and the α amino or carbamylated amino group, helping to stabilize the T structure and providing an explanation for the greater binding of CO_2 in the deoxy structure than in the oxy structure (42).

The breaking of the salt linkages in the transition from the deoxy to the oxy state leads to changes in the environment of the basic groups involved. This results in a lowering of their pK_a values and the release of protons that are involved in the conversion of HCO_3^- to CO_2 in the lungs. The reverse changes occur in the tissue where CO_2, HCO_3^-, and H^+ are being formed. For those hemoglobins (Hbs) that do not change their oxygen affinity with pH (i.e., no Bohr effect) there are changes in the residues involved in these salt linkages, such as acetylation of α amino groups in α chains or a C-terminal residue different from histidine in β chains.

Metabolic products with multiple phosphate groups, such as 2,3-diphosphoglycerate in mammals and inositol or purine phosphate esters in birds and fish, bind preferentially to the deoxy-conformation of tetrameric hemoglobin stabilizing the T state and lowering the oxygen affinity (8,64). The binding occurs in the central cavity of the molecule (4), which is larger in the deoxy-conformation. Transition to the oxy-conformation squeezes out the phosphate compound. For DPG there is binding of the negatively charged groups to the α- amino group of Val NA1, imidazole groups of His NA2 and His H21 of both β chains, and the ϵ amino group of Lys EF6 of one β chain. Inositol hexaphosphate combines with deoxyhemoglobin at the same site (5). Replacement of these residues by amino acids with different side chains will reduce the strength of binding of DPG and raise the oxygen affinity, as seen in human fetal hemoglobin where the γ chain residue H21 is serine.

Studies on Marsupial Globins

The marsupial globins sequenced and the number of differences from the corresponding human globins are shown in Table 3. These globins are characterized by normal chain length, that is 153, 141, and 146 residues for myo-

TABLE 3
Globin Chains Sequenced and Number of Differences from Human Globin

Animal	Myo	α	β	References
Marsupials				
Gray kangaroo *Macropus giganteus*		27	38	1,7
Red kangaroo *Megaleia rufa*	20		37	2,74
Potoroo *Potorous tridactylus*			42	74
Wallaroo *Macropus robustus*			37	74
Monotremes				
Echidna *Tachyglossus aculeatus* HbIB		37	31	82,83
Echidna *Tachyglossus aculeatus* HbIA		36	31	26
Echidna *Tachyglossus aculeatus* HbIIA		41	31	75
Platypus *Ornithorhyncus anatinus*	25	39	34	29,80,81
Cartilaginous fish				
Shark *Heterodontus portusjacksoni*	83[a]	82[a]	92[a]	30,31,54
Mollusk				
Bivalve *Anadara trapezia* HbI		109[a]		19

[a] Additions or deletions counted as 1 change.

globin, α, and β chains, respectively, and with few exceptions all the important contact sites discussed earlier are the same as in human globins. The exceptions are in the heme contacts: G4 in myoglobin, where Phe is present compared with the usual Tyr; in heme contact E11 in α-globin, where Ile is present compared with the usual Val. In addition there is one change in the important $\alpha_1\beta_2$ contact with C5β Ser, where Gln is the usual residue. These changes are all conservative.

The kangaroo is the largest marsupial with two closely related species, the gray and red kangaroo. There was no difference observed in α chains from these animals, but one consistent difference, β56 Ala or Gly, respectively, in the β chains. An additional site β2 varied between histidine, the usual residue, and glutamine in the gray kangaroo (76). This polymorphism has not been observed in a survey of 88 red kangaroos but was found to occur also in the wallaroo *Macropus robustus erubescens* with a gene frequency for Hbβ^{2His} from 0.1 to 0.4, depending on the area investigated (62,74). The β-globin of this animal had β56 Gly like the red kangaroo.

The selection of β2 Gln into two different species is probably related to the different oxygen affinities in the two polymorphic hemoglobins, since β2 His is a DPG-binding side chain. When stripped of DPG, the two hemoglobins of the gray kangaroo had the same oxygen affinity (26). With β2 Gln the effect of DPG will be less, resulting in an increased oxygen affinity. This is reflected in a lower P_{50} for this $\alpha_2\beta_2{}^{2Gln}$, 24.8 mm compared with 26.1 mm for $\alpha_2\beta_2{}^{2His}$ in blood at 36°C. The DPG level was also lower for the former variant compared with the latter or with the hemoglobin of the red kangaroo (12). It would be expected that there could be a biological advantage to each of these hemoglobins in different circumstances because of their different affinities for oxygen.

As yet we do not know in what situation one or the other hemoglobin would be advantageous. This polymorphism, with a reasonable gene frequency, represents a rare instance of functional differences in animal hemoglobins caused by a single amino acid substitution, and in this respect compares with some abnormal human hemoglobins.

Only a limited study of the hemoglobins of other macropods was possible. The results suggested only small differences in amino acid sequence consistent with a relatively recent adaptive radiation. The potoroo or rat kangaroo is considered to have diverged well before this radiation and showed the most differences, with multiple α and β chains. The most common β chain was purified and sequenced (74). It showed 15 changes from that of the gray kangaroo, suggesting a divergence of the potoroine and macropodine lines about 50 million years ago.

Studies on Monotreme Globins

The monotreme globins sequenced and the number of differences from the corresponding human globins are shown in Table 3. In addition, the sequence of echidna myoglobin has been reported (17). It shows nine differences from platypus myoglobin and 25 differences from human myoglobin. All the major globin chains of both echidna and platypus have now been sequenced, and all have the normal chain lengths.

Important contact sites have been preserved, with the exception of several heme contacts, namely:

Myoglobin	G4	Tyr → Phe	(also in marsupials)
	E10	Thr → Val	(echidna and horse only)
α-globin	CD4	Phe → Met	(echidna only)
	E10	Lys → Arg	(echidna major haemoglobin only)

The important $\alpha_1\beta_2$ contact sites are identical in platypus and echidna, both showing a single change at αCD2, a deoxy contact site, where Ser is present instead of the usual Pro.

Other sites are identical to those in human globins for both animals, which is perhaps surprising considering the different life-styles of the two animals and the measured differences in their oxygen saturation profiles (37). Echidna hemoglobin IB is 50% saturated at 12 mm at 25° pH 7.1 compared with a P_{50} of 18 mm for platypus. Contact sites associated with the Bohr effect were identical, except that the side chain of NA1 whose α-amino group (Fig. 4) contributes 20% of the effect (42) was Met in platypus and Val in echidna hemoglobin, yet the platypus hemoglobin is more sensitive to H^+ and should show a greater response to CO_2 produced by respiration. Both monotremes have the usual contact site residues involved in DPG interaction, but experimentally the removal of DPG caused a more marked increase in the affinity of platypus hemoglobin, with P_{50} falling to 4.5 mm, compared to echidna hemoglobin, where P_{50} fell to 3.3 mm. The binding of DPG in the red cell will therefore reduce the oxygen affinity of platypus hemoglobin more than echidna hemoglobin (37). No measured levels of DPG in the blood of these monotremes are at present available.

With 16 and 15 differences between the α- and β-globins, respectively, it is not possible to assign these differences in oxygenation characteristics of echidna and platypus hemoglobins to particular residues at present.

Both monotremes possess major (HbI) and minor (HbII) hemoglobins, the proportion of major hemoglobin being approximately 75% of the total. The echidna hemoglobin has been examined in most detail and genetic variation has been found in the electrophoretic mobility of the HbI and HbII (20). Polymorphism of HbI occurs in echidnas of the mainland of Australia with three common variants A, B, and C. HbII polymorphism with A and B variants occurs in Tasmania. The amino acid sequences of two of the major hemoglobins, HbIA and HbIB, have been determined, as well as one of the minor hemoglobins, HbIIA. In all these the variations in sequence were confined to the α chain. There are four amino acid differences between α-IA and α-IB chains, including one negatively charged glutamic acid for alanine substitution. The same variation occurs in α-IIA as well as an additional aspartic acid for alanine substitution. The changes account for the electrophoretic mobility differences.

The oxygen binding parameters of these major and minor hemoglobins have been measured (38). The minor hemoglobin has a markedly higher affinity for oxygen and slight differences were detected in the properties of HbIA and HbIB that could be interpreted in terms of selection for different temperature ranges in the geographical distribution of the animals. None of the residues that are different in HbIA, HbIB, or HbIIA α chains are contact amino acids involved in the $\alpha_1\beta_2$ interactions that could be expected to influence oxygen affinity.

Rates of Evolution

Calculation of the rates of evolution are made from the observed amino acid differences, adjusted to take account of events such as multiple mutations at particular sites and back-mutations (21), together with the estimates of the dates of divergence for different species taken from the palaeontological record (65,87). If an approximately constant rate is determined, then the timing of phylogenetic events and dates of divergence could utilize the "molecular clocks" provided by different protein molecules.

From the sequences of different globin chains from the monotremes we have calculated average rates of protein evolution and calculated dates of divergence for each globin. Agreement between the average rates of change would not be expected, since it can be argued (34,39) that genes for α-globin must change more slowly than those for β chains. This conclusion is based on the requirement that the α-globin combine with several different types of non-α-globins synthesized during the life of the animal. The echidna with a common β chain and variant α chains is an exception to this rule. It might be expected, however, that dates of divergence calculated for the separate globin chains would be similar if there was any constancy in the rate of protein evolution for a particular protein.

A computer program was used to compare amino acid sequences over 156 positions using the alignments with gaps as listed in Dayhoff (21). The palaentological estimates of dates of divergence used (84) were taken from the literature (65,66).

The known sequences of each globin chain were compared and the number of differences determined. A 10% difference in sequence cutoff was applied so that closely related groups that have many extant members or have been particularly popular with sequencers, such as the primate group, did not unduly influence the calculation of the mean rate.

The amino acid differences were converted to the number per 100 residues compared and the formula of Dayhoff (21) was applied to allow for possible multiple mutations. This gives PAM units (accepted point mutations) per 100 residues compared.

Protein evolutionary rates per arm (r) were determined from comparisons that did not involve the monotremes whose date of divergence was being calculated, using the formula $r = \text{PAM}/2t$, where t is the relevant palaeontological estimate of divergence date. Mean rates of protein mutation in PAM units per 100 residues per arm per million years were obtained by averaging the r values for all possible comparisons.

Dates of divergence of monotremes from other mammals were computed from the mean r values and the number of amino acid differences between the globin of the monotreme and those of other animals.

This method differed in several particulars from the earlier method used (3) with marsupial globins, when only a limited number of globin sequences were available (5 Mb, 8α, 6β chains), which appeared to indicate a statistically constant rate for each globin, with reasonable agreement in the marsupial-placental divergence dates.

The mean r values obtained in the monotreme study (84), with 95% confidence intervals for the means, were myoglobin, 0.108 ± 0.011 ($n = 92$); α-globin, 0.115 ± 0.007 ($n = 92$); β-globin, 0.133 ± 0.012 ($n = 60$).

The narrowness of the confidence intervals is suggestive of constancy of evolutionary rate. However, when monotreme divergence date calculations were carried out using these rates and comparisons involving the monotreme sequences, three varying estimates of the date of divergence from other mammals were obtained (84): myoglobin, 90 ± 16 million years ($n = 26$); α chain, 186 ± 38 million years ($n = 24$); β chain, 114 ± 15 million years ($n = 16$). Limits are $\pm$ 1 SD. Clearly at least one of the chains in both monotremes has mutated at a nonaverage rate.

Widely varying estimates of the echidna-platypus divergence date—namely, 28, 52, and 38 million years ago—were similarly calculated using myoglobin, α, and β chains. Monotreme-marsupial date of divergence estimates were 72, 234, and 122 million years ago for the myoglobin, α, and β chains, respectively. Clearly, these values are not sufficiently consistent to help resolve the question of the origin of monotremes relative to marsupials or placental mammals (70). There is insufficient constancy in the mutational rate to allow accurate divergence date estimates to be made directly from amino acid sequence data.

Since amino acid residues are the expression of a nucleic acid codon, it would be preferable to use the sequence of nucleotides in messenger RNA. Since this information has not been available some authors have used minimum base changes (32) and estimated base changes ["random evolutionary hits" (40)] instead of amino acid differences. Our results were not improved by consideration of minimum nucleotide changes required to give the observed sequence differences.

Rate variability between chains of different globin type is evident in comparison with species for which all three globin chain sequences are known (Table 4). This comparison, which is independent of paleontologically derived data, indicates variability in rates up to 2.6 times for α chains and 2.2 times for β chains relative to myoglobin. It seems likely (35) that globin genes have evolved at markedly nonconstant rates at different times, with more rapid changes occurring, after duplication of a gene, in the gene product that acquires a new function. Rates probably decelerate after functional opportunities have been realized. For example, the number of differences between α chains of platypus and echidna are indicative of four to five times the rate of change in the IIA gene product (Fig. 5).

Studies on Shark Globins

Since life arose in the sea it is reasonable to expect that studies of heme proteins in primitive marine animals could add to our knowledge of their evolution. Agnathans or jawless fish, such as the lamprey, are the oldest

TABLE 4
Variability of Mutation Rate between Chains[a]

	Myo	α	β	α/myo	β/myo
Horse	18	18	25	1.0	1.4
Ox	29	17	25	0.6	0.9
Kangaroo	20	27	38	1.4	1.9
Echidna	25	37	31	1.5	1.2
Platypus	25	39	34	1.6	1.4
Chicken	35	35	45	1.0	1.3

[a] Comparison of the amino acid differences for animals in which all three globin chains have been sequenced (comparison against human globins).

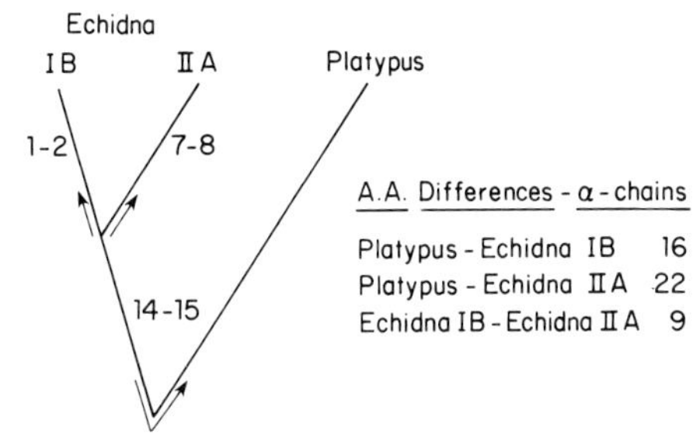

Fig. 5. Mutational rate variation between the α chains of the major and minor hemoglobin of echidna. Numbers are assigned to each arm to balance the observed number of amino acid sequence differences.

living vertebrates. They produce a monomeric hemoglobin that aggregates to dimers on deoxygenation (15). The amino acid sequence is known for two species (13,47).

Elasmobranchs, or cartilaginous fish, produce tetrameric hemoglobins and are generally considered to be a primitive class of animal. An example of this class is the shark *Heterodontus portusjacksoni.* The major hemoglobin and its α and β chains have been purified (53). These chains and the myoglobin from muscle have been sequenced to give the number of differences shown in Table 3.

There are many more changes in each globin chain than have hitherto been found for a vertebrate, with the exception of the lamprey. As in the lamprey there is an extended N-terminal NA region of seven residues in the α globin, and the α amino group is acetylated. With 148 residues this makes the chain the longest vertebrate α-globin (usually 141 residues), and there are four cysteinyl residues, the highest number in a vertebrate globin.

The β-globin, with 141 instead of the usual 146 residues, lacks five residues in the region responsible for the D-helix normally present in myoglobin and β chains but absent in α chains. The myoglobin is also shorter than normal, with 148 instead of 153 residues. Four of these residues are missing from the N-terminal region, which has an acetylated amino terminus corresponding to residue A2.

The C-terminal residues for the three globin chains are normal.

Many more changes are found in contact site residues associated with functional aspects of these globins. Heme contact changes are the following: (1) for myoglobin—F7 Thr instead of Ser, also in chicken; G4 Asn instead of Tyr, or Phe in marsupials and monotremes; G5 Phe instead of Leu, invariant; (2) for α-globin—CD3 Asp instead of His; CD4 Tyr instead of Phe, also in opossum; F7 Phe instead of Leu; (3) for β-globin—C7 Tyr instead of Phe; F7 Lys instead of Leu; G4 Ser instead of Asn, invariant; H19 Ile instead of Leu, invariant.

Although many of these changes are conservative and should preserve essential atomic contacts, the changes at G4 in myoglobin and β chain are interesting, since the replacement of Tyr by Asn in the former is rather radical because of the difference in size, and the change from Asn to Ser in the β chain is usually associated with reduced oxygen affinity (51). Residue F7 shows differences in each globin with the β-globin Lys being radically different in chemical nature to the usual hydrophobic residue Leu, although it is not an internal residue (60). Residue G5 has been invariantly Leu in myoglobins and is reported to play an important role in maintaining the angle of the heme group compared with the effect of Phe in hemoglobin α and β chains. Replacement of the Phe in β chain by Leu in the abnormal human hemoglobin Heathrow results in an abnormally high oxygen affinity (67).

Changes in the $\alpha_1\beta_2$ contact site are considerable and are illustrated in Fig. 6. Among so many changes the change at G4 is one that has a known effect in the abnormal human hemoglobin Beth Israel (51), where the same replacement of Asn by Ser occurs. As discussed earlier, this residue is involved in an important hydrogen bond that stabilizes the oxy conformation. In its absence the abnormal hemoglobins Beth Israel and Kansas, where Asn is replaced by Thr (10), have a very low oxygen affinity. Comparatively little is known about the oxygen binding characteristic of *H. portusjacksoni* hemoglobin, but the P_{50} and Hill number are not markedly different from those of other elasmobranch hemoglobins (52).

There is no change in oxygen affinity with pH, and of the residues known to be involved in the Bohr effect the free α amino group, NA1, is absent, because of the acetylated extended section mentioned previously. The

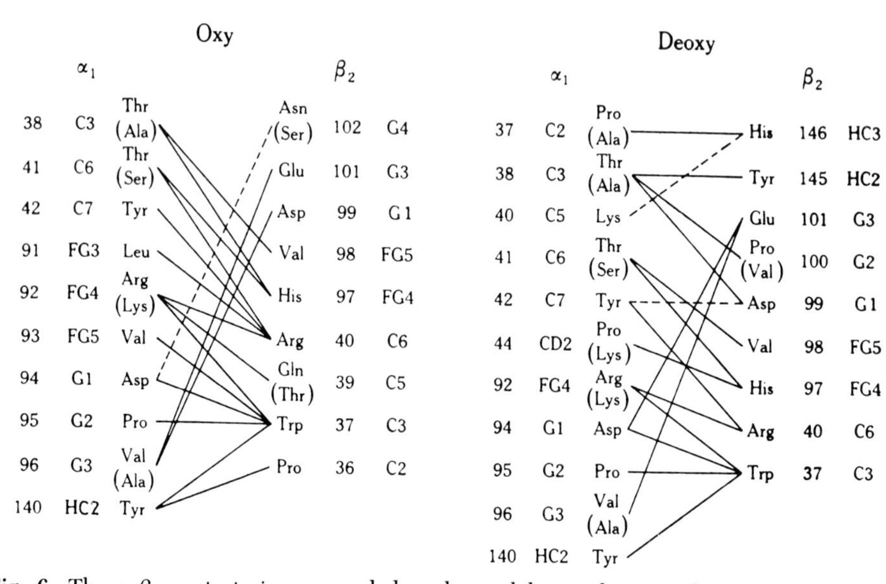

Fig. 6. The $\alpha_1\beta_2$ contacts in oxy- and deoxyhemoglobin in horse and shark hemoglobins. Residues in horse hemoglobin (9) are shown; any different residues in shark chains are shown in parentheses. Reproduced with permission (31).

C-terminal histidine on the β chain is present, but its partner in the usual salt link to FG1 is different, being Glu instead of Asp, which, although negatively charged, is of different dimensions and may not have the same effect on the pK_a value. The other group involved in the Bohr effect in human hemoglobin have not been clarified as yet (42). The binding sites for organic phosphate esters are present in this hemoglobin, with the difference that H21 is Lys rather than Arg.

The myoglobin of this shark shows many deficiencies in the inter- and intrasegmental hydrogen bonds and salt links that stabilize the myoglobin conformation in later animals (67,78). Uncharged side chains not capable of forming salt links are present in B8, G6, and G19, and residue A2 normally involved in such a link is missing. Similarly, many of the hydrogen-bonding-site residues have hydrophobic side chains, or side chains of different dimensions or charge to those normally present. This is probably best interpreted as indicating evolutionary changes have occurred to produce finer tuning and an increased stability of the myoglobin molecule, a view that would support selection processes (35).

Certain animals said to be "living fossils" retain morphologic characteristics over long periods of evolutionary time, and it is of interest to see if their biochemistry as reflected in the structure of protein molecules has also been maintained. As already mentioned, there are many changes in the sequences of the globin chains of *H. portusjacksoni* compared with human globins. Among those changes are many examples in myoglobin of residues that are unlike residues in those positions in most myoglobins but are the same as residues occurring in those same helical or interhelical positions in α and β chains. These positions are listed in Table 5. This strong conservation may have lasted since the duplication of the gene leading to myglobin and the separate precursor of present-day α- and β-globins, an event placed approximately 1000 million years ago (21) and have been retained in *H. portusjacksoni* since its divergence.

From the number of amino acid differences in the three globin chains investigated, we used the average rates of mutation listed earlier to calculate the time since divergence for this shark as follows: α chain, 465 ± 40 ($n = 15$) million years; β chain, 478 ± 19 ($n = 12$) million years; myoglobin, 450 ± 16 ($n = 15$) million years. These results show much more agreement than was obtained with monotreme globins, and probably reflect the long period during which these globins have evolved, with a consequent leveling in the number of mutational events despite differences in rate during some periods of time.

In any discussion of the reliability of protein clocks it must be borne in mind that there are perhaps 10^6 proteins in a mammal in a large number of superfamilies (6), and it would be expected that conclusions would be safer if derived from a large number of different proteins. Until the body of knowledge is available, Williams (85) is probably correct when he says, "unfortu-

TABLE 5
SHARK MYOGLOBIN RESIDUES TYPICAL OF α- OR β-GLOBINS[a]

Residue	Most animals	Shark
G2	Val/Ile	Pro as in all α's and β's (except Val Shark β)
G4	Tyr/Phe	Asn as in all α's and β's (except Ser Shark)
G5	Leu	Phe as in all α's and β's
H1	Glu	Thr as in 14/19 α's and 16/18 β's
C5	Leu	Lys as in 18 of 19 α's
CD2	Asp	Pro as in 13 of 19 α's
F6	Gln	Asp as in 15 of 19 α's
H7	Ala	Ser as in 15 of 19 α's
B5	His	Val as in 10 of 18 β's
E1	Ser	Asn as in 17 of 18 β's
G10	Glu/Asp	Asn as in 16 of 18 β's
H11	Ala	Val as in 11 of 18 β's

[a] The myoglobin sequences are shown in Table 1, and the α- and β-globin sequences are in Table 2.

nately therefore it must be concluded that proteins cannot be regarded as reliable clocks for timing past evolutionary events."

STUDIES ON MOLLUSK GLOBINS

The amino acid sequences of myoglobins from two gastropod mollusks, *Aplysia limacina* (73) and *Busycon canaliculatum* (11), have been reported. *Anadara trapezia*, which we have investigated, is a bivalve mollusk, one of four species of the genus *Anadara* found in Australian waters. *A. trapezia* has nucleated red cells with a major hemoglobin HbI that is tetrameric of the $\alpha_2\beta_2$ form and a minor component that is dimeric and polymorphic (55). Two different heme proteins have also been reported for *Anadara satowi* (56) and *A. broughtonii* (33). The polymorphic forms, HbIIa and HbIIb, of *A. trapezia* are composed of identical subunits very similar in amino acid composition. From the oxygen equilibrium curves (24) it has been shown that animals possessing HbIIb have an advantage over the HbIIa morph at temperatures between 15 and 25°C.

The amino acid sequence of one of the globin chains of the tetrameric hemoglobin has been determined (19). It is 153 residues in length, with additional residues at the amino terminus and an acetylated α amino group. In common with other invertebrate globin sequences, when compared with vertebrate globins it is necessary to insert extra gaps in the vertebrate alignment to preserve homology.

In Table 6 additional gaps in interhelical regions—4 in AB, 5 in EF, and 2 in FG—have been included to accommodate the longer invertebrate sequences. The alignment of Dayhoff (21) has been followed as far as possible, and Table 6 shows the sequences of human, shark, and lamprey globins for comparison.

The only common residues in these sequences are the F8 His and CD1

TABLE 6

Amino Acid Sequences of Seven Invertebrate Globins Aligned with Myoglobin, α and β Chains from Human and Shark; Lamprey Globin; and Leghemoglobin A from Kidney and Soy bean[a]

				NA	A															A	AB				AB	B									
Globin		1	2	3	1	2	3	4	5	6	7	8	9	10	11	12	13	14	15	16	1	2	3	4	5	1	2	3	4	5	6	7	8	9	10
Human Mb		G	—	L	S	D	G	E	W	Q	L	V	L	N	V	W	G	K	V	E	A	—	—	—	—	D	I	P	G	H	G	Q	E	V	L
Human α		V	—	L	S	P	A	D	K	T	N	V	K	A	A	W	G	K	V	G	A	—	—	—	—	H	A	G	E	Y	G	A	E	A	L
Human β		V	H	L	T	P	E	E	K	S	A	V	T	A	L	W	G	K	V	—	—	—	—	—	—	N	V	D	E	V	G	G	E	A	L
Shark Mb		—	—	—	—	Ac-	T	E	W	E	H	V	N	K	V	W	A	V	V	E	P	—	—	—	—	D	I	P	A	V	G	L	A	I	L
Shark α	Ac-S T S T S T S	D	—	Y	S	A	A	D	R	A	E	L	A	A	L	S	K	V	L	A	Q	—	—	—	—	N	A	E	A	F	G	A	E	A	L
Shark β		V	H	W	S	E	V	E	L	H	E	I	T	T	T	W	K	S	I	—	—	—	—	—	—	D	K	H	S	L	G	A	K	A	L
Lamprey	F- P I V D S G S V	A	P	L	S	A	A	E	K	T	K	I	R	S	A	W	A	P	V	Y	S	—	—	—	—	N	Y	E	T	S	G	V	D	I	L
Aplysia Mb	Ac-	S	—	L	S	A	A	E	A	D	L	A	G	K	S	W	A	P	V	F	A	—	—	—	—	N	K	N	A	N	G	A	D	F	L
Busycon Mb		G	—	L	D	G	A	Q	K	T	A	L	K	E	S	W	—	K	V	L	G	A	D	G	P	T	M	M	K	N	G	S	L	L	F
Anadara I-α	Ac-V A D A V A K	V	C	G	S	E	A	I	K	G	N	L	R	R	S	W	G	V	L	—	—	—	M	S	A	D	I	E	A	T	G	L	T	Y	L
Chironomus IIβ		A	P	L	S	A	D	E	A	S	L	V	R	G	S	W	A	N	V	K	—	—	—	—	—	—	—	—	H	S	E	V	D	I	L
Chironomus III		—	—	L	S	A	D	Q	I	S	X	V	Z	A	S	F	D	K	V	K	G	—	—	—	—	—	—	D	P	V	G	—	—	I	L
Chironomus VIIβ		S	P	L	T	A	D	E	A	S	L	V	Q	S	S	W	K	A	V	S	—	—	—	—	—	—	—	—	H	N	E	V	D	I	L
Glycera Hb		G	—	L	S	A	A	Q	R	Q	V	I	A	A	T	W	K	D	I	A	G	N	D	—	—	N	G	A	G	V	G	K	D	C	L
Leg Hb K. bean		G	A	F	T	E	K	Q	E	A	L	V	N	S	S	W	E	A	F	K	G	—	—	—	—	N	I	P	Q	Y	S	V	V	F	Y
Leg Hb S. bean		V	A	F	T	E	K	Q	D	A	L	V	S	S	S	F	E	A	F	K	A	—	—	—	—	N	I	P	Q	Y	S	V	V	F	Y

[a] The alignment and sequences are from the *Atlas of Protein Sequence and Structure* (21), but to allow for deletions and insertions in the invertebrate chains additional gaps have been necessary in the interhelical regions, 4 in AB, 5 in EF and 2 in FG. Ac- and F- represent an acetyl and formyl group respectively. The sequences have been published (11,21,30,31,44,54,71).

(continued)

TABLE 6—*Continued*

						B	C						C	CD							CD	D						D		E															
Globin	11	12	13	14	15	16	1	2	3	4	5	6	7	1	2	3	4	5	6	7	8	1	2	3	4	5	6	7	1	2	3	4	5	6	7	8	9	10	11	12		13	14	15	16
Human Mb	I	R	L	F	K	G	H	P	E	T	L	E	K	F	D	K	F	K	H	L	K	S	E	D	E	M	K	A	S	E	D	L	K	K	H	G	A	T	V	L	—	T	A	L	G
Human *α*	E	R	M	F	L	S	F	P	T	T	K	T	Y	F	P	H	F	—	D	L	S	H	—	—	—	—	—	G	S	A	Q	V	K	G	H	G	K	K	V	A	—	D	A	L	T
Human *β*	G	R	L	L	V	V	Y	P	W	T	Q	R	F	F	E	S	F	G	D	L	S	T	P	D	A	V	M	G	N	P	K	V	K	A	H	G	K	K	V	L	—	G	A	F	S
Shark Mb	L	R	L	F	K	E	H	K	E	T	K	D	L	F	P	K	F	K	E	I	—	P	V	Q	Q	L	G	N	N	E	D	L	R	K	H	G	V	T	V	L	—	R	A	L	G
Shark *α*	A	R	M	F	T	V	Y	A	A	T	K	S	Y	F	K	D	Y	K	D	F	T	A	—	—	—	—	—	A	A	P	S	I	K	A	H	G	A	K	V	V	—	T	A	L	A
Shark *β*	A	R	M	F	I	V	Y	P	W	T	T	R	Y	F	G	N	L	K	E	F	T	A	—	—	—	—	—	C	S	Y	H	—	K	E	H	A	K	K	V	T	—	G	A	L	G
Lamprey	V	K	F	F	T	S	T	P	A	A	Q	E	F	F	P	K	F	K	G	M	T	S	A	D	E	L	K	K	S	A	D	V	R	W	H	A	E	R	I	I	—	N	A	V	N
Aplysia Mb	V	A	L	F	E	K	F	P	D	S	A	N	F	F	A	D	F	K	G	—	K	S	V	A	D	I	K	A	S	P	K	L	R	D	V	S	S	R	I	F	—	T	R	L	N
Busycon Mb	G	L	L	F	K	T	Y	P	D	T	K	K	H	F	K	H	F	D	D	A	T	—	F	A	A	M	D	T	T	G	V	G	K	A	H	G	V	A	V	F	—	S	G	L	G
Anadara I-*α*	A	N	L	F	T	L	R	P	D	T	K	T	Y	F	T	R	L	G	D	V	Q	K	G	—	—	—	K	A	N	S	K	L	R	G	H	A	I	T	L	T	—	Y	A	L	D
Chironomus II*β*	Y	Y	I	F	K	A	N	P	D	I	M	A	K	F	P	N	F	A	G	—	K	D	L	E	T	L	K	G	T	G	Q	F	A	T	H	G	G	R	I	V	—	G	F	V	S
Chironomus III	Y	A	V	F	K	A	D	P	S	I	M	A	K	F	T	Q	F	A	G	—	K	D	L	E	S	I	K	G	T	A	P	F	H	T	E	A	N	R	I	V	—	G	F	F	S
Chironomus VII*β*	A	A	V	F	A	A	Y	P	D	I	Q M	A	K	F	P S	Q	F	A	G	—	K	D	L	A	S	I	K	D	T	G	A	F	A	T	H	A	T	R	I	V	—	S	F	L	S
Glycera Hb	I	K	H	L	S	A	H	P	Q	M	A	A	V	F	G	—	F	S	G	—	A	S	D	P	A	V	—	—	—	A	D	L	—	—	—	G	A	K	V	L	—	A	Z	I	G
Leg Hb K. bean	T	S	I	L	E	K	A	P	A	A	K	N	L	F	—	S	F	—	—	L	A	N	G	V	D	—	P	T	N	P	K	L	T	A	H	A	E	S	L`	F	G	L	V	R	D
Leg Hb S. bean	T	S	I	L	E	K	A	P	A	A	K	D	L	F	—	S	F	—	—	L	A	N	G	V	D	—	P	T	N	P	K	L	T	G	H	A	E	K	L	F	A	L	V	R	D

TABLE 6—*Continued*

				E	EF													EF	F								F	FG						FG	G										
Globin	17	18	19	20	1	2	3	4	5	6	7	8	9	10	11	12	13	14	1	2	3	4	5	6	7	8	9	1	2	3	4	5	6	7	1	2	3	4	5	6	7	8	9	10	11
Human Mb	G	I	L	K	K	K	G	H	—	—	—	H	E	A	—	—	—	E	I	K	P	L	A	Q	S	H	A	T	K	—	—	H	K	I	P	V	K	Y	L	E	F	I	S	E	C
Human α	N	A	V	A	H	V	D	D	—	—	—	M	P	N	—	—	—	A	L	S	A	L	S	D	L	H	A	H	K	—	—	L	R	V	D	P	V	N	F	K	L	L	S	H	C
Human β	D	G	L	A	H	L	D	N	—	—	—	L	K	G	—	—	—	T	F	A	T	L	S	E	L	H	C	D	K	—	—	L	H	V	D	P	E	N	F	R	L	L	G	N	V
Shark Mb	N	I	L	K	Q	K	G	K	—	—	—	H	S	T	—	—	—	N	V	K	E	L	A	D	T	H	I	N	K	—	—	H	K	I	P	P	K	N	F	V	L	I	T	N	I
Shark α	K	A	C	D	H	L	D	D	—	—	—	L	K	T	—	—	—	H	L	H	K	L	A	T	F	H	G	S	E	—	—	L	K	V	D	P	A	N	F	Q	Y	L	S	Y	C
Shark β	V	A	V	T	H	L	G	D	—	—	—	V	K	S	—	—	—	Q	F	T	D	L	S	K	K	H	A	E	E	—	—	L	H	V	D	V	E	S	F	K	L	L	A	K	C
Lamprey	D	A	V	A	S	M	D	D	T	E	K	M	S	—	—	—	—	—	M	K	D	L	S	G	K	H	A	K	S	—	—	F	Q	V	D	P	Q	Y	F	K	V	L	A	—	V
Aplysia Mb	E	F	V	N	D	A	A	N	A	G	K	M	S	A	—	—	—	M	L	S	Q	F	A	K	E	H	V	G	F	G	V	G	S	A	Q	F	E	N	V	R	S	M	F	P	G
Busycon Mb	S	M	I	C	S	I	D	D	D	D	C	V	B	G	—	—	—	L	A	K	K	L	S	R	N	H	L	A	—	—	—	R	G	V	S	A	A	D	F	K	L	L	—	E	A
Anadara I-α	W	F	V	D	S	L	D	D	P	S	R	L	K	C	—	—	—	V	V	E	K	F	A	V	N	H	I	N	R	—	—	—	K	I	S	G	D	A	F	G	S	I	I	P	E
Chironomus IIβ	E	I	V	A	L	M	G	N	S	A	N	M	P	A	M	E	T	L	I	K	D	M	A	A	N	H	—	K	A	—	—	R	G	I	P	K	A	Q	F	N	E	F	R	A	S
Chironomus III	K	L	I	G	E	I	P	E	—	—	N	I	A	D	—	—	—	V	N	T	F	P	A	S	K	H	V	—	R	G	V	T	H	—	—	B	Z	—	L	B	B	F	R	A	G
Chironomus VIIβ	E	V	I	A	L	S/M	G	N	E/A	S	N	A	A	A	V	Q	G	L	L	D	K	L	G	D	D	H	K	—	A	—	—	R	G	V	S	A	A	Q	F	G	E	F	R	L/T	A
Glycera Hb	V	A	V	S	H	L	G	D	Z	G	K	M	V	A	—	—	—	Q	M	K	A	V	G	V	R	H	K	G	Y	G	N	K	H	I	K	G	Q	Y	F	E	P	L	G	A	S
Leg Hb K. bean	S	A	A	Q	L	R	A	N	—	—	G	A	V	V	—	—	—	A	D	A	A	L	G	S	I	H	S	Q	K	G	V	S	N	—	—	D	Q	—	F	L	V	V	K	E	A
Leg Hb S. bean	S	A	G	Q	L	K	A	S	—	—	G	T	V	V	—	—	—	A	D	A	A	L	G	S	V	H	A	Q	K	A	V	T	N	—	—	P	Q	—	F	—	V	V	K	E	A

(*continued*)

TABLE 6—*Continued*

								G	GH				GH	H																				H	HC			(	H	HC		Mb	)
Globin	12	13	14	15	16	17	18	19	1	2	3	4	5	1	2	3	4	5	6	7	8	9	10	11	12	13	14	15	16	17	18	19	20	21	1	2	3	(25	26	1	2	3	4)
Human Mb	I	I	Q	V	L	Q	S	K	H	P	G	D	F	G	A	D	A	Q	G	A	M	N	K	A	L	E	L	F	R	K	D	M	A	S	N	Y	K	E	L	G	F	Q	G
Human α	L	L	V	T	L	A	A	H	L	P	A	E	F	T	P	A	V	H	A	S	L	D	K	F	L	A	S	V	S	T	V	L	T	S	K	Y	R						
Human β	L	V	C	V	L	A	H	H	F	G	K	E	F	T	P	P	V	Q	A	A	Y	Q	K	V	V	A	G	V	A	N	A	L	A	H	K	Y	H						
Shark Mb	A	V	K	V	L	T	E	M	Y	P	S	D	M	T	G	P	M	Q	E	S	F	S	K	V	F	T	V	I	C	S	D	L	E	T	L	Y	K	E	A	D	F	Q	G
Shark α	L	E	V	A	L	A	V	H	L	—	T	E	F	S	P	E	T	H	C	A	L	D	K	F	L	T	N	V	C	H	E	L	S	S	R	Y	R						
Shark β	F	V	V	E	L	G	I	L	L	K	D	K	F	A	P	Q	T	Q	A	I	W	E	K	Y	F	G	V	V	V	D	A	I	S	K	E	Y	H						
Lamprey	I	A	D	T	V	A	A	—	—	—	—	—	—	—	—	—	G	D	A	G	F	E	K	L	S	M	(I	C)	I	L	M	L	R	S	A	Y	—						
Aplysia Mb	F	V	A	S	V	A	A	—	—	—	—	—	—	P	P	A	G	A	D	A	W	T	K	L	F	G	L	I	I	D	A	L	K	A	A	G	K						
Busycon Mb	V	F	K	Z	F	L	—	—	—	—	D	E	A	T	Q	R	K	A	T	D	A	Q	K	D	A	D	G	A	L	L	T	M	L	I	K	A	H	V					
Anadara I-α	M	K	E	T	L	K	A	R	M	G	S	Y	S	D	D	V	G	—	A	A	W	V	Q	A	I	L	G	M	Q	N	A	V	L	S	A	L							
Chironomus IIβ	L	V	S	Y	L	Q	S	K	—	—	V	S	W	N	D	S	L	G	A	A	W	T	N	G	L	D	N	V	F	N	M	M	F	S	—	Y	L						
Chironomus III	F	V	S	Y	M	K	A	H	—	—	T	D	F	G	A	E	A	—	A	A	W	G	T	L	(F	D	T	F	A)	G	M	I	F	S	K	M	—						
Chironomus VIIβ	L	V	A	Y	L	Q	A	H	—	—	V	S	W	G	N	N	V	A	A	A	W	S	K	A	L	D	N	T	V F	A	I	V	V	P	R	L	_						
Glycera Hb	L	L	S	A	M	E	H	R	I	G	G	K	M	N	A	A	A	K	D	A	W	A	A	A	Y	A	D	I	S	G	A	L	I	S	—	—	—	—	—	G	L	Q	S
Leg Hb K. bean	L	L	K	T	L	K	Q	A	V	G	D	K	W	T	D	Q	L	S	T	A	L	E	L	A	Y	D	E	L	A	A	A	I	K	K	A	Y	A						
Leg Hb S. bean	L	L	K	I	I	K	A	A	V	G	D	K	W	S	D	E	L	S	R	A	W	E	V	A	Y	D	E	L	A	A	A	I	K	A	K	—							

Phe. *A. trapezia* exhibits 26 identical residues in comparison to *Busycon* and 31 in comparison to *Aplysia* myoglobin, whereas the two gastropod mollusks have only 19 identical residues.

The sequence of the *A. trapezia* globin has been designated α chain because the sequence in the region C2 to CD1 is strongly homologous to α chains rather than β chains. Three D-helical residues are missing, which is a characteristic of α chains and possibly of invertebrate globins, since X-ray studies have shown the absence of the D helix in *Glycera dibranchiata* hemoglobin (48). The sequence shows 31 ± 2 identical residues compared with vertebrate α, β, or myoglobin chains, and with the exception of *Aplysia l.* -globin, 23 ± 3 identical residues compared with invertebrate or plant globins out of a total of 136–153 residues.

Acknowledgments

This work has been supported by the Australian Research Grants Commitee. The author gratefully acknowledges the significant contributions of many co-workers and research assistants named in the references from his laboratory.

REFERENCES

1. Air, G.M., and Thompson, E.O.P., Studies on marsupial proteins. II. Amino acid sequence of the β-chain of haemoglobin from the grey kangaroo, *Macropus giganteus. Aust. J. Biol. Sci.*, 1969, **22**:1437–1454.
2. Air, G.M., and Thompson, E.O.P., Studies on marsupial proteins. IV. Amino acid sequence of myoglobin from the red kangaroo, *Megaleia rufa. Aust. J. Biol. Sci.*, 1971, **24**:75–95.
3. Air, G.M., Thompson, E.O.P., Richardson, B.J., and Sharman, G.B., Amino acid sequences of kangaroo myoglobin and haemoglobin and the date of marsupial-eutherian divergence. *Nature (London)*, 1971, **229**:391–394.
4. Arnone, A., X-ray diffraction study of binding of 2,3-diphosphoglycerate to human deoxyhaemoglobin. *Nature (London)*, 1972, **237**:146–149.
5. Arnone, A., and Perutz, M.F., Structure of inositol hexaphosphate—human deoxyhaemoglobin complex. *Nature (London)*, 1974, **249**:34–36.
6. Barker, W.C., and Dayhoff, M.O., Evolution of homologous physiological mechanisms based on protein sequence data. *Comp. Biochem. Physiol.* [*B*], 1979, **62**:1–5.
7. Beard, J.M., and Thompson, E.O.P., Studies on marsupial proteins. V. Amino acid sequence of the α-chain of haemoglobin from the grey kangaroo, *Macropus giganteus. Aust. J. Biol. Sci.*, 1971, **24**:765–786.
8. Benesch, R.E., and Benesch, R., The mechanism of interaction of red cell organic phosphates with hemoglobin. *Adv. Protein Chem.*, 1974, **28**:211–237.
9. Bolton, W., and Perutz, M.F., Three dimensional Fourier synthesis of horse deoxyhaemoglobin at 2.8 Å resolution. *Nature (London)*, 1970, **228**:551–552.
10. Bonaventura, J., and Riggs, A., Hemoglobin Kansas, a human hemoglobin with a neutral amino acid substitution and an abnormal oxygen equilibrium. *J. Biol. Chem.*, 1968, **243**:980–991.

11. Bonner, A.G., and Laursen, R.A., The amino acid sequence of a dimeric myoglobin from the gastropod mollusc, *Busycon canaliculatum L. FEBS Lett.*, 1977 **73**:201–203.
12. Brand, D.K., and Holland, R.A.B., Oxygen affinity and 2,3-diphosphoglycerate in blood of Australian marsupials of differing body size. *Respir. Physiol.* 1975, **31**:279–290.
13. Braunitzer, G., and Fujiki, H., Zur Evolutionen der Vertebraten. Die Konstitution und Tertiärstruktur des Hamoglobins des Flussneunauges. *Naturwiss enschaften*, 1969, **56**:322–323.
14. Braunitzer, G., Hilse, K., Rudloff, V., and Hilschmann, N., The hemoglobins *Adv. Protein Chem.*, 1964, **19**:1–71.
15. Briehl, R., The relation between the oxygen equilibrium and aggregation of subunits in lamprey hemoglobin. *J. Biol. Chem.*, 1963, **238**:2361–2366.
16. Castillo, O., and Lehmann, H., The myoglobin of the fruit-bat *(Rousettus aegyptiacus). Biochim. Biophys. Acta*, 1977, **492**:232–236.
17. Castillo, O., Jones, L.T., and Lehmann, H., The myoglobin of an echidna *(Tachyglossus aculeatus aculeatus). Biochim. Biophys. Acta*, 1978, **533**:289–292.
18. Castillo, O., Lehmann, H., and Jones, L.T., The myoglobin of the killer whale *(Orcinus orca). Biochim. Biophys. Acta*, 1977, **491**:23–28.
19. Como, P.F., *Multiple Haemoglobins of the Bivalve Mollusc* Anadara trapezia. Ph.D. Thesis, University of New South Wales, Kensington, 1978.
20. Cooper, D.W., VandeBurg, J.L., Griffiths, M.E., and Ealey, E.H.M., Haemoglobin polymorphisms in the echidna, *Tachyglossus aculeatus. Aust. J. Biol. Sci.*, 1973, **26**:605–612.
21. Dayhoff, M.O. (ed.), *Atlas of Protein Sequence and Structure*, National Biomedical Research Foundation, Washington D.C., Vol 5, 1972; Vol 5, Suppl 1, 1973; Vol 5, Suppl 2, 1976; Vol 5, Suppl 3, 1978.
22. Dickerson, R.E., X-ray analysis and protein structure, In: *The Proteins* (H. Neurath, ed.), 2nd ed., Vol 2. Academic Press, New York, 1964:603–778.
23. Dixon, G.H., Mechanisms of protein evolution. *Essays Biochem.* 1966, **2**:147–204.
24. Dixon, P.I., The biology of *Anadara trapezia* with particular reference to its haemoglobin. *Malacol. Rev.*, 1976, **9**:139.
25. Dodgson, S.J., Fisher, W.K., and Thompson, E.O.P., Studies on monotreme proteins. IV. Amino acid sequence of haemoglobin-IA of the echidna: A comparison of major haemoglobins from two geographical groups of echidnas. *Aust. J. Biol. Sci.*, 1973, **27**:111–115.
26. Dodgson-Schneider, S.J., and Holland, R.A.B., Equilibrium and kinetic properties of two kangaroo hemoglobins. *Physiologist*, 1976, **19**:170.
27. Edmundson, A.B., Amino-acid sequence of sperm whale myoglobin. *Nature (London)*, 1965, **205**:883–887.
28. Edsall, J.T., Blood and hemoglobin: The evolution of knowledge of functional adaptation in a biochemical system. *J. Hist. Biol.*, 1972, **5**:205–257.
29. Fisher, W.K., and Thompson, E.O.P., Studies on monotreme proteins. VII. Amino acid sequence of myoglobin from the platypus, *Ornithorhynchus anatinus. Aust. J. Biol. Sci.*, 1976, **29**:57–72.
30. Fisher, W.K., and Thompson, E.O.P., Myoglobin of the shark *Heterodontus*

portusjacksoni: Isolation and amino acid sequence. *Aust. J. Biol. Sci.*, 1979, **32**:277–294.

31. Fisher, W.K., Nash, A.R., and Thompson, E.O.P., Haemoglobins of the shark *Heterodontus portusjacksoni*. III. Amino acid sequence of the β-chain. *Aust. J. Biol. Sci.*, 1977, **30**:487–506.
32. Fitch, W.M., and Margoliash, E., Construction of phylogenetic trees. *Science*, 1967, **155**:276–284.
33. Furuta, H., Ohe, M., and Kajita, A., Subunit structure of hemoglobins from erythrocytes of the blood clam, *Anadara broughtonni. J. Biochem. (Tokyo)*, 1977, **82**:1723–1730.
34. Goodman, M., Barnabas, J., Matsuda, G., and Moore, G.W., Molecular evolution in the descent of man. *Nature (London)*, 1971, **233**:604–613.
35. Goodman, M., Moore, G.W., and Matsuda, G., Darwinian evolution in the genealogy of haemoglobin. *Nature (London)*, 1975, **253**:603–608.
36. Gratzer, W.B., and Allison, A.C., Multiple haemoglobins. *Biol. Rev. Cambridge Philos. Soc.*, 1960, **35**:459–506.
37. Hosken, R.W., Oxygen binding of monotreme haemoglobins. I. Oxygen binding parameters of platypus, *Ornithorhynchus anatinus* and echidna, *Tachyglossus aculeatus* haemoglobins. *Aust. J. Biol. Sci.*, 1978, **31**:327–336.
38. Hosken, R.W., and Boettcher, B., Oxygen binding of monotreme haemoglobins. II. Fixation of functionally non-equivalent haemoglobins in the echidna population. *Aust. J. Biol. Sci.*, 1978, **31**:337–344.
39. Ingram, V., Gene evolution and the haemoglobins. *Nature (London)*, **189**:704–708.
40. Jukes, T.H., and Holmquist, R., Estimation of evolutionary changes in certain homologous polypeptide chains. *J. Mol. Biol.*, 1972, **64**:163–179.
41. Kendrew, J.C., Dickerson, R.E., Strandberg, B.E., Hart, R.G., Davies, D.R., Phillips, D.C., and Shore, V.C., Structure of myoglobin. A three-dimensional Fourier synthesis at 2Å resolution. *Nature (London)*, 1960, **185**:422–427.
42. Kilmartin, J.V., Interaction of haemoglobin with protons, CO_2 and 2,3-diphosphoglycerate. *Br. Med. Bull.*, 1976, **32**:209–212.
43. Kitchen, H., Heterogeneity of animal hemoglobins. *Adv. Vet. Sci.*, 1969, **13**:247–330.
44. Kleinschmidt, T., and Braunitzer, G., Die Sequenz eines dimeren hämoglobins (komponente IIβ, *Chironomus thummi thummi, Diptera). Hoppe Seyler's Z. Physiol. Chem.*, 1976, **357**:1805–1808.
45. Lehmann, H., and Carrell, R.W., Variations in the structure of human haemoglobins. *Br. Med. Bull.*, 1969, **25**:14–23.
46. Lehmann, H., and Huntsman, R.G., *Man's Haemoglobins*. Elsevier, Amsterdam, 1974.
47. Li, S.L., and Riggs, A., The amino acid sequence of hemoglobin V from the lamprey *Petromyzon marinus. J. Biol. Chem.*, 1970, **245**:6149–6169.
48. Love, W.E., Klock, P.A., Lattman, E.E., Padlan, E.A., Ward, K.B., and Hendrickson, W.A., The structures of lamprey and bloodworm hemoglobins in relation to their evolution and function. *Cold Spring Harbor Symp. Quant. Biol.*, 1971, **36**:349–357.
49. Margoliash, E., and Smith, E.L., Structural and functional aspects of cyto-

chrome C in relation to evolution. In: *Evolving Genes and Proteins* (V. Bryson and H.J. Vogel, eds.). Academic Press, New York, 1965:221–242.
50. Morimoto, H., Lehmann, H., and Perutz, M.F., Molecular pathology of human haemoglobin: Sterochemical interpretation of abnormal oxygen affinities. *Nature (London),* 1971, **232**:408–413.
51. Nagel, R.L., Lynfield, J., Johnson, J., Landau, L., Bookchin, R.M., and Harris, M.B., Hemoglobin Beth Isreal. A mutant causing clinically apparent cyanosis. *N. Engl. J. Med.*, 1976, **295**:125–130.
52. Nash, A.R., *The Structure and Function of Shark Hemoglobins.* Ph.D. Thesis, University of New South Wales, Kensington, 1976.
53. Nash, A.R., and Thompson, E.O.P., Haemoglobins of the shark, *Heterodontus portusjacksoni. Aust. J. Biol. Sci.*, 1974, **27**:607–615.
54. Nash, A.R., Fisher, W.K., and Thompson, E.O.P., Haemoglobins of the shark, *Heterodontus portusjacksoni.* II. Amino acid sequence of the α-chain. *Aust. J. Biol. Sci.*, 1976, **29**:73–97.
55. Nicol, P.I., and O'Gower, A.K., Haemoglobin variation in *Anadara trapezia. Nature (London),* 1967, **216**:684.
56. Onoki, S., Mitoni, Y., Hata, R., and Satake, K., Heterogeneity of hemoglobin from Arca (*Anadara satowi*). Molecular weights and oxygen equilibria of Arca Hb I and II. *J. Biochem. (Tokyo),* 1973, **73**:717–725.
57. Perutz, M.F., The haemoglobin molecule. *Proc. R. Soc. London [B*], 1969, **173**:113–140.
58. Perutz, M.F., Structure and mechanism of haemoglobin. *Br. Med. Bull.*, 1976, **32**:195–208.
59. Perutz, M.F., and Lehmann, H., Molecular pathology of human haemoglobin. *Nature (London),* 1968, **219**:902–909.
60. Perutz, M.F., Kendrew, J.C., and Watson, H.C., Structure and function of haemoglobin. II. Some relations between polypeptide chain configuration and amino acid sequence. *J. Mol. Biol.*, 1969, **13**:669–678.
61. Perutz, M.F., Muirhead, H., Cox, J.M., and Goaman, L.C.G., Three dimensional Fourier synthesis of horse oxyhaemoglobin at 2.8 Å resolution: The atomic model. *Nature (London),* 1968, **219**:131–139.
62. Richardson, B.J., *A Morphological and Biochemical Study of Two Closely Related Marsupial Genera,* Megaleia *(the red kangaroo) and* Osphranter *(the Wallaroos).* Ph.D. Thesis, University of New South Wales, Kensington, 1970.
63. Riggs, A., Factors in the evolution of hemoglobin function. *Fed. Proc. Fed. Am. Soc. Exp. Biol.*, 1976, **35**:2115–2118.
64. Riggs, A., Studies of the hemoglobins of Amazonian fishes: An overview. *Comp. Biochem. Physiol.* [*A*], 1979, **62**:257–271.
65. Romer, A.S., *Vertebrate Palaeontology.* 3rd ed. Chicago University Press, Chicago, Illinois, 1966
66. Romero-Herrera, A.E., Lehmann, H., Joysey, K.A., and Friday, A.E., Molecular evolution of myoglobin and the fossil record: A phylogenetic synthesis. *Nature (London),* 1973, **246**:389–395.
67. Romero-Herrera, A.E., Lehmann, H., Joysey, K.A., and Friday, A.E., On the evolution of myoglobin. *Philos. Trans. R. Soc. London* [*B*], 1978, **283**:61–163.
68. Rucknagel, D.L., and Winter, W.P., Duplication of structural genes for hemoglobin α and β chains in man. *Ann. N.Y. Acad. Sci.*, 1974, **241**:80–92.

69. Schroeder, W.A., and Huisman, T.H.J., Multiple cistrons for fetal hemoglobin in man. *Ann. N.Y. Acad. Sci.*, 1974, **241**:70–79.
70. Sharman, G.B., Monotreme biology. Chairman's summary of the symposium. *Aust. Zool.*, 1978, **20**:1–7.
71. Sladic-Simic, D., Kleinschmidt, T., and Braunitzer, G., Die Sequenz eines dimeren Hämoglobins (Komponente VII B, *Chironomus thummi thummi, Diptera). Hoppe-Seylers Z. Physiol. Chem.*, 1977, **358**:591–594.
72. Stenzel, P., Opossum Hb chain sequence and neutral mutation theory. *Nature (London)*, 1974, **252**:62–63.
73. Tentori, L., Vivaldi, G., Carta, S., Marinucci, M., Massa, A., Antonini, E., and Brunori, M., The amino acid sequence of myoglobin from the mollusc *Aplysia limacina. Int. J. Pept. Protein Res.*, 1973, **5**:187–200.
74. Thompson, E.O.P., and Air, G.M., Studies on marsupial proteins. VI. Evolutionary changes in β-globins of the macropodidae and the amino acid sequence of β-globin from *Potorous tridactylus. Aust. J. Biol. Sci.*, 1971, **24**:1199–1217.
75. Thompson, E.O.P., Fisher, W.K., and Whittaker, R.G., Studies on monotreme proteins. III. Amino acid sequence of the α- and β-globin chains of the minor haemoglobin from the echidna, *Tachyglossus aculeatus aculeatus. Aust. J. Biol. Sci.*, 1973, **26**:1327–1335.
76. Thompson, E.O.P., Hosken, R., and Air, G.M., Studies on marsupial proteins. I. Polymorphism of haemoglobin of the grey kangaroo, *Macropus giganteus. Aust. J. Biol. Sci.*, 1969, **22**:449–462.
77. Vainshtein, B.K., Harutyunyan, E.H., Kuranova, I.P., Borisov, V.V., Sosfenov, N.I., Pavlovsky, A.G., Grebenko, A.O., and Konareva, N.N., Structure of leghaemoglobin from lupin root nodules at 5 Å resolution. *Nature (London)*, 1975, **254**:163–164.
78. Watson, H.C., The stereochemistry of the protein myoglobin. *Prog. Stereochem.*, 1969, **4**:299–333.
79. Weatherall, D.J., and Clegg, J.B., Molecular genetics of human hemoglobin. *Annu. Rev. Genet.*, 1976, **10**:157–178.
80. Whittaker, R.G., and Thompson, E.O.P., Studies on monotreme proteins. V. Amino acid sequence of the α-chain of haemoglobin from the platypus, *Ornithorhychus anatinus. Aust. J. Biol. Sci.*, 1974, **27**:591–605.
81. Whittaker, R.G., and Thompson, E.O.P., Studies on monotreme proteins. VI. Amino acid sequence of the β-chain of haemoglobin from the platypus, *Ornithorhynchus anatinus. Aust. J. Biol. Sci.*, 1975, **28**:353–365.
82. Whittaker, R.G., Fisher, W.K., and Thompson, E.O.P., Studies on monotreme proteins. I. Amino acid sequence of the β-chain of haemoglobin from the echidna, *Tachyglossus aculeatus aculeatus. Aust. J. Biol. Chem.*, 1972, **25**:989–1004.
83. Whittaker, R.G., Fisher, W.K., and Thompson, E.O.P., Studies on monotreme proteins. II. Amino acid sequence of the α-chain in haemoglobin from the echidna, *Tachyglossus aculeatus aculeatus. Aust. J. Biol. Sci.*, 1973, **26**:877–888.
84. Whittaker, R.G., Fisher, W.K., and Thompson, E.O.P., Monotreme haemoglobin and myoglobin amino acid sequences and their use in phylogenetic divergence point estimations. *Aust. Zool.*, 1978, **20**:57–68.
85. Williams, J., The primary structure of proteins in relation to evolution. *MTP In. Rev. Sci. Chem. Macromol. 1974,* **1**:1–57.

86. Wilson, A.C., Carlson, S.S., and White, T.J., Biochemical evolution. *Annu. Rev. Biochem.*, 1977, **46**:573–639.
87. Zuckerkandl, E., and Pauling, L., Molecular disease, evolution, and genic heterogeneity, In: *Horizons in Biochemistry* (M. Kasha and B. Pullman, eds.). Academic Press, New York, 1962:189–225.

Evolutionary Adaptation of Mitochondrial Cytochrome *c* to Its Functional Milieu

E. MARGOLIASH

Department of Biochemistry and Molecular Biology
Northwestern University
Evanston, Illinois

The early history of science is often as much the account of struggle against misconceptions as that of the achievement of progressively sharper insights into the nature of the subject. This is still particularly evident in biology, where so much remains unknown. Cytochrome *c* in its deceptive simplicity is no exception. The following short review attempts to document the present state of our understanding of the relation between the functional activities of the protein and the many transformations it has undergone during the course of long evolution in eukaryotes.

Neutral Mutations versus Selection in the Evolution of Cytochrome *c*

Since the first determination of the amino acid sequence of a cytochrome *c* in 1961 (59, 63, 70, 71, 79, 80) the properties of the protein have made it a prime object for molecular evolution studies, the major objective of which remains the relations between the primary structures and the evolutionary mechanisms that lead to each species synthesizing its own particular protein (3,5,6,19–21,30, 33–45,47,48,52–55,60,61,64–69,72,74–78,88,89,101,102,115,118). Mitochondrial cytochromes *c* are small, consisting of a single polypeptide chain just a little longer than 100 residues and their amino acid sequences are relatively easy to determine; they are ubiquitously present in eukaryotes and their commonly large net positive charge makes them easy to prepare and purify even from tissues that contain very little of the protein (8,69,73). Through the continued efforts of several groups of investigators, the amino acid sequences of about 80 different eukaryote cytochromes *c* have been determined. These are from species ranging all over the animal and plant kingdoms and represent

ISBN 0-12-643150-7

the widest toxonomic coverage for any protein (4,13–16). Thus it is not surprising that the first evidence for an apparently constant rate of fixation of mutations during the course of evolutionary descent was deduced for cytochrome *c* as early as 1963 from a comparison of the structures of the cytochromes *c* of four mammals (horse, man, pig, rabbit), one bird (chicken), one fish (tuna), and bakers' yeast (64). This in turn made it possible to propose the widely used measure of the rate of descent of proteins, termed in 1965 the *unit evolutionary period* (72). At that time, the implication derived from the apparent constancy of the rate of descent of the cytochrome *c* gene in all phylogenetic lineages examined, that all of those changes in amino acid sequence were evolutionarily neutral and fixed by some random nonselective mechanism, was supported by functional studies. Employing the spectrophotometric technique for following the oxidation of ferrocytochrome *c* by submitochondrial preparations, it was found that all the eukaryotic cytochromes *c* examined behaved in a quantitatively indentical fashion even for proteins that differed by a good proportion of their amino acid sequences (107). Cytrochrome *c* thus became the prime example of a so-called non-Darwinian evolutionary molecular system.

However, as more evidence accumulated, the foundations on which this concept was built began to disintegrate. In the first place, the development of a procedure for obtaining statistical phylogenetic trees from the amino acid sequences of a set of orthologous proteins in 1967 (42) led to the assignment of individual mutations to various lines of descent, in order to account for the structures of the present-day proteins (Fig. 1). With an early tree of only 23 cytochromes *c* it became apparent that for this gene the mammalian lines of descent had fixed about 39% more mutations than those of birds and that vertebrates as a whole had varied some 48% faster than invertebrates (77).

Second, the determination of the three-dimensional structure of the protein by X-ray crystallography made it possible to examine the spatial distribution of the positions at which amino acid sequence changes had occurred in the cytochrome *c* of various taxonomic groups (35,65,78). This clearly showed that for large groupings, such as mammals, insects, fungi, and higher plants, these locations were distinctly different, at a high level of significance, even though there was some overlap. This result did not fit into the neutral mutation hypothesis for the evolution of the protein, since if all the changes in amino acid sequence are of no functional significance, there is no reason that a particular change occurring in one set of species should not be just as likely to occur in another.

Third, with the accumulation of many cytochrome *c* primary structures and the development of procedures for dealing simultaneously with the phylogenetic relations derived from the structures of several proteins, the rates of molecular evolutionary descent could be estimated more accurately and determined over much shorter intervals than before (40,60,61). This led to the demonstration that, contrary to the earlier conclusion, the rate of

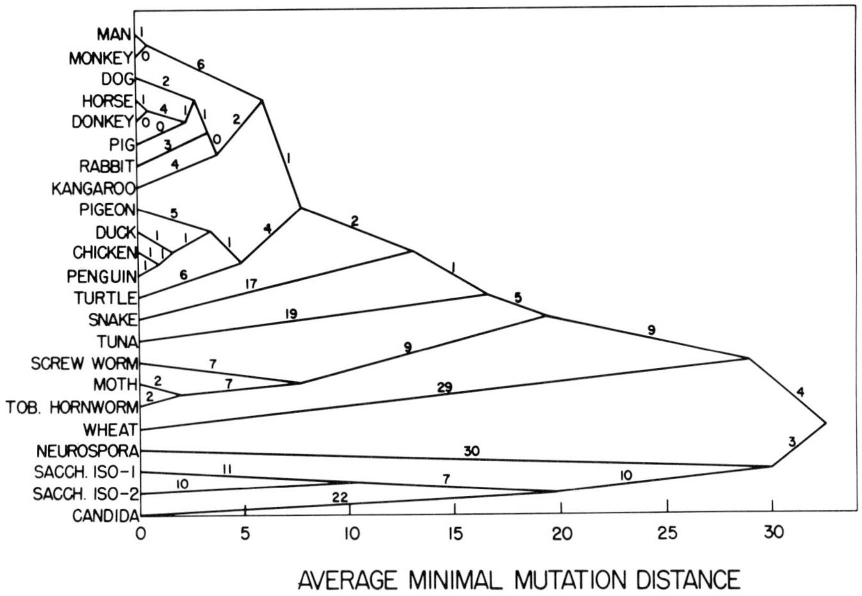

Fig. 1. Statistical phylogenetic tree based on the minimal mutation distances between the cytochromes *c* of the species listed, according to Fitch and Margoliash (42). Each number on the figure is the mutation distance along the line of descent as determined from the best fit of the data found. Each apex is placed at an ordinate value that is the average of the sums of all mutations in the lines of descent from that apex.

fixation of mutations is not constant for a given protein either over different evolutionary intervals in the same line of descent or over the same interval for different lines of descent. It became clear that the constancy of evolutionary rate of descent observed earlier was merely the result of a simple fallacy, namely, that if the rate of a process can vary only within finite limits, averaging it over a long enough period of time will make it appear to be constant (40). This is why constancy had been found for comparisons between the cytochromes *c* of large taxonomic groups only (64), such as mammals, birds, fish, vertebrates, invertebrates, and fungi, that had diverged from each other a long time ago, and not within any one group (64,66,72).

If one accepted these findings and interpretations, the hypothesis that the bulk of the changes that had taken place in the course of evolution in the cytochrome *c* gene were neutral mutations could no longer be entertained. That these variations must therefore, in some way or other, reflect changes in function led to a reexamination of the kinetics of reaction of cytochrome *c* with its reaction partners, particularly cytochrome oxidase, under conditions more closely approximating those in mitochondria (28,76). It was therefore satisfying to observe that under such conditions the cytochromes *c* of eukaryotes were not all functionally indistinguishable. Quite the contrary, some showed large functional differences, and the molecular changes responsible for these differences are beginning to be understood.

In general, cytochromes *c* that could be expected to be well adapted to the cytochrome oxidase with which they were tested showed biphasic kinetics of reaction, the high- and low-affinity phases having apparent K_m values of about $10^{-8}M$ and $10^{-6}M$, respectively (28) (Fig. 2). Cytochromes *c* not well adapted to the oxidase reacted either with uniformly high affinity, as in the case of bakers' yeast cytochrome *c* and beef oxidase, or low affinity, such as for euglena cytochrome *c* and beef oxidase. The high and low affinity reactions corresponded to two molecules of cytochrome *c* binding to one molecule of the enzyme (cytochrome aa_3), the apparent K_m values being close to directly determined K_D values (28,92).

These and related recent developments are the subject of this short review. However, it should not be overlooked that even after all this work the controversy between the selectionist and the neutralist points of view of the evolutionary transformations of cytochrome *c* has not yet been entirely settled. It has been unquestionably demonstrated that in a number of cases the cytochrome *c* having diverged away from an older structure, so that it reacts poorly with the earlier oxidase, is exquisitely adapted to the new form of the oxidase carried by the species that also makes the new form of the cytochrome *c* (74,75). Thus the enzyme can be considered to represent part of the adap-

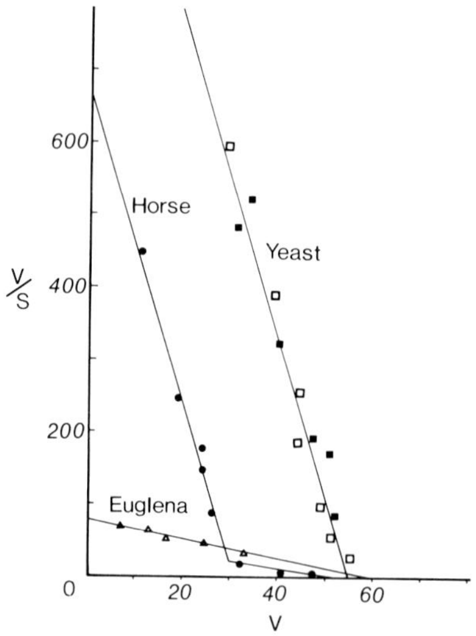

Fig. 2. Eadie–Hofstee representation of the kinetics of reaction of the various cytochromes *c* indicated in the figure with the cytochrome oxidase of a Keilin–Hartree particle preparation from beef heart (KHP) according to Ferguson-Miller *et al.* (28). Rates of oxygen consumption were measured polarographically with a Gilson oxygraph at 25°C in a medium containing 25 m*M* cacodylate (Tris), 250 m*M* sucrose, 0.7 m*M* *N,N,N′,N′*-tetramethylphenylenediamine (TMPD), 7 m*M* sodium ascorbate, at pH 7.8. The cytochrome *c* concentration range was 0.04–4 μM. The final KHP concentration was 0.07 mg protein per ml. Velocity (V) is in nmoles O_2 min^{-1} and the cytochrome *c* concentration (S) is μM. Results shown were obtained with cytochrome *c* from horse (●), euglena (△), yeast iso-1 (■), and yeast iso-2 (□).

tive environment for the cytochrome *c* and *vice versa*, cytochrome *c* must be part of the selective environment of the oxidase. To date, the best-documented series showing such a situation occurs within the primate line of descent. Prosimian cytochromes *c* react normally with a nonprimate oxidase, New World monkey cytochromes *c* react poorly, and Old World monkey and ape proteins react even less well. By contrast, the monkey and ape proteins react normally with oxidase prepared from an Old World monkey, and the prosimian cytochromes *c* react poorly with it, just as do nonprimate proteins (74,75) (Fig. 3). The evolutionary problem is that within the relatively crude representation of physiological parameters afforded by these tests of function, the reaction of the Old World monkey and ape cytochromes *c* with their own oxidase does not appear to be any better or more advantageous than the reaction of prosimian or nonprimate cytochromes *c* with their own oxidase. What, then, is the mechanism that initiated the series of evolutionary fixations leading to the higher primate system. Even if it is accepted that after that initial impetus all subsequent changes are adaptive in the ordinary Darwinian sense, one can still maintain that the first change was the result of a random nonselective evolutionary event.

This is a far cry from the initial interpretation that the bulk of evolutionary changes in cytochrome *c* are neutral in character. Nevertheless, in the present example, unless the selective advantage given by every stage of the transformation from the prosimian to the higher primate system can be uncovered, it will not be possible to maintain that the fixation of neutral mutations does not play a significant role. This author is encouraged by the fact that in the course of the development of this subject since 1963, as outlined above, every time a sizable advance was made in the understanding of either the structure or the function of cytochrome *c*, the domain within which hypothetical nonselective events could be considered possible became progressively more restricted.

Today it has become practical to use recombinant DNA technology to isolate cytochrome *c* genes and determine their sequences of nucleotides (83,84,108), and to transform such genes at will by various forms of directed mutagenesis and express them in appropriate systems. Thus one could obtain cytochromes *c* of any desirable amino acid sequence, so that a definitive attack can be mounted on problems of structure–function relations in the protein. It should be possible to employ the same approach to obtain ancestral cytochromes *c* representing any node in a statistical phylogenetic tree of the gene for a suitable group of species and examine directly the functional changes that may have attended every step of the evolutionary descent of the protein. There is hope that such a transformation of molecular evolution from a purely descriptive to an experimental science will, among numerous other questions, also settle the problem of whether the fixation of evolutionarily neutral mutations is or is not an important mechanism in the phylogenetic descent of cytochrome *c*.

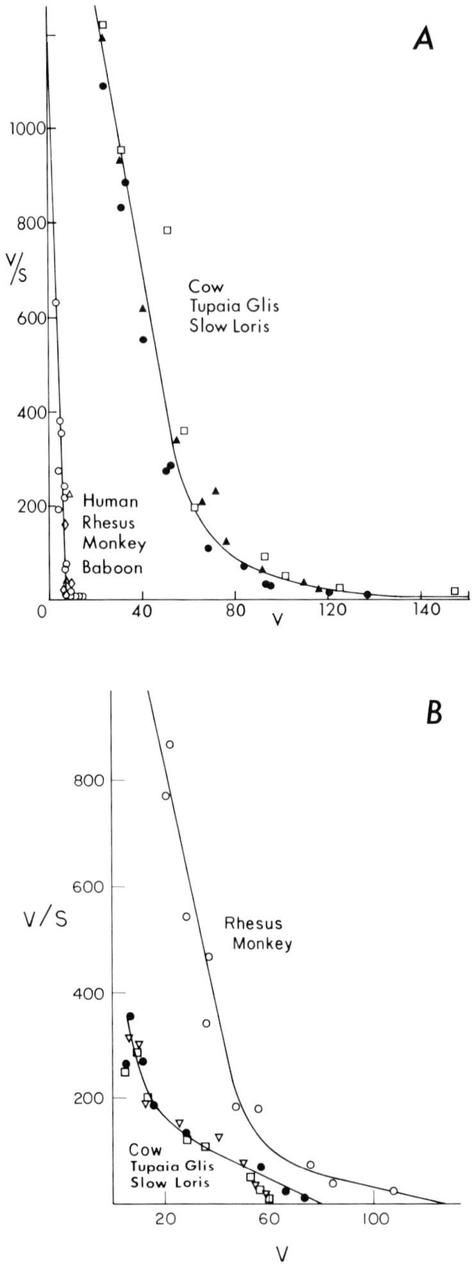

Fig. 3. Eadie–Hofstee representation of the kinetics of reaction of the cytochromes *c* of the various species indicated in the figure with the cytochrome oxidase of Keilin-Hartree particle preparation from (A) beef heart and (B) rhesus monkey heart, according to Margoliash *et al.* (75). Polarographic assay system, conditions, and units as in Fig. 2.

THE REACTION OF CYTOCHROME *c* WITH MITOCHONDRIAL ELECTRON DONORS AND ACCEPTORS

An assessment of the influence of evolutionary changes of the structure of a protein on its function ideally requires precise quantitative assays of function under conditions that approximate the physiological state of the system in which the protein operates. In the case of eukaryotic cytochromes *c* these conditions are particularly difficult to mimic (30,69). Cytochrome *c* is confined to the intermembrane space of mitochondria, an enclosure whose depth is less than three times the diameter of the protein under some conditions, but much larger under others (46); it obtains an electron from the terminal cytochrome c_1 component of the reductase segment of the respiratory chain and delivers it to cytochrome oxidase. Both the reductase and the oxidase are integral protein systems of the inner mitochondrial membrane. The surface of the inner mitochondrial membrane is negatively charged because of its cardiolipin and protein content, resulting in a tendency to hold cytochrome *c* electrostatically in the orientation dictated by the large dipole moment of the protein (58). Such an interaction is subject to the influence of the ionic strength of the medium. However, because the negatively charged membrane surface would tend to repel anions and because of a Donnan effect, anions may be partially excluded from the intermembrane space. Thus it is difficult to assess the effective anion concentration at the point at which cytochrome *c* reacts with its physiological electron exchange partners (93). Furthermore, there is little information concerning the relative spatial arrangement in the inner mitochondrial membrane, if any stable one exists, between cytochrome *c* reductase and cytochrome *c* oxidase. There is some experimental evidence to suggest that these systems tend to form stoichiometric complexes (90).

Finally, the various components of the respiratory chain in mitochondria are present in approximately equal amounts and at relatively high concentrations (117). For example, one commonly measures one to two molecules of cytochrome *c* for each molecule of cytochrome oxidase (cytochrome aa_3). Because of the very rapid reactions it is virtually impossible to set up an *in vitro* steady-state system that would reproduce these conditions at physiological temperatures and pH and permit the determination of the influence of varying the concentration of cytochrome *c*, as required to analyze the kinetics of the reactions.

It is important to bear these complexities in mind when attempting to estimate how and to what extent a change in cytochrome *c* function, observed by the available tests, bears on the physiological balance of the mitochondria carrying the variant protein.

ASSAY SYSTEMS

The earliest quantitative estimates of the reactions of cytochrome *c* with its mitochondrial electron exchange partners were manometric (99,100). The gas uptake resulting from the reaction of reduced cytochrome oxidase with oxygen to form water was measured, whether the electrons were supplied from succinate via the entire length of the respiratory chain or were supplied more directly from cytochrome *c* kept in the reduced form by ascorbate, which itself does not react at a significant rate with the oxidase.

Manometric assays are notoriously inconvenient, so that when spectrophotometers became common laboratory instruments in the 1950s, these procedures were supplanted by simple and accurate spectrophotometric assays (104). For example, in determining the activity with cytochrome oxidase, a small amount of the enzyme is added to fully reduced cytochrome *c* and the rate at which cytochrome *c* is oxidized is followed at appropriate wavelengths. These became the standard assays, even though the lowest concentrations of cytochrome *c* that could be employed with the spectrophotometers available at that time were at least an order of magnitude higher than the apparent K_m values for the reaction between cytochrome *c* and cytochrome oxidase determined manometrically. Another complication was that throughout its time course the reaction appeared to be first order, even though the manometric data had satisfied Michaelis–Menten kinetics. This was later explained by assuming that there is little difference between the binding affinity of the substrate (ferrocytochrome *c*) and the product (ferricytochrome *c*) to the enzyme under the relatively high ionic strength and low pH conditions employed, necessarily leading to a first-order time course (82; see also 25,30,86,87,119).

More recently a different type of assay has been developed. Oxygen uptake is followed polarographically and the cytochrome *c* is kept reduced by the dye *N*,*N*,*N'*,*N'*-tetramethylphenylenediamine (TMPD) and by ascorbate (28,29). It was shown that TMPD rapidly reduces cytochrome *c* bound to the enzyme, so that under relatively low ionic strength conditions the bulk of the electrons transfered to the oxidase come from cytochrome *c* that remains bound to the enzyme during turnover. Thus the measured initial steady-state rates, as one increases the concentration of cytochrome *c*, accurately reflect the proportion of the enzyme bound to cytochrome *c*; therefore an Eadie–Hofstee representation of such polarographic kinetics is in fact an Eadie–Scatchard plot (28,29). The apparent K_m is in fact a K_D (see Figs. 2 and 3). This has been confirmed by showing that direct measures of the K_D obtained with purified resting cytochrome oxidase and ferricytochrome *c* give values very similar to those yielded by the polarographic kinetics (28,92). Precisely what ratio of rate constants constitutes the kinetically determined K_D has not been demonstrated. However, since the rate of electron transfer from ferfocytochrome *c* in solution to the enzyme, as obtained by stopped-flow kinetics, is much higher than the rate of reduction of a ferricytochrome *c*–cytochrome oxidase complex by TMPD, it is likely that this K_D measures the ratio of k_3/k_1 (see Fig. 4), namely, the rate of dissociation of ferricytochrome *c* from the enzyme-product complex over the rate of association of ferrocytochrome *c* with the enzyme.

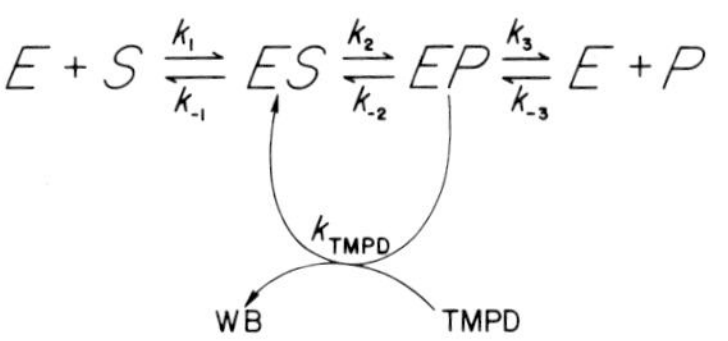

Fig. 4. Simplified scheme for the reaction of cytochrome *c* with cytochrome *c* oxidase (E). The substrate (S) is ferrocytochrome *c* and the product (P) is ferricytochrome *c*.

No mention has been made of the state of the oxidase that may be involved in these reactions. Considering the four electrons the enzyme can pick up with its two hemes and two copper atoms (62), the different spectrophotometric compounds of the oxidase known to occur during its action (10,11) and the apparent differences between resting enzyme and enzyme that has turned over (2), further complications of the interpretation of cytochrome *c* assays of function must be expected. Nevertheless, that the polarographic assay system measures accurately and sensitively some aspect of the binding of cytochrome *c* to cytochrome oxidase remains a particularly important advantage. Indeed, as outlined below, this permitted the localization of the enzymic interaction domain on the molecular surface of cytochrome *c* and the definition of the modes of reaction of the cytochromes *c* of various species with a variety of cytochrome oxidases (7,9,17,27,29,31,32,57,91,92,94,110). The main shortcoming of the polarographic assay system is that the reaction is most probably limited by the rate at which the dye can reduce enzyme-bound cytochrome *c*, and the V_{max} is unlikely to have any physiological significance.

In contrast, in the spectrophotometric assay, after a molecule of ferrocytochrome *c* has given up an electron to the oxidase, becoming ferricytochrome *c*, it has to dissociate from the enzyme before a second molecule can react to deliver an electron. Thus, ignoring any effects on the enzyme itself, maximal rates will be obtained at ionic strengths that are high enough that the reaction is not limited by the rate of dissociation of ferricytochrome *c* from the enzyme and low enough that the reaction is not limited by the rate at which ferrocytochrome associates with the enzyme. Since in mitochondria cytochrome *c* also has to dissociate and associate with the oxidase, at least to some extent, every time another electron is transported, one may consider the rates observed spectrophotometrically to be a closer representation of the physiological system than those given by the polarographic assay. However, the spectrophotometric assay presents a major difficulty stemming from the very high affinity of cytochrome *c* for cytochrome oxidase. Indeed, because these affinity constants range to about 10^{-8} *M* for the first molecule of cytochrome *c*, the concentration range required to explore the kinetics of the high-affinity reaction is so low that it is very difficult to attain with any degree of precision, even with the best present-day spectrophotometers. This is just the range over which the polarographic system yields data of excellent quality (see ref. 30).

Finally, it should be noted that (1) neither assay system measures the rate at which the electron is transferred to the enzyme once the ferrocytochrome *c*–cytochrome oxidase complex is formed, a rate that is surely faster than any observed by steady-state assays with either procedure; (2) the significance of the ionic strength dependence is quite different for the two assay systems; and (3) as already noted, it is not possible to assess what is the effective ionic strength at the distance from the inner mitochondrial membrane at which the charges on cytochrome *c* interact with those of the enzyme.

There are two other ways in which the functional activities of cytochrome *c* can be assessed. In one (10,12,27), the terminal respiratory chain carried by mitochondria or submitochondrial particles is fully reduced by a substrate, such as succinate, in the presence of carbon monoxide and in a medium containing ethylene glycol at temperatures down to about −50°C. Following the introduction of some oxygen, the CO-oxidase complex is dissociated by an appropriate laser flash, O_2 reacts with enzyme and one can follow spectrophotometrically a wave of oxidation traveling down the respiratory chain. The temperature can be adjusted to make the particular reaction to be observed conveniently slow. A difficulty with this system is that results may not be readily quantified. It has so far been used nearly exclusively with intact mitochondria and the little experience gained with submitochondrial particles and varying concentrations of cytochrome *c* is still too restricted to judge how useful it could eventually become in estimating cytochrome *c* functional parameters in media and at temperatures that more closely approximate physiological conditions.

Finally, using purified oxidase (cytochrome aa_3) or cytochrome c_1 preparations, it is possible to determine the rate of electron transfer from and to cytochrome *c* by stopped-flow experiments (see ref. 30). Because of the very rapid rates of these reactions under optimal conditions, putting them well beyond the time resolution of stopped-flow kinetic procedures, such experiments have so far been carried out under relatively high ionic strength conditions to slow the rate of association of the protein components appropriately (see, for example, ref. 94). Here again experience with the procedures as measures of changes in cytochrome *c* function, resulting from natural or chemically induced variations in structure, is yet too small to judge of their eventual general usefulness.

Steady-state kinetic assays of cytochrome *c* reactions with mitochondrial electron exchange partners other than cytochrome oxidase have in general been carried out spectrophotometrically (see reviews in refs. 8,30; also 1,26,49,50,56,57,74,75,81,85,103,105,106,109,110,112,113). Such tests of function have been used with the succinate–cytochrome *c* reductase system, the purified mitochondrial cytochrome *c* reductase (complex III; cytochrome bc_1 complex; reduced ubiquinone–cytochrome *c* reductase), yeast cytochrome *c* peroxidase, and sulfite oxidase. In these systems cytochrome *c* obtains an electron from the cytochrome c_1 of the reductase or from the cytochrome b_5 domain of sulfite oxidase, or delivers an electron to the hydrogen peroxide-oxidized form of the peroxidase. Each system has its own kinetic characteristics and, in general, is under the same types of constraint as the spectrophotometric assay for the reaction with cytochrome oxidase listed above. Considerable caution is required in attempting to relate the results of such tests to their possible significance in a physiological context. With electron exchange partners, such as cytochrome b_5, for which no steady-state kinetic systems have been developed, the electron transfer is measured by stopped-flow experiments (85).

THE ENZYMIC INTERACTION DOMAIN ON THE SURFACE OF CYTOCHROME *C*

With the demonstration that the polarographic assay afforded a sensitive measure of the binding of cytochrome *c* to cytochrome oxidase it became possible to examine the influence of perturbations on the surface of the molecule on its binding and reaction characteristics with the enzyme. To produce well-defined single-surface perturbations, a series of 12 4-carboxy-2,6-dinitrophenyl (CDNP) derivatives of single lysine residues of horse cytochrome *c* were purified to better than 99% homogeneity from reaction mixtures (7,9,92). These were the mono-CDNP derivatives of lysine-7, -8, -13, -25, -27, -39, -60, -72, -73, -86, -87, and -99. In each of these the positive charge of the unsubstituted ε-amino group is replaced by a large substituent that carries a negative charge. In every case it was shown by a variety of techniques, including proton NMR spectroscopy, that there was no detectable change in the spatial conformation or the usual properties of the protein. Presumably, the carboxyl group of the CDNP holds the substituent on the surface of the protein, leading to only a local surface perturbation of the protein topology. The relative K_m values for these modified cytochromes *c* reacting with the cytochrome oxidase of submitochondrial particles or the purified enzyme, indicated that modifications in the upper front of the protein—the front of the molecule being defined as the surface at which one edge of the square planar heme prosthetic group is at or close to the surface—are the most inhibited. Conversely, modifications at the back had little or no effect, whereas those located between the back and the upper front had intermediate effects (29,92) (Fig. 5).

It thus became obvious that cytochrome oxidase bound with high affinity to the area of cytochrome *c* that contained lysines-13, -72, -86, and -27 (Fig. 6). This is the area that is centered at the point at which the positive end of the dipole axis crosses the protein surface. Therefore it is likely that the relative orientation of the two proteins at the time of interaction is directed by electrostatic forces (57). This interaction domain also contains the majority, if not all, of the exposed heme edge, indicating that electron transfer occurs via that edge or a structure in close approximation to it (29). These observations appear to put to rest hypotheses of electron transfer through the protein.

The question as to how modifications outside the ostensible enzymic in-

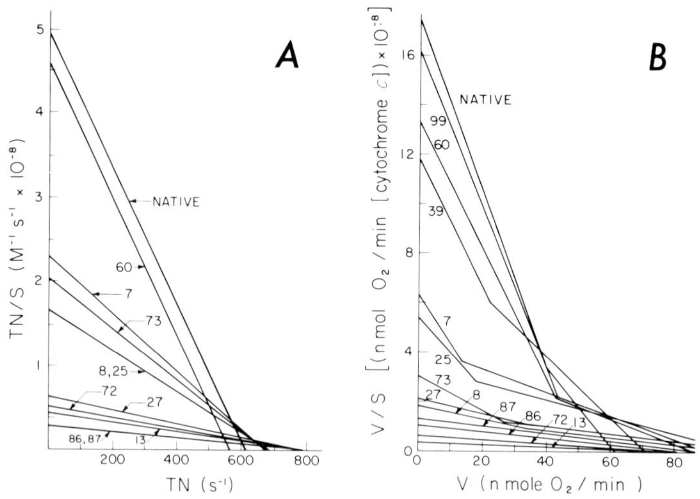

Fig. 5. Activity of various singly modified CDNP-horse cytochromes *c* in the polarographic cytochrome oxidase assay system (panel B), and with purified reduced ubiquinone–cytochrome *c* reductase (cytochrome bc_1) (panel A). Eadie–Hofstee representations. Conditions for the polarographic oxidase assay as in Fig. 2. Conditions for the spectrophotometric reductase assay according to Speck *et al.* (109), in which the velocities are given in turnover numbers. The numbers on the figures designate the amino acid sequence positions of the lysyl residues modified in the mono-CDNP cytochromes *c* studied.

teraction domain could affect activity at all was settled by calculating the dipole moments of all the CDNP modified cytochromes *c*. Assuming that when the dipole moment is changed by the chemical substitution, the protein will bind incorrectly to the oxidase so as to prevent electron transfer and that the work needed to move cytochrome *c* in the electric field of the enzyme to the correct position appears as extra activation energy for the reaction, it is possible to predict the effect of the changes in dipole on the rates of the reactions of the CDNP-modified cytochromes *c* relative to that for the unmodified protein (57,110). Only CDNP-cytochromes *c* that are more inhibited than would be expected on these bases can be considered to have a modification inside the enzymic interaction domain. This is how the cytochrome oxidase domain described earlier was delineated.

Another approach to the localization of enzymic interaction domains on the surface of cytochrome *c* is by differential chemical modification (95–97). One measures the relative chemical reactivities of the ε-amino groups of lysyl residues of the protein free in solution as compared to those on the protein bound to purified preparations of cytochrome oxidase, or for that matter of any other purified protein electron transfer partner that binds tightly to cytochrome *c*, such as cytochrome bc_1. The major advantage of this approach is its ease as compared to the difficult and lengthy preparation of highly purified and well-characterized monosubstituted chemical derivatives of cytochrome *c*. Its major disadvantage is that, unlike the chemical modification approach, it does not involve any measure of electron transfer activity and must assume rather than demonstrate that the complexes formed are the kinetically competent ones.

The location of binding domains by examination of the kinetics of reaction of nature cytochrome *c* of different species (26,106) is much less certain, mainly because conclusions reached from comparisons of proteins varying by numerous residues are at best indirect and generally tenuous. There are, however, some situations in which single-residue changes between cytochromes *c* are known, such as in the numerous mutant bakers' yeast cytochromes *c* (98). Apparently, these have not yet been employed for the thorough structure–function studies needed to decipher the significance of the residue substitutions. A case of a single-residue mutant butterfly cytochrome *c* is described later.

Remarkably, all these approaches to the localization of the interaction domain with cytochrome oxidase have yielded essentially the same result, namely, the area described earlier as the high-affinity enzyme binding site on the top front of cytochrome *c*. Even more remarkably, when the same techniques were employed with other protein electron exchange partners of cytochrome *c*, the same binding site was observed (see Fig. 6). These systems include purified cytochrome bc_1 complex (cytochrome reductase) (96,109), purified cytochrome c_1 (56), the succinate–cytochrome *c* reductase system (1), all of which presumably reflect the binding of cytochrome *c* to cytochrome c_1 located in the inner mitochondrial membrane, as well as cytochrome b_5 (85) and enzymes free in the intermembrane space of mitochondria, including yeast cytochrome *c* peroxidase (49,95) and vertebrate sulfite oxidase (18). Even though the kinetics of reaction of cytochrome *c* with these systems differ from one to another and are not the same as with cytochrome

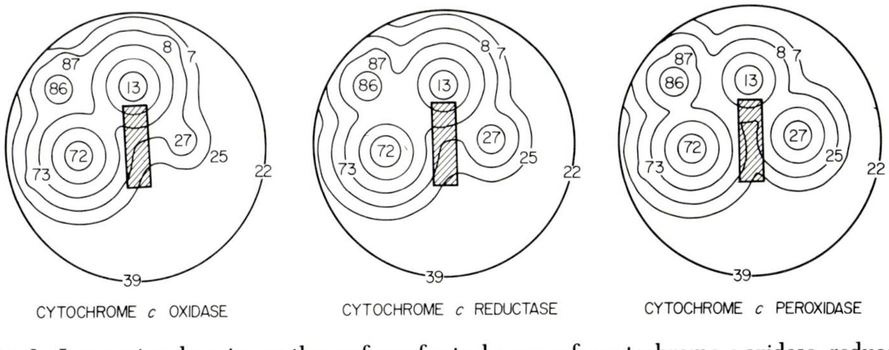

Fig. 6. Interaction domains on the surface of cytochrome *c* for cytochrome *c* oxidase, reduced ubiquinone–cytochrome *c* reductase and for yeast cytochrome *c* peroxidase. The shaded rectangle represents the exposed heme edge. The number of circles around a given lysine residue is proportional to the percentage of the inhibition of the CDNP–cytochrome *c* modified at that lysine that is unaccounted for by the alteration of the dipole moment. The radii of these circles are mutiples of an arbitrary value of 2.5 Å. The numbers indicate the relative positions of the β-carbons of the residues. According to Koppenol *et al.* (57).

oxidase, exhibiting different kinetic parameters, different dependencies on ionic strength, particular ions, pH, etc., the interaction domains on cytochrome *c* are essentially indistinguishable. This again emphasizes the paramount importance of electrostatic forces in aligning cytochrome *c* for reaction with every one of its protein electron exchange partners so far examined (57,110). Furthermore, dipole moment calculations for the CDNP-modified cytochromes *c*, employed as described previously for the reaction with cytochrome oxidase, again permitted a clear distinction between modified residues in the interaction domain and modified residues outside it, but influencing the reaction by virtue of their effect on the electrostatic field (57).

These results also provided some rationale for the presence of a high-affinity, $K_{m_1} \simeq 10^{-8}$ *M* and a low-affinity reaction site, $K_{m_2} \simeq 10^{-6}$ *M*, on cytochrome oxidase for cytochrome *c* (29), and an indication of the mobility of cytochrome *c* required for its respiratory chain electron transport function (109). The reaction of the CDNP-modified cytochromes *c* in the low-affinity phase of the kinetics with cytochrome oxidase are most easily interpreted by assuming that the second molecule of cytochrome *c* to bind productively to the enzyme approaches the first closely (Fig. 7). Each having a net positive charge would exert repulsion on the other, and that repulsion is diminished by modification with CDNP, particularly near the points of closest approach. Whether high- and low-affinity sites on the enzyme are well-defined stable structural entities, whether the first molecule of cytochrome *c* to bind to the oxidase is held with high affinity irrespective of which of two sites it occupies, the second being then fixed with low affinity merely because of close proximity, or whether the first molecule to bind to the oxidase allosterically transforms the second site to a low-affinity state, is not known. Other hypotheses may also explain the present results. In any case, the low-affinity site would appear to be physiologically essential. With one low-affinity and one high-affinity site, as compared to two high-affinity sites, a larger proportion of the cytochrome *c* is free in solution, presumably permitting enough of the protein to remain available for reaction with the enzymes in the intermembrane space, such as the peroxidase of yeast mitochondria and the sulfite oxidase of vertebrate mitochondria. Conversely, if both sites on the oxidase were of low affinity, the efficiency of electron transport in the respiratory chain may be diminished (29,30,74,75).

Finally, since both the mitochondrial reductase and oxidase interact with cytochrome *c* at the same area of its surface, the protein must move to transport electrons between them. It may do so by merely rotating or by two-dimensional diffusion on the surface of the inner mitochondrial membrane or by three-dimensional diffusion. A preliminary attempt to examine the extent of such possible motions in mitochondrial particle preparations, by an examination of the activity of succinate oxidase as a function of cytochrome *c* concentration, is described below.

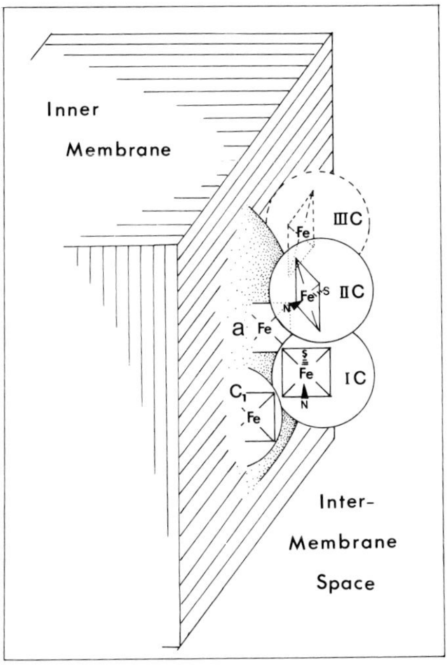

Fig. 7. Schematic representation of possible spatial relationships between cytochrome oxidase, cytochrome c_1 and cytochrome *c*, assuming the "solid-state" model of cytochrome *c* function. The hemes are indicated by the iron-centered squares and drawn to show their relative orientations with respect to the surface of the inner mitochondrial membrane. *N* and S indicate the histidine and methionine heme ligands of cytochrome *c*. Cytochrome oxidase (*a*) and cytochrome c_1 (c_1) are shown embedded in the membrane and associated, so that a molecule of cytochrome *c* (*IC*) can bind to both. This is assumed to allow electron transfer from the cytochrome c_1 heme to the bottom of the exposed heme edge of cytochrome *c*, and from the top of that heme edge to the bottom of the heme edge of cytochrome *a*. A second molecule of cytochrome *c* (*IIC*) can transfer electrons from the bottom of its exposed heme edge to the top of the heme edge of cytochrome *a*. It is shown with the exposed heme edge facing into the plane of the paper and against the mitochondrial membrane, with the bottom of the molecule adjacent to the top of the first cytochrome *c*. A third molecule of cytochrome *c* (*IIIC*) is drawn in dashed lines to represent a condition that occurs only at high concentrations of the protein. According to Ferguson-Miller *et al.* (29).

THE ACTIVITY OF SUCCINATE OXIDASE

There are three obvious models of the way cytochrome *c* may function in transporting electrons between the reductase and oxidase segments of the respiratory chain along the outer surface of the inner mitochondrial membrane (Fig. 8). It can (1) be wedged between closely and specifically related oxidase and reductase complexes, so as not to require any but minimal movement in transferring electrons, the so-called solid-state model; (2) move between more distant oxidase and reductase electron transfer sites by two-dimensional diffusion on the surface of the membrane, being held there

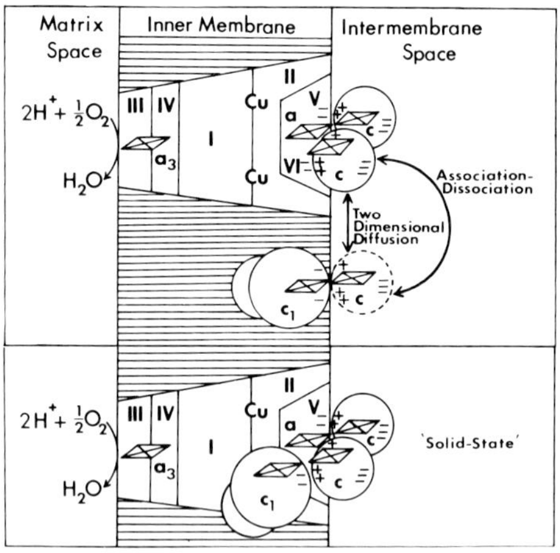

Fig. 8. Schematic representation of possible modes of interaction between cytochrome *c*, cytochrome oxidase, and cytochrome c_1 on the outer surface of the inner mitochondrial membrane. The diagrams illustrate how two molecules of cytochrome *c* may react simultaneously with one molecule of cytochrome aa_3 and may accomplish electron transfer between the oxidase and cytochrome c_1 by various possible mechanisms: two-dimensional diffusion on the surface of membrane, association-dissociation or three-dimensional diffusion via the aqueous phase, and in a "solid state" configuration that implies a ternary complex between cytochromes c_1, *c*, and aa_3. The roman numerals indicate the subunits of cytochrome oxidase where I $\simeq$ 40,000; II $\simeq$ 20,000; III $\simeq$ 15,000; IV $\simeq$ 13,000; V $\simeq$ 11,000; VI $\simeq$ 8,000. The relative positions of the heme groups of the cytochromes are schematically indicated. Positively and negatively charged areas of the proteins are shown by pluses and minuses. According to Ferguson-Miller *et al.* (30).

electrostatically, with the positive end of its dipole axis and the surrounding enzymic interaction domain against the negatively charged membrane surface, an orientation that should facilitate electron exchange with membrane-embedded respiratory chain components, the "two-dimensional diffusion" model; or (3) dissociate into solution in the intermembrane space, the "three-dimensional diffusion" model.

Figure 8 shows these various possibilities diagrammatically. In every case a second molecule of cytochrome *c* is depicted to account for the low-affinity binding. It is always assumed to dissociate from the enzyme into the intermembrane space, since its function is presumed to be reaction with electron exchange partners located in that space. The three models thus refer only to the mechanics of function of the molecule of cytochrome *c* binding with high affinity to the oxidase and taken to be mainly responsible for respiratory chain electron transport.

The succinate oxidase assay consists of two enzymic reactions coupled *via* cytochrome *c*. The kinetics it displays are biphasic with respect to cytochrome *c* concentration, and the apparent Michaelis constants for the high and low affinity phases are in good agreement with those determined for the oxidation of ferrocytochrome *c* by cytochrome oxidase (114). The kinetics of reduction of cytochrome *c* by succinate–cytochrome *c* reductase and purified ubiquinol–

cytochrome *c* reductase are monophasic, indicating that the enzymic reaction that is rate-limiting in the succinate oxidase assay, monitored polarographically, is the oxidation of ferrocytochrome *c* by cytochrome *c* oxidase. Additional support for this conclusion comes from the maximal turnover number for the high affinity phase of five electrons per second, under low ionic strength conditions. This value correlates very well with the turnover number observed for the high affinity phase of the reaction between cytochrome *c* and cytochrome oxidase monitored spectrophotometrically, in which the rate of electron transfer is limited by the dissociation of ferricytochrome *c* from the oxidase. That under any conditions succinate oxidase activity is limited by the rate of dissociation of ferri–cytochrome *c* from its high affinity binding site to cytochrome oxidase necessarily implies that cytochrome *c* must diffuse between oxidase and reductase in its physiological redox cycle (114). Furthermore, such a model provides a simple interpretation of the effect of ionic strength on this system. At low ionic strength the activity is low; as the ionic strength is increased it reaches on optimum at about 50 m*M* and then decreases. This can be explained by assuming that at low ionic strength cytochrome *c* is bound tightly to the membrane, limiting its rate of two-dimensional diffusion. At a higher ionic strength, two dimensional diffusion sustains an optimal rate of electron transport, while at still higher ionic strength three-dimensional diffusion with its decreased efficiency becomes dominant. (114).

This demonstration that cytochrome *c* is indeed mobile in the process of transferring electrons in the respiratory chain contradicts conclusions drawn by Erecińska and co-authors (22–24,116) from experiments in which cytochrome *c* is covalently bound to mitochondrial inner membranes or inner membrane fragments.

In these experiments the cytochrome *c* is modified such that it has one or more substituents carrying on arylazido moiety, which can then be activated on irradiation with light and bind covalently to organic material in its proximity. This photoaffinity labeling procedure is reported to have led to covalent complexes of cytochrome *c* with cytochrome oxidase in molar proportions of about 1:1 and which are competent to transfer electrons from externally added reducing agents or from succinate across the cytochrome reductase–cytochrome oxidase gap.

How these results fit with the observed functional mobility of cytochrome *c* is not immediately obvious. However, the contradiction may not be as absolute as it may appear. Some of the compounds by which cytochrome *c* is held are some 20 Å in length, giving the protein a considerable distance over which it remains free to move. Moreover, if the cytochrome *c* is anchored to lipid, the mobility of the lipid itself in the membrane would be sufficient to account for the rate of electron transfer observed. It can be expected that further work will resolve the apparent conflict.

Do *in Vitro* Tests of Function Reflect Physiological Activities in the Organism?

It has always been a difficult task to assess the physiological significance of a functional activity difference observed under artificial *in vitro* conditions. To further relate such a molecular change to its possible influence as a selective factor in an evolving population is an even more formidable enterprise. As far as this author is aware, it is still debatable whether an unequivocal demonstration has yet been obtained of the mechanism by which any protein molecular change may exert its evolutionary influence.

These difficulties apply in full to cytochrome *c*. When one considers the remaining uncertainties concerning its mechanics of function and the effective ionic strength under which it operates in the mitochondrial intermembrane space, its very high concentrations in the organelle and its very low high-affinity K_m values, one must wonder how any functional change, other than a very drastic one, could possibly be physiologically significant.

Two cases have so far been uncovered that may shed some light on this situation. One concerns primate cytochromes *c* and primate oxidases. The basic data are depicted in Fig. 3 (30,74,75). Old World monkey and ape cytochromes *c* bind with very high affinity to beef cytochrome oxidase, the entire polarographic kinetics consisting of one high-affinity phase. The low-affinity phase is lacking, indicating that only one molecule of cytochrome *c* can react with the enzyme. Prosimian cytochromes *c* react with the beef enzyme very much like the usual nonprimate proteins, whereas the New World monkey proteins show intermediate characteristics. By contrast, the situation is reversed when an Old World monkey oxidase is employed (30,74,75). The conclusions that these phenomena are due to a mutual adaptation of primate cytochrome *c* and primate oxidase in regard to mode of binding and binding affinity and that primate cytochromes *c* bind excessively tightly to the nonprimate enzyme were supported by direct determination of K_D values and stopped-flow kinetic experiments. The latter showed that the rate of oxidation of a primate ferrocytochrome *c* by beef cytochrome oxidase is at least as fast as the oxidation of horse cytochrome *c*, whereas the rate at which TMPD reduces a preformed cytochrome *c* – cytochrome oxidase complex is much slower if the complex is between a primate cytochrome *c* and beef oxidase than if it is between horse cytochrome *c* and beef oxidase (94). The tightness of binding presumably decreases the rate at which the dye can reduce the enzyme-bound cytochrome *c* leading to the small V_{max} values shown in Fig. 3.

Clearly, as discussed earlier, those mutations in the primate cytochrome *c* and primate oxidase lines of descent that have affected the way in which these proteins interact have resulted in functional changes that have evolved a system with new characteristics while maintaining the two components of the system well adapted to each other. Why this set of evolutionary changes was initiated at all is not obvious, since the final products do not seem to present any advantage over the presumed initial state, as far as our present tests of function can show. This set of proteins nevertheless presents an opportunity for unraveling the functional concomitants of evolutionary changes in precise molecular terms. Whether this will eventually lead to an understanding of what selective pressures underlie the process remains to be seen.

An even more interesting situation has been recently brought to light (51). A laboratory colony of the white cabbage butterfly, *Pieris brassicae*, was found to contain a strain carrying a cytochrome *c*, which, preliminary data indicate, differs from the wild-type protein by a change of the lysyl residue at position 39 into a glutemic acid. Though this residue is not in the enzyme interaction domain, the mutation has a large depressing effect on activity with both cytochrome oxidase and cytochrome reductase (111), whether the enzymes are from the same species, other related lepidoptera, or horse heart. The preliminary data obtained so far on small amounts of material also indicate that the concentration of cytochrome *c* in the thorax muscle mitochondria of the mutant butterfly is possibly several times higher than it is in the wild type. Interestingly, the first breeding experiments appear to indicate that it is the heterozygote butterflies that have the advantage over either of the homozygotes (51). Whether or not these early results will be fully substantiated by further experimentation, there is no doubt that this situation provides an unusually favorable system for the determination of the functional results of a spontaneous evolutionary change in cytochrome *c* structure and the possibility of uncovering the mechanisms of the related selective influences.

REFERENCES

1. Ahmed, A.J., Smith, H.T., Smith, M.B., and Millett, F.S., *Biochemistry*, 1978, **17**:2479.
2. Antonini, E., Brunori, M., Colosimo, A., Greenwood, C., and Wilson, M.T., *Proc. Natl. Acad. Sci. U.S.A.*, 1977, **74**:3128.
3. Arnheim, N., In: *The Antigens* (M. Sela, ed.). Academic Press, New York, 1973:377.
4. Borden, D., and Margoliash, E., In: *Handbook of Biochemistry* (G. Fasman, ed.), 3rd ed. Chemical Rubber Publishing Co., Cleveland, Ohio, 1976:268.
5. Boulter, D., *Syst. Zool.*, 1973, **22**:549.
6. Boulter, D., Ramshaw, J.A.M., Thompson, E.W., Richardson, M., and Brown, R.H., *Proc. R. Soc. London [B]*, 1972, **181**:441.
7. Brautigan, D.L., Ferguson-Miller, S., and Margoliash, E., *J. Biol. Chem.*, 1978, **253**:130.
8. Brautigan, D.L., Ferguson-Miller, S., and Margoliash, E., In: *Methods in Enzymology* (S. Fleischer and L. Packer, eds.), Vol. 53. Academic Press, New York, 1978:128–164.

9. Brautigan, D.L., Ferguson-Miller, S., Tarr, G.E., and Margoliash, E., *J. Biol. Chem.*, 1978, **253**:140.
10. Chance, B., and Leigh, J.S., *Proc. Natl. Acad. Sci. U.S.A.*, 1977, **74**:4777.
11. Chance, B., Saronio, C., and Leigh, J.S., *J. Biol. Chem.*, 1975, **250**:9926.
12. Chance, B., Saronio, C., and Leigh, J.S., *Proc. Natl. Acad. Sci. U.S.A.*, 1975, **72**:1635.
13. Dayhoff, M.O. (ed.), *Atlas of Protein Sequence and Structure*. Vol. 5. National Biomedical Research Foundation, Washington, D.C., 1972.
14. Dayhoff, M. O. (ed.), *Atlas of Protein Sequence and Structure*. Suppl. 1. National Biomedical Research Foundation, Washington, D.C., 1973.
15. Dayhoff, M.O. (ed.), *Atlas of Protein Sequence and Structure*. Suppl. 2. National Biomedical Research Foundation, Washington, D.C., 1976.
16. Dayhoff, M.O. (ed.), *Atlas of Protein Sequence and Structure*. Suppl. 3. National Biomedical Research Foundation, Washington, D.C., 1978.
17. Dethmers, J.K., Ferguson-Miller, S., and Margoliash, E., *J. Biol. Chem.*, 1979, **254**:11973–11981.
18. Dethmers, J.K., Margoliash, E., and Rajagopalan, K.V., *Proc. Int. Congr. Biochem., 11th, 1979* 1979, Abstract 04-2-S64.
19. Dickerson, R.E., *J. Mol. Evol.*, 1971, **1**:26.
20. Dickerson, R.E., *J. Mol. Biol.*, 1971, **57**:1.
21. Dickerson, R.E., Timkovich, R., and Almassy, R.J., *J. Mol. Biol.*, 1976, **100**:473.
22. Erecińska, M., *Biochem. Biophys. Res. Commun.*, 1977, **76**:495.
23. Erecińska, M., Oshino, R., and Wilson, D.F., *Biochem. Biophys. Res. Commun.*, 1980, **92**:743.
24. Erecińska, M., Vanderkooi, J.M., and Wilson, D.F., *Arch. Biochim. Biophys.*, 1975, **171**:108.
25. Errede, B., Haight, G.P., and Kamen, M.D., *Proc. Natl. Acad. Sci. U.S.A.*, 1976, **73**:113.
26. Errede, B., and Kamen, M.D., *Biochemistry*, 1978, **17**:1015.
27. Ferguson-Miller, S., Brautigan, D.L., Chance, B., Waring, A., and Margoliash, E., *Biochemistry*, 1978, **17**:2246.
28. Ferguson-Miller, S., Brautigan, D.L., and Margoliash, E., *J. Biol. Chem.*, 1976, **251**:1104.
29. Ferguson-Miller, S., Brautigan, D.L., and Margoliash, E., *J. Biol. Chem.*, 1978, **253**:149.
30. Ferguson-Miller, S., Brautigan, D.L., and Margoliash, E., In: *The Porphyrins* (D. Dolphin, ed.). Academic Press, New York, 1979:149–240.
31. Ferguson-Miller, S., Brautigan, D.L., and Margoliash, E., In: *Tunneling in Biological Systems* (B. Chance, R.A. Marcus, D.C. DeVault, J. Robert Schrieffer, H. Frauenfelder, and N. Sutin, eds.). Academic Press, New York, 1979:513–521.
32. Ferguson-Miller, S., Weiss, H., Speck, S., Brautigan, D.L., Osheroff, N., and Margoliash, E., In: *Cytochrome Oxidase* (B. Chance, T.E. King, K. Okunuki, and Y. Orii, eds.). Elsevier, Amsterdam, 1979:281–292.
33. Fitch, W.M., *J. Mol. Biol.*, 1970, **49**:1.
34. Fitch, W.M., *Syst. Zool.*, 1970, **19**:99.

35. Fitch, W.M., *Biochem. Genet.*, 1971, **5**:231.
36. Fitch, W.M., *Syst. Zool.*, 1971, **20**:406.
37. Fitch, W.M., *Annu. Rev. Genet.*, 1973, **7**:343.
38. Fitch, W.M., *J. Mol. Evol.*, 1973, **2**:123.
39. Fitch, W.M., *J. Mol. Evol.*, 1973, **2**:181.
40. Fitch, W.M., and Langley, C.H., *Fed. Proc., Fed. Am. Soc. Exp. Biol.*, 1976, **35**:2092.
41. Fitch, W.M., and Margoliash, E., *Biochem. Genet.*, 1967, **1**:65.
42. Fitch, W.M., and Margoliash, E., *Science*, 1967, **155**:279.
43. Fitch, W.M., and Margoliash, E., *Brookhaven Symp. Biol.*, 1968, **21**:217.
44. Fitch, W.M., and Margoliash, E., *Evol. Biol.*, 1970, **4**:67.
45. Fitch, W.M., and Markowitz, E., *Biochem Genet.*, 1970, **4**:579.
46. Hackenbrock, C.R., In: *Structure of Biological Membranes* (S. Abrahamsson and I. Pascher, eds.). Plenum, New York, 1976:199–234.
47. Jukes, T.H., and Holmquist, R., *Science*, 1972, **177**:530.
48. Jukes, T.H., and King, J.L., *Nature (London)*, 1971, **231**:114.
49. Kang, C.H., Brautigan, D.L., Osheroff, N., and Margoliash, E., *J. Biol. Chem.*, 1978, **253**:6502.
50. Kang, C.H., Ferguson-Miller, S., and Margoliash, E., *J. Biol. Chem.*, 1977, **252**:919.
51. Kayser, H., Unpublished results, 1980.
52. Kimura, M., *Nature (London)*, 1968, **217**:624.
53. Kimura, M., *Proc. Natl. Acad. Sci. U.S.A.*, 1969, **63**:1181.
54. Kimura, M., and Ohta, T., *J. Mol. Evol.*, 1971, **1**:1.
55. King, J.L., and Jukes, T.H., *Science*, 1969, **164**:788.
56. König, B.W., Osheroff, N., Wilms, J., Muijsers, A.O., Dikker, H.L., and Margoliash, E., *FEBS Lett.*, 1980, **111**:395.
57. Koppenol, W.H., Ferguson-Miller, S., Osheroff, N., Speck, S.H., and Margoliash, E., In: *Oxidases and Related Redox Systems* (T.E. King, H.S. Mason, and M. Morrison, eds.). University Park Press, Baltimore, Maryland, 1980 (in press).
58. Koppenol, W.H., Vroonland, C.A.J., and Braams, R., *Biochim. Biophys. Acta*, 1978, **503**:499.
59. Kreil, G., and Tuppy, H., *Nature (London)*, 1961, **192**:1123.
60. Langley, C.H., and Fitch, W.M., In: *Genetic Structure of Populations* (N.E. Morton, ed.). University of Hawaii Press, Honolulu, 1973:246.
61. Langley, C.H., and Fitch, W.M., *J. Mol. Evol.*, 1974, **3**:161.
62. Malmström, B.G., *Biochim. Biophys. Acta*, 1979, **549**:281.
63. Margoliash, E., *J. Biol. Chem.*, 1962, **237**:2161.
64. Margoliash, E., *Proc. Natl. Acad. Sci. U.S.A.*, 1963, **50**:672.
65. Margoliash, E., *Harvey Lect.*, 1972, **66**:177.
66. Margoliash, E., and Fitch, W.M., *Ann. N.Y. Acad. Sci.*, 1968, **151**:359.
67. Margoliash, E., and Fitch, W.M., In: *Homologies in Enzymes and Metabolic Pathways* (W. J. Whelan, ed.). North-Holland Publishers, Amsterdam, 1970:33.
68. Margoliash, E., and Fitch, W.M., *Taxon*, 1971, **20**:51.
69. Margoliash, E., and Schejter, A., *Adv. Protein Chem.*, 1966, **21**:113.
70. Margoliash, E., and Smith, E.L., *Nature (London)*, 1961, **192**:1121.

71. Margoliash, E., and Smith, E.L., *J. Biol. Chem.*, 1962, **237**:2151.
72. Margoliash, E., and Smith, E.L., In: *Evolving Genes and Proteins* (V. Bryson and H.J. Vogel, eds.). Academic Press, New York, 1965:221.
73. Margoliash, E., and Walasek, O., In: *Methods in Enzymology* (R.W. Estabrook and M.E. Pullman, eds.), Vol. 10. Academic Press, New York, 1967:339.
74. Margoliash, E., Ferguson-Miller, S., Brautigan, D.L., and Chaviano, A.H., In: *Structure-Function Relationships in Proteins* (R. Markham and R.W. Horne, eds.). Elsevier, Amsterdam, 1976:145.
75. Margoliash, E., Ferguson-Miller, S., Brautigan, D.L., Kang, C.H., and Dethmers, J.K., In: *Structure and Function of Energy-Transducing Membranes* (K. Van Dam and B.F. Van Gelder, eds.). Elsevier, Amsterdam, 1977:69.
76. Margoliash, E., Ferguson-Miller, S., Kang, C.H., and Brautigan, D.L., *Fed. Proc., Fed. Am. Soc. Exp. Biol.*, 1976, **35**:2124.
77. Margoliash, E., Fitch, W.M., and Dickerson, R.E., *Brookhaven Symp. Biol.*, 1968, **21**:259.
78. Margoliash, E., Fitch, W.M., Markowitz, E., and Dickerson, R.E., In: *Oxidation Reduction Enzymes* (A. Åkeson and A. Ehrenberg, eds.). Pergamon, Oxford, 1972:5.
79. Margoliash, E., Kimmel, J.R., Hill, R.L., and Schmidt, W.R., *J. Biol. Chem.*, 1962, **237**:2148.
80. Margoliash, E., Smith, E.L., Kreil, G., and Tuppy, H., *Nature (London)*, 1961, **192**:1125.
81. Millett, F.S., Ahmed, A.J., Smith, H.T., and Smith, M.B., *Fed. Proc., Fed. Am. Soc. Exp. Biol.*, 1978, **37**:1324.
82. Minnaert, K., *Biochim Biophys. Acta*, 1961, **50**:23.
83. Montgomery, D.L., Hall, B.D., Gillam, S., and Smith, M., *Cell*, 1978, **14**:673.
84. Montgomery, D.L., Leung, D.W., Smith, M., Shabit, P., Faye, G., and Hall, B.D., *Proc. Natl. Acad. Sci. U.S.A.*, 1980, **77**:541.
85. Ng, S., Smith, M.B., Smith, H.T., and Millett, F., *Biochemistry*, 1977, **16**:4975.
86. Nichols, P., In: *Oxidases and Related Redox Systems* (T.E. King, H.S. Mason, and M. Morrison, eds.). Wiley, New York, 1965:764.
87. Nicholls, P., and Kimelberg, H.K., In: *Biochemistry and Biophysics of Mitochondrial Membranes* (G.F. Azzone, E. Cargoli, A.L. Lehninger, E. Quagliariello, and N. Siliprandi, eds.). Academic Press, New York, 1972:17.
88. Nolan, C., and Margoliash, E., *Annu. Rev. Biochem.*, 1968, **37**:727.
89. Ohta, T., and Kimura, M., *J. Mol. Evol.*, 1971, **1**:18.
90. Orii, Y., and King, T.E., In: *Frontiers of Biological Energetics: Electrons to Tissues* (A. Scarpa, P.L. Dutton, and J.S. Leigh, eds.). Academic Press, New York, 1978:833.
91. Osheroff, N., Borden, D., Koppenol, W.H., and Margoliash, E., In: *Cytochrome Oxidase* (B. Chance, T.E. King, K. Okunuki, and Y. Orii, eds.). Elsevier, Amsterdam, 1979:385.
92. Osheroff, N., Brautigan, D.L., and Margoliash, E., *J. Biol. Chem.*, 1980, **255**:8245.
93. Osheroff, N., Koppenol, W.H., and Margoliash, E., In: *Frontiers of Biological Energetics* (P.L. Dutton, J.S. Leigh, and A. Scarpa, eds.), Vol. 1. Academic Press, New York, 1978:439.

94. Osheroff, N., Margoliash, E., Veerman, E.C.I., Wilms, J., and Muijsers, A.O., Unpublished results, 1980.
95. Pettigrew, G., *FEBS Lett.*, 1978, **86**:14.
96. Rieder, R., and Bosshard, H.R., *FEBS Lett.*, 1978, **92**:223.
97. Rieder, R., and Bosshard, H.R., *J. Biol. Chem.*, 1978, **253**:6045.
98. Sherman, F., Stewart, J.W., Parker, J.H., Putterman, J., Agrawal, B.B.L., and Margoliash, E., *Symp. Soc. Exp. Biol.*, 1970, **24**:85.
99. Slater, E.C., *Biochem. J.*, 1949, **44**:305.
100. Slater, E.C., *Biochem. J.*, 1949, **45**:1.
101. Smith, E.L., In: *The Enzymes* (P.D. Boyer, ed.), 3rd ed., Vol 1. Academic Press, New York, 1970:267.
102. Smith, E.L., and Margoliash, E., *Fed. Proc., Fed. Am. Soc. Exp. Biol.*, 1964, **23**:1243.
103. Smith, H.T., Staudenmayer, N., and Millett, F., *Biochemistry*, 1977, **16**:4971.
104. Smith, L., and Conrad, H., *Arch. Biochem. Biophys.*, 1965, **63**:403.
105. Smith, L., Davies, H.C., and Nava, M.E., *J. Biol. Chem.*, 1974, **249**:2904.
106. Smith, L., Davies, H.C., and Nava, M.E., *Biochemistry*, 1976, **15**:5827.
107. Smith, L., Nava, M.E., and Margoliash, E., In: *Oxidases and Related Redox Systems* (T.E. King, H.S. Mason, and M. Morrison, eds.). University Park Press, Baltimore, Maryland, 1973:629.
108. Smith, M., Leung, D.W., Gillam, S., Astell, C.R., Montgomery, D.L., and Hall, B.D., *Cell*, 1979, **16**:753.
109. Speck, S.H., Ferguson-Miller, S., Osheroff, N., and Margoliash, E., *Proc. Natl. Acad. Sci. U.S.A.*, 1979, **76**:155.
110. Speck, S.H., Koppenol, W.H., Osheroff, N., Dethmers, J.K., Kang, C.H., and Margoliash, E., In: *Membrane Bioenergetics* (C.P. Lee, G. Schatz, and L. Ernster, eds.). Addison-Wesley, Reading, Massachusetts, 1980:31.
111. Speck, S.H., Osheroff, N., Tarr, G.E., and Kayser, H., *Fed. Proc., Fed. Am. Soc. Exp. Biol.*, 1980, **39**:2005.
112. Staudenmayer, N., Ng, S., Smith, M.B., and Millett, F., *Biochemistry*, 1977, **16**:600.
113. Staudenmayer, N., Ng, S., Smith, M.B., Spies, F.J., Jr., and Millett, F., *Biochemistry*, 1976, **15**:3198.
114. Swanson, M., Speck, S.N., Koppenol, W.H., and Margoliash, E., Unpublished data, 1980.
115. Uzzell, T., and Corbin, K.W., *Science*, 1971, **172**:1089.
116. Waring, A., Chance, B., and Erecińska, M., *Fed. Proc., Fed. Am. Soc. Exp. Biol.*, 1979, **38**:576.
117. Williams, J.N., Jr., *Arch. Biochem. Biophys.*, 1964, **107**:537.
118. Wu, T.T., Fitch, W.M., and Margoliash, E., *Annu. Rev. Biochem.*, 1974, **43**:539.
119. Yonetani, T., and Ray, G.S., *J. Biol. Chem.*, 1965, **235**:3392.

Appendix I
Tributes to Professor Emil L. Smith by Symposium Participants

TRIBUTE BY ROBERT L. HILL

It is a pleasure to contribute to this volume the discussion of some recent work from my laboratory and to be together with friends and colleagues of the man we honor on this occasion. My remarks would be incomplete, however, without a few personal words about Emil L. Smith, since he has influenced so many, and especially myself, in so many different ways.

I first met Emil L. Smith in April, 1953, at the annual Federation meetings in Chicago, although I had come to know his science earlier from his publications on the metal-activated peptidases. Our meeting was arranged to consider whether I could do postdoctoral research in his laboratory, and in particular, work with him on the peptidases. I remember little I talked about at that first meeting, held at lunch at the Blackstone Hotel over corned beef and cabbage, but whatever was said did not jeopardize me in Emil's eyes, and a little over a year later I arrived in Salt Lake City with a two-year NIH postdoctoral fellowship to begin what proved to be a seven-year stay in his laboratory, the last five years as a junior faculty member in the Biochemistry Department at the University of Utah. There is no way for me to assess objectively the impact on my future of those seven years, although I know he influenced me greatly by his ideas, his scientific standards, his views of our work and that of others, and not in the least, our discussions of all matters of interest at the time. But whatever else was learned, I believe that I learned biochemistry, how to think about it, how to do it, how to communicate its substance to others, and even how to argue about it, for Emil loved even then, as he does today, a good, passionate dialogue. Perhaps of equal importance, he introduced me to those biochemists whose work and ideas I was to inherit if I too were to become a part of our scientific culture—Selig Hecht, David Keilin, Max Bergmann, who were Emil's own mentors—Willy Kuhne, Otto Meyerhof, Emil Fischer, Frederick Sanger, Francis Crick, John Edsall,

ISBN 0-12-643150-7

and dozens of others. Thus, on leaving his laboratory to begin my own, I was equipped by Emil, perhaps unconsciously so at the time, although retrospectively perhaps far better than many my age, to try my hand at biochemistry. The discussion given of recent research by students and fellows in my laboratory I consider a progress report to Emil, a statement for him alone to evaluate whether his efforts with me were worthwhile. Irrespective of his decision, I sincerely hope those younger aspirants to biochemistry, who call me now their mentor, have learned something of Emil's gifts to me and will carry them on through me to their own. They join me in honoring Emil L. Smith on this occasion of his advancement to Professor Emeritus of Biological Chemistry, and they thank him, just as I do, for his many contributions to his science, to his friends, and to his students. Thus, with Thomas Beyer, Lowrie Glasgow, Earl Guthrow, Michael Imber, Mark Lehrman, James Paulson, Jean-Paul Prieels, Salvatore Pizzo, James Rearick, and Evan Sadler, whose studies I describe in this volume, I wish my mentor, Emil Smith, the very best as he continues his newly chosen tasks in biochemistry.

Tribute by Bo G. Malmström

When I worked as a graduate student in the University of Minnesota from 1948 to 1951, I participated in the evening seminars on enzyme chemistry organized by another contributor to this symposium, Professor Paul D. Boyer. These seminars made me decide that after completion of my Ph.D. I would change my field of research and try to investigate enzyme mechanisms. In search for a specific problem to tackle I read the enzyme literature thoroughly, and I was fascinated by a review entitled "Role of Metal Ions in Enzyme Systems" that was published by A. L. Lehninger (1) in 1950. Through this I discovered an extensive series of papers by Emil Smith on the metal ion-activation of peptidases. These studies led him to suggest that one role of the metal ion is to form a bridge between the protein and substrate in formation of the enzyme–substrate complex (7). He realized furthermore that the ability of the metal ion to participate in chelate formation may contribute to the destabilization of the substrate in this complex.

Emil Smith's conclusions were based largely on kinetic measurements. He had also studied (8) complex formation between activating ions and the substrate, but the lability of the peptidases precluded a characterization of their metal-binding properties. It appeared to me that a rigorous test of his ideas should include the correlation of the kinetic results with direct measurements of the binding of the activating ions to a homogeneous, well-defined enzyme. After studying the literature I decided that only two enzymes, enolase and phosphoglycerate kinase, were suitable for my project unless I wanted to develop new purification procedures.

Fortunately, I was soon able to try out my ideas in the laboratory. At a congress in New York I met Professor Arne Tiselius, who offered me a position in his department in Uppsala. Over Christmas 1951, I returned by ship to my native Sweden, setting to work in Tiselius' laboratory in January of 1952. I was lucky in my experimental efforts, and in about one month I had prepared a sufficient amount of crystalline enolase to initiate metal-binding studies with the equilibrium-dialysis technique. My first results were published as a note in "Nature" in February 1953 (2). In the following year I was extremely happy to find that Emil Smith, whom I had never met but whom I still considered my mentor, judged my experiments significant enough to describe them in his contribution (10) to the McCollum-Pratt Symposium on "The Mechanism of Enzyme Action." The next year he referred to them again (9) at the Faraday Society discussion in Oxford on "The Physical Chemistry of Enzymes," in response to a comment from R. J. P. Williams and B. L. Vallee, who alleged that metal-ion activation represents nonspecific effects that must be distinguished from the role of metals in metalloenzymes. I was even more thrilled by a letter from Emil Smith asking if he could visit me in Uppsala on his return from the International Congress in Brussels and the Oxford meeting. I think he would agree with me that we had a delightful time during this visit, not the least because we discovered that we had similar tastes not only in biochemistry but also in literature and music, and this became the beginning of a life-long friendship.

In 1957 Emil Smith and I met again, this time at the "Symposium on Protein Structure" held at the Sorbonne in Paris. By then I had published a series of papers on the kinetic and metal ion-binding properties of enolase (3) and also had initiated similar studies on phosphoglycerate kinase together with a student, M. Larsson-Raźnikiewics (4). I had become convinced that to understand the catalytic action of these enzymes I would have to supplement the approach used until then with structural studies. As Emil Smith had already established himself as one of the leading investigators of protein primary structure, I asked if I could visit his laboratory in Salt Lake City to apply his techniques to enolase. I was accepted and a Rockefeller Foundation fellowship made it possible for me to spend most of the spring and early summer of 1958 at the University of Utah. My work there resulted in a joint publication describing the amino acid composition and amino-terminal sequence of yeast enolase (5).

This was my first paper on primary structure, and it turned out to be my only one. [Before going to Utah I had been fascinated by a new physical technique, now well known as electron paramagnetic resonance (EPR), that I wanted to use to characterize the interaction between the enzyme and the activating ion(6).] I did set up the techniques I had learned at Utah in the Uppsala laboratory and two of my former students, P. O. Nyman and L. Rydén, have carried out extensive structural investigations on metalloenzymes. Both moved with me when I left Uppsala in 1963 to set up a new Biochemistry Department in Göteborg, and Numan is still associated with this department as an independent investigator and faculty member.

References to Tribute by Bo G. Malmström

1. Lehninger, A.L., Role of metal ions in enzyme systems. *Physiol. Rev.*, 1950, **30**:393–429.
2. Malmström, B.G., Interaction of manganous ions with enolase. *Nature (London)*, 1953, **171**:392–393.
3. Malmström, B.G., *The Mechanism of Metal-ion Activation of Enzymes. Studies on Enolase.* Inaugural dissertation, Uppsala, 1956.
4. Malmström, B.G., and Larsson-Raźnikiewicz, M., 3-phosphoglycerate kinase. In: *The Enzymes* (P.D. Boyer, H.A. Lardy, and K. Myrback, eds.) 2nd ed., Vol. 6. Academic Press, New York, 1961:85–94.
5. Malmström, B.G., Kimmel, J.R., and Smith, E.L., Amino acid composition and amino-terminal sequence of yeast enolase. *J. Biol. Chem.*, 1959, **234**:1108–1111.
6. Malmström, B.G., Vänngård, T., and Larsson, M., An electron-spin-resonance study of the interaction of manganous ions with enolase and its substrate. *Biochim. Biophys. Acta*, 1958, **30**:1–5.
7. Smith, E.L., The mode of action of the metal-peptidases. *Proc. Natl. Acad. Sci. U.S.A.*, 1949, **35**:80–90.
8. Smith, E.L., Aspects of the specificity and mode of action of some peptidases. In: *Enzymes and Enzyme Systems—Their State in Nature* (J.T. Edsall, ed.), No. 1. Harvard University Press, Cambridge, Massachusetts, 1951:47–76.
9. Smith, E.L., The physical chemistry of enzymes. *Faraday Soc.*, 1955, **20**: 262.
10. Smith, E.L., Davis, N.C., Adams, E., and Spackman, D.H., The specificity and mode of action of two metal-peptidases. In: *The Mechanism of Enzyme Action* (W.D. McElroy and B. Glass, eds.). Johns Hopkins Press, Baltimore, Maryland, 1954:291–312.

Tribute by Thomas H. Jukes

It is just fifty years since I was given my first laboratory job, which was to analyze egg proteins for tyrosine and tryptophan. For about half this time, I have enjoyed Emil's friendship.

Before meeting him, I was aware of his presence in protein biochemistry, and later I found that he combined the roles of scientific father and brother to his colleagues and students. Of all my friends, he is the one from whom I learn most by arguing. He loves to hold forth on a variety of scientific subjects. The impact of his logical and incisive mind is compelling, and his smile is captivating. His laboratory research in the field of proteins has been prodigious, and his standards of excellence are the highest.

Scientists gladly come to this meeting from four continents to pay a tribute to Emil Smith, the man, and to his scientific leadership. I am indeed grateful for the opportunity to participate and to wish him continued happiness in the years ahead.

Appendix II
Representative Publications of Emil L. Smith (1935–1979)

1. Hecht, S., Shlaer, S., and Smith, E. L., Intermittent light stimulation and the duplicity theory of vision. *Cold Spring Harbor Symp. Quant. Biol.*, 1935, **3**:237.
2. Smith, E.L., The influence of light and carbon dioxide on photosynthesis. *J. Gen. Physiol.*, 1937, **20**:807.
3. Smith, E.L., Solution of chlorophyll-protein compounds (phyllochlorine) extracted from spinach. *Science*, 1938, **88**:170.
4. Chase, A.M., and Smith, E.L., Regeneration of visual purple in solution. *J. Gen. Physiol.*, 1939, **23**:21.
5. Keilin, D., and Smith, E.L., Direct perception of pigment in the nerve tissue of human retina. *Nature (London)*, 1939, **143**:333.
6. Smith, E.L., Chlorophyll as the prosthetic group of a protein in the green leaf. *Science*, 1940, **91**:199.
7. Smith, E.L., and Pickels, E.G., Micelle formation in aqueous solutions of digitonin. *Proc. Natl. Acad. Sci. U.S.A.*, 1940, **26**:272.
8. Smith, E.L., The chlorophyll-protein compound of the green leaf. *J. Gen. Physiol.*, 1941, **24**:565.
9. Smith, E.L., The action of sodium dodecyl sulfate on the chlorophyll-protein compound of the spinach leaf. *J. Gen. Physiol.*, 1941, **24**:583.
10. Smith, E.L., and Pickels, E.G., The effect of detergents of the chlorophyll-protein compound of spinach as studied in the ultracentrifuge. *J. Gen. Physiol.*, 1941, **24**:753.
11. Smith, E.L., and Bergmann, M., The peptidases of intestinal mucosa. *J. Biol. Chem.*, 1944, **153**:627.
12. Smith, E.L., Manganese and l-leucine-aminoexopeptidase. *J. Biol. Chem.*, 1946, **163**:15.
13. Smith, E.L., The immune proteins of bovine colostrum and plasma. *J. Biol. Chem.*, 1946, **164**:345.
14. Smith, E.L., Isolation and properties of immune lactoglobulins from bovine whey. *J. Biol. Chem.*, 1946, **165**:665.
15. Smith, E.L., and Gerlough, T.D., The isolation and properties of the proteins

associated with tetanus antitoxic activity in equine plasma. *J. Biol. Chem.*, 1947, **167**:679.
16. Hanson, H.T., and Smith, E.L., The application of peptides containing β-alanine to the study of the specificity of various peptidases. *J. Biol. Chem.*, 1948, **175**:833.
17. Smith, E.L., The isolation and properties of the immune proteins of bovine milk and colostrum and their role in immunity: A review. *J. Dairy Sci.*, 1948, **31**:127.
18. Smith, E.L., Action of carboxypeptidase on peptide derivatives of L-tryptophan. *J. Biol. Chem.*, 1948, **175**:39.
19. Smith, E.L., and Holm, A., The transfer of immunity to the new born calf from colostrum. *J. Biol. Chem.*, 1948, **175**:349.
20. Hanson, H.T., and Smith, E.L., Carnosinase: An enzyme of swine kidney. *J. Biol. Chem.*, 1949, **179**:789.
21. Smith, E.L., The mode of action of the metal-peptidases. *Proc. Natl. Acad. Sci. U.S.A.*, 1949, **35**:80.
22. Smith, E.L., and Polglase, W.J., The specificity of leucine amino-peptidase. II. Optical and side chain specificity. *J. Biol. Chem.*, 1949, **180**:1209.
23. Smith, E.L., and Lumry, R., Some consideration of the interaction of the metal-peptidases with their substrates. *Cold Spring Harbor Symp. Quant. Biol.*, 1950, **14**:168.
24. Adams, E., and Smith, E.L., Proteolytic activity of pituitary extracts. *J. Biol. Chem.*, 1951, **191**:651.
25. Lumry, R., Smith, E.L., and Glantz, R.R., Kinetics of carboxypeptidase action I. Effect of various extrinsic factors on kinetic parameters. *J. Am. Chem. Soc.*, 1951, **73**:4330.
26. Smith, E.L., The specificity of certain peptidases. *Adv. Enzymol.*, 1951, **12**:191.
27. Smith, E.L., Proteolytic enzymes. In *The Enzymes* (K. Myrback and J.B. Sumner, eds.). Academic Press, New York, 1951:793–872.
28. Smith, E.L., Lumry, R., and Polglase, W.J., The Van der Waals factor in carboxypeptidase interaction with inhibitors and substrates. *J. Phys. Colloid Chem.*, 1951, **55**:125.
29. Kimmel, J.R., and Smith, E.L., Crystalline papain. I. Preparation, specificity, and activation. *J. Biol. Chem.*, 1954, **207**:515.
30. Smith, E.L., Kimmel, J.R., and Brown, D.M., Crystalline papain. II. Physical studies; the mercury complex. *J. Biol. Chem.*, 1954, **207**:533.
31. White, A., Handler, P., Smith, E.L., and Stetten, D., Jr., *Principles of Biochemistry*. McGraw-Hill, New York, 1954:XIII, 1117.
32. Smith, E.L., Kimmel, J.R., Brown, D.M., and Thompson, E.O.P., Isolation and properties of a crystalline mercury derivative of a lysozyme from papaya latex. *J. Biol. Chem.*, 1955, **215**:67.
33. Spackman, D.H., Smith, E.L., and Brown, D.M., Leucine aminopeptidase. IV. Isolation and properties of the enzyme from swine kidney. *J. Biol. Chem.*, 1955, **212**:255.
34. Davis, N.C., and Smith, E.L., Purification and some properties of prolidase of swine kidney. *J. Biol. Chem.*, 1957, **224**:261.
35. Hill, R.L., and Smith, E.L., Leucine aminopeptidase. VI. Inhibition by alcohols and other compounds. *J. Biol. Chem.*, 1957, **224**:209.

36. Stockell, A., and Smith, E.L., Kinetics of papain action. I. Hydrolysis of benzoyl-L-argininamide. *J. Biol. Chem.*, 1957, **227**:1.
37. Rosevear, J.W., and Smith, E.L., Structure of glycopeptides from a human γ-globulin. *J. Am. Chem. Soc.*, 1958, **80**:250.
38. Malmstrom, B.G., Kimmel, J.R., and Smith, E.L., Amino acid composition and N-terminal sequence of yeast enolase. *J. Biol. Chem.*, 1959, **234**:1108.
39. Glazer, A. N., and Smith, E. L., Studies on the ultraviolet difference spectra of proteins and polypeptides. *J. Biol. Chem.*, 1961, **236**:2942.
40. Margoliash, E., Smith, E.L., Kreil, G., and Tuppy, H., Amino acid sequence of cytochrome *c*; The complete amino acid sequence. *Nature (London)*, 1961, **192**:1125.
41. Rosevear, J.W., and Smith, E.L., Glycopeptides. I. Isolation and properties of glycopeptides from a fraction of human γ-globulin. *J. Biol. Chem.*, 1961, **236**:425.
42. Nolan, C., and Smith, E.L., Glycopeptides. III. Isolation and properties of glycopeptides from rabbit γ-globulin. *J. Biol. Chem.*, 1962, **237**:446.
43. Nolan, C., and Smith, E.L., Glycopeptides. III. Isolation and properties of glycopeptides from a bovine globulin of colostrum and from fraction II-3 of human globulin. *J. Biol. Chem.*, 1962, **237**:453.
44. Matsubara, H., and Smith, E.L., Human heart cytochrome *c*: Chymotryptic peptides, tryptic peptides, and the complete amino acid sequence. *J. Biol. Chem.*, 1963, **238**:2732.
45. Light, A., Frater, R., Kimmel, J.R., and Smith, E.L., Current status of the structure of papain: The linear sequence, active sulfhydryl group, and the disulfide bridges. *Proc. Natl. Soc. Acad. Sci. U.S.A.*, 1964, **52**:1276.
46. Smith, E.L., and Margoliash, E., Evolution of cytochrome *c*. *Fed. Proc., Fed. Am. Soc. Exp. Biol.*, 1964, **23**:1276.
47. Margoliash, E., and Smith, E.L., Structural and functional aspects of cytochrome *c* in relation to evolution, In *Evolving Genes and Proteins* (V. Bryson and H.J. Vogel, eds.). Academic Press, New York, 1965:221.
48. Smith, E.L., Markland, F.S., Kasper, C.B., DeLange, R.J., Landon, M., and Evans, W.H., The complete amino acid sequences of two types of subtilisin: BPN′ and Carlsberg. *J. Biol. Chem.*, 1966, **241**:5974.
49. DeLange, R.J., Smith, E.L., Fambrough, D.M., and Bonner, J., Amino acid sequence of histone IV: Presence of ϵ-*N*-acetyllysine. *Proc. Natl. Acad. Sci. U.S.A.*, 1968, **61**(3):1145.
50. Smith, E.L., The evolution of proteins. *Harvey Lect.*, 1968, **62**:231.
51. Smith, E.L., Structural Variations in Cytochrome c, in *Structure and Function of Cytochromes* (K. Okunuki, M.D. Kamen, and I. Sekuzu, eds.). University of Tokyo Press, Tokyo, 1968:282–288.
52. Smith, E.L., Delange, R.J., Evans, W.H., Landon, M., and Markland, F.S., Subtilisin Carlsberg. V. The complete sequence; comparison with subtilisin BPN′; evolutionary relationships. *J. Biol. Chem.*, 1968, **243**:2184.
53. DeLange, R.J., Glazer, A.N., and Smith, E.L., Presence and location of an unusual amino acid, ϵ-*N*-trimethyllysine, in cytochrome *c* of wheat germ and neurospora. *J. Biol. Chem.*, 1969, **244**:1385.
54. DeLange, R.J., Glazer, A.N., and Smith, E.L., Identification and location of

ε-*N*-trimethyllysine in yeast cytochromes *c*. *J. Biol. Chem.*, 1970, **245**:3325.
55. Piszkiewicz, D., Landon, M., and Smith, E.L., Anomalous cleavage of aspartyl-proline peptide bonds during amino acid sequence determinations. *Biochem. Biophys. Res. Commun.*, 1970, **40**:1173.
56. Piszkiewicz, D., Landon, M., and Smith, E.L., Bovine liver glutamate dehydrogenase. Sequence of a hexadecapeptide containing a lysyl residue reactive with pyridoxal 5′-phosphate. *J. Biol. Chem.*, 1970, **245**:2622.
57. Smith, E.L., Evolution of enzymes. In *The Enzymes* (P.D. Boyer, ed.), 3rd ed., vol. 1. Academic Press, New York, 1970:267–339.
58. Smith, E.L., DeLange, R.J., and Bonner, J., Chemistry and biology of the histones. *Physiol. Rev.*, 1970, **50**:159.
59. Smith, E.L., Landon, M., Piszkiewicz, D., Brattin, W.J., Langley, T.J., and Melamed, M.D., Bovine liver glutamate dehydrogenase: Tentative amino acid sequence; indentification of a reactive lysine; nitration of a specific tyrosine and loss of allosteric inhibition of guanosine triphosphate. *Proc. Natl. Acad. Sci. U.S.A.*, 1970, **67**:724.
60. Smith, E.L., Markland, F.S., and Glazer, A.N., Some structure-function relationships in the subtilisins. In *Structure-Function Relationships of Proteolytic Enzymes* (P. Desnuelle, H. Neurath, and M. Ottesen, eds.). Munksgaard, Copenhagen, 1970:160–172.
61. Piszkiewicz, D., and Smith, E.L., Bovine liver glutamate dehydrogenase. Equilibria and kinetics of inactivation by pyridoxal. *Biochemistry*, 1971, **10**:4538.
62. Piszkiewicz, D., and Smith, E.L., Bovine liver glutamate dehydrogenase. Equilibria and kinetics of imine formation by lysine-97 with pyridoxal 5′-phosphate. *Biochemistry*, 1971, **10**:4544.
63. Kurihara, M., Markland, F.S., and Smith, E.L., Subtilisin amylosacchariticus. III. Isolation and sequence of the chymotryptic peptides and the complete amino acid sequence. *J. Biol. Chem.*, 1972, **247**:5619.
64. Moon, K., Piszkiewicz, D., and Smith, E.L., Glutamate dehydrogenase: Amino-acid sequence of the bovine enzyme and comparison with that from chicken liver. *Proc. Natl. Acad. Sci. U.S.A.*, 1972, **69**(6):1380.
65. Blumenthal, K.M., and Smith, E.L., Nicotinamide adenine dinucleotide phosphate-specific glutamate dehydrogenase of *Neurospora*. I. Isolation, subunits, amino acid composition, sulfhydryl groups, and identification of a lysine residue reactive with pyridoxal phosphate and *N*-ethylmaleimide. *J. Biol. Chem.*, 1973, **248**:6002.
66. DeLange, R.J., Hooper, J.A., and Smith, E.L., Histone III. III. Sequence studies on the cyanogen bromide peptides; complete amino acid sequence of calf thymus histone III. *J. Biol. Chem.*, 1973, **248**:3261.
67. Moon, K., and Smith, E.L., Sequence of bovine liver glutamate dehydrogenase. VIII. Peptides produced by specific chemical cleavages; the complete sequence of the protein. *J. Biol. Chem.*, 1973, **248**:3082.
68. Moon, K., Piszkiewicz, D., and Smith, E.L., Amino acid sequence of chicken liver glutamate dehydrogenase. *J. Biol. Chem.*, 1973, **248**:3093.
69. Patthy, L., Smith, E.L., and Johnson, J., Histone III. V. The amino acid sequence of pea embryo histone III. *J. Biol. Chem.*, 1973, **248**:6834.
70. Wootton, J.C., Chambers, G.K., Holder, A.A., Baron, A.J., Taylor, J.G., Fincham, J.R.S., Blumenthal, K.M., Moon, K., and Smith, E.L., Amino-acid se-

quence of NADP-specific glutamate dehydrogenase of *Neurospora crassa*. *Proc. Natl. Acad. Sci. U.S.A.*, 1974, **71**:4361.

71. Blumenthal, K.M., and Smith, E.L., Nicotinamide adenine dinucleotide phosphate-specific glutamate dehydrogenase of *Neurospora*. III. Inactivation by nitration of a tyrosine residue involved in coenzyme binding. *J. Biol. Chem.*, 1975, **250**:6560.
72. Blumenthal, K.M., Moon, K., and Smith, E.L., Nicotinamide adenine dinucleotide phosphate-specific glutamate dehydrogenase of *Neurospora*. II. Sequences of the cyanogen bromide peptides; The complete sequence of the enzyme. *J. Biol. Chem.*, 1975, **250**:3644.
73. Patthy, L., and Smith, E.L., Reversible modification of arginine residues. Application to sequence studies by restriction of tryptic hydrolysis to lysine residues. *J. Biol. Chem.*, 1975, **250**:557.
74. Patthy, L., and Smith, E.L., Identification of functional arginine residues in ribonuclease A and lysozyme. *J. Biol. Chem.*, 1975, **250**:565.
75. Smith, E.L., Austen, B.M., Blumenthal, K.M., and Nyc, J.F., Glutamate dehydrogenases. In *The Enzymes* (P.D. Boyer, ed.), Vol. 11, Part A. Academic Press, New York, 1975:293–367.
76. Austen, B.M., and Smith, E.L., Identification of a functional arginine residue involved in coenzyme binding by the NADP-specific glutamate dehydrogenase of *Neurospora*. *J. Biol. Chem.*, 1976, **251**:5835.
77. Austen, B.M., Haberland, M.E., Nyc, J.F., and Smith, E.L., Nicotinamide adenine dinucleotide-specific glutamate dehydrogenase of *Neurospora*. IV. The COOH-terminal 669 residues of the peptide chain; comparison with other glutamate dehydrogenases. *J. Biol. Chem.*, 1977, **252**:8142.
78. White, A., Handler, P., Smith, E.L., Hill, R.L., and Lehman, I.R., *Principles of Biochemistry*. 6th ed. McGraw-Hill, New York, 1978:xi, 1492.
79. Smith, E.L., The evolution of glutamate dehydrogenases and a hypothesis for the insertion or deletion of multiple residues in the interior of polypeptide chains. *Proc. Philos. Soc.*, 1979, **123**(1).

Author Index

A

Aasa, R., 89(8), 94(1), *95, 96*
Adams, E., 325(10), *326, 328*
Adams, M. J., 163(26), *169*
Adelberg, E. A., 174(65), 196(65), *202*
Adler, A. J., 163(10), *168*
Adman, E. T., 90(2), *95,* 162(1), 163(1), *167,* 246(1), 247(1), 249(1), 259(1), *262*
Agaki, J. M., 163(27, 28), *169*
Aggarwal, S. J., 247(2), *262*
Agosto, G., 109(57), *116*
Agrawal, B. B. L., 311(98), *321*
Ahmed, A. J., 307(1), 308(1, 21), 311(1), 313(1), *317, 320*
Air, G. M., 207(26), *219,* 280 (1, 2, 3, 74, 76), 281(74), 285(1), *293, 297*
Alberte, R. S., 223(79), 224(79), *243*
Alden, R. A., 174(59), 176(59), *202*
Allfrey, V. G., 134(2), 144, 146(42, 43), *147,* 154(19), *149, 157*
Allison, A. C., 268(36), *295*
Almassy, R. J., 174(1, 23), 176(1), 180(1), 194(1, 23), *199, 200,* 299(21), *318*
Aloj, S. M., 108(58), 109(35, 43, 44, 55, 58), *114, 115, 116*
Altosaar, I., 41(7), 44, 46(1), *47*
Ambesi-Impiomato, F. S., 109(35), *114, 115*
Ambler, R. P., 174(7, 8), 182, 183(7, 8), 190(7, 8), 191(2), 197(7, 8), *199*
Amelunxen, R. E., 162(2), 163(2), 165(2), *167*
Andersson, B., 88(4), 89(3), *95*
Anderson, J. M., 223(5), 224(1), *239*
Andrew, P. W., 249(59), 250(3), 252(3), *262, 265*
Andréasson, L.-E., 89(16), 90(16), *96*
Andreesen, J. R., 164(18), *168*
Anfinsen, C. B., 6(45, 46, 47), *17*
Ansell, N. J., 80(41), *86*
Antonini, E., 288(73), *297,* 307(2), 308(2), 313(2), *317*
Apell, G. S., 233(20), 234(20, 39), *240, 241*
Archer, O. K., 120, *130*
Arnheim, N., 299(3), 308(3), 314(3), *317*
Arnon, D. I., 245(57), 246(57), 256(72), 259(73), *265, 266*
Arnone, A., 279(4, 5), *293*
Arshad, J. H., 224(50), *241*
Ascoli, M., 105(1), *112*
Ashida, T., 173(9), 176(9), *199*
Ashwell, G., 10, 11(84), *19,* 77(25), *85*
Ashworth, J. N., 124(8), *130*
Astell, C. R., 303(108), *321*
Atassi, M. Z., 105(59), *116*
Austen, B. M., *331*
Axel, R., 144, *148*
Azari, P., 162(3), 163(3), *167*

B

Bachmann, B. J., 190(10), *199*
Bach, J.-F., 121(2, 4), 123, *130, 132*
Bach, M.-A., 121(4), 123(5, 6), *130*
Bäckström, G., 64(22), *84, 85*
Baffa, L., 146(42, 43), *149*
Bagi, G., 143(19), *148*
Bahl, O. P., 105(7), 106(7), *113*
Bahramian, M. B., 43(8), *47, 48*
Bailey, J. L., 6(40), *16*
Bakke, A., 134(3), 136(3), *148*
Bandoni, R. J., *243*

Barbour, S. D., 23(1), *36*
Barber, J., 236(69), 238(73), *242, 243*
Barham, S. S., 154(21), *157*
Barker, W. C., 246(5, 6), 249(6), *262*, 287(6), *293*
Barnabas, J., 180(49), *201*, 282(34), *295*
Barnard, E. A., 7, 9, *17, 18*
Barnes, E. M., 163(27), *169*
Baron, A. J., *330, 331*
Barrell, B. G., 207(26), 208(3), 212, *218, 219*
Barrett, J., 224(1), *239*
Bartholeyns, J., 10, *19*
Bartsch, R. G., 174(7), 176(11, 15), 183(7), 186(6), 190(7), 194(11), 197(7), *199*, 249(10), 259(10), *262*
Baudhuin, P., 10, *19*
Beard, J. M., 280(7), *293*
Beckwith, J. R., 51(31), *62*
Béguin, S., 229(14), 231(14), *239*
Beintema, J. J., 9, *18*
Benesch, R., 279(8), *293*
Benesch, R. E., 279(8), *293*
Bennett, A., 224(49), *241*
Bennett, G., 212(19), *218*
Benson, A. M., 163(29), 169, 245(62), *265, 266*
Beranek, W. E., 67(23), *85*
Berg, B., 143(19), *148*
Berger, A., 6(47), *17*
Bergman, C., 90(5), *95*
Bergmann, M., 1(5), 2(5, 6), *15, 327*
Berns, D. S., 229(3), 231(2), 235(60), 238(4), *239, 242*
Berry, J. A., 175(61), *202*
Bertrand, K., 212(19), *218*
Beyer, T. A., 65, 67(6), 68, 69(4, 5), 70(4), 71, 72, 75, 76(24 *83, 84, 85*
Bhorjee, J. S., 54(1), *60*
Biesecker, G. 159(4), 160, 163(4), 165(4), *167*
Billing, R., 134(4), 135(4), 143(4), *148*
Birken, S., 109(57), *116*
Birktoft, J. J., 40(2), 46(2), *47*
Biserte, G., 136(32), *149*
Blackburn, P., 9(78), 11, 12, 13, 14, *18, 19*
Blair, T. T., 180(50), *201*
Blarot, D., 121(4), *130*
Blouquit, Y., 123(24), *132*
Bloomfield, G. A., 104(69, 70), *117*
Blow, D. M., 40(2), 46(2), *47*
Blumenthal, K. M., *330, 331*
Boardman, N. K., 223(5), *239*
Boccù, E., 165(34), *169*
Böhlen, P., 5(30), *16*
Boettcher, B., 282(38), *295*
Bogorad, L., 229(7), 231(6), *239*
Boime, I., 100(42), *115*
Bolinger, R. E., 163(25), *169*
Bolivar, F., 56(13), *61*
Bolton, W., 271(9), 286(9), *293*
Bolwar, F., 56(18), *61*
Bonaventura, J., 286(10), *293*
Bonner, A. G., 288(11), 289(11), *294*
Bonner, J., 133(6, 26), 133(6), 134(3, 4, 6, 14, 15, 16), 135(4, 13), 136(3, 12, 23), 137(8, 16), 141(34), 143(4, 19), 144(45), 145(5, 7, 20, 45), 147(20), *148, 149, 150*, 151(9, 10, 14, 17, 31), 152(9, 10), 153(9, 10), 154(9, 14), *156, 157, 329, 330*
Bookchin, R. M., 286(51), *296*
Boquet, P., 108, 111(3), *112*
Borden, D., 300(4), 307(91), *317, 320*
Borisov, V. V., 269(77), *297*
Bosshard, H. R., 310(96, 97), 311(96), *321*
Bothe, H., 259(70), *266*
Boulter, D., 247(58), 249(58, 59), 250(3), 252(3), 256(29), *262, 264, 265*, 299(5, 6), *317*
Bowen, C. C., 225(85), *243*
Boyer, H. W., 56(18), *61*
Braams, R., 305(58), *319*
Bradbury, E. M., 146(42), *149*
Brady, R. O., 108(58), 109(43, 55, 58, 64, 65), *115, 116*
Brake, A. J., 56(2), 58(3), *60, 61*
Brand, D. K., 280(12), *294*
Brand, E., 3(8), *15*
Brandt, W. F., 152(34), 153(35), *158*
Brattin, W. J., *330*
Braunitzer, G., 268(14), 285(13), 289(44, 71), *294, 295, 297*
Brautigan, D. L., 299(8, 30, 74, 75, 76), 301(28, 76), 302(28, 74, 75, 92), 303(74, 75), 304(75), 305(30), 306(28, 29, 30), 307(7, 9, 27, 29, 30, 31, 32, 92), 308(8, 27, 30, 49, 74, 75), 309(7, 9, 29, 92), 311(49), 312(29, 30, 74, 75), 313, 314(30), 316(30, 74, 75), *317, 318, 320*
Brenner, S., 51(31), *62*
Brew, K., 66(31), *85*
Bricas, E., 121(4), *130*
Briehl, R., 285(15), *294*
Briving, *95*
Broda, E., 192(12), 196(12), *199*
Brodbeck, U., 66(7), *84*
Brooks, C., 235(8), *239*
Brown, A. S., 232(80), 234(81), *243*
Brown, D. M., 6(43), *17*, 229(14, 42), 231(14), 234(39), *239, 241, 328*
Brown, F. F., 104(4), *112*
Brown, J. L., 40(9), *48*, 51(4), *61*
Brown, N. L., 207(26), *219*
Brown, R. H., 299(6), *317*
Brownlee, G. G., 207(22), *219*
Brunori, M., 288(73), *297*, 307(2), 308(2), 313(2), *317*
Bruschi, M., 246(7), 249(7), *262*
Bustin, M., 145(35), *149*
Brutlag, D., 137, *148*

Bryant, D. A., 234(9, 39), 237, 238(10), *239, 241*
Buchanan, B. B., 175(61), *202*, 245(35), *264*
Buchanan, R. E., 163(5), *167*, 182(13), 196(13), *199*
Buehner, M., 159(6), 163(6, 26), *167, 169*
Burger, M. M., 83(42), *86*
Burgoyne, L., 137, *149*
Burleigh, B. D., 41(17), 42(17), 43(17), *48*, 101(55a), 104(5a), 105(5), 106(5), *112, 116*
Burris, R. H., 256(55), *265*
Burton, P., 123(7, 47), 124, 125, 126, *130, 132*
Burzawa-Gerard, E., 99(25), 100(24, 25), *114*
Busch, H., 136(48), *150*, 151(25), *157*
Buse, G., 92, *96*
Butler, W. L., 223(11), *239*
Buttin, G., 50(33), *62*

C

Cahan, L., 83(26), *85*
Cahan, L. D., 65(8), 82(8), *84*
Cammack, R., 245(16), 246(15, 16, 17), 247(33), 249(15, 16, 17), 256(29), *263, 264*
Cancedda, F., 109(40), *115*
Candido, E., 145, *148*
Canfield, R. E., 109(57), *112, 113, 116*
Cantell, K., 81(20), *84*
Cantor, C. R., 246(43), 252(43), 256(43), *264*
Carithers, R. P., 259(73), *266*
Carlsen, R. B., 105(7), 106(7), *113*
Carlson, G., 173(42), 176(42), *201*
Carlson, S. S., 204(29), 208(29), *219*, 268(86), *298*
Carnahan, J. E., 245(48), *265*
Carnaud, C., *130*
Carrasco, L., *116*
Carrell, R. W., 269(45), *295*
Carroll, W. R., 6(45), *17*
Carta, S., 288(73), *297*
Cassel, D., 108(8), *113*
Castenholz, R. E., 176(57), 192(57), *202*
Castets, A.-M., 237, 238(10), *239*
Castillo, O., 272(16, 17, 18), 280(16), 281(17), *294*
Catt, K. J., 100(82), 111(9, 82), *113, 117*
Celada, F., 56(2, 5, 6), 57, 58, *60, 61*
Cerami, A., 8(60), *18*
Chain, R. K., 246(44), *264*
Chalkley, R., 136(25), 146, 151(27), *149, 157*
Chambers, G. K., *330, 331*
Chan, S. H. P., 92(28), *96*
Chance, B., 307(10, 11, 27), 308(10, 12, 27), 315(116), *318, 321*
Chang, A. C. Y., 216(24), *219*
Chang, D., 109(40), *115*, 123(20), *131*
Chang, T., 111(10), *113*
Charnetzky, W. T., 43, *47*
Charreire, J., 121(4), *130*
Chase, A. M., *327*
Chaviano, A. H., 299(74), 302(74), 303(74), 308(74), 312(74), 316(74), *320*
Cheng, K.-W., 104(34), 105(13, 14), 106(11, 12, 13), 107(11, 12, 34, 73), *113, 114, 117*
Cheron, A., 68(40), 73(40), *86*
Chin, W. W., 100(15), *113*
Cho, F., 227(12), *239*
Choate, W. L., 6(45), *17*
Chou, P. Y., 163, *167, 168*
Clark, A. J., 23(1), *36*
Clark, R. W., 23, *36*
Clayton, R. K., 175, *199*
Clegg, J. B., 268(79), *297*
Cohen-Bazire, G., 50(29), *62*, 227(57, 77), 229(14), 231(14, 33, 40), 235(34, 41), 237, 238(10), *239, 240, 241, 242, 243*
Cohen, G. H., 123(19), *131*
Cohen, G. N., 50(33), *62*
Cohen, H., 110(16, 53), *113, 115, 116*
Cohen, S. N., 216(24), *219*
Cohn, E. J., 124, *130*
Cohn, M., 49(7), 50(16, 29), *61, 62*
Cole, R. D., 163(25), *169*
Collier, J. R., 112(54), *116*
Collier, R. J., 108(17), 111(28), *113, 114*
Colman, P. M., 90(7), 91, *95*, 159(20), *168*
Colosimo, A., 307(2), 308(2), 313(2), *317*
Combarnous, Y., 105(19), 106, 107, *113*
Como, P. F., 280(19), 288(19), *294*
Comsa, J., 123(9, 10), *130*
Conrad, H., 306(104), *321*
Conti, S. F., 224(24), 225(23, 25), *240*
Cooke, J., 6(47), *17*
Coon, H. G., 109(35), *114, 115*
Cooper, A., 173(22), 176(22), *200*
Cooper, D. W., 282(20), *294*
Corbin, K. W., 299(115), *321*
Coulson, A. R., 207(26), *219*
Cox, J. C., 89(8), *95, 96*
Craig, L. C., 3(10), *15*
Crews, D., 99(45), 100(45), *115*
Crea, R., 56(13, 18), *61*
Crestfield, A. M., 7(55, 56), *17*
Crick, F. H. C., 204, *218*
Crivellaro, O., 129(14), *130*
Crothers, D. M., 154(22), *157*
Cusanovich, M. A., 176(15), *199*

D

Dahr, W., 65(9), 82(9), *84*
Dairman, W., 5(30), *16*
Dale, R. E., 238(15), *239*
D'Alessio, G., 9(69, 70), *18*
D'Anna, J. A., 138, 139, *148*, 154(21), *157*
Daniel, M., 174(7), 183(7), 190(7), 197(7), *199*

Dardenne, M., 121(4), 123, *130, 131, 132*
Dattagupta, N., 154(22), *157*
Davidson, B. E., 159(8), 163(8), *168*
Davidson, N., 27(3), *36*
Davie, J., 145, *148*
Davies, D. R., 268(41), 269(41), 270(41), *295*
Davies, H. C., 308(105, 106), 311(106), *321*
Davies, T. F., 100(82), 111(82), *117*
Davis, K. A., 194(15a), *199*
Davis, N. C., 325(10), *326, 328*
Davis, R. W., 27, *36*
Dayhoff, M. O., 180, 181, *199*, 246(5, 8, 9), 249(9), *262*, 267(21), 268(21), 269(21), 270(21), 272(21), 275(21), 282(21), 283, 287(6, 21), 288, 289(21), *293, 294*, 300(13, 14, 15, 16), *318*
Dazord, A., 111(10), *113*
Deaven, L. L., 154(21), *157*
Degelaen, J., 104(52), *115*
Deinum, J., 92(9), 94(1), *95, 96*
DeKlerk, H., 249(10), 259(10), *262*
de la Llosa, P., 104(22), 105(20), 106(21), 107, *113, 114*
DeLange, R. J., 135(13), 136(12), *148*, 151(4, 5, 6, 7, 8, 9, 10, 11, 12, 14, 26, 31, 32), 152, 153(9, 10, 13, 15), 155(3a), *156, 157*, 229(43), *241, 329, 330*
DePrisco, R., 9(69), *18*
Dethmers, J. K., 299(75), 302(75), 303(75), 304(75), 307(17, 110), 308(75, 110), 312(75, 110), 316(75), 310(110), 311(18), *318, 320, 321*
Devanathan, T., 163(27), *169*
Dickerson, R. E., 40(20), *48*, 173(22, 42), 174(1, 23, 67), 176(1, 20, 22, 42, 66), 180(1, 18), 182(66), 194(1, 19, 21, 23), 197, *199, 200, 201, 202*, 268(41), 269(41), 270, *294, 295*, 299(19, 21, 77, 78), 300(77, 78), *318, 320*
Dikker, H. L., 308(56), 311(56), *319*
Dixon, G. H., 146(31), *149*, 267(23), *294*
Dixon, P. I., 288(24), *294*
Dodge, J. D., 235(16), *239*
Dodgson, S. J., 280(25), *294*
Dodgson-Schneider, S. J., *294*
Dohen, G. H., 123(23), *131*
Dorland, L., 64(38), 68(40), 73(40), *86*
Dothie, J. M., 41(3a, 7), 44(7), *47*
Doudoroff, M., 174(65), 196(65), *202*
Drzeniek, Z., *86*
Dube, S. K., 145(7), *148, 150*
Duckworth, W. L. H., 119(13), *130*
Dufau, M. L., 111(9), *112*
Dumont, J. E., 110(78), *117*
Dupourque, D., 159(20), *168*
Durosay, M., 105(20), *113*
Dus, K., 249(10), 259(10), *262*
Dutton, P. L., 192(58), *202*
du Vigneaud, V., 3, *15*

E

Ealey, E. H. M., 282(20), *294*
Ebner, K. E., 66(7), *84*
Eder, J., 231(17), *240*
Edman, P., 6, *16*
Edmundson, A. B., 268(27), 270(27), *294*
Edsall, J. T., 269(28), *294*
Edwards, M. R., 224(27), 235(27), 238(4), *239, 240*
Efstratiadis, A., 212(13), 217(13), *218*
Eisenberg, D., 173(22), 176(22), *200*
Eiserling, F. A., 238(18), *240*
Elgin, S. C. R., 151(16), 152, 154(16), *157*
Eliezer, N., 163(35), *169*
Epstein, W., 24(9), *37*
Erecińska, M., 315(22, 23, 24, 116), *318, 321*
Ericsson, L. H., *169*
Errede, B., 194(23a), *200*, 306(25), 308(26), 311(26), *318*
Esfahani, M., 107(48), *115*
Evans, M. C. W., 249(18), 256(61), 258(18, 20, 45, 60, 61), *263, 264, 265*
Evans, W. H., 151(32), *157, 329*
Eventoff, W., 163(9), 164, *168*
Eylar, E. H., 63(13), 65(13), *84*

F

Faith, M. R., 104(71), 105(23), *114, 117*
Fambrough, D. M., 134(14, 15, 16), 135(13), 136(12), 137, *148*, 151(9, 10, 14, 17), 152(9, 10), 153(9, 10), 154(9, 14), *156, 157*
Fang, S., 229(35, 42), 231(36), 238(35), *240, 241*
Farmer, S. W., 99(45), 100(45), *115*
Fasman, G. D., 163, *167, 168*
Faulkner, R., 134(2), 144(2), *147*
Fee, J. A., 89(10), *96*
Feingold, D. S., 64(22), *84, 85*
Felsenfeld, G., 152(18), 153(18), *157*
Ferguson-Miller, S., 299(8, 30, 74, 75, 76), 301(28, 76), 302, 303(74, 75), 304(75), 305(30), 306(28, 29, 30), 307(7, 9, 17, 27, 29, 30, 31, 32, 57), 308(8, 27, 30, 50, 57, 74, 75, 109), 309(7, 9, 29, 57), 310(57, 109), 311(57, 109), 312(29, 30, 57, 74, 75, 109), 313, 314, 316(30, 74, 75), *317, 318, 320, 321*
Fernandez-Puentes, C., *116*
Fiddes, J. C., 207(26), 208(3), 212, *218, 219*
Fields, S., 207(22), *219*
Fincham, J. R. S., *330, 331*
Finne, J., 64(10), 65(10), 74(17), *84*
Fisher, L. A., 206, 217(25), *219*
Fisher, W. K., 272(29), 275(31), 280(25, 29, 30, 31, 75, 82, 83), 283(84), 286(31), 289(30, 31), *294, 295, 297*
Fishman, P. H., 108(58), 109(55, 58, 64, 65), *116*

Fitch, W. M., 109(37), *115*, 246(11), 252(12), 255(12), *263*, 299(33, 34, 35, 36, 37, 38, 39, 40, 41, 42, 43, 44, 45, 46, 60, 61, 66, 67, 68, 77, 78, 118), 300(35, 40, 42, 60, 61, 77, 78), 301, *318, 319, 320, 321*
Fitch, W. N., 180, *200*
Floridi, A., 9, *18*
Fontana, A., 165(34), *169*
Fontaine, Y. A., 99(25), 100(24, 25), *114*
Ford, G. C., 159(6), 163(6, 26), *167, 169*
Forget, B. G., 212(13), 217(13), *218*
Forget, B. J., 211(20), 212(20), *218*
Fossitt, D. D., 40(13), *48*
Foster, J. A., 234(81), *243*
Fothergill, J. E., 207(22), *218, 219*
Fournet, B., 64(38), 68(40), 73(40), *86*
Fournier, C., 121(4), *130*
Fowler, A. V., 49(10, 11), 51(39), 54(1, 9), 55(8, 11), 56(2, 6, 20, 24, 25), 57(6), 58(3,6), 59(12,17), *60,61, 62*
Fox, G. E., 197(26, 27), *200*, 256(68), *266*
Frambrough, D. M., *329*
Frank, G., 229(19), 232(19), 233(19), 234(19), *240*
Franzblau, C., 234(81), *243*
Frater, R., *329*
Freer, S. T., 174(59), 176(59), *202*
Freedman, M., 159(11), *168*
Freeman, H. C., 90(7), 91(7), *95*
Freidenreich, P., 233(20), 234(20, 87), *240, 243, 244*
Freire, M., 129(14), *130*
Friday, A. E., 283(66), *296*
Friedman, H., 121(15), *130*
Fry, I., 249(25), *263*
Fry, K. T., 245(13), *263*
Fujiki, H., 285(13), *294*
Fujiki, Y., 102, *114*
Fujimura, F., 134(16), 137(16), *148*
Fukuyama, K., 247(14, 64), 249(14), 250(14), *262, 266*
Furia, A., 9(74), *18*
Furlong, N. B., 106(49), *115*
Furthmayr, H., 65(11), 66(11), 82(11), *84*
Furuta, H., 288(33), *295*

G

Gandvik, E.-K., 89(5a), 90(5), 91(5a), *95*
Gantt, E., 224(24, 27), 225, 235(8, 27), 236, 237(22, 25, 26, 29), 238(21, 46), *239, 240, 241*
Garel, A., 144(17), *148*
Geren, C. R., *84*
Gerlough, T. D., *327*
Gershey, E. L., 154, *157*
Gest, H., 192(29, 41, 46, 47), 193(25), *200, 201*
Gething, M. J., 44(18), *48*
Gibbons, N. E., 163(5), *167*, 182(13), 196(13), *199*
Gibson, J., 197(26, 27), *200*
Gill, D. M., 108(26, 27), 109(26), *114*
Gillam, S., 303(83, 108), *320, 321*
Gilliland, D. G., 111(28), *114*
Giraud, G., 225(58), 229(58), *242*
Giudice, L. C., 100(29, 30, 33), 101, 102, 104(31, 32, 34, 68, 71), 106(32), 107(34), *114, 116*
Givol, D., 207(22), *218, 219*
Gjerset, R., 145(30), *149*
Glantz, M., 162(3), 163(3), *167*
Glantz, R. R., *328*
Glasgow, L. R., 65(14), 68(14), 76(24), 77(14, 27, 32), 78(14), 80(32), *84, 85*
Glazer, A. N., 105(14), *113*, 151(11, 12), *156*, 227(89), 229(14, 30, 35, 37, 42, 43, 86), 230, 231(14, 30, 33, 36, 40), 233(20), 234(20, 39, 87), 235(34, 41), 238(18, 35, 38), *239, 240, 241, 243, 244, 329, 330*
Glover, C. V. C., 152(20), 152(20), 154(20), *157*
Godson, G. N., 208(3), 212, *218*
Goeddel, D. V., 56(13), *61*
Goedheer, J. C., 227(44), *241*
Göthe, P. O., 89(3), *95*
Goldmark, P. J., 23(4), *36*
Goldstein, A. L., 120(6, 17, 50), 121(17, 25, 42, 50), 123(18, 19, 20, 21, 23, 49, 50), 128(16, 21), 129(48), *130, 131, 132*
Goldwater, W. H., 3(8), *15*
Goodchild, D. J., 223(5), *239*
Goodman, M., 209(4), *218*, 271(35), 282(34), 284(35), 287(35), *295*
Goodman, M. A., 180, *201*
Goodwin, G., 146(27), *149*
Gorovsky, M. A., 152(20), 154(20), *157*
Gottesfeld, J., 143, 144, 145(20), 147, *148*
Govindjee, *239*
Grabowski, J., 225(28), 236, 238(46), *240, 241*
Grandi, C., 165(34), *169*
Gratzer, W. B., 268(36), *295*
Graves, D. J., 65(12), 67(12), *84*
Gray, B. H., 225(47), *241*
Gray, G. R., 10, 11, *19*
Gray, H. B., 90(25), *96*
Grebenko, A. O., 269(77), *297*
Greenwood, C., 307(2), 308(2), 313(2), *317*
Griesbach, M., 163(16), *168*
Griffith, J., 137(22), 138, 139(21), *148*
Griffiths, M. E., 282(20), *294*
Grollman, E. F., 109(35), *114, 115*
Gromet-Elhanan, Z., 175, 192(29), *200*
Gross, E., 7, *17*
Groudine, M., 144(46), *150*
Grütter, M. K., 159(12), 160(12), 166, *168*
Gudas, L. J., 24(5), *36*
Guglielmi, G., 237, 238(10), *239*

Guha, A., 123(18), *131*
Gundlach, H. G., 7(54), *17*
Gurley, L. R., 154(21), *157*
Guss, J. M., 90(7), 91(7), *95*
Guthrow, C. E., 65(14), 68(14), 77(14), 78(14), *84*
Gysi, J., 249(24), *263*

H

Habig, W., 235(60), *242*
Habener, J. F., 100(15), *113*
Haberland, M. E., *331*
Hackenbrock, C. R., 305(46), *319*
Hackert, M. L., 163(26), *169*
Haddock, B. A., 190(30), *200*
Hagopian, A., 63(13), 65(13), *84*
Haight, G. P., 306(25), *318*
Hall, B. D., 303(83, 108), *320, 321*
Hall, D. O., 245(16), 246(15, 16, 17), 247(33), 249(15, 16, 17, 19, 22, 24, 25), 258(19), 259(19), *263, 264*
Halldal, P., 225(48), *241*
Hamilton, P. B., 5(26), *16*
Hamlin, J., 60(14), *61*
Handler, P., *328, 331*
Haniu, M., 163(27, 28), *169*, 256(61), 258(60, 61), *265*
Hanson, H. T., *328*
Hardy, M. A., 123(18), *131*
Hare, J. W., 90(25), *96*
Harris, J. I., 159(4, 8, 14), 160(4), 163(4, 8, 13, 14, 15), 164(22), 165(4), *167, 168*
Harris, M. B., 286(51), *296*
Hart, R. G., 268(41), 269(41), 270(41), *295*
Hartley, B. S., 40(4, 5, 9), 41(3a, 7, 17), 42(6, 17), 43(8, 17), 44, 46(1, 14a, 14b, 15), 47(12, 15), *47, 48*, 59(15), *61*
Hartman, P. E., 190(60), *202*
Harutyunyan, E. H., 269(77), *297*
Harvey, C., 123(20), *131*
Hase, T., 209(21), *218*, 246(42), 249, 256(66), 258, 259(19), 260(21, 23), *263, 264, 266*
Haslett, B. G., 249(59), 250(3), 252(3), 256(29), *262, 264, 265*
Hata, R., 288(56), *296*
Hatefi, Y., 194(15a), *199*
Hauschka, P. V., 5(28), *16*
Haverkamp, J., 64(38), *86*
Hawkes, R. B., 159(12), 160(12), 166(12), *168*
Haxo, F. T., 224(49, 70), 227(71), *241, 242, 243*
Hayakawa, K., 123(32), *131*
Hayashi, H., 136(23), *148*
Heath, E. C., 110, 111, *116*
Hecht, S., *327*
Heindell, H. C., 212, *218*
Henderson, L. E., 88(11), *96*
Hendrickson, W. A., 293(48), *295*
Hennen, G., 107, *113*
Henriksson, D., 88(11), *96*
Hermier, C., 104(22), *114*
Hermodson, M. A., *169*
Hermoso, J., 174(7), 183(7), 190(7), 197(7), *199*
Hesse, J. E., 24(9), *37*
Hewish, D., 137, *149*
Heyneker, H. L., 56(13, 18), *61*
Higa, H. H., 83(26), *85*
Hill, R. L., 64(34), 65(4, 5, 14), 67(6, 23, 43), 68(5, 6, 14, 35), 69(4, 5), 70(4), 71(4), 72(4), 75(4), 77(14, 25, 27, 28, 32), 78(14), 80(32), 81(34), 82(29, 34), *83, 84, 85, 86*, 299(79), *320, 328*
Hilschmann, N., 268(14), *294*
Hilse, K., 268(14), *294*
Himes, R. H., 163(27, 28), *169*
Hirose, T., 56(13, 18), *61*
Hirs, C. H. W., 5(34), 6, *16*
Hixon, C. S., 229(37, 43), 234(39), 238(38), *241*
Höök, M., 64(22), *84, 85*
Hogan, M., 154(22), *157*
Hogness, D. S., 50(16), *61*
Holder, A. A., *330, 331*
Holdsworth, E. S., 224(50), *241*
Holland, R. A. B., 280(12), *294*
Holloway, B. W., 191(31), *200*
Holm, A., *328*
Holmquist, R., 284(40), *295*, 299(47), *319*
Holmquist, R., 205(12), 206(12), 209(10), 213, 216(10, 11, 12), *218*, 299(47), *319*
Holroyde, M. J., 76(24), *85*
Hood, J. M., 59(17), *61*
Hooper, J. A., 123(19, 23), *131*, 136(25), *149*, 152(13), 153(13), *156, 330*
Horecker, B. L., 129(14), *130*
Hori, H., 249(26), 256(26), *263*
Horowitz, M. I., 66(33), *85*
Hosken, R. W., 281(37), 282(37, 38), *295*
Hosken, R., 280(76), *297*
Howard, J. B., 259(27), *263, 264*
Huang, R. C. C., 133, *149*
Hubbard, A. L., 11, *19*
Hubbell, R. B., 1(4), 2(4), *14, 15*
Hughes, W. L., Jr., 124(8), *130*
Hugli, T. E., 5(27), *16*
Huisman, T. H. J., 268(69), *297*
Huisman, J. G., 256(28), *264*
Hum, V. G., 105(36), 106(36), *115*
Hunter, F. A., 223(79), 224(79), *243*
Huntsman, R. G., 268(46), *295*
Hutchinson, C. A., III., 207(26), *219*
Hutson, K. G., 256(29), *264*

I

Iden, S., 123(7), 124(7), 125(7), 126(7), *130*
Imai, T., 249(21), 258(21), 260(21), *263*

Imber, M., 65(14), 68(14), 77(14), 78(14), *84*
Ingram, V., 146(39), *149*, 268(39), 282(39), *295*
Inoue, A., 216(24), *219*
Isaacs, C., 129(14), *130*
Isenberg, I., 138, 139, *148*
Itakura, K., 56(13, 18), *61*
Ito-Mega, S., 78(46), *86*
Ivatt, R. J., 65(15), 74(15), *84*
Iwai, K., 136(23), *148*

J

Jacob, J., 50(19, 32), 55(35), *61, 62*
Jacobsson, I., 64(22), *84, 85*
Jailkhani, 13, *19*
Jansonius, J. N., 159(20), *168*
Janssen, E., 65(9), 82(9), *84*
Jennings, J. V., 258(20), *263*
Jensen, L. H., 90(2), *95*, 162(1), 163(1), *167*, 246(1), 247(1), 249(1), 259(1), *262*
Jörnwall, H., 56(20), *61*
John, P., 196(32, 33), *200*
Johns, E., 146, *149*
Johnson, J., 136(23, 37), *148, 149, 152(28), 153(28), 157*, 286(51), *296, 330*
Jones, C. W., 190(30), *200*
Jones, L. T., 272(17, 18), 281(17), *294*
Jones, G. M. T., 163(15), *168*
Jordon, J., 136(48), *150*, 151(25), *157*
Joysey, K. A., 283(66), *296*
Jüttner, F., 249(25), *263*
Jukes, T. H., 203, 204, 205(12, 17), 206, 208(8), 209(10), 213, 216(7, 10, 11, 12), 217(17), *218*, 246(30, 43), 252(43), 256(43), 284(40), 299(47, 48, 55), *264, 295, 319*
Junge, W., 222(51), *241*
Jutisz, M., 105(20), *113*

K

Kaback, H. R., 109(35), *114, 115*
Kafatos, F. C., 212, 217(13), *218*
Kajita, A., 288(33), *295*
Kakudo, M., 173(9), 176(9), *199*, 247(64), *266*
Kallai, O. B., 173(22), 176(22), *200*
Kamen, M. D., 174(8), 183(8), 190(8), 194(15a, 23a), 197(8), *199, 200*, 306(25), 308(26), 311(26), *318*
Kang, C. H., 299(75, 76), 301(76), 302(75), 303(75), 304(75), 307(110), 308(49, 50, 75, 110), 310(110), 311(49), 312(75, 110), 316(75), *319, 320, 321*
Kammerman, S., *112, 113*
Kan, K.-S., 223(79), 224(79), *243*
Kania, J., 58(3, 30), *60, 61, 62*
Karlsson, U., 49(21), *61*
Karstrom, H., 49, *61*
Kasper, C. B., *329*
Kassel, R. L., 9(78), *18, 19*
Kassell, B., 3(8), *15*
Katoh, S., 245(31), *264*
Katsube, Y., 247(64), *266*
Katze, J., 51(4), *61*
Kauffman, D. L., 40(9), *48*
Kay, L. M., 40(20), *48*
Kayser, H., 317(51, 111), *319, 321*
Keil, W., 163(9, 16), 164(9), *168*
Keilen, D., *200, 327*
Keister, D. L., 192(35), *200*
Kellenberger, G., 23, *37*
Kendrew, J. C., 268(41), 269(41), 270(41), 286(60), *295, 296*
Kepes, A., 50(40), *62*
Keresztes-Nagy, S., 246(32), *264*
Kerscher, L., 247(33), 249(22), *263, 264*
Ketcham, L. K., *262*
Kieffer, J. D., 100(15), *113*
Kilmărtin, J. V., 269(42), 279(42), 281(42), 287(42), *295*
Kiltz, H. H., 163(9, 16), 164(9), *168*
Kimelberg, H. K., 306(87), *320*
Kimmel, J. R., 3(17), 4(17), *15*, 299(79), *320*, 325(5), *326, 328, 329*
Kimura, M., 204, 208, 209, 210, *218*, 299(52, 53, 54, 89), *319, 320*
King, J. L., 204, 205(17), 213(17), 217(17), *218*, 299(48, 55), *319*
King, T. E., 305(90), *320*
Kingston, B., 89(12), *96*
Kingston, B. L., 89(12), *96*
Kipling, R., 40(10), *48*
Kirk, J. T. O., 221(53, 54), 222, *241, 242*
Kita, T., 216(24), *219*
Kitazume, Y., 256(53), *265*
Kitchen, H., 268(43), *295*
Kleid, D. G., 24(7), *37*, 56(13), *61*
Kleinschmidt, T., 289(44, 71), *295, 297*
Klevan, L., 154(22), *157*
Klock, P. A., 293(48), *295*
Knaff, D. B., 192(36), 193(36), *200*
Knipfel, J. E., 105(36), 106(36), *115*
Knox, R. S., 238(55), *242*
Koller, K.-P., 237(64), 238(56), *242*
Kobata, A., 78(46), *86*
Kohn, L., 109(44), *115*
Kohn, L. D., 108(58), 109(35, 43, 55, 58), *114, 115, 116*
Koide, S. S., 110(53), *115, 116*
Konareva, N. N., 269(77), *297*
Konig, B. W., 308(56), 311(56), *319*
Konkel, D. A., 211(18), *218*
Kook, A. I., 123(24, 52), *131, 132*
Koorajian, S., 49(21), 51(23), *61, 62*
Koorajian, S., 51(4), *61*

Koppenol, W. H., 305(58, 93), 307(57, 91, 110), 308(57, 110), 309(57, 110), 309(57), 310(57), 311, 312(57, 110), 315(114), *319, 320, 321*
Kornberg, R. D., 137, 141, *149*, 152(23), 153(23), *157*
Kornfeld, R., 64(16), *84*
Kornfeld, S., 10, *19*, 64(16), *84*
Korszun, Z. R., 174(37), 176(37), *201*
Krantz, M. J., 78(18), *84*
Kraszewski, A., 56(13), *61*
Kraut, J., 174(59), 176(59), *202*
Kreil, G., 176(44), *201*, 299(59, 80), *319, 320, 329*
Krishnapillai, V., 191(31), *200*
Kristiansen, T., 67(30), *85*
Krusius, T., 64(10), 65(10), 74(17), *84*
Kunitz, M., 5(33), *16*
Kuo, P.-F., 122(31), *131*
Kurosky, A., 109(37), 110, *115*
Kuranova, I. P., 269(77), *297*
Kurihara, M., *330*
Kycia, J. H., 224(49), *241*

L

Lai, C.-Y., 109(39, 40), *115*, 123(20), *131*
LaMonica, R. F., 192(39), 193(39), *201*
Landau, L., 286(51), *296*
Landefeld, T. D., 100(41, 42), 103(41), *115*
Landon, M., 151(32), *157, 329, 330*
Langley, C. H., 299 (40, 60, 61), 300(40, 60, 61), 301(40), *319*
Langley, K. E., 56(24, 25), *62*
Langley, T. J., *330*
Larsson, M., 325(6), *326*
Larsson-Raźnikiewicz, M., 325, *326*
Lattman, E. E., 293(48), *295*
Laughlin, S. B., *243*
Laursen, R. A., 288(11), 289(11), *294*
Lazarow, P. B., 14, *19*
Leder, P., 211(18), *218*
Lederberg, K., 49(26), *62*
Lee, F., 212, *218*
Lee, G., 108(58), 109(43, 44, 58), *115, 116*
Lee, Y. C., 78(18), *84*
Leed, G., 109(35), *114, 115*
Lefort-Tran, M., 237(13), *239*
LeGall, J., 246(7), 249(7), 258(63), 259(63), *266*
Lehman, I. R., 24(10, 13), *37*
Lehmann, H., 268(46), 269(46, 50, 59), 271(67), 272(16, 17, 18, 67), 278(50), 280(16), 281(17), 286(66, 67), 287(67), *294, 295, 296*
Lehninger, A. L., 324, *326*
Lehrman, M., 65(14), 68(14), 77(14), 78(14), *84*
Leigh, J. S., 307(10, 11), 308(10, 12), *318*
Leimgruber, W., 5(30), *16*
Lemasson, C., 227, *242*
Lentz, P. J., Jr., 163(26), *169*
Leone, E., 9(69), *18*
Leonhardt, H., 123(10), *130*
Lerch, K., 92(9), *96*
Lerner, S. A., 40(11), *48*
Leung, D. W., 303(84, 108), *320, 321*
Levine, R., 121(51), *132*
Levitt, M., 163, *168*
Levy-Wilson, B., 145, 146, *149*
Ley, A. C., 227(71), *243*
Li, C.-H., 105(74), 106(75), 107(74), *117*
Li, S. L., 285(47), *295*
Liao, T.-H., 8(65, 66), *18*
Libonati, M., 9, *18*
Licht, P., 99(45), 100(45), *115*
Lichtlé, C., 225(58), 229(58), *242*
Light, A., *329*
Lin, E. C. C., 40, *48*
Lin, E. E. C., 41(22), *48*
Lin, S., 55(27), *62*
Lindahl, U., 64(22), *84, 85*
Linkskog, S., 88(19), *96*
Linn, S., 23(4), *36*
Lipschultz, C. A., 225(26, 28, 47), 237(25, 26, 29), *240, 241*
Lisowska, E., *86*
Little, J. W., 24(7), *37*
Liu, A., 212(5), *218*
Liu, T.-Y., 8(63), *18*
Liu, W.-K., 104(5a), 105(5, 47, 50), 106(5, 47, 49), *112, 115*
Livingston, A. M., 110(51), *115*
Livingston, V. W.-C., 110(51), *115*
Ljungdahl, L. G., 164(18), *168*
Loeblich, A. R., Jr., *242*
Lorsbach, T., 259(27), *263, 264*
Low, K. B., 190(10), *199*
Low, T. L. K., 121(25), 123(20), 128(21), *131*
Lowe, D. J., 259(70), *266*
Lowman, C., 76(24), *85*
Love, W. E., 293(48), *295*
Lovenberg, W., 245(34, 35, 54), *264, 265*
Luckey, T. D., 121(26), 123(27), *131*
Ludwig, M., 176(40), 180(40), *201*
Luehrsen, K. R., 252(36), *264*
Lumry, R., *328*
Lundgren, J.-O., 89(22, 23), 91, *96*
Lynfield, J., 286(51), *296*

M

McAdoo, M., 123(20), *131*
McCarthy, B. J., 145(30), *149*, 153(24), *157*
McClure, J., 123(20), *131*
MacColl, R., 235(60), *242*
McDaniel, M. C., 123(19, 23), *131*

McEntee, K., 24(8, 9, 10, 13), *37*
McGuire, E. J., 74(21, 37), *84, 86*
McKee, P. A., 77(39), *86*
Mckerns, K. W., 100(53a), *116*
McKnight, S., 145, *149*
McLachan, A. D., 180(48), *201*, 246(47), *265*
McLaughlin, P. C., 246(5), *262*
McPherson, A., Jr., 163(26), *169*
McReynolds, L., 207, *218, 219*
McWilliams, D., 100(42), *115*
Madigan, M. T., 192(41), *201*
Mäkelä, O., 81(20), *84*
Magee, S. C., *84*
Maghuin-Rogister, G., 104, 105(19), 106, *113*
Magrum, L. J., 252(36), *264*
Makert, C. L., 88(17), *96*
Malkin, R., 245(37), *264*
Malmström, B. G., 88(13, 14), 89(8, 16), 90(16), 92(15), 94(1), *95, 96*, 307(62), *319*, 325(2, 3), *326, 329*
Maloof, F., 100(15), *113*
Malorni, M. C., 9(70), *18*
Man, N. K., 123(12), *130*
Mandel, G., 173(42), 176(42), *201*
Mandel, N., 173(42), 176(42), *201*
Mangeat, P., 136(32), *149*
Mann, R., 146(43), *149*
Manning, J. M., 5(31), 8(61), *16, 18*
Manning, J. M., 8, *18*
Marchis-Mouren, G., 136(32), *149*
Margoliash, E., 173(22), 176(22, 44), 180, 182(43), *200, 201*, 246(32), 252(12), 255(12), *263, 264*, 267, 284(32), *295, 296*, 299(30, 41, 42, 43, 44, 63, 64, 65, 66, 67, 68, 69, 70, 72, 73, 74, 75, 76, 77, 78, 79, 80, 88, 102, 118), 300(4, 42, 64, 65, 77, 78, 102, 107), 301, 302(74, 75, 92), 303(74, 75), 304, 305(30, 69, 93), 306(29, 30, 31, 32), 307(7, 9, 17, 27, 57, 91, 92, 94, 110), 308(27, 30, 49, 50, 56, 57, 74, 75, 94, 109, 110), 309(7, 9, 29, 57, 74, 75, 92), 310(57, 109, 110), 311(18, 49, 56, 57, 98, 109), 312(29, 30, 57, 74, 75, 109, 110), 313, 314(30), 315(114), 316(30, 74, 75, 94), *317, 318, 319, 320, 321, 329*
Margulies, A. D., 23, *36*
Marinucci, M., 288(73), *297*
Markel, D. E., 109(37), 110(38), *115*
Markowitz, E., 299(45, 78), 300(78), *319, 320*
Markland, F. S., 151(32), *157, 329, 330*
Markwell, M. A. K., 83(26), *85*
Marotta, C. A., 211(20), 212, *218*
Marrs, B. L., 190(45, 63), 192(39, 46, 47, 70), 193(39), *201, 202*
Martin, A. J. P., 3, *15*
Martinage, A., 136(32), *149*
Martinson, H. G., 153(15, 24), 154(15), *157*
Maruo, T., 110(53), *115, 116*
Marushige, K., 141, 142, 144, 145, *149*
Marxhall, L., 83(26), *85*
Masaki, R., 252(38, 46), *264, 265*
Massa, A., 288(73), *297*
Matheson, A. T., 246(39), 256(39), *264*
Matousek, J., 9(77), *18*
Matsubara, H., 209(21), *218*, 246(40, 42, 43, 44, 45), 247(2, 64), 249(18, 19, 21, 22, 23, 24, 25), 252(38, 46), 256(66), 258, 259(19), 260(21, 23), *262, 263, 264, 265, 266, 329*
Matsuda, G., 209(4), *218*, 271(35), 282(34), 284(35), 287(35), *295*
Matsueda, G., 163(27), *169*
Matsumoto, T., 249(21), 258(21), 260(21), *263*
Matsushita, Y., 123(30, 32), *131*
Matsuura, Y., 247(64), *266*
Matthews, B. W., 159(12, 19, 20), 160(12), 166(12), *168*
Meinhofer, J., 123(20), *131*
Melin, M., 124(8), *130*
Mekalanos, J. J., 112(54), *116*
Melamed, M. D., *330*
Meldolesi, M. F., 109(35, 55), *114, 115, 116*
Meren, R., 108(27, 63), *114, 116*
Mereschkovsky, C., 228(61), *242*
Meselson, M., 23, *37*
Meyer, H., 163(9, 16), 164(9), *168*
Meyer, T. E., 174(7, 8), 183(7, 8), 190(7, 8), 197(7, 8), *199*
Midwinter, G. G., 47(12), *48*
Milcu, S. M., 122(36), *132*
Miller, A., 49(21), *61*
Miller, C., 145(35), *149*
Millett, F. S., 307(1), 308(1, 21, 85, 112), 309(85), 311(1, 85), 313(1), *317, 320, 321*
Miller, J. F. A. P., 120, *130, 131*
Minnaert, K., 306(82), *320*
Miozzarri, G. F., 212, *219*
Mitchell, K., 123(7), 124(7), 125(7), 126(7), *130*
Mitoni, Y., 288(56), *296*
Mitsky, A., 134(2), 144(2), *147*
Mizutani, A., 122(31, 33), 123(30, 32), *131*
Mizutani, T., 122(31), 123(30), *131*
Moehring, J. M., 111(28), *114*
Mörschel, E., 235(63), 237(64), 238(56), *242*
Møller, F., 88(17), *96*
Molmström, A., 64(22), *84, 85*
Monod, J., 49(7), 50(16, 19, 29, 32, 33, 40), 55(35), *61, 62*
Montgomery, D. L., 303(83, 84, 108), *320, 321*
Montreuil, J., 68(40), 73(40), *86*
Moon, K., *330, 331*
Moore, G. W., 180(49), *201*, 209(4), *218*, 271(35), 282(34), 284(35), 287(35), *295*
Moore, M. R., 164(18), *168*

Moore, S., 2(7), 3(11, 12, 13, 14, 16), 4(19, 20, 21, 22, 23), 5(25, 27, 31, 34), 6(34, 38, 39, 40, 41), 7(49, 54, 55, 56, 58), 8(58, 63, 64, 66), 9(71, 75), 11(87), 12(87), 13(87), *15, 16, 17, 18, 19*
Moore, W. T., Jr., 101, *116*
Moras, D., 159(6), 163(6), 164(22), *167, 168*
Morell, A. G., 10, *19*
Morgan, A. F., 191(31), *200*
Morgan, F. J., *112, 113*
Mori, K. E., 105(36), 106(36), *115*
Morimoto, H., 269(50), 278, *296*
Morris, H. R., 47(12), *48*
Morteson, L. E., 245(48), *265*
Mortlock, R. P., 40, 43, *47, 48*
Moshera, J., 129(14), *130*
Moss, J., 108(56), 109(56, 57), *116*
Mount, D. W., 24(5), *36*
Mountain, I. M., 9(78), *18, 19*
Mower, H. F., 163(29), *169*, 245(62), *265, 266*
Muckle, G., 230(65, 66), *242*
Müller-Hill, B., 58(3, 30), *61, 62*
Muijsers, A. O., 256(28), *264*, 308(56), 311(56), *319*
Mulford, D. J., 124(8), *130*
Muller, C. H., 99(45), 100(45), *115*
Mullin, B. R., 108(58), 109(58), *116*
Mullinger, R. N., 249(19), 258(19), 259(19), *263*
Murata, M., 90(7), 91(7), *95*
Murphy, R., 144(45), 145(7, 20, 45), 147(20), *148, 150*

N

Nagar, R. H., 246(39), 256(39), *264*
Nagel, R. L., 286(51), *296*
Nakagawa, Y., 105(50), 107(50), *115*
Nakamura, M., 216(24), *219*
Nakanishi, S., 216, *219*
Nakashima, T., 163(29), *169*, 245(62), *265, 266*
Nash, A. R., 275(31, 54), 280(31, 54), 285(53), 286(31, 52), 289(31, 54), *295, 296*
Nava, M. E., 300(107), 308(105, 106), 311(106), *321*
Needleman, S. B., 180(50), *201*
Neuberger, M. S., 41(7), 43(8), 44(7, 14), 46(14a, 14b, 15), 47(15), *47, 48*
Neurath, H., 40(21), *48, 169*
Neville, D. M., Jr., 111(10), *113*
Newman, D. J., 258(63), 259(63), *266*
Newmann, J., 146(39), *149*
Newton, W. A., 51(31), *62*
Ng, S., 308(85, 112, 113), 309(85), 311(85), *320, 321*
Nichols, P., 306(86, 87), *320*
Nicol, P. I., 288(55), *296*
Nigen, A. M., 8(61), *18*
Nilsson, T., 89(3), *95*
Nisbet, A. D., 207(22), *218, 219*
Nolan, C., 299(88), *320, 329*
Nolan, L. S., 1(4), 2(4), *14, 15*
Noller, H. E., 159(8), 163(8), *168*
Norris, V. A., 90(7), 91(7), *95*
Nozaki, Y., 163, *168*
Numa, S., 216(24), *219*
Nureddin, A., 105(59), *116*
Nye, J. F., *331*
Nyman, P. O., 88, 89, 91(5a), *95, 96*

O

O'Carra, P., 224(68), 229(67), *242*
Oda, *96*
O'Donnell, M. K., 259(70), *266*
Oeltmann, T. N., 110, 111, *116*
Oesterhelt, D., 247(33), 249(22), *263, 264*
Ogawa, Y., 151(25), *157*
O'Gower, A. K., 288(55), *296*
Ohe, M., 288(33), *295*
O'h Eocha, C., 224(68), 229(67), *242*
Ohmiya, M., 249(19), 258(19), 259(19), *263*
Ohmori, D., 249(23), 258(23), 259(49), 260(23), *263, 265*
Ohta, T., 204(16), *218*, 299(54, 89), *319, 320*
Okada, Y., 111(76), *117*
Olaitan, S. A., 151(26), *157*
Olins, A., 139, *149*
Olsen, K. W., 159(6), 163(6), 164, *167, 168*
Olsnes, S., 108, *116*
Olson, M., 136(48), *150*
Olson, J. M., 175, 182(52), *201*
O'Malley, B. W., 207(22), *218, 219*
Omodeo-Sale, F., 109(64), *116*
Onoki, S., 288(56), *296*
Orii, Y., 305(90), *320*
Orme-Johnson, W. H., 245(50, 51), 246(50, 51), 256(55), *265*
Osawa, S., 249(26), 256(26), *263*
Osborne, J. C., Jr., 109(65), *116*
Osheroff, N., 302(92), 305(93), 307(57, 91, 92, 94, 110), 308(49, 56, 57, 94, 109, 110), 309(57, 92), 310(57, 109, 110), 311(49, 56, 57, 109), 312(57, 109, 110), 316(94), 317(111), *319, 320, 321*
Oshino, R., 256(53), *265*, 315(23), *318*
Osoba, D., 120(29), *131*
Otto, J., 230(66), *242*

P

Pacuszka, T., 109(65), *116*
Paddock, W. V., 212(5), *218*
Padlan, E. A., 293(48), *295*
Page, J., 6(45), *17*
Palmer, G., 252(52), *265*
Panyim, S., 151(27), *157*

Papkoff, H., 99(45), 100(45), 106(75), *115* · *117*
Pappenheimer, A. M., Jr., 108, 111(3), *112, 116*
Pardee, A. B., 50(32), *62*
Parente, A., 9(70), *18*
Parker, J. H., 311(98), *321*
Parsons, T. F., 104(4, 66, 70), 105(66), 106(66), *112, 116, 117*
Partington, G., 144(18), *148*
Patterson, R. Q., 46(15), 47(15), *48*
Patthy, L., 14, *19*, 136(37), *149*, 152(28), 153(28), *157, 330, 331*
Pauling, L., 267(87), 268, 282(87), *298*
Paulson, J. C., 65(4, 8, 14), 67(6, 23), 68(6, 14), 69(4), 70(4), 71(4), 72(4), 75(4), 76(24), 77(14, 25, 27, 28, 32, 39), 78(14), 80(32), 81(34), 82(8, 29, 34), 83(26), *83, 84, 85, 86*
Pavlovsky, A. G., 269(77), *297*
Peck, H. D., Jr., 258(63), 259(63), *266*
Perham, R. N., 163(13), *168*
Perimi, F., 246(32), *264*
Perutz, M. F., 159, *168*, 268(61), 269, 270, 271(9, 57, 58, 61), 278(50), 279(5), 286(9, 60), *293, 296*
Petering, D. H., 252(52), *265*
Peters, E., 146(31), *149*
Peterson, E. A., 5, *16*
Peterson, J. W., 110(38), *115*
Petry, K., 163(16), *168*
Pettigrew, G., 310(95), 311(95), *321*
Pfenning, N., 174(54, 68), 175, 176(55), 192(53, 55, 68), 197(56), *201, 202*
Pfeuffer, T., 108(8), *113*
Pfleiderer, G., 160(33), *169*
Phillips, D. C., 268(41), 269(41), 270(41), *295*
Phillips, D. M., 134(38), *149*
Pickels, E. G., 327
Pierce, J. C., 120, *130*
Pierce, J. G., 3, *15*, 100(29, 30, 82), 101, 102, 104(4, 31, 32, 34, 66, 68, 69, 70, 71), 105(13, 14, 23, 66), 106(13, 32, 66), 107(34, 73), 111(82), *112, 113, 114, 116, 117*
Pierson, B. K., 176(57), 192(57), *202*
Pignero, A., 9(69), *18*
Pinder, P. B., 194(62), *202*
Piszkiewiez, D., 8(62), *18, 30*
Pizzo, S. V., 65(14), 68(14), 77(14, 32), 78(14), 80(32), *84, 85*
Pleau, J.-M., 121(4), 123(6, 12, 34), *130, 132*
Plummer, T. H., Jr., 5(35), 6, *16*
Pochet, R., 110(78), *117*
Polglase, W. J., *328*
Porath, J., 67(30), *85*
Porter, G., 236, 238(73), *242, 243*
Potop, I., 122(35, 36), *132*
Potts, J. T., 6(47), *17*
Powell, J. T., 66(31), *85*
Prézelin, B. B., 224(70), 227, *242, 243*
Price, P. A., 8(63), *18*
Prieels, J.-P., 65(4, 14), 68(14), 69(4), 70(4), 71(4), 72(4), 75(4), 77(14, 27, 32), 78(14), 80(32), *83, 84, 85*
Prince, R. C., 192(58), *202*
Provasoli, L., 224(27), 235(27), *240*
Puett, D., 105(1), *112*
Putnam, F. W., 89(12), *96*
Putterman, J., 311(98), *321*

Q

Quagliarotti, G., 151(25), *157*
Que, L., 259(27), *263, 264*

R

Rabinowitz, J. C., 245(35, 37), 259(56), *264, 265*
Radding, C. M., 36(12), *37*
Raidt, H., 159, *168*
Rajagopalan, K. V., 311(18), *318*
Rall, S. C., 163(25), *169*
Ramel, A. H., 123(20), *131*
Ramshaw, J. A. M., 90(7), 91(7), *95*, 299(6), *317*
Rao, K. K., 245(16), 246(15, 16, 17), 247(2), 249(15, 16, 17, 19, 22, 24, 25), 256(61), 258(19, 60, 61), 259(19), *263, 265, 266*
Rathman, P., 102(25a), 104(72), 107(72), *114, 117*
Rauvalla, H., 64(10), 65(10), *84*
Ray, G. S., 306(119), *321*
Rearick, J. I., 65(4), 67(6), 68(6, 35), 69(4), 70(4), 71(4), 72(4), 75(4), 77(28), *83, 84, 85, 86*
Redfield, R. R., 6(45, 46), *17*
Reeve, J. R., 104(71), 107(73), *117*
Reichert, L. E., Jr., 106(80, 81), *117*
Reick, G. R., 206, 217(25), *219*
Reider, R., 310(96, 97), 311(96), *321*
Reiner, A. M., 44, *48*
Reinhammar, B., 89(16), 90(16), 91(21), 92(9), 94(1), *95, 96*
Remington, S. J., 159(19), *168*
Restelli, A. D., 120, *132*
Richards, F. M., 7, *17*
Richardson, B. J., 280(62), 283(8), *293, 296*
Richardson, M., 299(6), *317*
Rickenberg, H. W., 50(33), *62*
Rigby, P. W. J., 41(17), 42(17), 43(17), 44(18), *48*
Riggs, A., 269(63, 64), 279(64), 286(10), 287(47), *293, 295, 296*
Riggs, A. D., 56(13, 18), *61*
Riggs, M., 146, *149*
Rimon, S., 229(14), 231(14), 234(39), *239, 241*
Robbi, M., 14, *19*
Roberts, G. C. K., 104(52), *115*
Robertson, M., 207(22), *219*
Robey, W. G., 123(38), *132*
Rodén, L., 64(22), 66(33), *84, 85*

Rogers, J. C., 10, *19*, 249(59), 250(3), 252(3), 256(29), *262, 264, 265*
Rolleston, H. D., 119(39), *132*
Romer, A. S., 282(65), 283(65), *296*
Romero-Herrera, A. E., 271(67), 272(67), 283(66), 286(67), 287(67), *296*
Romig, W. R., 112(54), *116*
Roseman, S., 74(37), *86*
Rosenberg, J., 173(42), 176(42), *201*
Rosevear, J. W., *329*
Ross, G. T., *112, 113*
Ross, P. S., 109(57), *116*
Rossio, J. L., 123(20), 128(21), *131*
Rossmann, M. G., 159(6), 163(6, 9, 26), 164(9, 22), *167, 168, 169*
Roth, J. S., 11, *19*
Roth, M., 5(29), *16*
Rouse, G. E., *243*
Rucknagel, D. L., 268(68), *296*
Rudloff, V., 268(14), *294*
Rüdiger, W., 230(65, 66), 231(17), *240, 242*
Ryan, F. J., 3(8), *15*
Ryan, R. J., 104(34), 107(34), *114*
Rydén, L., 89(22, 23), 91, *96*

S

Sadler, J. E., 65(4, 5), 67(6), 68(6, 35), 69(4, 5), 70(4), 71(4), 72(4), 75(4), 81(34), 82(29, 34), *83, 84, 85, 86*
Saidel, L. J., 3(8), *15*
Sairam, M. R., 105(74), 107(74), 106(75), *117*
Sajgó, M., 159(8), 163(8), *168*
Salemme, F. R., 174(37, 59), 176(37, 59), 194(15a), *199, 201, 202*
Salniknow, J., 8(64, 65, 66), *18*
Salser, W. A., 212(5), *218*
Samson, L., 173(22), 176(22), *200*
Sanders, C., 146(27), *149*
Sanderson, K. E., 190(60), *202*
Sanger, F., 5(32, 37), *16*, 207, *219*
San Pietro, A., 245(13), *263*
Sargent, T., 144(45), 145(7, 45), *148, 150*
Saronio, C., 307(11), 308(12), *318*
Sasaki, R. M., 246(44, 45), 258(45), *264*
Satake, K., 288(56), *296*
Sautiere, P., 136(32, 40), *149*
Saxena, B. B., 102(25a), 104(72), 107(72), *114, 117*
Scagel, R. F., *243*
Schachter, H., 74(37), *86*
Scheele, G., 14, *19*
Schejter, A., 299(69), 305(69), *319*
Schevitz, R. W., 163(26), *169*
Schlehuber, C., 137(8), *148*
Schmalisch, R., 65(9), 82(9), *84*
Schmidt, W. R., 299(79), *320*
Schneider, H., 237(64), *242*
Schoenborn, B. P., 159(20), *168*
Schofield, W. B., *243*
Schroeder, W. A., 268(69), *297*
Schulof, R. S., 123(19, 23), *131*
Schut, B. L., 64(38), *86*
Schwartz, R. M., 246(6, 9), 249(6, 9), *262*
Schwarz, J. A., 123(10), *130*
Sealy, L., 146, *149*
Searcy, D. G., 155(3a, 28a, 33), *156, 157*
Searle, G. F. W., 236(69), 238, *242, 243*
Selset, R., 234(84), *243*
Sharman, G. B., 283(3), 284(70), *293, 297*
Sherman, F., 311(98), *321*
Sherod, D. W., 164(18), *168*
Shetlar, M. D., 153(24), *157*
Shin, M., 256(53), *265*
Shiozawa, J. A., 223(79), 224(79), *243*
Shlaer, S., *327*
Shore, V. C., 268(41), 269(41), 270(41), *295*
Shotton, D. M., 40(19), *48*
Sidler, W., 229(19), 232(19), 233(19), 234(19), *240*
Siegelman, H. W., 224(49), *241*
Sieker, L. C., 90(2), *95*, 162(1), 163(1), *167*, 261(1), 247(1), 249(1), 259(1), *262*
Sigman, D. S., 104(4), *112*
Silcox, M. A., 41(3a), *47*
Silverman, M. S., 111(3), *112*
Simha, R., 163(35), *169*
Simon, M., 27(3), *36*
Singleton, R., Jr., 162(2), 163(2), 165(2), *167*
Sinnott, M. L., 59(12, 34), *61, 62*
Sirevag, R., 175(61), *202*
Sistrom, W. R., 175, *199*
Sjöstrand, F. S., 49(21), *61*
Skotnicki, A. B., 123(40), *132*
Sladic-Simic, D., 289(71), *297*
Slater, E. C., 305(99, 100), *321*
Sletton, K., 234(84), *243*, 249(10), 259(10), *262*
Slocombe, P. M., 207(26), *219*
Smiley, I. E., 163(26), *169*
Smillie, L. B., 40(9), *48*
Smith, E. L., 1(1, 2, 3, 4, 5), 2(4, 5), 3(17), 4, 6(43, 44), 8(62), 14, *14, 15, 17, 18, 19*, 88, *96*, 135(13), 136(12, 25, 37), *148, 149*, 151, 152, 153 (9, 10, 13, 28), 154(5, 7, 8, 9, 14), *156, 157*, 176(44), *201*, 204(27), *219*, 267, *295, 296*, 299(70, 71, 72, 80, 101, 102), 300(102), 301(72), *319, 320, 321*, 324(7, 8), 325, *326, 327, 328, 329, 330, 331*
Smith, L., 194(62), *202*, 300(107), 306(104), 308(105, 106), 311(106), *321*
Smith, H. T., 307(1), 308(1, 81), 311(1), 313(1), *317, 320*
Smith, M., 207(26), *219*, 303(83, 84, 108), *320, 321*
Smith, M. B., 307(1), 308(1, 81, 85, 112, 113), 309(85), 311(1, 85), 313(1), *317, 320, 321*

Smith, M. H., 206, *219*
Smith, P. J., 59(12, 34), *61, 62*
Smyth, D. G., 7, *17*
Snyder, A. W., 238(74), *243*
Sobel, B. E., 245(54), *265*
Sober, H. A., 5, *16*
Sodetz, J. M., 77(39), *86*
Solioz, M., 190(63), *202*
Soloman, E. I., 90(25), *96*
Somer, K. R., 136(25), *149*
Somers, G. F., 224(49), *241*
Sommer, K. R., 151(27), *157*
Song, P.-S., 225(75), *243*
Sosfenov, N. I., 269(77), *297*
Spackman, D. H., 4(23, 24), 6(41, 43), *16, 17,* 325(10), *326, 328*
Speck, S. H., 307(57, 110), 308(57, 109, 110), 309(57), 310, 311(57, 109), 312(57, 109, 110), 315(114), 317(111), *319, 321*
Spies, F. J., Jr., 308(113), *321*
Spik, G., 68(40), 73(40), *86*
Springer, G. F., 80(41), *86*
Stackebrandt, E., 197(27), *200*
Staden, R., 208(3), 212, *218*
Stanier, R. Y., 174(65), 196(65), 198(64), *202,* 228(76), 231(40), 235(41, 76), *241, 243*
Stapel, S., 256(28), *264*
Starbuck, W. C., 136(48), *150,* 151(25), *157*
Stark, G. W., 8, *17*
Staudenmayer, N., 308(112, 113), *321*
Steffens, G. J., 92, *96*
Stein, D. B., 155(28a, 33), *157*
Stein, J. R., *243*
Stein, S., 5(30), *16*
Stein, W. D., 7, 10(53), *17*
Stein, W. H., 2(6, 7), 3(11, 12, 13, 14, 16), 4(18, 19, 20, 21, 22, 23), 5(34), 6(34, 38, 39, 40, 41), 7(49, 54, 55, 56, 58), 8(58, 63, 64, 65, 66), 9(71), *15, 16, 17, 18*
Stenkamp, R. E., 90(2), *95*
Stenzel, P., 275(72), *297*
Stetten, D., Jr., *328*
Stewart, J. W., 311(98), *321*
Stockell, A., 3(17), 4(17), *15, 329*
Stombaugh, N. A., 265(55), *265*
Stout, C. D., 259(55a), *265*
Stowell, C. P., 78(18), *84*
Strampp, A., 110(16), *113*
Strandberg, B. E., 268(41), 269(41), 270(41), *295*
Strecker, G., 68(40), 73(40), *86*
Strickland, M., 152(34), *158*
Strickland, M. S., 152(35), 153(35), *158*
Strickland, W. N., 152(34), 153(35), *158*
Strid, L., 88(4), 89(3, 20), 90(5), *95, 96*
Strong, L. E., 124(8), *130*
Stroud, R. M., 40(20), *48*
Studnicka, G. U., 212(5), *218*
Stukenbrok, R., 11(84), *19*
Sugano, N., 136(48), *150*
Sukenobu, M., 256(53), *265*
Suzuki, I., 122(31, 33), 123(32), *131*
Suzuki, K., 249(23), 256(66), 258(23), 260(23), *263, 266*
Swanson, M., 315(114), *321*
Sweeney, W. V., 259(56), *265*
Synge, R. L. M., 3, *15*

T

Tagawa, K., 245(57), 246(57), *265*
Tahara, H., 247(64), *266*
Takaichi, S., 256(66), *266*
Takamiya, A., 245(31), *264*
Takano, T., 173(22), 176(22), *200*
Takasaki, S., 78(46), *86*
Takuri, I., 247(58), 249, *265*
Tallan, H. H., 4(22), *15*
Talmadge, K. W., 83(42), *86*
Tanabe, T., 77(25), *85*
Tanaka, N., 173(9), 176(9), *199,* 247(64), *266*
Tanaka, M., 163(27, 28, 29), *169,* 245(62), 246, 256(61), 258, *265, 266*
Tanaka, S., 41(22), *48*
Tandeau de Marsac, N., 227(57, 77), 237, 238(10), *239, 242, 243*
Tanford, C., 163, *168*
Tarnowski, G. S., 9(78), *18, 19*
Tarr, G. E., 307(9), 309(9), 317(111), *318, 321*
Tashian, R. E., 89, *96*
Taylor, A. L., 190(10), *199*
Taylor, C. W., 136(48), *150,* 151(25), *157*
Taylor, F. J. R., 235(78), *243*
Taylor, H. L., 124(8), *130*
Taylor, J. G., *330, 331*
Taylor, S. S., 163(9, 30), 164(9), *168, 169*
Taylor, T. M. C., *243*
Teale, F. W. J., 238(15), *239*
Tentori, L., 288(73), *297*
Tetrin-Clary, C., 105(20), *113*
Thierry, J. C., 159(4), 160(4), 163(4), 165(4), *167*
Thompson, E. W., 299(6), *317*
Thompson, E. O. P., 272(29), 275(31, 80, 81), 280(1, 2, 3, 7, 25, 29, 30, 31, 74, 75, 76, 80, 81, 82, 83), 281(74), 283(84), 285(1, 53), 286(31), 289(30, 31), *293, 294, 295, 296, 297, 328*
Thornber, J. P., 223(79), 224(79), *243*
Thurman, G. B., 123(19, 20, 23), 128(21), *131*
Tilghman, S. M., 211(18), *218*
Timkovich, R., 174(23, 67), 176(66), 182(66), 194(21, 23), 196(21), *200, 202,* 299(21), *318*
Titani, K., *169*
Tobari, J., 249(21), 258(21), 260(21), *263*

Tobey, R. A., 154(21), *157*
Torff, H. J., 163(9), 164(9), *168*
Tracy, R. P., 92(28), *96*
Trainin, N., 120(41), 123(24, 52), *131, 132*
Travis, J., 258, 259(63), *266*
Trayer, I. P., 64(43), 67(43), *86*
Tredwell, C. J., 236(69, 73), 238(73), *242, 243*
Trivers, G. E., 121(42), 128(21), *131*
Troughton, J. H., 175(61), *202*
Troxler, R. F., 232(80), 234(81), *243*
Trüper, H. G., 174(68), 175(68), 192(68), 197(56), *201, 202*
Trus, B. L., 173(42), 176(42), *201*
T'so, P. O. P., 133(6), 134(6), *148*
Tsuchiya, D. K., 246(45), 258(45), *264*
Tsui, H.-W., 99(45), 100(45), *115*
Tsukihara, T., 173(9), 176(9), *199*, 247(64), *266*
Tsunoda, J. N., 159(32), 162(3), 163(3, 32), *167, 169*
Tuengler, P., 160(33), *169*
Tuppy, H., 176(44), *201*, 299(59, 80), *319, 320, 329*

U

Uchida, T., 111(76), *117*
Udenfriend, S., 5(30), *16*
Uhlenbruck, G., 65(9), 82(9), *84*
Ullmann, A., 55, *62*
Uzzell, T., 299(115), *321*

V

Vänngård, T., 94(1), *95*, 325(6), *326*
Vainshtein, B. K., 269(77), *297*
Valentine, R. C., 245(48), *265*
Van Bekkum, D. W., 121(43), *132*
VandeBurg, J. L., 282(20), *294*
van den Berg, A., 9(68), *18*
Vanderkooi, J. M., 315(24), *318*
Van Heyningen, S., 112(77), *117*
Van Lente, F., 153(36), *158*
Van Sande, J., 110(78), *117*
Vaughan, M., 108(56), 109(56, 57), *116*
Vaughan, M. H., Jr., 231(82), *243*
Vazquez, D., *116*
Venkatappa, M. P., 90(7), 91(7), *95*
Vernon, W. B., 111(3), *112*
Vernotte, C., 238(83), *243*
Veronese, F. M., 8(62), *18*, 165(34), *169*
Venugopal, B., 123(27), *131*
Visentin, L. P., 246(39), 256(39), *264*
Vickery, H. B., 1(4), 2(4), *14, 15*
Vidali, G., 146, *149*, 154(19), *157*
Villarejo, M. R., 55(36), 56(25), *62*
Vivaldi, G., 288(73), *297*
Vliegenthart, J. F. G., 64(38), 68(40), 73(40), *86*
von Holt, C., 152, 153(35), *158*
Vroonland, C. A. J., 305(58), *319*

W

Wada, K., 209(21), *218*, 246(42), 247(64), 249(25), 252(38, 46, 65), 256(66), *263, 264, 265, 266*
Wagenmann, R., 231(17), *240*
Wakabayashi, S., 209(21), *218*, 246(42), 249(21, 22, 23, 24, 25), 258(20, 21, 23), 260(21, 23), *263, 264*
Walasek, O., 299(73), *320*
Walker, J. M., 146(27), *149*
Walker, J. E., 159(4, 14), 160(4), 163(4, 14), 165(4), *167, 168*
Walker, J. D., 47(14b), *48*
Wallace, R. B., 144, 145(7, 45), *148, 150*
Wallenfels, K., 49, *62*
Wallin, R., 234(84), *243*
Walpole, H., 173(69), *202*
Walsh, K. A., 40(21), *48, 169*
Wang, D., 9, *18, 19*
Wang, S.-S., 123(20), *131*
Ward, D. N., 101(55a), 104(5a, 79), 105(5, 47, 50), 106(5, 47, 49, 83), 107(46, 48, 50), *115, 116, 117*
Ward, K. B., 293(48), *295*
Waring, A., 307(27), 308(27), 315(116), *318, 321*
Wasniowska, K., *86*
Watkins, W. M., 81(45), *86*
Watson, D., 146(31), *149*
Watson, H. C., 40(19), *48*, 268(78), 271(78), 286(60), 287(78), *296, 297*
Weare, J. A., 106(80, 81), *117*
Weatherall, D. J., 268(79), *297*
Wehrmeyer, W., 235(63), 237(64), 238(56), *242*
Weigle, J. J., 23, *37*
Weigele, M., 5(30), *16*
Weinstein, B., 246(67), *266*
Weinstein, J., 83(26), *85*
Weinstock, G. M., 24(10, 13), *37*
Weintraub, B. D., 100(33), *114*
Weintraub, H., 144, 146, *150*, 151(16), 152, 153(36), 154(16), *157, 158*
Weisbrod, S., 146(47), *150*
Weissman, S. M., 211(20), 212(13, 20), 217(13), *218*
Wellick, G. E., 246(39), 256(39), *264*
Westermark, G., 89(20), *96*
Whatley, F. R., 196(32, 33), *200*
White, A., 120(17, 46, 50), 121(17, 50), 122(46, 51), 123(7, 18, 19, 47, 49, 50), 124(7), 125(7), 126(7), 129(44, 45, 48), *130, 131, 132, 328, 331*
White, T. J., 204(29), 208(29), *219*, 268(86), *298*
Whitley, R., 104(34), 107(34), *114*
Whiteley, H. R., 159(32), 163(32), *169*

Whittaker, R. G., 146(39), *149*, 275(80, 81), 280(75, 80, 81, 82, 83), 283(84), *297*
Widmer, H., 229(19), 232(19), 233(19), 234(19), *240*, 249(24), *263*
Wildman, R. B., 225(85), *243*
Williams, D. M., 47(12), *48*
Williams, J., 287, *297*
Williams, J. F., 100(82), 111(82), *117*
Williams, J. N., Jr., 305(117), *321*
Williams, L. C., 153(15), 154(15), *157*
Williams, R. C., 227(89), *244*
Williams, V. P., 229(86), 234(87), *243*, *244*
Wilms, J., 308(56), 311(56), *319*
Wilson, A. G., 204(29), 208, *219*, 268(86), *298*
Wilson, D. F., 315(23, 24), *318*
Wilson, G., 9, 10, 11, 12, *18*, *19*
Wilson, J. T., 211(20), 212(20), *218*
Wilson, M. T., 307(2), 308(2), 313(2), *317*
Winter, W. P., 268(68), *296*
Witkin, E., 24, *37*
Witkop, B., 7, *17*
Wöhler, F., 7, *17*
Woese, C. R., 197(26, 27), *200*, 252(36), 256(68), *264*, *266*
Wollman, F.-A., 225(88), *244*
Wonacott, A. J., 159(4), 160(4), 163(4), 165(4), *167*
Wong, N. C., 146(31), *149*
Wood, W. A., 40(13), *48*
Wootton, J. C., *330*, *331*
Wu, T. T., 40(11), 41(22), *48*, 299(118), *321*
Wu, Y.-T., 65(12), 67(12), *84*
Wyckoff, H. W., 7, *17*

X

Xuong, Ng. H., 174(59), 176(59), *202*

Y

Yaguchi, M., 246(39), 256(39), *264*
Yakir, Y., 123(24, 52), *131*, *132*
Yamamoto, H., 122(33), *131*
Yamanaka, G., 227(89), *244*
Yamane, T., 173(9), 176(9), *199*
Yamizumi, M., 111(76), *117*
Yang, K.-P., 105(50), 106(83), 107(50), *115*, *117*
Yanofsky, C., 212, *218*, *219*
Yansura, D. G., 56(13), *61*
Yasunobu, K. T., 159(32), 162(3), 163(3, 27, 28, 29, 32), *167*, *169*, 245(62), 246, 256(61), 258, *265*, *266*
Yates, M. G., 259(70), *266*
Yčas, M., 206, *219*
Yen, H.-C., 192(70), *202*
Yeoman, L. C., 136(48), *150*
Yoch, D. C., 256(71, 72), 259(56, 73), *265*, *266*
Yonetani, T., 306(119), *321*
Yoshima, H., 78(46), *86*

Z

Zabin, I., 49(10, 11, 21), 50(38, 40), 51(4, 23, 39), 54(1, 9), 55(11, 36), 56(2, 5, 6, 20, 24, 25, 28), 57(6), 58(3, 6), 59(12, 17), 60(14), *60*, *61*, *62*
Zablen, L. B., 197(27), *200*
Zamenhof, P. J., 56(25), *62*
Zarnitz, M. L., 49(37), *62*
Zata, M., 123(18), *131*
Zichichi, M. L., 23, *37*
Zilinskas, E., 237(29), *240*
Zipser, D., 51(31), *62*
Zimmerman, B. K., 225(28), *240*
Zimmerman, J. M., 163, *169*
Zittle, C. A., 7(51), *17*
Zuber, H., 229(19), 232(19), 233(19), 234(19), 249(24), *240*, *263*
Zuckerkandl, E., 267, 268(87), 282(87), *298*

Subject Index

A

Amino acid sequence, 3, 5–7, 13, 39–41, 43, 46, 47, 49–54, 56, 58, 59, 77, 82, 88, 91, 92, 99–102, 109–112, 122, 126, 135, 136, 151–155, 159-167, 174–181, 191, 192, 196, 197, 203–208, 217, 231, 235, 245–262, 267-293, 299, 300, 303, 317
Azathioprine-sensitive E-rosette assay, 124–129

B

Bilins, 225–238
Blue copper proteins, 88–95
  binuclear metal complexes, 95
  electron paramagnetic resonance (EPR), 87, 90, 94
  reduction of dioxygen, 89, 90, 92, 94
  spectroscopic measurements, 90

C

Carbonic anhydrase, 88, 89
Chemical modifications, 104–107
  measurements,
    circular dichroism, 104–106
    proton nuclear magnetic resonance, 104
    radioimmunoassay, 104, 106
  methods,
    carbodiimide-mediated coupling, 104–106
    carboxymethylation of methionine, 105–107
    carboxypeptidase treatment, 105, 106
    cyanogen bromide treatment, 106, 107
    dinitrophenylation of tyrosine, 107
    iodination of tyrosine, 105, 106
    methylation of lysines, 105
    nitration of tyrosine, 105, 106
    treatments on lysine, 107
Cytochrome *c*, 152, 173–198, 299–317,
  assays, 305–307
    carbon monoxide method, 308
    manometric, 306
    polarographic, 306–309, 315, 316
    TMPD dye, 306–308
    spectrophotometric, 300, 306–309
    steady state kinetic, 308, 309
  cytochrome oxidase. *See* Cytochrome oxidase
  cytochrome reductase, 305, 308, 309, 313–315
  electron transfer (*see* Electron transfer system)
  enzymic interaction domain, 309–313
    differential chemical modification, 310, 311
    kinetics studies, 311, 312, 316, 317
      butterfly cytochrome *c*, 311, 317
    use of mono-CDNP derivatives of lysine, 309–312
      dipole moments, 305, 309, 310, 312
      heme edge, 309
  gene transfer, 181–192, 197
  mutations, 299–305, 311, 316, 317
  orientation in membrane, 305, 309
  sequence, 174–181, 192, 196, 197, 299, 300, 303, 317
  subjects,
    ape, 303, 316
    butterfly, 311, 317
    euglena, 302
    eukaryotic, 176–181, 192, 299, 301, 305
    horse, 302, 309, 316
    monkey, 303, 316
    photosynthetic, 174–180, 192–197

Cytochrome *c* (*continued*)
prokaryotic, 180, 181, 192
prosimian, 303, 316
respiratory, 180, 181, 191–197, 305, 306, 308, 312–315
yeast, 302
x-ray structure analyses, 174–181, 196, 197
Cytochrome *c* oxidase, 88, 92–95, 301, 302, 305–317
binuclear metal complexes, 94, 95
kinetics of reaction, 301–303, 305–309, 311, 316
ferricytochrome *c*, 300, 316
high and low affinity reactions, 302, 307, 312–314, 316
K_D values, 302, 306, 316
stopped flow experiments, 306, 308, 309, 316
reduction of dioxygen, 89, 92, 94
Cytosine–guanine (CG) nucleotide pairs, 214, 215

D

Duplex DNA, 24–34, 36
D-loops, 29–36
nitrocellulose filters, 28, 29

E

Electron transfer system, 89, 222, 245, 246, 252, 306, 308–315
influence of viscosity on cytochrome *c*, 312–315
mobility, 314, 315
comparison with photoaffinity labelling experiments, 315
kinetics of reaction. *See* Cytochrome oxidase
models of cytochrome *c* function, 313–315
Electron transport, 173–180, 190–193
Escherichia coli, 23–36, 41–44, 49–60
Evolution, 23, 40–42, 44–47, 58, 59, 87, 88–92, 95, 99, 104, 110–112, 136, 151–155, 165, 166, 173–197, 204–217, 225–238, 245, 246, 253, 257, 262, 267–270, 282–284, 288, 299–303, 316, 317. *See also* Cytochrome *c*; Ferredoxin; β-Galactoside; Genetic recombination; Globins; Glycoprotein hormones; Histones; Messenger RNA; Pentitol operons; Metalloenzymes; Phycobiliproteins; Respiration; Ribitol dehydrogenases; Thermal stability; Xylitol dehydrogenases
Evolutionary clock, 203, 204, 208, 209, 282, 288
Evolutionary divergence, 89, 203, 204, 209
Exchanges, 161–167
internal hydrophobicity, 163–165, 167
stabilizing effect, 162–167
statistical observations, 161–167
Experimental enzyme evolution, 9, 10, 39–47

F

Ferredoxin, 159, 160, 162–167, 209, 245–247, 253, 257, 260–262
absorption spectra, 245, 257, 259
bacterial-type, 245, 246, 257–262
chloroplast-type, 246–257, 262
electron paramagnetic resonance (EPR), 245, 259
electron transfer, 222, 245, 246, 251
eukaryotic, 246, 256, 257
evolution, 245, 246, 253
halobacterial, 246, 247, 253, 256
iron clusters, 245–253, 257–261
prokaryotic, 246, 256
redox potentials, 245, 257–261
sequence, 159–167
sites, 247–259

G

β-Galactoside, 49–60
active site, 59
hybrids, 58
immunological cross-reactivity, 51, 55, 56
lactose operon, 50–58
Lac Z, 54–59
comparison with dihydrofolate reductase, 59
sequence, 49–54, 56, 58
cyanogen bromide, 54
tryptic digest, 54, 56
nonsense mutants, 51, 55, 60
topography, 56–60
antibodies, 56–60
complementation studies, 56–60
Gene duplication, 40, 46, 59, 89, 90, 92, 246, 253–260, 268, 284, 287
Gene expression, 141, 147
Genetic code, 205–217
Genetic recombination, 23, 24, 29, 34–36
Escherichia coli, 2
rec A protein, 23–36
Gene transfer, 181–192, 197
Geological period, 235
Globins, 209, 210, 213, 267–293
accepted point mutations (PAM) units, 283
contact sites, 270, 271, 280, 281, 285
dates of divergence, 282–284, 287
interactions of hemoglobin, 269–282
protein variability, 268, 281
rates of protein evolution, 209, 282–284
subjects,
agnathan, 284, 288
elasmobranch, 284–288
human, 279–281, 287, 288
marsupial, 279, 280, 283, 284

mollusk, 288, 293
monotreme, 281–284
Glyceraldehyde-3-phosphate (GAPDH), 159, 162, 164–167
sequence, 159–167
Glycoprotein hormones, 99–112
chemical modifications, 99–107, 109 (*see also* Chemical modifications)
comparison with toxins, 104, 108–110
receptor binding, 100–108, 111, 112
sequence, 99–108, 109–112
subjects,
bovine lutropin, 99, 104, 105–108
human choriagonadotropin, 110, 111
pregnant mare serum gonadotropin (PMSG), 99, 105
subunit interaction, 103–107, 109
α subunits, 99–112
β subunits, 99–112
Glycoproteins, 63–83
Glycosyltransferases, 63–83
oligosaccharide biosynthesis, 63–83
purification, 66–68
sialyltransferases, 70, 76–82
hepatic receptors, 77, 78
M/N blood group, 80–82
viral adherence, 82, 83
substrate specificity, 69–76

H

Histones, 133–141, 145–147, 151–155
chromatin, 133–147, 152–154
acetylation, 144–147, 154
cross-linking techniques, 153
phosphorylation, 154
polyacrylamide gel electrophoresis, 134, 145, 147
trypsin digestion, 134, 143, 153
sequence, 135–137, 151–155
subjects,
calf thymus, 135, 136, 151
eukaryote, 136, 141, 155
pea seedlings, 135–137, 151
prokaryote, 136, 155

I

Immunological response, 122–129
human plasma, 124–126
macrophage inhibitory factor (MIF), 122
T and B cells, 122
Iron chelation, 246–250, 257–261

L

Lactate dehydrogenase (LDH), 160, 162, 166, 167
sequence, 160–167
Lactose operon, 50–59

M

Messenger RNA (mRNA), 205–217
Metabolic tree, 194
Metalloenzymes, 87–95
metal coordination, 88–90
sequence 88, 90–92
substances,
blue oxidases, 88–94
carbonic anhydrase, 88, 89
copper proteins, 88–92
cytochrome *c* oxidase, 88, 92–95
enolase, 88, 92–95
Myoglobin fold, 269–271

N

Neutral theory, 204–217, 299–303
DNA of bacteriophage Φx-174 and G-4, 207, 208, 212, 213
frameshift readings, 207, 208, 212, 213, 217
mRNA, 205, 206, 210–217
silent changes, 210–213, 217
Nucleoside triphosphates, 24–28, 34, 36
activity, 24–28
Nucleotide replacements, 208–215

P

Pancreatic ribonuclease. *See* Ribonuclease
Pentitol operons, 41, 43–47
DNA sequencing, 46, 47
Escherichia coli C, 41, 43, 44
gene duplication, 46
Klebsiella aerogenes, 43, 44, 46
plasmids, 46, 47
transducing phages, 44–46
Photosynthesis, 221–238
accessory complexes, 223, 225–238
reaction centers, 223, 225
subjects,
brown algae, 221, 225
chromophores, 221–223
cryptomonads, 231, 235
cyanobacteria, 221, 225, 227–231
green algae, 221–225
red algae, 221, 225–231
thylakoids, 222–227, 231
Phycobiliproteins, 225–238, 253
electron microscopy, 227, 231–237, 238
fluorescence, 235–238

Phycobiliproteins (*continued*)
spectroscopic properties, 228, 229, 236
subjects,
cyanobacteria, 227-238
cryptomonads, 231, 235
red algae, 227-238
Phycobilisome, 225, 227, 231, 236-238
Phylogenetic tree, 180, 192, 197, 210, 246, 253, 256, 260, 262, 300, 301, 303
Polymorphism, 89, 280, 282, 288
Prealbumin, 124-126
comparison with fraction, human plasma, 124-126

R

Rec A protein, 23-36
assimilation, 29, 31, 34
binding to duplex DNA, 25-36
DNA aggregation, 25, 28, 34, 36
nucleoside triphosphatase activity, 24-28
rec A mutants, 23, 24, 35
renaturation, 31-35
Respiration, 176-196
bacteria,
blue-green, 174, 176, 192-196
green nonsulfur, 174-176, 194-196
green sulfur, 174-176, 181, 192, 194-196
purple nonsulfur, *Rhodospirillaceae*, 174, 176, 180-197
purple sulfur, *Chromatiaceae*, 174, 176, 192-197
evolution, 181-197
genes for, 190
Ribitol dehydrogenase, 41, 42
gene multiplication, 42, 43
Klebsiella aerogenes, 40-42
mutants, 41, 42
substrate specificity, 40, 41
Ribonuclease, 5-11
activity toward poly A·poly U, 9, 10
coupling of carbohydrates, 10, 11
cross-linking, 9, 10
hybridization, 10, 11
sequence, 5-8
Ribonuclease inhibitor, 11-14
kinetics, 13
purification from human placenta, 11-14
translation systems, 14

S

Selectionism, 40-44
chemostat, 41-44,
Silent changes, 210-213, 217

T

Thermal stability
comparison of sequences, 159-167
mesophiles, 160-162
thermophiles, 160-162
Thymic hormone-like activity, 121-129
comparisons, 122-126
human plasma, 124-128
prealbumin, 124-126
rosette assay, 124-129 (*see also* Azathioprine-sensitive E-rosette assay)
Thymus hormones, 119-129
designation as, 121-123, 126-129
history, 120
immunological impairment, 122, 123
immunological response, 121-127
restoration of immunity 126, 127
putative, 121, 122
Toxins, 104, 105

X

Xylitol dehydrogenases, 41
Escherichia coli, 41
Klebsiella aerogenes, 41
mutants, 41, 42
substrate specificity, 40-42